# Introduction to Linear System Theory

Holt, Rinehart and Winston Series
in Electrical Engineering, Electronics, and Systems

Other Books in the Series:

**Shu-Park Chan,** Introductory Topological Analysis of Electrical Networks
**George R. Cooper and Clare D. McGillem,** Methods of Signal and System Analysis
**Mohammed S. Ghausi and John J. Kelly,** Introduction to Distributed-Parameter Network: With Application to Integrated Circuits
**Roger A. Holmes,** Physical Principles of Solid State Devices
**Benjamin J. Leon,** Lumped Systems
**Benjamin J. Leon and Paul A. Wintz,** Basic Linear Networks for Electrical and Electronics Engineers
**Samuel Seely,** Electronic Circuits

# Introduction to Linear System Theory


CHI-TSONG CHEN

Associate Professor, Department of Electrical Sciences

*State University of New York at Stony Brook*

HOLT, RINEHART AND WINSTON, INC.

New York Chicago San Francisco Atlanta
Dallas Montreal Toronto London Sydney

To Beatrice


Library of Congress Catalog Card Number: 77-126528
**SBN: 03-077155-2**
Printed in the United States of America
0 1 2 3 22 9 8 7 6 5 4 3 2 1

# Preface

The text is intended for use at the senior–graduate level for university courses on the analysis and introductory design of linear systems. It is also intended for independent study and reference by engineers and applied mathematicians. The mathematical background assumed for this book is a working knowledge of matrix manipulation and an elementary knowledge of differential equations. The unstarred sections of this book were developed from the lecture notes for the first graduate course in system theory taught for four years at The State University of New York at Stony Brook.

The book is mainly concerned with the analysis and introductory design of linear systems. Over the past decade, engineers have been interested in designing not merely a workable system, but the best available system. However because of the structure, a system often has some inherent limitation, and the best the system can achieve is not unlimited. Without knowing this limitation, a designer might be wasting his time and energy in trying to design an impossible system. Hence, a thorough study of all the properties of a system is essential in the design. In fact, many design techniques are evolved from this study.

In the conventional control theory, analysis and design are carried out mostly by using transfer functions—an input–output description.

On the other hand, the state variable description is mainly used in modern control theory. In this text, these two mathematical descriptions are developed from the concepts of linearity, causality, relaxedness, time invariance and the state. They are introduced with equal emphasis and their relationships are stressed.

The digital computer is a very powerful tool in engineering design. As engineers, we shall make full use of digital computers. Aiming at this end, most of the results in this book are developed in a manner suitable for digital computer programming. The detailed programmings, however, are not discussed; they can be found in the books on numerical computations.

In Chapter 2, the mathematical background required is introduced. The basic concepts of system theory are introduced in Chapter 3; its presentation is intended to be intuitively appealing; the mathematical rigor is not a primary concern. The solutions of state-variable equations are developed in Chapter 4. The concepts of controllability and observability are introduced in Chapter 5. Their practical implications are studied in Chapter 7. In Chapter 6, the realization problem is covered. Stability and composite systems are studied in the last two chapters.

The text is flexible permitting use in one- or two-semester course. By skipping Theorems 3-1, 4-5, 4-11, 5-2, 5-5, 5-11, 8-3, 8-7, and Chapter 9; and skipping the proofs of Theorems 2-6, 2-8, 2-10 to 2-13, 4-1, 4-2, 4-20, 4-21 and 8-10, the unstarred sections and Appendix A are covered in a one-semester course at Stony Brook. Since the logical dependence among Chapter 6, Chapter 7 and Chapters 8, 9 are rather loose (see the end of the Preface) the material in the unstarred sections could be easily adopted in a one-semester or one-quarter course. By covering the starred sections, the book could be used in a two semester course.

The problem sets form an integral part of the book. They are designed to help the reader understand and utilize the concepts and results covered. In order to retain the continuity of the main text, some important results are listed in the problem sets.

I owe a great deal to many people in writing this book. Kalman's work and Zadeh and Desoer's book *Linear System Theory* form the foundation of this book. I have benefitted immensely in my learning from Professor C. A. Desoer, for this I could never express enough my gratitude. I also wish to thank him for his constant encouragement and his critical scrutiny of the manuscript. To Professors B. J. Leon, E. J. Craig, and I. B. Rhodes I wish to express my appreciation for their review and valuable suggestions. Thanks are due to Professor P. E. Barry for many stimulating discussions, and to Professor M. Y. Wu for his help in proofreading. Thanks are also due to Professor S. S. L. Chang, former chairman of the Department of Electrical Sciences, SUNY at

Stony Brook, for providing the departmental assistance. I wish to thank many of my students, especially Messrs. C. Waters and C. H. Hsu for their suggestions in improving the presentation. I am indebted to Mrs. R. Lustig, V. Donahue, C. Calomiros and Misses B. Martin, J. Gould for typing various drafts of the manuscript. I would also like to thank Mrs. L. Liang and Miss L. Banks of Holt, Rinehart and Winston, Inc. for their assistance in the production of the book. Finally my special thanks go to my wife Beatrice for her patience and assistance in making the writing of this book possible.

Logical Dependence of Chapters.

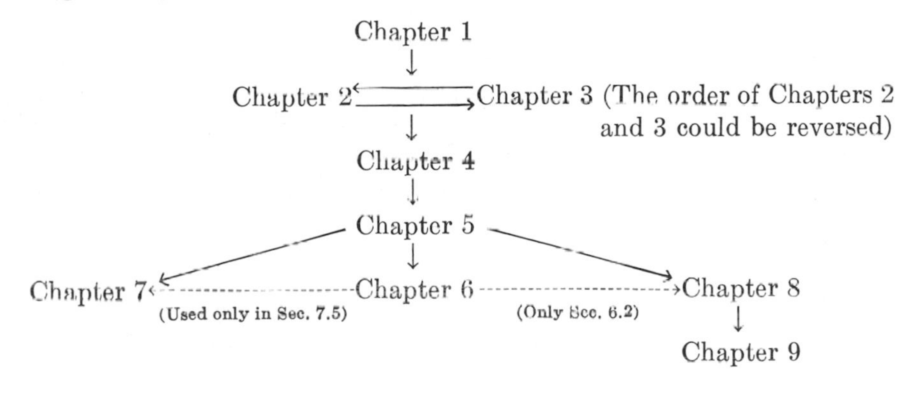


*Stony Brook, New York*
*March, 1970*

CHI-TSONG CHEN

# Contents

**Preface** v

**Glossary of Symbols** xv

**Chapter 1 Introduction** 1

1-1 The Study of Systems/1
1-2 The Scope of the Book/3
1-3 Chapter Description/4

**Chapter 2 Linear Spaces and Linear Operators** 7

2-1 Introduction/7
2-2 Linear Spaces Over a Field/8
2-3 Linear Independence, Bases and Representations/13
Change of Basis/20
2-4 Linear Operators and Their Representations/21
Matrix Representations of a Linear Operator/23

2-5 Systems of Linear Algebraic Equations/29
2-6 Eigenvectors, Generalized Eigenvectors and Jordan-Form Representations of a Linear Operator/35
A Procedure for Computing a Jordan-Form Representation/44
2-7 Functions of a Square Matrix/49
Polynomials of a Matrix/49
Functions of a Matrix/55
Functions of a Matrix defined by Means of Power Series/58
2-8 Norms and Inner Product/61
2-9 Concluding Remarks/64
Problems/65

**Chapter 3 Mathematical Descriptions of Systems 72**

3-1 Introduction/72
3-2 The Input–Output Description/74
Linearity/75
Causality/79
Relaxedness/79
Time-Invariance/82
Transfer-Function Matrix/84
3-3 The State-Variable Description/86
The Concept of State/86
Dynamical Equations/89
Analog Computer Simulations of Linear Dynamical Equations/92
3-4 Examples/94
Dynamical Equations for *RLC* Networks/99
3-5 Comparison of the Input–Output Description and the State-Variable Description/104
*3-6 Mathematical Descriptions of Composite Systems/107
Time-Varying Case/107
Time-Invariant Case/110
*3-7 Discrete-Time Systems/113
3-8 Concluding Remarks/116
Problems/117

* May be omitted without loss of continuity.

**Chapter 4** **Linear Dynamical Equations and Impulse-Response Matrices** **125**

4-1 Introduction/125
4-2 Solutions of a Dynamical Equation/126
Time-Varying Case/126
Time-Invariant Case/133
4-3 Equivalent Dynamical Equations/138
Time-Invariant Case/138
*Time-Varying Case/143
Linear Dynamical Equation with Periodic **A** (·)/144
4-4 Impulse Response Matrices and Dynamical Equations/146
Time-Varying Case/146
Time-Invariant Case/149
4-5 Concluding Remarks/154
Problems/155

**Chapter 5** **Controllability and Observability of Linear Dynamical Equations** **161**

5-1 Introduction/161
5-2 Linear Independence of Time Functions/163
5-3 Controllability of Linear Dynamical Equations/169
Time-Varying Case/169
*Differential Controllability, Instantaneous Controllability and Uniform Controllability/174
Time-Invariant Case/177
*Simplified Controllability Condition/179
5-4 Observability of Linear Dynamical Equations/182
Time-Varying Case/182
*Differential Observability, Instantaneous Observability and Uniform Observability/186
Time-Invariant Case/187
*Simplified Observability Condition/188
*5-5 Controllability and Observability of Jordan-Form Dynamical Equations/189

* May be omitted without loss of continuity.

5-6 Canonical Decomposition of a Linear Time-Invariant Dynamical Equation/197
Irreducible Dynamical Equations/203
*5-7 Output Controllability and Output-Function Controllability/206
5-8 Concluding Remarks/209
Problems/210

**Chapter 6 Irreducible Realizations of Rational Transfer-Function Matrices 216**

6-1 Introduction/216
6-2 The Characteristic Polynomial and the Degree of a Proper Rational Matrix/218
6-3 Irreducible Realizations of a Scalar Rational Transfer Function/221
Irreducible Realizations of $\beta/D(s)$/221
Irreducible Realizations of $\hat{g}(s) = N(s)/D(s)$/225
Observable Canonical-Form Dynamical-Equation Realization/225
Controllable Canonical-Form Dynamical-Equation Realization/227
Jordan Canonical-Form Dynamical-Equation Realization/229
*Realization of $N(s)/D(s)$ Where $N(s)$ and $D(s)$ Have Common Factors/232
*Realization of Linear Time-Varying Differential Equations/234
*6-4 An Irreducible Realization of a Proper Rational Transfer-Function Matrix/235
*6-5 Irreducible Jordan-Form Realization of a Proper Rational Transfer-Function Matrix/240
Controllable Jordan-Form Dynamical-Equation Realization of $\hat{\mathbf{G}}(s)$/240
Reduction of Jordan-Form Dynamical Equations/246
6-6 Concluding Remarks/252
Problems/254

* May be omitted without loss of continuity.

**Chapter 7 Canonical Forms, State Feedback, and State Estimators 259**

7-1 Introduction/259
7-2 Canonical-Form Dynamical Equations/260
Single-Variable Case/260
*Multivariable Case/268
7-3 State Feedback/271
Single-Variable Case/271
*Multivariable Case/277
7-4 State Estimators/281
Single-Variable Case/281
The $n$-Dimensional Estimator/281
The Separation Property/287
The $(n-1)$-Dimensional Estimator/289
*Multivariable Case/292
7-5 Design of Feedback Systems: An Example/296
*7-6 Decoupling by State Feedback/298
7-7 Concluding Remarks/305
Problems/305

**Chapter 8 Stability of Linear Systems 311**

8-1 Introduction/311
8-2 Stability Criteria in Terms of the Input–Output Description/312
Time-Varying Case/312
Time-Invariant Case/315
8-3 Routh–Hurwitz Criterion and Liénard–Chipart Criterion/322
8-4 Stability of Linear Dynamical Equations/328
Time-Varying Case/328
Time-Invariant Case/335
*8-5 A Proof of the Routh–Hurwitz Criterion/339
8-6 Concluding Remarks/347
Problems/348

**Chapter 9 Linear Time-Invariant Composite Systems 354**

9-1 Introduction/354
9-2 Transfer-Function Descriptions of Composite Systems/356

* May be omitted without loss of continuity.

*9-3 Controllability and Observability of Composite Systems/361
Parallel Connection/361
Tandem Connection/363
Feedback Connection/364
9-4 Stability of Linear, Time-Invariant Feedback Systems/367
Single-Variable Feedback Systems/368
Multivariable Feedback Systems/375
9-5 Design of Pole-Placement Compensators/382
Two Algebraic Theorems/383
Design of Pole-Placement Compensators for Single-Input Two-Output Systems/388
Discussion and Extension/393
9-6 Concluding Remarks/394
Problems/395

**Appendix A** **Analytic Functions of a Real Variable** **399**

**Appendix B** **Minimum Energy Control** **401**

**Appendix C** **Controllability after the Introduction of Sampling** **403**

**Appendix D** **Hermitian Forms** **409**

**Appendix E** **On the Matrix Equation AM + MB = N** **414**

**References** **417**

**Index** **425**

* May be omitted without loss of continuity.

# Glossary of Symbols

| | |
|---|---|
| ▮ | This symbol denotes the end of a statement or an example wherever the ending is not clear. |
| $\mathbf{A},\mathbf{B},\mathbf{P}, \ldots$ | Capital boldface letters denote matrices. |
| $\mathbf{u},\mathbf{y},\boldsymbol{\alpha}, \ldots$ | Lowercase boldface letters denote vectors. |
| $u,y,\alpha, \ldots$ | Lowercase italic and Greek type denote scalar valued functions or scalars. |
| $\mathcal{L}$ | Laplace transform. |
| $\hat{\mathbf{u}}(s),\hat{\mathbf{G}}(s), \ldots$ | A letter with circumflex denotes the Laplace transform of the letter (e.g., $\hat{\mathbf{u}}(s) \triangleq \mathcal{L}[\mathbf{u}]$) except two cases: (1) $\hat{\mathbf{A}}$ denotes the Jordan-form matrix, (2) $\hat{\mathbf{x}}$, $\hat{\bar{\mathbf{x}}}$ denote estimates of $\mathbf{x}$ and $\bar{\mathbf{x}}$. |
| $\rho(\mathbf{A}), \ldots$ | The rank of the constant matrix $\mathbf{A}$. |
| $\nu(\mathbf{A}), \ldots$ | The nullity of the constant matrix $\mathbf{A}$. |
| $\delta(\hat{\mathbf{G}}(s)), \ldots$ | The degree of the rational matrix $\hat{\mathbf{G}}(s)$. |
| $\mathbf{A}',\mathbf{x}', \ldots$ | The transpose of the matrix $\mathbf{A}$ and the vector $\mathbf{x}$. |
| $\mathbf{A}^*,\mathbf{x}^*, \ldots$ | The complex-conjugate transpose of the matrix $\mathbf{A}$ and the vector $\mathbf{x}$. |
| $\det \mathbf{A}, \ldots$ | The determinant of $\mathbf{A}$. |
| $\mathbb{C}$ | The field of complex numbers. |
| $\mathbb{R}$ | The field of real numbers. |

$\triangleq$ Equals by definition.

$\frac{d}{dt}\mathbf{A} \triangleq \left(\frac{d}{dt}a_{ij}\right)$,

$\mathfrak{L}[\mathbf{A}] \triangleq (\mathfrak{L}[a_{ij}])$, . . . When an operator is applied to a matrix or a vector, it means that the operator is applied to every entry of the matrix or the vector.

$\dot{\mathbf{x}} \triangleq \frac{d}{dt}\mathbf{x}$

CHAPTER

# 1 Introduction

## 1-1 The Study of Systems

The study and design of a physical system often consists of four steps:

1. Modeling.
2. Setting up mathematical descriptions.
3. Analyses.
4. Design.

The first step, modeling, involves searching for a model that resembles the physical system in its salient features but is easier to study. A physical system is an object existing in the real world; its precise characteristics are often unknown to us. We can, however, apply to it all kinds of testing signals, and from the measured data, determine its characteristics. If we want to study it analytically, a model that resembles the physical system has to be determined from the measured characteristics. The process of determining a model is called modeling. As engineers, we are all familiar with the distinction between a physical system and a model. In electrical engineering, for example, an amplifier circuit, as shown in

Figure 1-1(a), may be modeled as shown in Figure 1-1(b); in mechanical engineering, an automobile suspension system may be modeled as shown in Figure 1-2. Modeling is a very important problem because the success of the design depends upon whether or not the model is properly chosen.

Depending on different questions asked, or depending on different operating ranges, a physical system may have many different models. For example, an electrical amplifier may have different models at high and low frequencies. A spaceship may be modeled as a particle in the

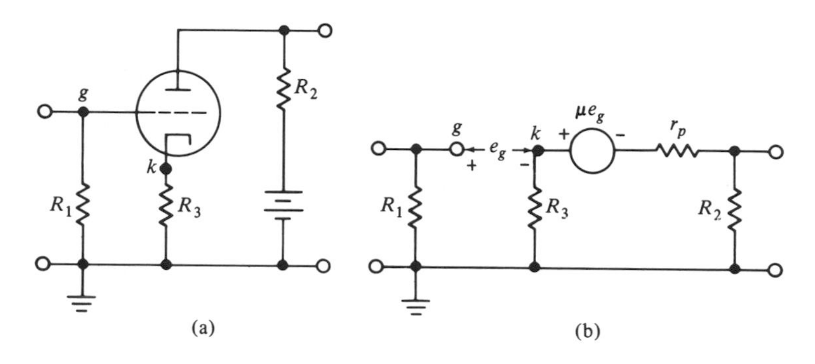


**Figure 1-1** An amplifier. (a) Circuit. (b) Model.

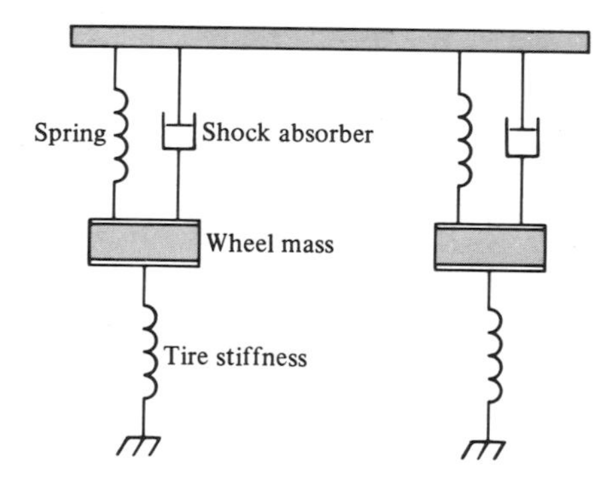


**Figure 1-2** Model of an automobile suspension system.

study of trajectory; however, it has to be modeled as a rigid body in studies of maneuvering. In order to develop a suitable model of a physical system, a thorough understanding of the physical system and its operating range is essential. In this book, we shall refer to models of physical systems as *systems*. Hence, a physical system is a device or a collection of devices existing in the real world; a system is a model of a physical system.

After a system (a model) is found for a physical system, the next step in the study is to find mathematical equations to describe the system. Depending on different systems, we apply different physical laws such as Kirchhoff's voltage and current laws in electrical systems, or Newton's laws in mechanical systems, to set up the equations. The equations that describe systems may assume many forms; they may be linear equations, nonlinear equations, integral equations, difference equations, differential equations, or others. Depending on the question asked, one form of equation might be preferable to another in describing the same system. In conclusion, a system may have many different mathematical-equation descriptions, just as a physical system may have many different models.

Once a mathematical description of a system is obtained, the next step in the study involves analyses–quantitative and/or qualitative. In the quantitative analysis, we are interested in the exact response of the system to certain input and initial conditions. This part of the analysis can be easily achieved by using a digital or an analog computer. In the qualitative analysis, we are interested in the general properties of the system, such as stability, controllability, and observability. This part of analysis is very important, because design techniques may often evolve from this study.

If the response of a system is found to be unsatisfactory, the system has to be improved or optimized. In some cases, the responses of systems can be improved by adjusting certain parameters of the systems; in other cases, compensators have to be introduced. Note that the design is carried out on the model of a physical system. However, if the model is properly chosen, the performance of the physical system should be correspondingly improved by introducing the required adjustments or compensators.

## 1-2 The Scope of the Book

As discussed in the previous section, the study of systems may be divided into four parts: modeling, setting up mathematical descriptions, analyses, and design. In this book, we study only the setting up of mathematical descriptions and analyses of systems. Modeling problems will not be covered; design problem will be touched upon only slightly. This choice of material is based on the following reasoning. The development of models for physical systems requires knowledge of each particular field and certain instruments. For example, to develop models for transistors requires not only the knowledge of quantum physics but also laboratory experimentation techniques. Developing models for automobile suspension systems requires actual testing and measuring; it cannot be achieved by use of pencil and paper alone. The modeling problem should be studied in connection with each specific field and therefore will not be included here. We devote the main part of this book on the development and analysis of mathematical equations for systems. With regard to design, we limit ourselves to those techniques that are deducible directly from analyses. Although the design techniques developed from optimal control theory are not discussed, this book will provide the necessary background for the study of the subject.

The mathematical equations used to describe systems can be classified into linear equations and nonlinear equations. In this book we study only linear equations for the following reasons: (1) A great number of physical systems over their normal operating ranges can be modeled by systems that are describable by linear equations; (2) the theory of linear equations is complete and well-organized; and (3) the theory is the basis for the study of nonlinear equations.

The two linear equations that are basic and most important in the study of network and control systems are of the forms

$$\mathbf{y}(t) = \int_{t_0}^{t} \mathbf{G}(t, \tau)\mathbf{u}(\tau)\, d\tau \tag{1-1}$$

and

$$\dot{\mathbf{x}}(t) = \mathbf{A}(t)\mathbf{x}(t) + \mathbf{B}(t)\mathbf{u}(t) \tag{1-2a}$$
$$\mathbf{y}(t) = \mathbf{C}(t)\mathbf{x}(t) + \mathbf{D}(t)\mathbf{u}(t) \tag{1-2b}$$

Equation (1-1) describes the relation between the input $\mathbf{u}$ and the output $\mathbf{y}$ of a system and is called the *input–output description* or the *external description* of the system. The transfer function is a special case of this description. The set of two equations of the form (1-2) is called a dynamical equation. If it is used to describe a system, it is called the *dynamical-equation description* or the *state-variable description* of the system. These two equations will be developed in this book from a very general setting. They will be thoroughly investigated, their relationship will be established, and various concepts and techniques associated with them will be introduced.

## 1-3 Chapter Description

In this section we give a brief description of the contents of each chapter.

We review in Chapter 2 a number of concepts and results in linear algebra. The objective of this chapter is to enable the reader to carry out similarity transformations, to transform a matrix into a Jordan canonical form and to compute functions of a matrix. These techniques are very important, if not indispensable, in analysis and design of linear systems.

In Chapter 3 we develop systematically the input–output description and the state-variable description of linear systems. These descriptions are developed from the concepts of linearity, relaxedness, causality, and time-invariance. We also show, by the use of examples, how these descriptions can be set up for systems. Mathematical descriptions of composite systems and discrete-time equations are also introduced.

In Chapter 4 we study the solutions of linear dynamical equations. We also show that different analysis often leads to different dynamical-equation descriptions of the same system. The relation between the input–output description and the state-variable description is also established.

We introduce in Chapter 5 the concepts of controllability and observability. The importance of introducing these concepts can be seen from the following example. Consider the networks shown in Figure 1-3, in which the transfer functions are both equal to 1. There is no doubt about the transfer function of the network shown in Figure 1-3(b); however, we may ask why the capacitor in Figure 1-3(a) does not play any role in the transfer function. In order to answer this question, the concepts of controllability and observability are needed. These two concepts are also essential in optimal control theory, stability studies, and the prediction or

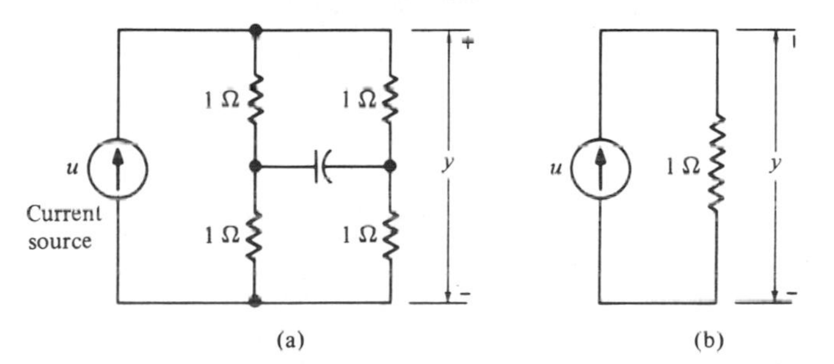


**Figure 1-3** Two different networks with the same transfer function, 1.

filtering of signals. Various necessary and sufficient conditions for a dynamical equation to be controllable and observable are derived.

In Chapter 6 we study irreducible realizations of rational transfer-function matrices. The problem is to find a controllable and observable linear time-invariant dynamical equation that has a prescribed rational matrix. Its solution is very important because it renders a means of simulating a system in an analog or a digital computer. It also offers a method of synthesizing a rational matrix by using operational-amplifier circuits.

The practical implications of the concepts of controllability and observability are studied in Chapter 7. We show what can be achieved under the assumption of controllability and/or observability. Some design techniques are introduced in this chapter.

We study in Chapter 8 a qualitative property of linear systems. This comes under the heading of stability, which is always the first requirement to be met in the design of a system. We introduce the concepts of bounded-input bounded-output stability, stability in the sense of Lyapunov, asymptotic stability, and total stability. Their characterizations and relationships are studied.

In the last chapter we study various problems associated with linear, time-invariant composite systems. One of them is to study the implication of pole-zero cancellation of transfer functions. For example, consider two systems with the transfer functions $1/(s-1)$ and $(s-1)/(s+1)$

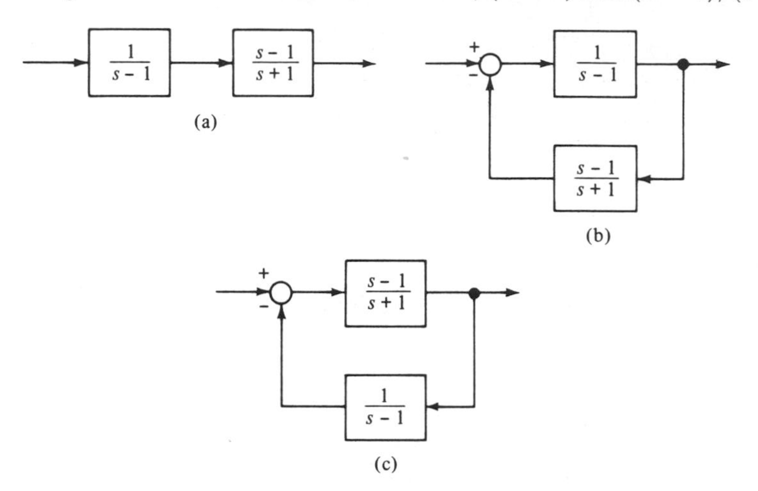


**Figure 1-4** Three different connections of $1/(s-1)$ and $(s-1)/(s+1)$.

connected in three different ways, as shown in Figure 1-4. We show why the system in Figure 1-4(b) can be studied from its composite transfer function, but the systems in Figure 1-4(a) and (c) cannot. We also study stabilities of single-variable and multivariable feedback systems. Finally, we study the design of feedback systems by using an algebraic method.

CHAPTER

# 2
# Linear Spaces and Linear Operators

## 2-1 Introduction

In this chapter we shall review a number of concepts and results in linear algebra that are essential in the development of linear system theory. The topics are carefully selected and only those which will be subsequently used are introduced. The purpose of this chapter is to enable the reader to understand the mechanism of similarity transformation, to find Jordan-form representations of square matrices, and to compute functions of a matrix, in particular, exponential functions of a matrix. (See Section 2-9, Concluding Remarks.[1])

In Section 2-2 we introduce the concepts of field and linear space over a field. The fields we shall encounter in this book are the field of real numbers, the field of complex numbers, and the field of rational functions. In order to have a representation of a vector in a linear space, we introduce, in Section 2-3, the concept of basis. The relationship between dif-

[1] It is recommended that the reader keeps the concluding remarks in mind, for they provide the reader with motivations for studying the mathematical theorems introduced in this chapter.

ferent representations of the same vector is established. In Section 2-4, we study linear operators and their representations. The concept of similarity transformation is embedded here. In Section 2-5, the solutions of a set of linear algebraic equations are studied. The concepts of rank and nullity are essential here. In Section 2-6, we show that every square matrix has a Jordan-form representation; this is achieved by introducing eigenvectors and generalized eigenvectors as basis vectors. In Section 2-7, we study functions of a square matrix. The minimal polynomial and the Cayley–Hamilton theorem are introduced. In the last section, the concepts of inner product and norm are introduced.

This chapter is intended to be self-contained. The reader is assumed to have some basic knowledge of matrix theory (determinants, matrix addition, multiplication, and inversion). The matrix identities introduced below will also be used. Let **A**, **B**, **C**, **D** be $n \times m$, $m \times r$, $l \times n$ and $r \times p$ constant matrices, respectively. Let $\mathbf{a}^i$ be the $i$th *column* of **A**, and let $\mathbf{b}^j$ be the $j$th *row* of **B**. Then we have

$$\mathbf{AB} = [\mathbf{a}^1 \quad \mathbf{a}^2 \quad \cdots \quad \mathbf{a}^m] \begin{bmatrix} \mathbf{b}^1 \\ \mathbf{b}^2 \\ \cdot \\ \cdot \\ \cdot \\ \mathbf{b}^m \end{bmatrix} = \mathbf{a}^1\mathbf{b}^1 + \mathbf{a}^2\mathbf{b}^2 + \cdots + \mathbf{a}^m\mathbf{b}^m \tag{2-1}$$

$$\mathbf{CA} = \mathbf{C}[\mathbf{a}^1 \quad \mathbf{a}^2 \quad \cdots \quad \mathbf{a}^m] = [\mathbf{Ca}^1 \quad \mathbf{Ca}^2 \quad \cdots \quad \mathbf{Ca}^m] \tag{2-2}$$

$$\mathbf{BD} = \begin{bmatrix} \mathbf{b}^1 \\ \mathbf{b}^2 \\ \cdot \\ \cdot \\ \cdot \\ \mathbf{b}^m \end{bmatrix} \mathbf{D} = \begin{bmatrix} \mathbf{b}^1\mathbf{D} \\ \mathbf{b}^2\mathbf{D} \\ \cdot \\ \cdot \\ \cdot \\ \mathbf{b}^m\mathbf{D} \end{bmatrix} \tag{2-3}$$

These identities can be easily checked. Note that $\mathbf{a}^i\mathbf{b}^i$ is an $n \times r$ matrix, it is the product of an $n \times 1$ matrix $\mathbf{a}^i$ and a $1 \times r$ matrix $\mathbf{b}^i$.

The material presented here is well known and can be found in References [5], [38], [39], [43]–[45], [77], [86], and [116].[2] However our presentation is different. We emphasized the difference between a vector and its representations [see Equation (2-12) and Definition 2-7]. After stressing this distinction, the concepts of matrix representation of an operator and of the similarity transformation follow naturally.

## 2-2 Linear Spaces Over a Field

In the study of mathematics we must first specify a collection of objects that forms the center of study. This collection of objects or elements is

[2] Bracketed numbers correspond to the References at the end of the book.

called a *set*. For example, in arithmetic, we study the set of real numbers. In Boolean algebra, we study the set $\{0, 1\}$, which consists of only two elements. Other examples of sets include the set of complex numbers, the set of positive integers, the set of all polynomials of degree less than 5, the set of all $2 \times 2$ real constant matrices. In this section when we discuss a set of objects, the set could be any one of those just mentioned or any other the reader wishes to specify.

Consider the set of real numbers. It is clear that the set of real numbers has the following properties: the sum, difference, product, and quotient (except division by zero) of any two real numbers are real numbers. Any set with these properties is called a *field*. We give a formal definition of a *field* in the following.

### Definition 2-1

A field consists of a set, denoted by $\mathfrak{F}$, of elements called *scalars* and two operations called addition "+" and multiplication "·"; the two operations are defined over $\mathfrak{F}$ such that satisfying the following conditions:

**1.** To every pair of elements $\alpha$ and $\beta$ in $\mathfrak{F}$, there correspond an element $\alpha + \beta$ in $\mathfrak{F}$ called the sum of $\alpha$ and $\beta$, and an element $\alpha \cdot \beta$ or $\alpha\beta$ in $\mathfrak{F}$, called the product of $\alpha$ and $\beta$.
**2.** Addition and multiplication are respectively commutative: For any $\alpha$, $\beta$ in $\mathfrak{F}$,

$$\alpha + \beta = \beta + \alpha \qquad \alpha \cdot \beta = \beta \cdot \alpha$$

**3.** Addition and multiplication are respectively associative: For any $\alpha$, $\beta$ $\gamma$ in $\mathfrak{F}$,

$$(\alpha + \beta) + \gamma = \alpha + (\beta + \gamma) \qquad (\alpha \cdot \beta) \cdot \gamma = \alpha \cdot (\beta \cdot \gamma)$$

**4.** Multiplication is distributive with respect to addition: For any $\alpha$, $\beta$ $\gamma$ in $\mathfrak{F}$,

$$\alpha \cdot (\beta + \gamma) = (\alpha \cdot \beta) + (\alpha \cdot \gamma)$$

**5.** $\mathfrak{F}$ contains an element, denoted by 0 and an element, denoted by 1, such that $\alpha + 0 = \alpha$, $1 \cdot \alpha = \alpha$ for every $\alpha$ in $\mathfrak{F}$.
**6.** To every $\alpha$ in $\mathfrak{F}$, there is an element $\beta$ in $\mathfrak{F}$ such that $\alpha + \beta = 0$.
**7.** To every $\alpha$ in $\mathfrak{F}$ which is not the element 0, there is an element $\gamma$ in $\mathfrak{F}$ such that $\alpha \cdot \gamma = 1$. ∎

We give some examples to illustrate this concept.

### Example 1

Consider the set of numbers that consists of 0 and 1. The set $\{0, 1\}$ does not form a field if we use the usual definition of addition and multiplication, because the element $1 + 1 = 2$ is not in the set $\{0, 1\}$. However if we *define* $0 + 0 = 1 + 1 = 0$, $1 + 0 = 1$; $0 \cdot 1 = 0 \cdot 0 = 0$, $1 \cdot 1 = 1$, then it can be verified that $\{0, 1\}$ with the defined addition and multiplication satisfies all the conditions listed for a field. Hence the set $\{0, 1\}$ with the *defined operations* forms a field. It is called the field of *binary numbers*.

### Example 2

Consider the set of all $2 \times 2$ matrices of the form

$$\begin{bmatrix} x & -y \\ y & x \end{bmatrix}$$

where $x$ and $y$ are arbitrary real numbers. The set with the usual definitions of matrix addition and multiplication forms a field. The element 0 and 1 of the field are, respectively,

$$\begin{bmatrix} 0 & 0 \\ 0 & 0 \end{bmatrix} \quad \text{and} \quad \begin{bmatrix} 1 & 0 \\ 0 & 1 \end{bmatrix}$$

Note that the set of all $2 \times 2$ matrices does not form a field. ▮

From the foregoing examples, we see that the set of objects that forms a field could be anything so long as the two operations can be defined for these objects. The fields we shall encounter in this book are fortunately the most familiar ones: the field of real numbers, the field of complex numbers, and the field of rational functions with real coefficients. The additions and multiplications of these fields are defined in the usual ways. The reader is advised to show that they satisfy all the conditions required for a field. We use $\mathbb{R}$ and $\mathbb{C}$ to denote the field of real numbers and the field of complex numbers, respectively. Note that the set of positive real numbers does not form a field.

Before introducing the concept of vector space, let us consider the ordinary two-dimensional geometric plane. If the origin is chosen, then every point in the plane can be considered as a vector: it has direction as well as magnitude. A vector can be shrunk or extended. Any two vectors can be added, but the product of two points or vectors is *not* defined. Such a plane, in the mathematical terminology is called a *linear space*, or a *vector space*, or a *linear vector space*.

### Definition 2-2

A linear space over a field $\mathcal{F}$, denoted by $(\mathcal{X}, \mathcal{F})$, consists of a set, denoted by $\mathcal{X}$, of elements called *vectors*, a field $\mathcal{F}$, and two operations called vector addition and scalar multiplication. The two operations are defined over $\mathcal{X}$ and $\mathcal{F}$ such that satisfying all the following conditions:

1. To every pair of vectors $\mathbf{x}^1$ and $\mathbf{x}^2$ in $\mathcal{X}$, there corresponds a vector $\mathbf{x}^1 + \mathbf{x}^2$ in $\mathcal{X}$, called the sum of $\mathbf{x}^1$ and $\mathbf{x}^2$.

**2.** Addition is commutative: For any $\mathbf{x}^1$, $\mathbf{x}^2$ in $\mathfrak{X}$, $\mathbf{x}^1 + \mathbf{x}^2 = \mathbf{x}^2 + \mathbf{x}^1$.
**3.** Addition is associative: For any $\mathbf{x}^1$, $\mathbf{x}^2$, and $\mathbf{x}^3$ in $\mathfrak{X}$, $(\mathbf{x}^1 + \mathbf{x}^2) + \mathbf{x}^3 = \mathbf{x}^1 + (\mathbf{x}^2 + \mathbf{x}^3)$.
**4.** $\mathfrak{X}$ contains a vector, denoted by $\mathbf{0}$, such that $\mathbf{0} + \mathbf{x} = \mathbf{x}$ for every $\mathbf{x}$ in $\mathfrak{X}$. The vector $\mathbf{0}$ is called the zero vector or the origin.
**5.** To every $\mathbf{x}$ in $\mathfrak{X}$, there is a vector $\bar{\mathbf{x}}$ in $\mathfrak{X}$, such that $\mathbf{x} + \bar{\mathbf{x}} = \mathbf{0}$.
**6.** To every $\alpha$ in $\mathfrak{F}$, and every $\mathbf{x}$ in $\mathfrak{X}$, there corresponds a vector $\alpha\mathbf{x}$ in $\mathfrak{X}$ called the scalar product of $\alpha$ and $\mathbf{x}$.
**7.** Scalar multiplication is associative: For any $\alpha$, $\beta$, in $\mathfrak{F}$ and any $\mathbf{x}$ in $\mathfrak{X}$, $\alpha(\beta\mathbf{x}) = (\alpha\beta)\mathbf{x}$.
**8.** Scalar multiplication is distributive with respect to vector addition: For any $\alpha$ in $\mathfrak{F}$, and any $\mathbf{x}^1$, $\mathbf{x}^2$ in $\mathfrak{X}$, $\alpha(\mathbf{x}^1 + \mathbf{x}^2) = \alpha\mathbf{x}^1 + \alpha\mathbf{x}^2$.
**9.** Scalar multiplication is distributive with respect to scalar addition: For any $\alpha$, $\beta$ in $\mathfrak{F}$ and any $\mathbf{x}$ in $\mathfrak{X}$, $(\alpha + \beta)\mathbf{x} = \alpha\mathbf{x} + \beta\mathbf{x}$.
**10.** For any $\mathbf{x}$ in $\mathfrak{X}$, $1\mathbf{x} = \mathbf{x}$ where 1 is the element 1 in $\mathfrak{F}$. ▮

### Example 1

A field forms a vector space over itself with the vector addition and scalar multiplication defined as the corresponding operations in the field. For example, $(\mathbb{R}, \mathbb{R})$ and $(\mathbb{C}, \mathbb{C})$ are vector spaces. Note that $(\mathbb{C}, \mathbb{R})$ is a vector space but not $(\mathbb{R}, \mathbb{C})$. (Why?)

### Example 2

The set of all real-valued piecewise continuous functions defined over $(-\infty, \infty)$ forms a linear space over the field of real numbers. The addition and scalar multiplication are defined in the usual way. It is called a *function space.*

### Example 3

Given a field $\mathfrak{F}$. Let $\mathfrak{F}^n$ be all $n$-tuples of scalars written as columns

$$\mathbf{x}^i = \begin{bmatrix} x_1^i \\ x_2^i \\ \cdot \\ \cdot \\ \cdot \\ x_n^i \end{bmatrix} \tag{2-4}$$

where the superscript $i$ denotes different vectors in $\mathfrak{F}^n$, the subscript denotes various components of $\mathbf{x}^i$. If the vector addition and the scalar multiplication are defined in the following way

$$\mathbf{x}^i + \mathbf{x}^j = \begin{bmatrix} x_1{}^i + x_1{}^j \\ x_2{}^i + x_2{}^j \\ \cdot \\ \cdot \\ \cdot \\ x_n{}^i + x_n{}^j \end{bmatrix} \qquad \alpha\mathbf{x}^i = \begin{bmatrix} \alpha x_1{}^i \\ \alpha x_2{}^i \\ \cdot \\ \cdot \\ \cdot \\ \alpha x_n{}^i \end{bmatrix} \tag{2-5}$$

then $(\mathcal{F}^n, \mathcal{F})$ is a vector space. If $\mathcal{F} = \mathbb{R}$, $(\mathbb{R}^n, \mathbb{R})$ is called the $n$-dimensional *real vector space;* if $\mathcal{F} = \mathbb{C}$, $(\mathbb{C}^n, \mathbb{C})$ is called the $n$-dimensional *complex vector space.*

### Example 4

Consider the set $P_n[t]$ of all polynomials of degree less than $n$ with real coefficients.

$$\sum_{i=0}^{n-1} \alpha_i t^i$$

Let the vector addition and the scalar multiplication be defined as

$$\sum_{i=0}^{n-1} \alpha_i t^i + \sum_{i=0}^{n-1} \beta_i t^i = \sum_{i=0}^{n-1} (\alpha_i + \beta_i) t^i$$

$$\alpha \left( \sum_{i=0}^{n-1} \alpha_i t^i \right) = \sum_{i=0}^{n-1} (\alpha\alpha_i) t^i$$

It is easy to verify that $(P_n[t], \mathbb{R})$ is a linear space.

### Example 5

Let $\mathcal{X}$ denote the set of all solutions of the homogeneous differential equation $\ddot{x} + 2\dot{x} + 3x = 0$. Then $(\mathcal{X}, \mathbb{R})$ is a linear space with the vector addition and the scalar multiplication defined in the usual way. If the differential equation is not homogeneous, then $(\mathcal{X}, \mathbb{R})$ is not a linear space. (Why?) ∎

We introduce one more concept to conclude this section.

### Definition 2-3

Let $(\mathcal{X}, \mathcal{F})$ be a linear space and let $\mathcal{Y}$ be a subset of $\mathcal{X}$. Then $(\mathcal{Y}, \mathcal{F})$ is said to be a *subspace* of $(\mathcal{X}, \mathcal{F})$ if under the operations of $(\mathcal{X}, \mathcal{F})$, $\mathcal{Y}$ itself forms a vector space over $\mathcal{F}$. ∎

We remark on the conditions for a subset of $\mathcal{X}$ to form a subspace. Since the vector addition and scalar multiplication have been defined for the linear space $(\mathfrak{X}, \mathfrak{F})$, they satisfy conditions 2, 3 and 7 through 10 listed in Definition 2-2. Hence we need to check only conditions 1 and 4 through 6 to determine whether a set $\mathcal{Y}$ is a subspace of $(\mathfrak{X}, \mathfrak{F})$. It is easy to verify that if $\alpha_1\mathbf{y}^1 + \alpha_2\mathbf{y}^2$ is in $\mathcal{Y}$ for any $\mathbf{y}^1$, $\mathbf{y}^2$ in $\mathcal{Y}$ and any $\alpha_1$, $\alpha_2$ in $\mathfrak{F}$, then conditions 1 and 4 through 6 are satisfied. Hence we conclude that *a set $\mathcal{Y}$ is a subspace of $(\mathfrak{X}, \mathfrak{F})$ if $\alpha_1\mathbf{y}^1 + \alpha_2\mathbf{y}^2$ is in $\mathcal{Y}$, for any $\mathbf{y}^1$, $\mathbf{y}^2$ in $\mathcal{Y}$ and any $\alpha_1$, $\alpha_2$ in $\mathfrak{F}$.*

### Example 6

In the two-dimensional real vector space $(\mathbb{R}^2, \mathbb{R})$, every straight line passing through the origin is a subspace of $(\mathbb{R}^2, \mathbb{R})$. That is, the set

$$\begin{bmatrix} x_1 \\ \alpha x_1 \end{bmatrix}$$

for any fixed real $\alpha$ is a subspace of $(\mathbb{R}^2, \mathbb{R})$.

### Example 7

The real vector space $(\mathbb{R}^n, \mathbb{R})$ is a subspace of the complex vector space $(\mathbb{C}^n, \mathbb{C})$.

## 2-3 Linear Independence, Bases, and Representations

Every geometric plane has two coordinate axes, which are mutually perpendicular and of the same scale. The reason for having a coordinate system is to have some reference or standard to specify a point or vector in the plane. In this section, we will extend this concept of coordinate to general linear spaces. In linear spaces a coordinate system is called a *basis.* The basis vectors are generally not perpendicular to each other and have different scales. Before proceeding, we need the concept of linear independence of vectors.

### Definition 2-4

A set of vectors $\mathbf{x}^1, \mathbf{x}^2, \cdots, \mathbf{x}^n$ in a linear space over a field $\mathfrak{F}$, $(\mathfrak{X}, \mathfrak{F})$, is said to be *linearly dependent* if and only if there exist scalars $\alpha_1, \alpha_2, \cdots, \alpha_n$ in $\mathfrak{F}$, not all zero, such that

$$\alpha_1\mathbf{x}^1 + \alpha_2\mathbf{x}^2 + \cdots + \alpha_n\mathbf{x}^n = \mathbf{0} \tag{2-6}$$

If the only set of $\alpha_i$ for which (2-6) holds is $\alpha_1 = 0, \alpha_2 = 0, \cdots, \alpha_n = 0$, then the set of vectors $\mathbf{x}^1, \mathbf{x}^2, \cdots, \mathbf{x}^n$ is said to be *linearly independent.* ▌

Given any set of vectors, Equation (2-6) always holds for $\alpha_1 = 0$, $\alpha_2 = 0, \cdots, \alpha_n = 0$. Therefore, in order to show the linear independence of the set, we have to show that $\alpha_1 = 0, \alpha_2 = 0, \cdots, \alpha_n = 0$ is the *only* set of $\alpha_i$ for which (2-6) holds; that is, if any one of the $\alpha_i$'s is different from zero, then the right-hand side of (2-6) cannot be a zero vector. If a set of vectors is linearly dependent, there are generally infinitely many sets of $\alpha_i$, not all zero, that satisfy Equation (2-6). However, it is sufficient to find one set of $\alpha_i$, not all zero, to conclude the linear dependence of the set of vectors.

### Example 1

Consider the set of vectors $\mathbf{x}^1, \mathbf{x}^2, \cdots, \mathbf{x}^n$ in which $\mathbf{x}^1 = \mathbf{0}$. This set of vectors is always linearly dependent, because we may choose $\alpha_1 = 1$, $\alpha_2 = 0, \alpha_3 = 0, \cdots, \alpha_n = 0$, and Equation (2-6) holds.

### Example 2

Consider the set of vector $\mathbf{x}^1$ which consists of only one vector. The set of vector $\mathbf{x}^1$ is linearly independent if and only if $\mathbf{x}^1 \neq \mathbf{0}$. If $\mathbf{x}^1 \neq \mathbf{0}$, the only way to have $\alpha_1\mathbf{x}^1 = \mathbf{0}$ is $\alpha_1 = 0$. If $\mathbf{x}^1 = \mathbf{0}$, we may choose $\alpha_1 = 1$. ▮

If we introduce the notation

$$\alpha_1\mathbf{x}^1 + \alpha_2\mathbf{x}^2 + \cdots + \alpha_n\mathbf{x}^n \triangleq [\mathbf{x}^1 \quad \mathbf{x}^2 \quad \cdots \quad \mathbf{x}^n]\begin{bmatrix} \alpha_1 \\ \alpha_2 \\ \cdot \\ \cdot \\ \cdot \\ \alpha_n \end{bmatrix}$$

$$\triangleq [\mathbf{x}^1 \quad \mathbf{x}^2 \quad \cdots \quad \mathbf{x}^n]\boldsymbol{\alpha} \tag{2-7}$$

then the linear independence of a set of vectors can also be stated in the following definition.

### Definition 2-4′

A set of vectors $\mathbf{x}^1, \mathbf{x}^2, \cdots, \mathbf{x}^n$ in $(\mathcal{X}, \mathcal{F})$ is said to be linearly independent if and only if the equation

$$[\mathbf{x}^1 \quad \mathbf{x}^2 \quad \cdots \quad \mathbf{x}^n]\boldsymbol{\alpha} = \mathbf{0}$$

implies $\boldsymbol{\alpha} = \mathbf{0}$, where every component of $\boldsymbol{\alpha}$ is an element of $\mathcal{F}$ or, correspondingly, $\boldsymbol{\alpha}$ can be considered as a vector in $\mathcal{F}^n$. ▮

Observe that linear dependence depends not only on the set of vectors but also on the field. For example, the set of vectors $\{\mathbf{x}^1, \mathbf{x}^2\}$, where

$$\mathbf{x}^1 \triangleq \begin{bmatrix} \dfrac{1}{s+1} \\ \dfrac{1}{s+2} \end{bmatrix} \qquad \mathbf{x}^2 \triangleq \begin{bmatrix} \dfrac{s+2}{(s+1)(s+3)} \\ \dfrac{1}{s+3} \end{bmatrix}$$

is linearly dependent in the field of rational functions with real coefficients. Indeed, if we choose

$$\alpha_1 = -1 \qquad \text{and} \qquad \alpha_2 = \frac{s+3}{s+2}$$

then $\alpha_1 \mathbf{x}^1 + \alpha_2 \mathbf{x}^2 = \mathbf{0}$. However this set of vectors is linearly independent in the field of real numbers. Since there exist no $\alpha_1$ and $\alpha_2$ in $\mathbb{R}$ that are different from zero, such that $\alpha_1 \mathbf{x}^1 + \alpha_2 \mathbf{x}^2 = \mathbf{0}$.

It is clear from the definition of linear dependence that if the vectors $\mathbf{x}^1, \mathbf{x}^2, \cdots, \mathbf{x}^n$ are linearly dependent, then at least one of them can be written as a linear combination of the others. However, it is not necessarily true that every one of them can be expressed as a linear combination of the others.

### Definition 2-5

The maximal number of linearly independent vectors in a linear space $(\mathcal{X}, \mathcal{F})$ is called the *dimension* of the linear space $(\mathcal{X}, \mathcal{F})$. ▮

In the previous section we introduced the $n$-dimensional real vector space $(\mathbb{R}^n, \mathbb{R})$. The meaning of $n$-dimensional is now clear. It means that in $(\mathbb{R}^n, \mathbb{R})$, there are, at most, $n$ linearly independent vectors (over the field $\mathbb{R}$). In the two-dimensional real vector space $(\mathbb{R}^2, \mathbb{R})$, one cannot find three linearly independent vectors. (Try!)

### Example 3

Consider the function space that consists of all real-valued piecewise continuous functions defined over $(-\infty, \infty)$. The zero vector in this space is the one which is identically zero on $(-\infty, \infty)$. The following functions, with $-\infty < t < \infty$,

$$t, t^2, t^3, \cdots$$

are clearly elements of the function space. This set of functions $\{t^n, n = 1, 2, \cdots\}$ is linearly independent, because there exist no real constants, $\alpha_i$'s, not all zero, such that

$$\sum_{i=1} \alpha_i t^i \equiv 0$$

There are infinitely many of these functions; therefore, the dimension of this space is infinity. ▮

We assume that all the linear spaces we shall encounter are of finite dimensions unless stated otherwise.

### Definition 2-6

A set of linearly independent vectors of a linear space $(\mathfrak{X}, \mathfrak{F})$ is said to be a *basis* of $\mathfrak{X}$ if every vector in $\mathfrak{X}$ can be expressed as a unique linear combination of these vectors. ∎

### Theorem 2-1

In an $n$-dimensional vector space, *any* set of $n$ linearly independent vectors qualifies as a basis.

### Proof

Let $\mathbf{e}^1, \mathbf{e}^2, \cdots, \mathbf{e}^n$ be any $n$ linearly independent vectors in $\mathfrak{X}$, and let $\mathbf{x}$ be an arbitrary vector in $\mathfrak{X}$. Then the set of $(n + 1)$ vectors $\mathbf{x}, \mathbf{e}^1, \mathbf{e}^2, \cdots, \mathbf{e}^n$ is linearly dependent (since, by the definition of dimension, $n$ is the maximum number of linearly independent vectors we can have in the space). Consequently, there exist $\alpha_0, \alpha_1, \cdots, \alpha_n$ in $\mathfrak{F}$, not all zero, such that

$$\alpha_0\mathbf{x} + \alpha_1\mathbf{e}^1 + \alpha_2\mathbf{e}^2 + \cdots + \alpha_n\mathbf{e}^n = \mathbf{0} \tag{2-8}$$

We claim that $\alpha_0 \neq 0$. If $\alpha_0 = 0$, Equation (2-8) reduces to

$$\alpha_1\mathbf{e}^1 + \alpha_2\mathbf{e}^2 + \cdots + \alpha_n\mathbf{e}^n = \mathbf{0} \tag{2-9}$$

which, together with the linear independence assumption of $\mathbf{e}^1, \mathbf{e}^2, \cdots, \mathbf{e}^n$, implies that $\alpha_1 = 0, \alpha_2 = 0, \cdots, \alpha_n = 0$. This contradicts the assumption that not all $\alpha_0, \alpha_1, \cdots, \alpha_n$ are zero. If we define $\beta_i \triangleq -\alpha_i/\alpha_0$, for $i = 1, 2, \cdots, n$, then (2-8) becomes

$$\mathbf{x} = \beta_1\mathbf{e}^1 + \beta_2\mathbf{e}^2 + \cdots + \beta_n\mathbf{e}^n \tag{2-10}$$

This shows that every vector $\mathbf{x}$ in $\mathfrak{X}$ can be expressed as a linear combination of $\mathbf{e}^1, \mathbf{e}^2, \cdots, \mathbf{e}^n$. Now we show that this combination is unique. Suppose there is another linear combination, say

$$\mathbf{x} = \tilde{\beta}_1\mathbf{e}^1 + \tilde{\beta}_2\mathbf{e}^2 + \cdots + \tilde{\beta}_n\mathbf{e}^n \tag{2-11}$$

Then by subtracting (2-11) from (2-10), we obtain

$$\mathbf{0} = (\beta_1 - \tilde{\beta}_1)\mathbf{e}^1 + (\beta_2 - \tilde{\beta}_2)\mathbf{e}^2 + \cdots + (\beta_n - \tilde{\beta}_n)\mathbf{e}^n$$

which, together with the linear independence of $\{\mathbf{e}^i\}$, implies that

$$\beta_i = \tilde{\beta}_i \qquad i = 1, 2, \cdots, n \qquad \text{Q.E.D.}$$

This theorem has a very important implication. In an $n$-dimensional vector space $(\mathfrak{X}, \mathfrak{F})$, if a basis is chosen, then every vector in $\mathfrak{X}$ can be uniquely represented by a set of $n$ scalars $\beta_1, \beta_2, \cdots, \beta_n$ in $\mathfrak{F}$. If we use the notation of (2-7), we may write (2-10) as

$$\mathbf{x} = [\mathbf{e}^1 \quad \mathbf{e}^2 \quad \cdots \quad \mathbf{e}^n]\boldsymbol{\beta} \tag{2-12}$$

where $\boldsymbol{\beta} = [\beta_1 \quad \beta_2 \quad \cdots \quad \beta_n]'$ and the prime denotes the transpose. The $n \times 1$ vector $\boldsymbol{\beta}$ can be considered as a vector in $(\mathfrak{F}^n, \mathfrak{F})$. Consequently, *there is a one-to-one correspondence between any n-dimensional vector space* $(\mathfrak{X}, \mathfrak{F})$ *and the same dimensional linear space* $(\mathfrak{F}^n, \mathfrak{F})$ *if a basis is chosen for* $(\mathfrak{X}, \mathfrak{F})$.

### Definition 2-7

In an $n$-dimensional vector space $(\mathfrak{X}, \mathfrak{F})$, if a basis $\{\mathbf{e}^1, \mathbf{e}^2, \cdots, \mathbf{e}^n\}$ is chosen, then every vector $\mathbf{x}$ in $\mathfrak{X}$ can be uniquely written in the form of (2-12). $\boldsymbol{\beta}$ is called the *representation* of $\mathbf{x}$ with respect to the basis $\{\mathbf{e}^1, \mathbf{e}^2, \cdots, \mathbf{e}^n\}$.

### Example 4

The geometric plane shown in Figure 2-1 can be considered as a two-dimensional real vector space. Any point in the plane is a vector.

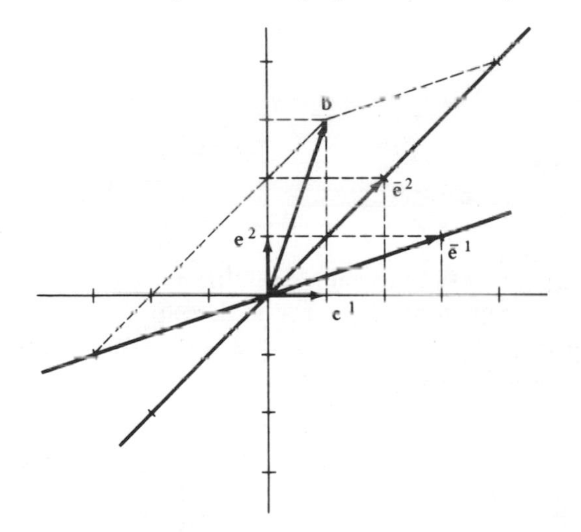


**Figure 2-1** A two-dimensional real vector space.

Theorem 2-1 states that any set of two linearly independent vectors forms a basis. Observe that we have not only the freedom in choosing the directions of the basis vectors (as long as they do not lie in the same line) but also the magnitude (scale) of these vectors. Therefore, given a vector in $(\mathbb{R}^2, \mathbb{R})$, for different bases we have different representations of the *same* vector. For example, the representations of the vector $\mathbf{b}$ in

Figure 2-1 with respect to the basis $\{\mathbf{e}^1, \mathbf{e}^2\}$ and the basis $\{\bar{\mathbf{e}}^1, \bar{\mathbf{e}}^2\}$ are, respectively, $[1 \quad 3]'$ and $[-1 \quad 2]'$ (where the "prime" symbol denotes the transpose). We summarize the representations of the vectors $\mathbf{b}$, $\bar{\mathbf{e}}^1$, $\bar{\mathbf{e}}^2$, $\mathbf{e}^1$, and $\mathbf{e}^2$ with respect to the bases $\{\mathbf{e}^1, \mathbf{e}^2\}$, $\{\bar{\mathbf{e}}^1, \bar{\mathbf{e}}^2\}$ in Table 2-1:

**Table 2-1** DIFFERENT REPRESENTATIONS OF VECTORS

| Bases \ Vectors | $\mathbf{b}$ | $\bar{\mathbf{e}}^1$ | $\bar{\mathbf{e}}^2$ | $\mathbf{e}^1$ | $\mathbf{e}^2$ |
|---|---|---|---|---|---|
| $[\mathbf{e}^1 \quad \mathbf{e}^2]$ | $\begin{bmatrix} 1 \\ 3 \end{bmatrix}$ | $\begin{bmatrix} 3 \\ 1 \end{bmatrix}$ | $\begin{bmatrix} 2 \\ 2 \end{bmatrix}$ | $\begin{bmatrix} 1 \\ 0 \end{bmatrix}$ | $\begin{bmatrix} 0 \\ 1 \end{bmatrix}$ |
| $[\bar{\mathbf{e}}^1 \quad \bar{\mathbf{e}}^2]$ | $\begin{bmatrix} -1 \\ 2 \end{bmatrix}$ | $\begin{bmatrix} 1 \\ 0 \end{bmatrix}$ | $\begin{bmatrix} 0 \\ 1 \end{bmatrix}$ | $\begin{bmatrix} \frac{1}{2} \\ -\frac{1}{4} \end{bmatrix}$ | $\begin{bmatrix} -\frac{1}{2} \\ \frac{3}{4} \end{bmatrix}$ |

### Example 5

Consider the linear space $(P_n[t], \mathbb{R})$, where $P_n[t]$ is the set of all polynomials of degree less than $n$ with coefficients in $\mathbb{R}$. Let $\mathbf{x} = \beta_1 t^{n-1} + \beta_2 t^{n-2} + \cdots + \beta_n$ be an arbitrary vector in $P_n[t]$. Let $\mathbf{e}^1 = t^{n-1}$, $\mathbf{e}^2 = t^{n-2}, \ldots, \mathbf{e}^{n-1} = t$, and $\mathbf{e}^n = 1$. Clearly, the vectors $\mathbf{e}^1, \mathbf{e}^2, \ldots, \mathbf{e}^n$ are linearly independent and qualify as basis vectors. With respect to these basis vectors, $\mathbf{x}$ can be written as

$$\mathbf{x} = [\mathbf{e}^1 \quad \mathbf{e}^2 \quad \cdots \quad \mathbf{e}^n] \begin{bmatrix} \beta_1 \\ \beta_2 \\ \cdot \\ \cdot \\ \cdot \\ \beta_n \end{bmatrix}$$

hence $[\beta_1 \quad \beta_2 \quad \cdots \quad \beta_n]'$ (where the "prime" symbol stands for the transpose) is the representation of $\mathbf{x}$ with respect to $\{\mathbf{e}^1, \mathbf{e}^2, \ldots, \mathbf{e}^n\}$. If we choose $\bar{\mathbf{e}}^1 = t^{n-1} - t^{n-2}$, $\bar{\mathbf{e}}^2 = t^{n-2} - t^{n-3}$, $\bar{\mathbf{e}}^3 = t^{n-3} - t^{n-4}, \ldots$, $\bar{\mathbf{e}}^{n-1} = t - 1$, and $\bar{\mathbf{e}}^n = 1$ as the basis vectors, then

$$\mathbf{x} = [\bar{\mathbf{e}}^1 \quad \bar{\mathbf{e}}^2 \quad \cdots \quad \bar{\mathbf{e}}^n] \begin{bmatrix} \beta_1 \\ \sum_{i=1}^{2} \beta_i \\ \sum_{i=1}^{3} \beta_i \\ \cdot \\ \cdot \\ \cdot \\ \sum_{i=1}^{n} \beta_i \end{bmatrix}$$

Hence the representation of $\mathbf{x}$ with respect to $\{\bar{\mathbf{e}}^1, \bar{\mathbf{e}}^2, \cdots, \bar{\mathbf{e}}^n\}$ is

$$\left[\beta_1 \quad \sum_{i=1}^{2} \beta_i \quad \sum_{i=1}^{3} \beta_i \quad \cdots \quad \sum_{i=1}^{n} \beta_i\right]'$$

Clearly, $(P_n[t], \mathbb{R})$ has dimension $n$. ∎

In this example, there is a sharp distinction between vectors and representations. However, this is not always the case. Let us consider the $n$-dimensional real vector space $(\mathbb{R}^n, \mathbb{R})$ or complex vector space $(\mathbb{C}^n, \mathbb{C})$, a vector is an $n$-tuple of real or complex numbers, written as

$$\mathbf{x} = \begin{bmatrix} \beta_1 \\ \beta_2 \\ \cdot \\ \cdot \\ \cdot \\ \beta_n \end{bmatrix}$$

This array of $n$ numbers can be interpreted in two ways: (1) It is defined as such; that is, it is a vector and is independent of basis. (2) It is a representation of a vector with respect to some fixed unknown basis. Given an array of numbers, unless it is tied up with some basis, we shall always consider it as a vector. However we shall also introduce, unless stated otherwise, the following vectors:[3]

$$\mathbf{n}^1 = \begin{bmatrix} 1 \\ 0 \\ 0 \\ \cdot \\ \cdot \\ \cdot \\ 0 \\ 0 \end{bmatrix}, \mathbf{n}^2 = \begin{bmatrix} 0 \\ 1 \\ 0 \\ \cdot \\ \cdot \\ \cdot \\ 0 \\ 0 \end{bmatrix}, \cdots, \mathbf{n}^{n-1} = \begin{bmatrix} 0 \\ 0 \\ 0 \\ \cdot \\ \cdot \\ \cdot \\ 1 \\ 0 \end{bmatrix}, \mathbf{n}^n = \begin{bmatrix} 0 \\ 0 \\ 0 \\ \cdot \\ \cdot \\ \cdot \\ 0 \\ 1 \end{bmatrix} \tag{2-13}$$

as the basis of $(\mathbb{R}^n, \mathbb{R})$ and $(\mathbb{C}^n, \mathbb{C})$. In this case, an array of numbers can be interpreted as a vector or the representation of a vector with respect to the basis $\{\mathbf{n}^1, \mathbf{n}^2, \cdots, \mathbf{n}^n\}$, because with respect to this particular set of bases, the representation and the vector itself are identical; that is,

$$\begin{bmatrix} \beta_1 \\ \beta_2 \\ \cdot \\ \cdot \\ \cdot \\ \beta_n \end{bmatrix} = [\mathbf{n}^1 \quad \mathbf{n}^2 \quad \cdots \quad \mathbf{n}^n] \begin{bmatrix} \beta_1 \\ \beta_2 \\ \cdot \\ \cdot \\ \cdot \\ \beta_n \end{bmatrix} \tag{2-14}$$

[3] This set of vectors is called an *orthonormal* set.

**Change of basis.** We have shown that a vector $\mathbf{x}$ in $(\mathcal{X}, \mathcal{F})$ has different representations with respect to different bases. It is natural to ask what the relationships are between these different representations of the same vector. In this subsection, this problem will be studied.

Let the representations of a vector $\mathbf{x}$ in $(\mathcal{X}, \mathcal{F})$ with respect to $\{\mathbf{e}^1, \mathbf{e}^2, \cdots, \mathbf{e}^n\}$ and $\{\bar{\mathbf{e}}^1, \bar{\mathbf{e}}^2, \cdots, \bar{\mathbf{e}}^n\}$ be $\boldsymbol{\beta}$ and $\bar{\boldsymbol{\beta}}$, respectively; that is,

$$\mathbf{x} = [\mathbf{e}^1 \quad \mathbf{e}^2 \quad \cdots \quad \mathbf{e}^n]\boldsymbol{\beta} = [\bar{\mathbf{e}}^1 \quad \bar{\mathbf{e}}^2 \quad \cdots \quad \bar{\mathbf{e}}^n]\bar{\boldsymbol{\beta}} \tag{2-15}$$

In order to derive the relationship between $\boldsymbol{\beta}$ and $\bar{\boldsymbol{\beta}}$, we need either the information of the representations of $\bar{\mathbf{e}}^i$, for $i = 1, 2, \cdots, n$, with respect to the basis $\{\mathbf{e}^1, \mathbf{e}^2, \cdots, \mathbf{e}^n\}$, or the information of the representations of $\mathbf{e}^i$, for $i = 1, 2, \cdots, n$, with respect to the basis $\{\bar{\mathbf{e}}^1, \bar{\mathbf{e}}^2, \cdots, \bar{\mathbf{e}}^n\}$. Let the representation of $\mathbf{e}^i$ with respect to $\{\bar{\mathbf{e}}^1, \bar{\mathbf{e}}^2, \cdots, \bar{\mathbf{e}}^n\}$ be $[p_{1i} \quad p_{2i} \quad \cdots \quad p_{ni}]'$; that is,

$$\mathbf{e}^i = [\bar{\mathbf{e}}^1 \quad \bar{\mathbf{e}}^2 \quad \cdots \quad \bar{\mathbf{e}}^n]\begin{bmatrix} p_{1i} \\ p_{2i} \\ \cdot \\ \cdot \\ \cdot \\ p_{ni} \end{bmatrix} \triangleq \mathbf{E}\mathbf{p}^i \qquad i = 1, 2, \cdots, n \tag{2-16}$$

where $\mathbf{E} \triangleq [\bar{\mathbf{e}}^1 \quad \bar{\mathbf{e}}^2 \quad \cdots \quad \bar{\mathbf{e}}^n]$, $\mathbf{p}^i \triangleq [p_{1i} \quad p_{2i} \quad \cdots \quad p_{ni}]'$. Using matrix notation, we write

$$[\mathbf{e}^1 \quad \mathbf{e}^2 \quad \cdots \quad \mathbf{e}^n] = [\mathbf{E}\mathbf{p}^1 \quad \mathbf{E}\mathbf{p}^2 \quad \cdots \quad \mathbf{E}\mathbf{p}^n] \tag{2-17}$$

which, by using (2-2), can be written as

$$\begin{aligned}
[\mathbf{e}^1 \quad \mathbf{e}^2 \quad \cdots \quad \mathbf{e}^n] &= \mathbf{E}[\mathbf{p}^1 \quad \mathbf{p}^2 \quad \cdots \quad \mathbf{p}^n] \\
&= [\bar{\mathbf{e}}^1 \quad \bar{\mathbf{e}}^2 \quad \cdots \quad \bar{\mathbf{e}}^n]\begin{bmatrix} p_{11} & p_{12} & \cdots & p_{1n} \\ p_{21} & p_{22} & \cdots & p_{2n} \\ \cdot & \cdot & & \cdot \\ \cdot & \cdot & & \cdot \\ \cdot & \cdot & & \cdot \\ p_{n1} & p_{n2} & \cdots & p_{nn} \end{bmatrix} \\
&\triangleq [\bar{\mathbf{e}}^1 \quad \bar{\mathbf{e}}^2 \quad \cdots \quad \bar{\mathbf{e}}^n]\mathbf{P}
\end{aligned} \tag{2-18}$$

Substituting (2-18) into (2-15), we obtain

$$\mathbf{x} = [\bar{\mathbf{e}}^1 \quad \bar{\mathbf{e}}^2 \quad \cdots \quad \bar{\mathbf{e}}^n]\mathbf{P}\boldsymbol{\beta} = [\bar{\mathbf{e}}^1 \quad \bar{\mathbf{e}}^2 \quad \cdots \quad \bar{\mathbf{e}}^n]\bar{\boldsymbol{\beta}} \tag{2-19}$$

Since the representation of $\mathbf{x}$ with respect to the basis $\{\bar{\mathbf{e}}^1, \bar{\mathbf{e}}^2, \cdots, \bar{\mathbf{e}}^n\}$ is unique, (2-19) implies

$$\bar{\boldsymbol{\beta}} = \mathbf{P}\boldsymbol{\beta} \tag{2-20}$$

where

$$\mathbf{P} = \begin{bmatrix} i\text{th column: the} \\ \text{representation of} \\ \mathbf{e}^i \text{ with respect to} \\ \{\bar{\mathbf{e}}^1, \bar{\mathbf{e}}^2, \cdots, \bar{\mathbf{e}}^n\} \end{bmatrix} \tag{2-21}$$

This establishes the relationship between $\bar{\beta}$ and $\beta$. In (2-16), if the representation of $\bar{\mathbf{e}}^i$ with respect to $\{\mathbf{e}^1, \mathbf{e}^2, \cdots, \mathbf{e}^n\}$ is used, then we shall obtain

$$\beta = \mathbf{Q}\bar{\beta} \tag{2-22}$$

where

$$\mathbf{Q} = \begin{bmatrix} i\text{th column: the} \\ \text{representation of} \\ \bar{\mathbf{e}}^i \text{ with respect to} \\ \{\mathbf{e}^1, \mathbf{e}^2, \cdots, \mathbf{e}^n\} \end{bmatrix} \tag{2-23}$$

Different representations of a vector are related by (2-20) or (2-22). Therefore, given two sets of bases, if the representation of a vector with respect to one set of bases is known, the representation of the same vector with respect to the other set of bases can be computed by using either (2-20) or (2-22). Since $\bar{\beta} = \mathbf{P}\beta$ and $\beta = \mathbf{Q}\bar{\beta}$, we have $\bar{\beta} = \mathbf{PQ}\bar{\beta}$, for all $\bar{\beta}$, hence we may conclude that

$$\mathbf{PQ} = \mathbf{I} \quad \text{or} \quad \mathbf{P} = \mathbf{Q}^{-1} \tag{2-24}$$

## 2-4 Linear Operators and Their Representations

The concept of a function is basic to all parts of analysis. Given two sets $\mathfrak{X}$ and $\mathcal{Y}$, if we assign to each element of $\mathfrak{X}$ one and only one element of $\mathcal{Y}$, then the rule of assignments is called a *function*. For example, the rule of assignments in Figure 2-2(a) is a function, but not the one in

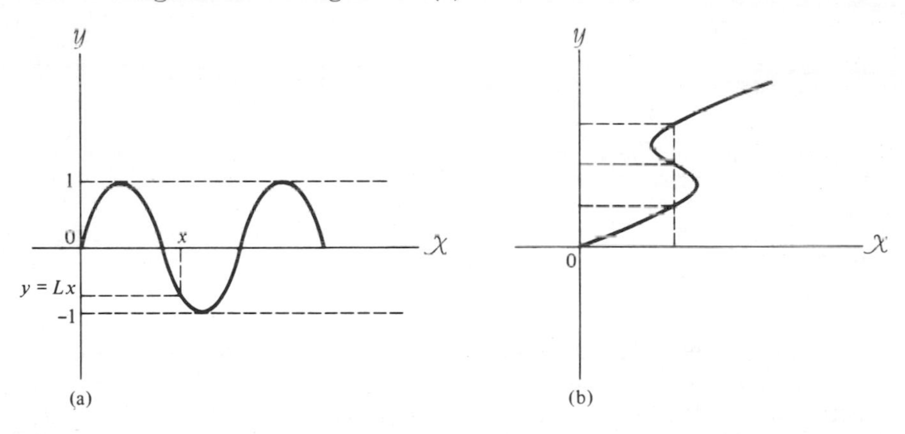


**Figure 2-2** Examples in which (a) the curve represents a function and (b) the curve does not represent a function.

Figure 2-2(b). A function is usually denoted by the notation $f\colon \mathfrak{X} \to \mathcal{Y}$, and the element of $\mathcal{Y}$ that is assigned to the element $x$ of $\mathfrak{X}$ is denoted by $y = f(x)$. The set $\mathfrak{X}$ on which a function is defined is called the *domain* of the function. The subset of $\mathcal{Y}$ that is assigned to some element of $\mathfrak{X}$ is called the *range* of the function. For example, the domain of the function shown in Figure 2-2(a) is the positive real line, the range of the function is the set $[-1, 1]$ which is a subset of the entire real line $\mathcal{Y}$.

The functions we shall study in this section is a restricted class of functions, called *linear functions,* or more often called *linear operators, linear mappings,* or *linear transformations.* The sets associated with linear operators are required to be linear spaces over the same field, say $(\mathcal{X}, \mathcal{F})$ and $(\mathcal{Y}, \mathcal{F})$. A linear operator is denoted by $L: (\mathcal{X}, \mathcal{F}) \to (\mathcal{Y}, \mathcal{F})$. In words, $L$ maps $(\mathcal{X}, \mathcal{F})$ into $(\mathcal{Y}, \mathcal{F})$.

### Definition 2-8

A function $L$ that maps $(\mathcal{X}, \mathcal{F})$ into $(\mathcal{Y}, \mathcal{F})$ is said to be a *linear operator* if and only if

$$L(\alpha_1\mathbf{x}^1 + \alpha_2\mathbf{x}^2) = \alpha_1 L\mathbf{x}^1 + \alpha_2 L\mathbf{x}^2$$

for any vectors $\mathbf{x}^1$, $\mathbf{x}^2$ in $\mathcal{X}$ and any scalars $\alpha_1$, $\alpha_2$ in $\mathcal{F}$. ∎

Note that the vectors $L\mathbf{x}^1$ and $L\mathbf{x}^2$ are elements of $\mathcal{Y}$. The reason for requiring that $\mathcal{Y}$ be defined over the same field as $\mathcal{X}$ is to ensure that $\alpha_1 L\mathbf{x}^1$ and $\alpha_2 L\mathbf{x}^2$ be defined.

### Example 1

Consider the transformation that rotates a point in a geometric plane counterclockwise 90° with respect to the origin as shown in Figure 2-3.

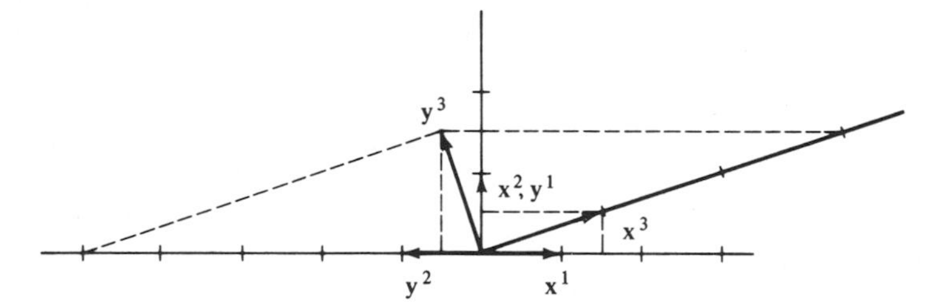


**Figure 2-3** The transformation that rotates a vector counterclockwise 90 degrees.

Given any two vectors in the plane, it is easy to verify that the vector that is the sum of the two vectors after rotation is equal to the rotation of the vector that is the sum of the two vectors before rotation. Hence the transformation is a linear transformation. The spaces $(\mathcal{X}, \mathcal{F})$ and $(\mathcal{Y}, \mathcal{F})$ of this example are all equal to $(\mathbb{R}^2, \mathbb{R})$.

### Example 2

Let $\mathcal{U}$ be the set of all real-valued piecewise continuous functions defined over $[0, \infty)$. It is clear that $(\mathcal{U}, \mathbb{R})$ is a linear space whose dimension is infinity (see Example 3, Section 2-3). Let $g$ be a continuous function defined over $[0, \infty)$. Then the transformation

$$y(t) = \int_0^\infty g(t - \tau)u(\tau)\, d\tau \tag{2-25}$$

is a linear transformation. The spaces $(\mathcal{X}, \mathcal{F})$ and $(\mathcal{Y}, \mathcal{F})$ of this example are all equal to $(\mathcal{U}, \mathbb{R})$.

**Matrix representations of a linear operator.** We see from the above two examples that the spaces $(\mathcal{X}, \mathcal{F})$ and $(\mathcal{Y}, \mathcal{F})$ on which a linear operator is defined may be of finite or infinite dimension. We show in the following that every linear operator that maps finite dimensional $(\mathcal{X}, \mathcal{F})$ into finite dimensional $(\mathcal{Y}, \mathcal{F})$ has matrix representations with coefficients in the field $\mathcal{F}$. If $(\mathcal{X}, \mathcal{F})$ and $(\mathcal{Y}, \mathcal{F})$ are of infinite dimension, a representation of a linear operator can still be found. However, the representation will not be in a matrix form; it will be of a form similar to (2-25). This will not be discussed in this chapter.

Theorem 2-2

Let $(\mathcal{X}, \mathcal{F})$ and $(\mathcal{Y}, \mathcal{F})$ be $n$- and $m$-dimensional vector spaces over the same field. Let $\mathbf{x}^1, \mathbf{x}^2, \cdots, \mathbf{x}^n$ be a set of linearly independent vectors in $\mathcal{X}$. Then the linear operator $L: (\mathcal{X}, \mathcal{F}) \to (\mathcal{Y}, \mathcal{F})$ is uniquely determined by the $n$-pairs of mappings $\mathbf{y}^i = L\mathbf{x}^i$, for $i = 1, 2, \cdots, n$. Furthermore, with respect to the basis $\{\mathbf{x}^1, \mathbf{x}^2, \cdots, \mathbf{x}^n\}$ of $\mathcal{X}$ and a basis $\{\mathbf{u}^1, \mathbf{u}^2, \cdots, \mathbf{u}^m\}$ of $\mathcal{Y}$, $L$ can be represented by an $m \times n$ matrix $\mathbf{A}$ with coefficients in the field $\mathcal{F}$. The $i$th column of $\mathbf{A}$ is the representation of $\mathbf{y}^i$ with respect to the basis $\{\mathbf{u}^1, \mathbf{u}^2, \cdots, \mathbf{u}^m\}$.

Proof

Let $\mathbf{x}$ be an arbitrary vector in $\mathcal{X}$. Since $\mathbf{x}^1, \mathbf{x}^2, \cdots, \mathbf{x}^n$ are linearly independent, the set of vectors qualifies as a basis. Consequently, the vector $\mathbf{x}$ can be expressed uniquely as $\alpha_1\mathbf{x}^1 + \alpha_2\mathbf{x}^2 + \cdots + \alpha_n\mathbf{x}^n$ (Theorem 2-1). By the linearity of $L$, we have

$$\begin{aligned} L\mathbf{x} &= \alpha_1 L\mathbf{x}^1 + \alpha_2 L\mathbf{x}^2 + \cdots + \alpha_n L\mathbf{x}^n \\ &= \alpha_1\mathbf{y}^1 + \alpha_2\mathbf{y}^2 + \cdots + \alpha_n\mathbf{y}^n \end{aligned}$$

which implies that for any $\mathbf{x}$ in $\mathcal{X}$, $L\mathbf{x}$ is uniquely determined by $\mathbf{y}^i = L\mathbf{x}^i$, for $i = 1, 2, \cdots, n$. This proves the first part of the theorem.

Let the representation of $\mathbf{y}^i$ with respect to $\{\mathbf{u}^1, \mathbf{u}^2, \cdots, \mathbf{u}^m\}$ be $[a_{1i} \quad a_{2i} \quad \cdots \quad a_{mi}]'$; that is,

$$\mathbf{y}^i = [\mathbf{u}^1 \quad \mathbf{u}^2 \quad \cdots \quad \mathbf{u}^m] \begin{bmatrix} a_{1i} \\ a_{2i} \\ \cdot \\ \cdot \\ \cdot \\ a_{mi} \end{bmatrix} \qquad i = 1, 2, \cdots, n \tag{2-26}$$

where the $a_{ij}$'s are elements of $\mathfrak{F}$. Let us write, as in (2-17) and (2-18),

$$L[\mathbf{x}^1 \quad \mathbf{x}^2 \quad \cdots \quad \mathbf{x}^n] = [\mathbf{y}^1 \quad \mathbf{y}^2 \quad \cdots \quad \mathbf{y}^n]$$

$$= [\mathbf{u}^1 \quad \mathbf{u}^2 \quad \cdots \quad \mathbf{u}^m] \begin{bmatrix} a_{11} & a_{12} & \cdots & a_{1n} \\ a_{21} & a_{22} & \cdots & a_{2n} \\ \cdot & \cdot & & \cdot \\ \cdot & \cdot & & \cdot \\ \cdot & \cdot & & \cdot \\ a_{m1} & a_{m2} & \cdots & a_{mn} \end{bmatrix}$$

$$\triangleq [\mathbf{u}^1 \quad \mathbf{u}^2 \quad \cdots \quad \mathbf{u}^m]\mathbf{A} \tag{2-27}$$

Note that the elements of $\mathbf{A}$ are in the field $\mathfrak{F}$ and the $i$th column of $\mathbf{A}$ is the representation of $\mathbf{y}^i$ with respect to the basis of $\mathcal{Y}$. With respect to the basis $\{\mathbf{x}^1, \mathbf{x}^2, \cdots, \mathbf{x}^n\}$ of $(\mathfrak{X}, \mathfrak{F})$ and the basis $\{\mathbf{u}^1, \mathbf{u}^2, \cdots, \mathbf{u}^m\}$ of $(\mathcal{Y}, \mathfrak{F})$, the linear operator $\mathbf{y} = L\mathbf{x}$ can be written as

$$[\mathbf{u}^1 \quad \mathbf{u}^2 \quad \cdots \quad \mathbf{u}^m]\boldsymbol{\beta} = L[\mathbf{x}^1 \quad \mathbf{x}^2 \quad \cdots \quad \mathbf{x}^n]\boldsymbol{\alpha} \tag{2-28}$$

where $\boldsymbol{\beta} \triangleq [\beta_1 \quad \beta_2 \quad \cdots \quad \beta_m]'$ and $\boldsymbol{\alpha} \triangleq [\alpha_1 \quad \alpha_2 \quad \cdots \quad \alpha_n]'$ are the representations of $\mathbf{y}$ and $\mathbf{x}$, respectively. After the bases are chosen, there are no differences between specifying $\mathbf{x}$, $\mathbf{y}$, and $\boldsymbol{\alpha}$, $\boldsymbol{\beta}$; hence in studying $\mathbf{y} = L\mathbf{x}$, we may just study the relationship between $\boldsymbol{\beta}$ and $\boldsymbol{\alpha}$. By substituting (2-27) into (2-28), we obtain

$$[\mathbf{u}^1 \quad \mathbf{u}^2 \quad \cdots \quad \mathbf{u}^m]\boldsymbol{\beta} = [\mathbf{u}^1 \quad \mathbf{u}^2 \quad \cdots \quad \mathbf{u}^m]\mathbf{A}\boldsymbol{\alpha} \tag{2-29}$$

which, together with the uniqueness of a representation, implies that

$$\boldsymbol{\beta} = \mathbf{A}\boldsymbol{\alpha} \tag{2-30}$$

Hence we conclude that if the bases of $(\mathfrak{X}, \mathfrak{F})$ and $(\mathcal{Y}, \mathfrak{F})$ are chosen, the operator can be represented by a matrix with coefficients in $\mathfrak{F}$. Q.E.D.

We see from (2-30) that the matrix $\mathbf{A}$ gives the relation between the representations $\boldsymbol{\alpha}$ and $\boldsymbol{\beta}$, not the vectors $\mathbf{x}$ and $\mathbf{y}$. We also see that $\mathbf{A}$ depends on the basis chosen. Hence, for different bases, we have different representations of the same operator.

We study in the following an important subclass of linear operators that maps a linear space $(\mathfrak{X}, \mathfrak{F})$ into itself; that is, $L: (\mathfrak{X}, \mathfrak{F}) \to (\mathfrak{X}, \mathfrak{F})$. In this case, the same basis is always used for these two linear spaces. If a basis of $\mathfrak{X}$, say $\{\mathbf{e}^1 \quad \mathbf{e}^2 \quad \cdots \quad \mathbf{e}^n\}$ is chosen then a matrix representation $\mathbf{A}$ of the linear operator $L$ can be obtained by using Theorem 2-2. For a different basis $\{\bar{\mathbf{e}}^1, \bar{\mathbf{e}}^2, \cdots, \bar{\mathbf{e}}^n\}$, we shall obtain a different representation $\bar{\mathbf{A}}$ of the same operator $L$. We shall now establish the relationship between $\mathbf{A}$ and $\bar{\mathbf{A}}$. Consider Figure 2-4; $\mathbf{x}$ is an arbitrary vector in $\mathfrak{X}$, $\boldsymbol{\alpha}$ and $\bar{\boldsymbol{\alpha}}$ are the representations of $\mathbf{x}$ with respect to the basis $\{\mathbf{e}^1, \mathbf{e}^2, \cdots, \mathbf{e}^n\}$ and the basis $\{\bar{\mathbf{e}}^1, \bar{\mathbf{e}}^2, \cdots, \bar{\mathbf{e}}^n\}$, respectively. Since the vector $\mathbf{y} = L\mathbf{x}$ is in the same space, its representations with respect to the bases chosen, say $\boldsymbol{\beta}$ and $\bar{\boldsymbol{\beta}}$, can also be found. The matrix representations $\mathbf{A}$ and $\bar{\mathbf{A}}$ can be computed by using Theorem 2-2. The relationships

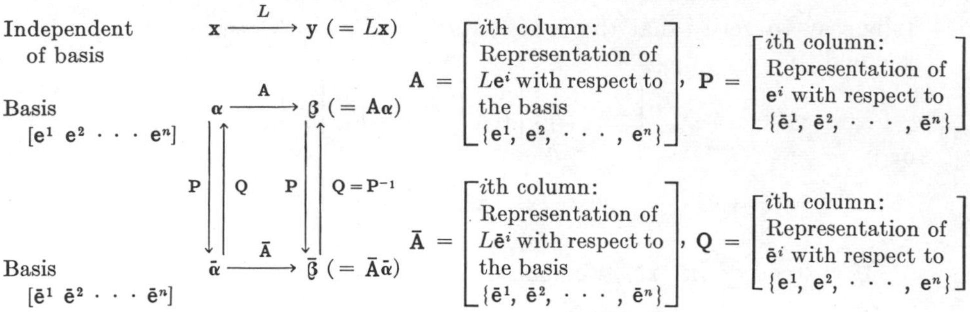

Independent of basis: $\mathbf{x} \xrightarrow{L} \mathbf{y}\ (= L\mathbf{x})$

Basis $[\mathbf{e}^1\ \mathbf{e}^2\ \cdots\ \mathbf{e}^n]$: $\boldsymbol{\alpha} \xrightarrow{\mathbf{A}} \boldsymbol{\beta}\ (= \mathbf{A}\boldsymbol{\alpha})$

$\mathbf{P}\downarrow\ \uparrow\mathbf{Q} \qquad \mathbf{P}\downarrow\ \uparrow\mathbf{Q} = \mathbf{P}^{-1}$

Basis $[\bar{\mathbf{e}}^1\ \bar{\mathbf{e}}^2\ \cdots\ \bar{\mathbf{e}}^n]$: $\bar{\boldsymbol{\alpha}} \xrightarrow{\bar{\mathbf{A}}} \bar{\boldsymbol{\beta}}\ (= \bar{\mathbf{A}}\bar{\boldsymbol{\alpha}})$

$\mathbf{A} =$ [ $i$th column: Representation of $L\mathbf{e}^i$ with respect to the basis $\{\mathbf{e}^1, \mathbf{e}^2, \cdots, \mathbf{e}^n\}$ ], $\mathbf{P} =$ [ $i$th column: Representation of $\mathbf{e}^i$ with respect to $\{\bar{\mathbf{e}}^1, \bar{\mathbf{e}}^2, \cdots, \bar{\mathbf{e}}^n\}$ ]

$\bar{\mathbf{A}} =$ [ $i$th column: Representation of $L\bar{\mathbf{e}}^i$ with respect to the basis $\{\bar{\mathbf{e}}^1, \bar{\mathbf{e}}^2, \cdots, \bar{\mathbf{e}}^n\}$ ], $\mathbf{Q} =$ [ $i$th column: Representation of $\bar{\mathbf{e}}^i$ with respect to $\{\mathbf{e}^1, \mathbf{e}^2, \cdots, \mathbf{e}^n\}$ ]

**Figure 2-4** Relationships between different representations of the same operator.

between $\boldsymbol{\alpha}$ and $\bar{\boldsymbol{\alpha}}$ and between $\boldsymbol{\beta}$ and $\bar{\boldsymbol{\beta}}$ have been established in (2-20); they are related by $\bar{\boldsymbol{\alpha}} = \mathbf{P}\boldsymbol{\alpha}$ and $\bar{\boldsymbol{\beta}} = \mathbf{P}\boldsymbol{\beta}$, where $\mathbf{P}$ is a nonsingular matrix with coefficients in the field $\mathcal{F}$ and the $i$th column of $\mathbf{P}$ is the representation of $\mathbf{e}^i$ with respect to the basis $\{\bar{\mathbf{e}}^1, \bar{\mathbf{e}}^2, \cdots, \bar{\mathbf{e}}^n\}$. From Figure 2-4, we have

$$\bar{\boldsymbol{\beta}} = \bar{\mathbf{A}}\bar{\boldsymbol{\alpha}} \qquad \text{and} \qquad \bar{\boldsymbol{\beta}} = \mathbf{P}\boldsymbol{\beta} = \mathbf{P}\mathbf{A}\boldsymbol{\alpha} = \mathbf{P}\mathbf{A}\mathbf{P}^{-1}\bar{\boldsymbol{\alpha}}$$

Hence, by the uniqueness of a representation with respect to a specific basis, we have $\bar{\mathbf{A}}\bar{\boldsymbol{\alpha}} = \mathbf{P}\mathbf{A}\mathbf{P}^{-1}\bar{\boldsymbol{\alpha}}$. Since the relation holds for any $\bar{\boldsymbol{\alpha}}$, we conclude that

$$\bar{\mathbf{A}} = \mathbf{P}\mathbf{A}\mathbf{P}^{-1} = \mathbf{Q}^{-1}\mathbf{A}\mathbf{Q} \tag{2-31a}$$

or

$$\mathbf{A} = \mathbf{P}^{-1}\bar{\mathbf{A}}\mathbf{P} = \mathbf{Q}\bar{\mathbf{A}}\mathbf{Q}^{-1} \tag{2-31b}$$

where $\mathbf{Q} \triangleq \mathbf{P}^{-1}$.

Two matrices $\mathbf{A}$ and $\bar{\mathbf{A}}$ are said to be *similar* if there exists a nonsingular matrix $\mathbf{P}$ satisfying (2-31). The transformation defined in (2-31) is called a *similarity transformation*. Clearly, *all the matrix representations (with respect to different bases) of the same operator are similar.*

### Example 3

Consider the linear operator $L$ of Example 1 shown in Figure 2-3. If we choose $\{\mathbf{x}^1, \mathbf{x}^2\}$ as a basis, then

$$\mathbf{y}^1 = L\mathbf{x}^1 = [\mathbf{x}^1\ \ \mathbf{x}^2]\begin{bmatrix} 0 \\ 1 \end{bmatrix} \qquad \text{and} \qquad \mathbf{y}^2 = L\mathbf{x}^2 = [\mathbf{x}^1\ \ \mathbf{x}^2]\begin{bmatrix} -1 \\ 0 \end{bmatrix}$$

Hence the representation of $L$ with respect to the basis $\{\mathbf{x}^1, \mathbf{x}^2\}$ is

$$\begin{bmatrix} 0 & -1 \\ 1 & 0 \end{bmatrix}$$

The representation of $\mathbf{x}^3$ is

$$\begin{bmatrix} 1.5 \\ 0.5 \end{bmatrix}$$

It is easy to verify that the representation of $\mathbf{y}^3$ with respect to $\{\mathbf{x}^1, \mathbf{x}^2\}$ is equal to

$$\begin{bmatrix} 0 & -1 \\ 1 & 0 \end{bmatrix} \begin{bmatrix} 1.5 \\ 0.5 \end{bmatrix} = \begin{bmatrix} -0.5 \\ 1.5 \end{bmatrix}$$

or

$$\mathbf{y}^3 = [\mathbf{x}^1 \quad \mathbf{x}^2] \begin{bmatrix} -0.5 \\ 1.5 \end{bmatrix}$$

If, instead of $\{\mathbf{x}^1, \mathbf{x}^2\}$, we choose $\{\mathbf{x}^1, \mathbf{x}^3\}$ as a basis, then

$$\mathbf{y}^1 = L\mathbf{x}^1 = [\mathbf{x}^1 \quad \mathbf{x}^3] \begin{bmatrix} -3 \\ 2 \end{bmatrix} \qquad \text{and} \qquad \mathbf{y}^3 = L\mathbf{x}^3 = [\mathbf{x}^1 \quad \mathbf{x}^3] \begin{bmatrix} -5 \\ 3 \end{bmatrix}$$

Hence the representation of $L$ with respect to the basis $\{\mathbf{x}^1, \mathbf{x}^3\}$ is

$$\begin{bmatrix} -3 & -5 \\ 2 & 3 \end{bmatrix}$$

The reader is advised to find the **P** matrix for this example and verify $\bar{\mathbf{A}} = \mathbf{PAP}^{-1}$. ▮

In matrix theory, a matrix is introduced as an array of numbers. With the concepts of linear operator and representation, we shall now give a new interpretation of a matrix. Given an $n \times n$ matrix **A** with coefficients in a field $\mathfrak{F}$, if it is not specified to be a representation of some operator we shall consider it as a linear operator that maps $(\mathfrak{F}^n, \mathfrak{F})$ into itself.[4] The matrix **A** is independent of the basis chosen for $(\mathfrak{F}^n, \mathfrak{F})$. However, if the set of the vectors $\mathbf{n}^1, \mathbf{n}^2, \cdots, \mathbf{n}^n$ in Equation (2-13) is chosen as a basis of $(\mathfrak{F}^n, \mathfrak{F})$, then the representation of the linear operator **A** is identical to the linear operator **A** (a matrix) itself. This can be checked by using the fact that the $i$th column of the representation is equal to the representation of $\mathbf{An}^i$ with respect to the basis $\{\mathbf{n}^1, \mathbf{n}^2, \cdots, \mathbf{n}^n\}$. If $\mathbf{a}^i$ is the $i$th column of **A**, then $\mathbf{An}^i = \mathbf{a}^i$. Now the representation of $\mathbf{a}^i$ with respect to the basis (2-13) is identical to itself. Therefore we conclude that the representation of a matrix (a linear operator) with respect to the basis (2-13) is identical to itself. For a matrix (an operator), Figure 2-4 can be modified as in Figure 2-5. The equation $\mathbf{Q} =$

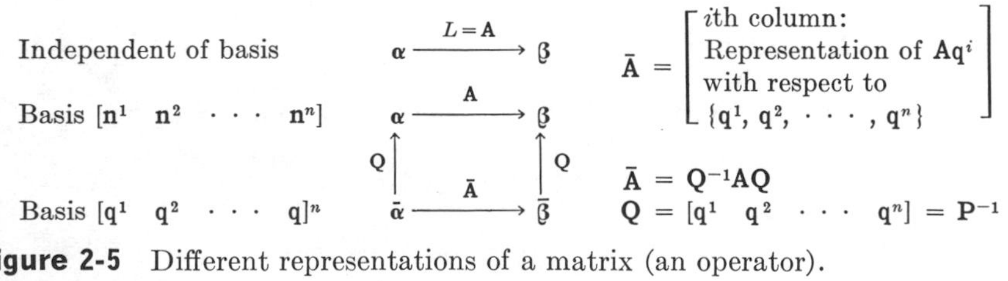

Independent of basis $\qquad \alpha \xrightarrow{L=\mathbf{A}} \beta$

Basis $[\mathbf{n}^1 \quad \mathbf{n}^2 \quad \cdots \quad \mathbf{n}^n] \qquad \alpha \xrightarrow{\mathbf{A}} \beta$

$\mathbf{Q} \uparrow \qquad \uparrow \mathbf{Q}$

Basis $[\mathbf{q}^1 \quad \mathbf{q}^2 \quad \cdots \quad \mathbf{q}]^n \qquad \bar{\alpha} \xrightarrow{\bar{\mathbf{A}}} \bar{\beta}$

$$\bar{\mathbf{A}} = \begin{bmatrix} i\text{th column:} \\ \text{Representation of } \mathbf{Aq}^i \\ \text{with respect to} \\ \{\mathbf{q}^1, \mathbf{q}^2, \cdots, \mathbf{q}^n\} \end{bmatrix}$$

$$\bar{\mathbf{A}} = \mathbf{Q}^{-1}\mathbf{AQ}$$

$$\mathbf{Q} = [\mathbf{q}^1 \quad \mathbf{q}^2 \quad \cdots \quad \mathbf{q}^n] = \mathbf{P}^{-1}$$

**Figure 2-5** Different representations of a matrix (an operator).

[4] This interpretation can be extended to nonsquare matrices.

$[\mathbf{q}^1 \quad \mathbf{q}^2 \quad \cdots \quad \mathbf{q}^n]$ follows from the fact that the $i$th column of $\mathbf{Q}$ is the representation of $\mathbf{q}^i$ with respect to the basis $\{\mathbf{n}^1, \mathbf{n}^2, \cdots, \mathbf{n}^n\}$.

If a basis $\{\mathbf{q}^1, \mathbf{q}^2, \cdots, \mathbf{q}^n\}$ is chosen for $(\mathfrak{F}^n, \mathfrak{F})$, a matrix $\mathbf{A}$ has a representation $\bar{\mathbf{A}}$. From Figure 2-5, we see that the matrix representation $\bar{\mathbf{A}}$ may be computed either from Theorem 2-2 or from a similarity transformation. In most of the problems encountered in this book, it is always much easier to compute $\bar{\mathbf{A}}$ from Theorem 2-2 than from using a similarity transformation.

### Example 4

Consider the following matrix with coefficients in $\mathbb{R}$:

$$L = \mathbf{A} = \begin{bmatrix} 3 & 2 & -1 \\ -2 & 1 & 0 \\ 4 & 3 & 1 \end{bmatrix}$$

Let

$$\mathbf{b} = \begin{bmatrix} 0 \\ 0 \\ 1 \end{bmatrix}$$

Then[5]

$$\mathbf{Ab} = \begin{bmatrix} -1 \\ 0 \\ 1 \end{bmatrix} \qquad \mathbf{A}^2\mathbf{b} = \begin{bmatrix} -4 \\ 2 \\ -3 \end{bmatrix} \qquad \mathbf{A}^3\mathbf{b} = \begin{bmatrix} -5 \\ 10 \\ -13 \end{bmatrix}$$

It can be shown that the following relation holds (check!):

$$\mathbf{A}^3\mathbf{b} = 5\mathbf{A}^2\mathbf{b} - 15\mathbf{Ab} + 17\mathbf{b} \tag{2-32}$$

Since the set of vectors $\mathbf{b}$, $\mathbf{Ab}$, and $\mathbf{A}^2\mathbf{b}$ are linearly independent, it qualifies as a basis. We compute now the representation of $\mathbf{A}$ with respect to this basis. It is clear that

$$\mathbf{A}(\mathbf{b}) = [\mathbf{b} \quad \mathbf{Ab} \quad \mathbf{A}^2\mathbf{b}] \begin{bmatrix} 0 \\ 1 \\ 0 \end{bmatrix}$$

$$\mathbf{A}(\mathbf{Ab}) = [\mathbf{b} \quad \mathbf{Ab} \quad \mathbf{A}^2\mathbf{b}] \begin{bmatrix} 0 \\ 0 \\ 1 \end{bmatrix}$$

and

$$\mathbf{A}(\mathbf{A}^2\mathbf{b}) = [\mathbf{b} \quad \mathbf{Ab} \quad \mathbf{A}^2\mathbf{b}] \begin{bmatrix} 17 \\ -15 \\ 5 \end{bmatrix}$$

---

[5] $\mathbf{A}^2 \triangleq \mathbf{AA}$, $\mathbf{A}^3 \triangleq \mathbf{AAA}$.

The last equation is obtained from (2-32). Hence the representation of $\mathbf{A}$ with respect to the basis $\{\mathbf{b}, \mathbf{Ab}, \mathbf{A}^2\mathbf{b}\}$ is

$$\bar{\mathbf{A}} = \left[\begin{array}{c:c:c} 0 & 0 & 17 \\ 1 & 0 & -15 \\ 0 & 1 & 5 \end{array}\right]$$

The matrix $\bar{\mathbf{A}}$ can also be obtained from $\mathbf{Q}^{-1}\mathbf{AQ}$, but it requires an inversion of a matrix and $n^3$ multiplications. However, we may use $\bar{\mathbf{A}} = \mathbf{Q}^{-1}\mathbf{AQ}$, or more easily, $\mathbf{Q}\bar{\mathbf{A}} = \mathbf{AQ}$ to check our result. The reader is asked to verify

$$\begin{bmatrix} 0 & -1 & -4 \\ 0 & 0 & 2 \\ 1 & 1 & -3 \end{bmatrix} \begin{bmatrix} 0 & 0 & 17 \\ 1 & 0 & -15 \\ 0 & 1 & 5 \end{bmatrix} = \begin{bmatrix} 3 & 2 & -1 \\ -2 & 1 & 0 \\ 4 & 3 & 1 \end{bmatrix} \begin{bmatrix} 0 & -1 & -4 \\ 0 & 0 & 2 \\ 1 & 1 & -3 \end{bmatrix}$$

### Example 5

We extend Example 4 to the general case. Let $\mathbf{A}$ be an $n \times n$ square matrix with real coefficients. If there exists a real vector $\mathbf{b}$ such that the vectors $\mathbf{b}, \mathbf{Ab}, \cdots, \mathbf{A}^{n-1}\mathbf{b}$ are linearly independent and if $\mathbf{A}^n\mathbf{b} = -\alpha_n\mathbf{b} - \alpha_{n-1}\mathbf{Ab} - \cdots - \alpha_1\mathbf{A}^{n-1}\mathbf{b}$ (see Section 2-7), then the representation of $\mathbf{A}$ with respect to the basis $\{\mathbf{b}, \mathbf{Ab}, \cdots, \mathbf{A}^{n-1}\mathbf{b}\}$ is

$$\bar{\mathbf{A}} = \begin{bmatrix} 0 & 0 & \cdots & 0 & -\alpha_n \\ 1 & 0 & \cdots & 0 & -\alpha_{n-1} \\ 0 & 1 & \cdots & 0 & -\alpha_{n-2} \\ \cdot & \cdot & & \cdot & \cdot \\ \cdot & \cdot & & \cdot & \cdot \\ \cdot & \cdot & & \cdot & \cdot \\ 0 & 0 & \cdots & 0 & -\alpha_2 \\ 0 & 0 & \cdots & 1 & -\alpha_1 \end{bmatrix} \tag{2-33}$$

∎

As an aid in memorizing Figure 2-5, we write $\bar{\mathbf{A}} = \mathbf{Q}^{-1}\mathbf{AQ}$ as

$$\mathbf{Q}\bar{\mathbf{A}} = \mathbf{AQ}$$

Since $\mathbf{Q} = [\mathbf{q}^1 \quad \mathbf{q}^2 \quad \cdots \quad \mathbf{q}^n]$, it can be further written as

$$[\mathbf{q}^1 \quad \mathbf{q}^2 \quad \cdots \quad \mathbf{q}^n]\bar{\mathbf{A}} = [\mathbf{Aq}^1 \quad \mathbf{Aq}^2 \quad \cdots \quad \mathbf{Aq}^n] \tag{2-34}$$

From (2-34), we see that the $i$th column of $\bar{\mathbf{A}}$ is indeed the representation of $\mathbf{Aq}^i$ with respect to the basis $\{\mathbf{q}^1, \mathbf{q}^2, \cdots, \mathbf{q}^n\}$.

We pose the following question to conclude this section: Since a linear operator has many representations, is it possible to choose one set of basis vectors such that the representation is nice and simple? The answer is yes. This will be studied in Section 2-6.

## 2-5 Systems of Linear Algebraic Equations

Consider the set of linear equations:

$$\begin{aligned} a_{11}x_1 + a_{12}x_2 + \cdots + a_{1n}x_n &= y_1 \\ a_{21}x_1 + a_{22}x_2 + \cdots + a_{2n}x_n &= y_2 \\ &\vdots \\ a_{m1}x_1 + a_{m2}x_2 + \cdots + a_{mn}x_n &= y_m \end{aligned} \tag{2-35}$$

where the given $a_{ij}$'s and $y_i$'s are assumed to be elements of a field $\mathfrak{F}$, the unknown $x_i$'s are also required to be in the same field $\mathfrak{F}$. This set of equations can be written in matrix form as

$$\mathbf{Ax} = \mathbf{y} \tag{2-36}$$

where

$$\mathbf{A} \triangleq \begin{bmatrix} a_{11} & a_{12} & \cdots & a_{1n} \\ a_{21} & a_{22} & \cdots & a_{2n} \\ \cdot & & & \cdot \\ \cdot & \cdot & & \cdot \\ \cdot & \cdot & & \cdot \\ a_{m1} & a_{m2} & \cdots & a_{mn} \end{bmatrix} \qquad \mathbf{x} \triangleq \begin{bmatrix} x_1 \\ x_2 \\ \cdot \\ \cdot \\ \cdot \\ x_n \end{bmatrix} \qquad \mathbf{y} \triangleq \begin{bmatrix} y_1 \\ y_2 \\ \cdot \\ \cdot \\ \cdot \\ y_m \end{bmatrix}$$

Clearly, $\mathbf{A}$ is an $m \times n$ matrix, $\mathbf{x}$ is an $n \times 1$ vector and $\mathbf{y}$ is an $m \times 1$ vector. No restriction is made on the integer $m$; it may be larger than, equal to, or smaller than the integer $n$. Two questions can be raised in regard to this set of equations: first, the existence of a solution and, second, the number of solutions. More specifically, suppose the matrix $\mathbf{A}$ and the vector $\mathbf{y}$ in Equation (2-36) are given, the first equation is concerned with the condition on $\mathbf{A}$ and $\mathbf{y}$ under which at least one vector $\mathbf{x}$ exists such that $\mathbf{Ax} = \mathbf{y}$. If solutions exist, then the second question is concerned with the number of linearly independent vectors $\mathbf{x}$ such that $\mathbf{Ax} = \mathbf{y}$. In order to answer these questions, the rank and the nullity of the matrix $\mathbf{A}$ have to be introduced.

We have agreed in the previous section to consider the matrix $\mathbf{A}$ as a linear operator which maps $(\mathfrak{F}^n, \mathfrak{F})$ into $(\mathfrak{F}^m, \mathfrak{F})$. Recall that the linear space $(\mathfrak{F}^n, \mathfrak{F})$ that undergoes transformation is called the domain of $\mathbf{A}$.

### Definition 2-9

The *range* of a linear operator $\mathbf{A}$ is the set $\mathcal{R}(\mathbf{A})$ defined by

$\mathcal{R}(\mathbf{A}) = \{$all the elements $\mathbf{y}$ of $(\mathfrak{F}^m, \mathfrak{F})$ for which there exists at least one vector $\mathbf{x}$ in $(\mathfrak{F}^n, \mathfrak{F})$ such that $\mathbf{y} = \mathbf{Ax}\}$

### Theorem 2-3

The range of a linear operator **A** is a subspace of $(\mathfrak{F}^m, \mathfrak{F})$.

### Proof

If $\mathbf{y}^1$ and $\mathbf{y}^2$ are elements of $\mathcal{R}(\mathbf{A})$, then by definition there exist vectors $\mathbf{x}^1$ and $\mathbf{x}^2$ in $(\mathfrak{F}^n, \mathfrak{F})$ such that $\mathbf{y}^1 = \mathbf{A}\mathbf{x}^1$, $\mathbf{y}^2 = \mathbf{A}\mathbf{x}^2$. We claim that for any $\alpha_1$ and $\alpha_2$ in $\mathfrak{F}$, the vector $\alpha_1\mathbf{y}^1 + \alpha_2\mathbf{y}^2$ is also an element of $\mathcal{R}(\mathbf{A})$. Indeed, by the linearity of **A**, it is easy to show that $\alpha_1\mathbf{y}^1 + \alpha_2\mathbf{y}^2 = \mathbf{A}(\alpha_1\mathbf{x}^1 + \alpha_2\mathbf{x}^2)$ and the vector $\alpha_1\mathbf{x}^1 + \alpha_2\mathbf{x}^2$ is an element of $(\mathfrak{F}^n, \mathfrak{F})$. Hence the range $\mathcal{R}(\mathbf{A})$ is a subspace of $(\mathfrak{F}^m, \mathfrak{F})$ (see the remark following Definition 2-3).
Q.E.D.

Let the $i$th column of **A** be denoted by $\mathbf{a}^i$; that is, $\mathbf{A} = [\mathbf{a}^1 \quad \mathbf{a}^2 \quad \cdots \quad \mathbf{a}^n]$, then the matrix equation (2-36) can be written as

$$\mathbf{y} = x_1\mathbf{a}^1 + x_2\mathbf{a}^2 + \cdots + x_n\mathbf{a}^n \tag{2-37}$$

where $x_i$, for $i = 1, 2, \cdots, n$, are components of **x** and are elements of $\mathfrak{F}$. The range space $\mathcal{R}(\mathbf{A})$ is, by definition, the set of **y** such that $\mathbf{y} = \mathbf{A}\mathbf{x}$ for some **x** in $(\mathfrak{F}^n, \mathfrak{F})$. It is the same as saying that $\mathcal{R}(\mathbf{A})$ is the set of **y** with $x_1, x_2, \cdots, x_n$ in (2-37) ranging through all the possible values of $\mathfrak{F}$. Therefore we conclude that *$\mathcal{R}(\mathbf{A})$ is the set of all the possible linear combinations of the columns of* **A**. Since $\mathcal{R}(\mathbf{A})$ is a linear space, its dimension is defined and is equal to the maximum number of linearly independent vectors in $\mathcal{R}(\mathbf{A})$. Hence, *the dimension of $\mathcal{R}(\mathbf{A})$ is the maximum number of linearly independent columns in* **A**.

### Definition 2-10

The *rank* of a matrix **A**, denoted by $\rho(\mathbf{A})$, is the maximum number of linearly independent columns in **A**, or, equivalently, the dimension of the range space of **A**. ∎

### Example 1

Consider the matrix

$$\mathbf{A} = \begin{bmatrix} 0 & 1 & 1 & 2 \\ 1 & 2 & 3 & 4 \\ 2 & 0 & 2 & 0 \end{bmatrix}$$

The range space of **A** is all the possible linear combinations of all the columns of **A**, or correspondingly, all the possible linear combinations of the first two columns of **A**, because the third and the fourth columns of **A** are linearly dependent on the first two columns. Hence the rank of **A** is 2. ∎

In matrix theory, the rank of a matrix is defined as the largest order of all nonvanishing minors of **A**. In other words, the matrix **A** has rank $k$ if and only if there is at least one minor of order $k$ in $\mathbf{A}$ that does not vanish and every minor of order higher than $k$ vanishes. This definition and Definition 2-10 are, in fact, equivalent; the proof can be found, for example, in Reference [43]. A consequence is that a square matrix has full rank if and only if the determinant of the matrix is different from zero; or correspondingly, *a matrix is nonsingular if and only if all the columns of the matrix are linearly independent.* It can also be shown that *the maximum number of linearly independent columns of a matrix is equal to the maximum number of linearly independent rows.* Therefore, the rank of a matrix can be checked either from the row vectors or from the column vectors. Another consequence of this is that if $\mathbf{A}$ is an $n \times m$ matrix, then

$$\rho(\mathbf{A}) \leq \min(n, m) \tag{2-38}$$

With the concepts of range space and rank, we are ready to study the existence problem of the solutions of $\mathbf{Ax} = \mathbf{y}$.

### Theorem 2-4

Consider the matrix equation $\mathbf{Ax} = \mathbf{y}$, where the $m \times n$ matrix $\mathbf{A}$ maps $(\mathfrak{F}^n, \mathfrak{F})$ into $(\mathfrak{F}^m, \mathfrak{F})$.

1. Given $\mathbf{A}$ and given a vector $\mathbf{y}$ in $(\mathfrak{F}^m, \mathfrak{F})$, there exists a vector $\mathbf{x}$ such that $\mathbf{Ax} = \mathbf{y}$ if and only if the vector $\mathbf{y}$ is an element of $\mathcal{R}(\mathbf{A})$.
2. Given $\mathbf{A}$ and for *any* $\mathbf{y}$ in $(\mathfrak{F}^m, \mathfrak{F})$, there exists a vector $\mathbf{x}$ such that $\mathbf{Ax} = \mathbf{y}$ if and only if $\mathcal{R}(\mathbf{A}) = (\mathfrak{F}^m, \mathfrak{F})$ or, equivalently, $\rho(\mathbf{A}) = m$.

### Proof

1. It follows immediately from the definition of the range space of $\mathbf{A}$. If the vector $\mathbf{y}$ is not an element of $\mathcal{R}(\mathbf{A})$, the equation $\mathbf{Ax} = \mathbf{y}$ has no solution and is said to be *inconsistent.*
2. The rank of $\mathbf{A}$, $\rho(\mathbf{A})$, is by definition the dimension of $\mathcal{R}(\mathbf{A})$. Since $\mathcal{R}(\mathbf{A})$ is a subspace of $(\mathfrak{F}^m, \mathfrak{F})$, if $\rho(\mathbf{A}) = m$, then $\mathcal{R}(\mathbf{A}) = (\mathfrak{F}^m, \mathfrak{F})$. If $\mathcal{R}(\mathbf{A}) = (\mathfrak{F}^m, \mathfrak{F})$, then for any $\mathbf{y}$ in $(\mathfrak{F}^m, \mathfrak{F})$, there exists a vector $\mathbf{x}$ such that $\mathbf{Ax} = \mathbf{y}$. If $\rho(\mathbf{A}) < m$, there exists at least one nonzero vector $\mathbf{y}$ in $(\mathfrak{F}^m, \mathfrak{F})$, but not in $\mathcal{R}(\mathbf{A})$, for which there exists no $\mathbf{x}$ such that $\mathbf{Ax} = \mathbf{y}$. Q.E.D.

After we find out that a linear equation has at least one solution, it is natural to ask how many solutions it may have. Instead of studying the general case, we discuss here only the homogeneous linear equation $\mathbf{Ax} = \mathbf{0}$.

### Definition 2-11

The *null space* of a linear operator **A** is the set $\mathfrak{N}(\mathbf{A})$ defined by

$$\mathfrak{N}(\mathbf{A}) = \{\text{all the elements } \mathbf{x} \text{ of } (\mathfrak{F}^n, \mathfrak{F}) \text{ for which } \mathbf{A}\mathbf{x} = \mathbf{0}\}$$

The dimension of $\mathfrak{N}(\mathbf{A})$ is called the *nullity* of **A** and is denoted by $\nu(\mathbf{A})$. ▮

In other words, the null space $\mathfrak{N}(\mathbf{A})$ is the set of all solutions of $\mathbf{A}\mathbf{x} = \mathbf{0}$. It is easy to show that $\mathfrak{N}(\mathbf{A})$ is indeed a linear space. If the dimension of $\mathfrak{N}(\mathbf{A})$, $\nu(\mathbf{A})$, is 0, then $\mathfrak{N}(\mathbf{A})$ consists of only the zero vector, and the only solution of $\mathbf{A}\mathbf{x} = \mathbf{0}$ is $\mathbf{x} = \mathbf{0}$. If $\nu(\mathbf{A}) = \nu$, then the equation $\mathbf{A}\mathbf{x} = \mathbf{0}$ has $\nu$ linearly independent vector solutions.

Note that the null space is a subspace of the domain $(\mathfrak{F}^n, \mathfrak{F})$, whereas the range space is a subspace of $(\mathfrak{F}^m, \mathfrak{F})$.

### Example 2

Consider the matrix

$$\mathbf{A} = \begin{bmatrix} 0 & 1 & 1 & 2 & -1 \\ 1 & 2 & 3 & 4 & -1 \\ 2 & 0 & 2 & 0 & 2 \end{bmatrix}$$

which maps $(\mathbb{R}^5, \mathbb{R})$ into $(\mathbb{R}^3, \mathbb{R})$. It is easy to check that the last three columns of **A** are linearly dependent on the first two columns of **A**. Hence the rank of **A**, $\rho(\mathbf{A})$, is equal to 2. Let $\mathbf{x} = [x_1 \quad x_2 \quad x_3 \quad x_4 \quad x_5]'$. Then

$$\begin{aligned} \mathbf{A}\mathbf{x} &= x_1 \begin{bmatrix} 0 \\ 1 \\ 2 \end{bmatrix} + x_2 \begin{bmatrix} 1 \\ 2 \\ 0 \end{bmatrix} + x_3 \begin{bmatrix} 1 \\ 3 \\ 2 \end{bmatrix} + x_4 \begin{bmatrix} 2 \\ 4 \\ 0 \end{bmatrix} + x_5 \begin{bmatrix} -1 \\ -1 \\ 2 \end{bmatrix} \\ &= (x_1 + x_3 + x_5) \begin{bmatrix} 0 \\ 1 \\ 2 \end{bmatrix} + (x_2 + x_3 + 2x_4 - x_5) \begin{bmatrix} 1 \\ 2 \\ 0 \end{bmatrix} \end{aligned} \quad \text{(2-39)}$$

Since the vectors $[0 \quad 1 \quad 2]'$ and $[1 \quad 2 \quad 0]'$ are linearly independent, we conclude from (2-39) that a vector **x** satisfies $\mathbf{A}\mathbf{x} = \mathbf{0}$ if and only if

$$\begin{aligned} x_1 + x_3 + x_5 &= 0 \\ x_2 + x_3 + 2x_4 - x_5 &= 0 \end{aligned}$$

Note that the number of equations here is equal to the rank of **A**, $\rho(\mathbf{A})$. There are five components in **x** but there are only two equations governing them; therefore three of the five components can be arbitrarily chosen. Let $x_3 = 1$, $x_4 = 0$, $x_5 = 0$; then $x_1 = -1$ and $x_2 = -1$. Let $x_3 = 0$, $x_4 = 1$, $x_5 = 0$; then $x_1 = 0$, $x_2 = -2$. Let $x_3 = 0$, $x_4 = 0$, $x_5 = 1$; then

$x_1 = -1, x_2 = 1$. It is clear that the three vectors

$$\begin{bmatrix} -1 \\ -1 \\ 1 \\ 0 \\ 0 \end{bmatrix} \quad \begin{bmatrix} 0 \\ -2 \\ 0 \\ 1 \\ 0 \end{bmatrix} \quad \begin{bmatrix} -1 \\ 1 \\ 0 \\ 0 \\ 1 \end{bmatrix}$$

are linearly independent, and that every solution of $\mathbf{Ax} = \mathbf{0}$ must be a linear combination of these three vectors. Therefore the set of vectors form a basis of $\mathfrak{N}(\mathbf{A})$ and $\nu(\mathbf{A}) = 3$. ▮

We see from this example that the number of equations that the vectors of $\mathfrak{N}(\mathbf{A})$ should obey is equal to $\rho(\mathbf{A})$ and that there are $n$ components in every vector of $\mathfrak{N}(\mathbf{A})$. Therefore $(n - \rho(\mathbf{A}))$ components of the vectors of $\mathfrak{N}(\mathbf{A})$ can be arbitrarily chosen. Consequently, there are $(n - \rho(\mathbf{A}))$ linearly independent vectors in $\mathfrak{N}(\mathbf{A})$. Hence we conclude that $n - \rho(\mathbf{A}) = \nu(\mathbf{A})$. We state this as a theorem; its formal proof can be found, for example, in References [43] and [86].

#### Theorem 2-5

Let $\mathbf{A}$ be an $m \times n$ matrix; then

$$\rho(\mathbf{A}) + \nu(\mathbf{A}) = n$$

#### Corollary 2-5

The number of linearly independent vector solutions of $\mathbf{Ax} = \mathbf{0}$ is equal to $n - \rho(\mathbf{A})$, where $n$ is the number of columns in $\mathbf{A}$, and $\rho(\mathbf{A})$ is the number of linearly independent columns in $\mathbf{A}$. ▮

This corollary follows directly from Theorem 2-5 and the definition of the null space of $\mathbf{A}$. It is clear that if $\rho(\mathbf{A}) = n$, then the only solution of $\mathbf{Ax} = \mathbf{0}$ is $\mathbf{x} = \mathbf{0}$, which is called the trivial solution. If $\rho(\mathbf{A}) < n$, then we can always find a nonzero vector $\mathbf{x}$ such that $\mathbf{Ax} = \mathbf{0}$. In particular, if $\mathbf{A}$ is a square matrix, then $\mathbf{Ax} = \mathbf{0}$ *has a nontrivial solution if and only if* $\rho(\mathbf{A}) < n$, *or equivalently*, $\det(\mathbf{A}) = 0$.

We introduce the following three very useful theorems.

#### Theorem 2-6 (Sylvester's inequality)

Let $\mathbf{A}, \mathbf{B}$ be $q \times n$ and $n \times p$ matrices with coefficients in the same field. Then

$$\rho(\mathbf{A}) + \rho(\mathbf{B}) - n \leq \rho(\mathbf{AB}) \leq \min(\rho(\mathbf{A}), \rho(\mathbf{B}))$$

#### Proof

The composite matrix $\mathbf{AB}$ can be considered as two linear transformations applied successively to $(\mathfrak{F}^p, \mathfrak{F})$ as shown in Figure 2-6. Since the domain

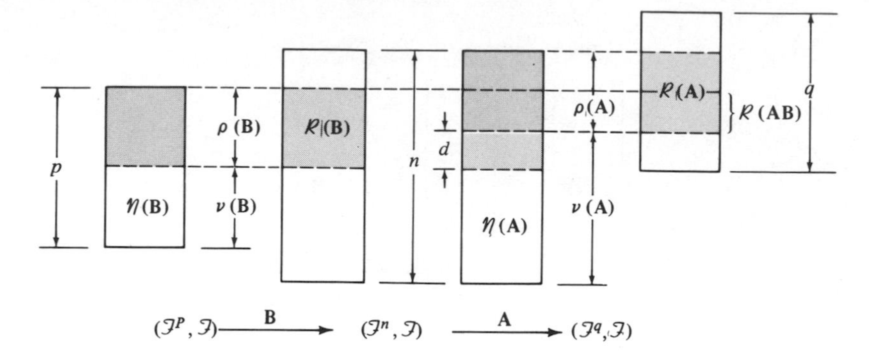


**Figure 2-6** A composite transformation.

of **AB** is $\mathcal{R}(\mathbf{B})$ and the range of **AB** is a subspace of $\mathcal{R}(\mathbf{A})$, we have immediately $\rho(\mathbf{AB}) \leq \min(\rho(\mathbf{A}), \rho(\mathbf{B}))$ by using (2-38). From Figure 2-6 we have $\rho(\mathbf{AB}) = \rho(\mathbf{B}) - d$, where $d$ is the dimension of the intersection of $\mathcal{R}(\mathbf{B})$ and $\mathcal{N}(\mathbf{A})$.[6] The dimension of $\mathcal{N}(\mathbf{A})$ is $n - \rho(\mathbf{A})$; hence, $d \leq n - \rho(\mathbf{A})$. Consequently, $\rho(\mathbf{AB}) \geq \rho(\mathbf{B}) - n + \rho(\mathbf{A})$. Q.E.D.

If **B** is an $n \times n$ matrix and nonsingular, then

$$\rho(\mathbf{A}) + \rho(\mathbf{B}) - n = \rho(\mathbf{A}) \leq \rho(\mathbf{AB}) \leq \min(\rho(\mathbf{A}), n) = \rho(\mathbf{A})$$

Hence we have the following important theorem:

### Theorem 2-7

Let **A** be an $m \times n$ matrix. Then

$$\rho(\mathbf{AC}) = \rho(\mathbf{A}) \qquad \text{and} \qquad \rho(\mathbf{DA}) = \rho(\mathbf{A})$$

for any $n \times n$ and $m \times m$ nonsingular matrices **C** and **D**. ▮

In words, the rank of a matrix does not change after premultiplying or postmultiplying a nonsingular matrix.

### Theorem 2-8

Let **A** be an $m \times n$ matrix with coefficients in a field, and let $\mathbf{A}^*$ be the complex conjugate transpose of **A**. Then

1. $\rho(\mathbf{A}) = n$ if and only if $\rho(\mathbf{A}^*\mathbf{A}) = n$, or equivalently,

$$\det(\mathbf{A}^*\mathbf{A}) \neq 0$$

2. $\rho(\mathbf{A}) = m$ if and only if $\rho(\mathbf{AA}^*) = m$, or equivalently,

$$\det(\mathbf{AA}^*) \neq 0$$

▮

[6] The intersection of two linear spaces is a linear space.

Note that $\mathbf{A}^*\mathbf{A}$ is an $n \times n$ matrix and $\mathbf{A}\mathbf{A}^*$ is an $m \times m$ matrix. In order to have $\rho(\mathbf{A}) = n$, it is necessary to have $n \leq m$. This theorem will be proved by using Definition 2-4′. More specifically, we use the fact that if $\mathbf{A}\boldsymbol{\alpha} = \mathbf{0}$ implies $\boldsymbol{\alpha} = \mathbf{0}$, then all the columns of $\mathbf{A}$ are linearly independent, and $\rho(\mathbf{A}) = n$, where $n$ is the number of the columns of $\mathbf{A}$.

### Proof

**1.** *Sufficiency:* $\rho(\mathbf{A}^*\mathbf{A}) = n$ implies $\rho(\mathbf{A}) = n$. We show that $\mathbf{A}\boldsymbol{\alpha} = \mathbf{0}$ implies $\boldsymbol{\alpha} = \mathbf{0}$ under the assumption of $\rho(\mathbf{A}^*\mathbf{A}) = n$, where $\boldsymbol{\alpha}$ is an $n \times 1$ vector. If $\mathbf{A}\boldsymbol{\alpha} = \mathbf{0}$, then $\mathbf{A}^*\mathbf{A}\boldsymbol{\alpha} = \mathbf{0}$ which, with the assumption $\rho(\mathbf{A}^*\mathbf{A}) = n$, implies $\boldsymbol{\alpha} = \mathbf{0}$. Hence we conclude that $\rho(\mathbf{A}) = n$. *Necessity:* $\rho(\mathbf{A}) = n$ implies $\rho(\mathbf{A}^*\mathbf{A}) = n$. Let $\boldsymbol{\alpha}$ be an $n \times 1$ vector. We show that $\mathbf{A}^*\mathbf{A}\boldsymbol{\alpha} = \mathbf{0}$ implies $\boldsymbol{\alpha} = \mathbf{0}$ under the assumption of $\rho(\mathbf{A}) = n$. The equality $\mathbf{A}^*\mathbf{A}\boldsymbol{\alpha} = \mathbf{0}$ implies $\boldsymbol{\alpha}^*\mathbf{A}^*\mathbf{A}\boldsymbol{\alpha} = 0$. Let $\mathbf{A}\boldsymbol{\alpha} = [\beta_1 \quad \beta_2 \quad \cdots \quad \beta_n]'$. Then

$$\boldsymbol{\alpha}^*\mathbf{A}^* = [\beta_1^* \quad \beta_2^* \quad \cdots \quad \beta_n^*]$$

and

$$\boldsymbol{\alpha}^*\mathbf{A}^*\mathbf{A}\boldsymbol{\alpha} = |\beta_1|^2 + |\beta_2|^2 + \cdots + |\beta_n|^2$$

Hence $\boldsymbol{\alpha}^*\mathbf{A}^*\mathbf{A}\boldsymbol{\alpha} = 0$ implies $\beta_i = 0$, for $i = 1, 2, \cdots, n$; or, equivalently, $\mathbf{A}\boldsymbol{\alpha} = \mathbf{0}$ which, in turn, implies $\boldsymbol{\alpha} = \mathbf{0}$ from the assumption of $\rho(\mathbf{A}) = n$. Therefore we conclude that $\rho(\mathbf{A}^*\mathbf{A}) = n$.
**2.** This part can be similarly proved or directly deduced from the foregoing by using the fact $\rho(\mathbf{A}) = \rho(\mathbf{A}^*)$. Q.E.D.

## 2-6 Eigenvectors, Generalized Eigenvectors and Jordan-Form Representations of a Linear Operator

With the background of Section 2-5, we are now capable of studying the problem posed at the end of Section 2-4. We discuss in this section only linear operators that map $(\mathbb{C}^n, \mathbb{C})$ into itself with the understanding that the results are applicable to any operator that maps a finite dimensional linear space over $\mathbb{C}$ into itself. The reason for restricting the field to the field of complex numbers will be seen immediately.

Let $\mathbf{A}$ be an $n \times n$ matrix with coefficients in the field $\mathbb{C}$. We have agreed to consider $\mathbf{A}$ as a linear operator that maps $(\mathbb{C}^n, \mathbb{C})$ into $(\mathbb{C}^n, \mathbb{C})$.

### Definition 2-12

Let $\mathbf{A}$ be a linear operator that maps $(\mathbb{C}^n, \mathbb{C})$ into itself. Then a scalar $\lambda$ in $\mathbb{C}$ is called an *eigenvalue* of $\mathbf{A}$ if there exists a nonzero vector $\mathbf{x}$ in $\mathbb{C}^n$ such that $\mathbf{A}\mathbf{x} = \lambda\mathbf{x}$. Any nonzero vector $\mathbf{x}$ satisfying $\mathbf{A}\mathbf{x} = \lambda\mathbf{x}$ is called an *eigenvector* of $\mathbf{A}$ associated with the eigenvalue $\lambda$. ▮

In order to find an eigenvalue of $\mathbf{A}$, we write $\mathbf{Ax} = \lambda\mathbf{x}$ as

$$(\mathbf{A} - \lambda\mathbf{I})\mathbf{x} = \mathbf{0} \tag{2-40}$$

where $\mathbf{I}$ is a unit matrix of order $n$. We see that for any fixed $\lambda$ in $\mathbb{C}$, Equation (2-40) is a set of homogeneous linear equations. The matrix $(\mathbf{A} - \lambda\mathbf{I})$ is an $n \times n$ square matrix. From Corollary 2-5, we know that Equation (2-40) has a nontrivial solution if and only if $\det(\mathbf{A} - \lambda\mathbf{I}) = 0$. It follows that *a scalar $\lambda$ is an eigenvalue of $\mathbf{A}$ if and only if it is a solution of* $\Delta(\lambda) \triangleq \det(\lambda\mathbf{I} - \mathbf{A}) = 0$. $\Delta(\lambda)$ is a polynomial of degree $n$ in $\lambda$ and is called the *characteristic polynomial* of $\mathbf{A}$. Since $\Delta(\lambda)$ is of degree $n$, the $n \times n$ matrix $\mathbf{A}$ has $n$ eigenvalues (not necessarily all distinct).

### Example 1

Consider the matrix

$$\mathbf{A} = \begin{bmatrix} 1 & -1 \\ 2 & -1 \end{bmatrix} \tag{2-41}$$

which maps $(\mathbb{R}^2, \mathbb{R})$ into itself. We like to check whether Definition 2-12 can be modified and applied to a linear operator that maps $(\mathbb{R}^n, \mathbb{R})$ into $(\mathbb{R}^n, \mathbb{R})$. A modified version of Definition 2-12 reads as a scalar $\lambda$ in $\mathbb{R}$ is an eigenvalue of $\mathbf{A}$ if there exists a nonzero vector $\mathbf{x}$ such that $\mathbf{Ax} = \lambda\mathbf{x}$. Clearly $\lambda$ is an eigenvalue of $\mathbf{A}$ if and only if it is a solution of $\det(\lambda\mathbf{I} - \mathbf{A}) = 0$. Now

$$\det(\lambda\mathbf{I} - \mathbf{A}) = \det\begin{bmatrix} \lambda - 1 & 1 \\ -2 & \lambda + 1 \end{bmatrix} = \lambda^2 + 1$$

which has no real-valued solution. Consequently the matrix $\mathbf{A}$ has no eigenvalue in $\mathbb{R}$.

Since the set of real numbers is a part of the field of complex numbers, there is no reason that we cannot consider the matrix $\mathbf{A}$ in (2-41) as a linear operator that maps $(\mathbb{C}^2, \mathbb{C})$ into itself. In so doing, then the matrix $\mathbf{A}$ has eigenvalues $+i$ and $-i$ where $i \triangleq \sqrt{-1}$. ▮

The constant matrices we shall encounter in this book are all real-valued. However in order to insure the existence of eigenvalues, we shall consider them as linear operators under the field of complex numbers.

With these preliminaries, we are ready to introduce a set of basis vectors such that a linear operator has a diagonal or almost diagonal representation. We study first the case in which all the eigenvalues of $\mathbf{A}$ are distinct; the case where $\mathbf{A}$ has repeated eigenvalues will then be studied.

### Case 1: All the eigenvalues of A are distinct

Let $\lambda_1, \lambda_2, \cdots, \lambda_n$ be the eigenvalues of $\mathbf{A}$ and let $\mathbf{v}^i$ be an eigenvector of $\mathbf{A}$ associated with $\lambda_i$, for $i = 1, 2, \cdots, n$; that is, $\mathbf{Av}^i = \lambda_i\mathbf{v}^i$. We shall use the set of vectors $\{\mathbf{v}^1, \mathbf{v}^2, \cdots, \mathbf{v}^n\}$ as a basis of $(\mathbb{C}^n, \mathbb{C})$. In order to do so, we have to show that the set is linearly independent and qualifies as a basis.

### Theorem 2-9

Let $\lambda_1, \lambda_2, \cdots, \lambda_n$ be the distinct eigenvalues of $\mathbf{A}$ and let $\mathbf{v}^i$ be an eigenvector of $\mathbf{A}$ associated with $\lambda_i$, for $i = 1, 2, \cdots, n$. Then the set $\{\mathbf{v}^1, \mathbf{v}^2, \cdots, \mathbf{v}^n\}$ is linearly independent (over $\mathbb{C}$).

### Proof

We prove the theorem by contradiction. Suppose $\mathbf{v}^1, \mathbf{v}^2, \cdots, \mathbf{v}^n$ are linearly dependent; then there exist $\alpha_1, \alpha_2, \cdots, \alpha_n$ (not all zero) in $\mathbb{C}$ such that

$$\alpha_1\mathbf{v}^1 + \alpha_2\mathbf{v}^2 + \cdots + \alpha_n\mathbf{v}^n = \mathbf{0} \tag{2-42}$$

We assume $\alpha_1 \neq 0$. If $\alpha_1 = 0$, we may reorder $\lambda_i$ in such a way that $\alpha_1 \neq 0$. Equation (2-42) implies that

$$(\mathbf{A} - \lambda_2\mathbf{I})(\mathbf{A} - \lambda_3\mathbf{I}) \cdots (\mathbf{A} - \lambda_n\mathbf{I}) \left(\sum_{i=1}^{n} \alpha_i\mathbf{v}^i\right) = \mathbf{0} \tag{2-43}$$

Since

$$(\mathbf{A} - \lambda_j\mathbf{I})\mathbf{v}^i = (\lambda_i - \lambda_j)\mathbf{v}^i \qquad \text{if } i \neq j$$

and

$$(\mathbf{A} - \lambda_i\mathbf{I})\mathbf{v}^i = \mathbf{0}$$

the left-hand side of (2-43) can be reduced to

$$\alpha_1(\lambda_1 - \lambda_2)(\lambda_1 - \lambda_3) \cdots (\lambda_1 - \lambda_n)\mathbf{v}^1 = \mathbf{0}$$

By assumption the $\lambda_i$'s, for $i = 1, 2, \cdots, n$, are all distinct; hence the equation

$$\alpha_1 \prod_{i=2}^{n} (\lambda_1 - \lambda_i)\mathbf{v}^1 = \mathbf{0}$$

implies $\alpha_1 = 0$. This is a contradiction. Thus, the set of vectors $\{\mathbf{v}^1, \mathbf{v}^2, \cdots, \mathbf{v}^n\}$ is linearly independent and qualifies as a basis.

Q.E.D.

Let $\hat{\mathbf{A}}$ be the representation of $\mathbf{A}$ with respect to the basis $\{\mathbf{v}^1, \mathbf{v}^2, \cdots, \mathbf{v}^n\}$. Recall from Figure 2-5 that the $i$th column of $\hat{\mathbf{A}}$ is the representation of $\mathbf{A}\mathbf{v}^i = \lambda_i\mathbf{v}^i$ with respect to $\{\mathbf{v}^1, \mathbf{v}^2, \cdots, \mathbf{v}^n\}$—that is, $[0 \ \cdots \ 0 \ \lambda_i \ 0 \ \cdots \ 0]'$, where $\lambda_i$ stands at the $i$th component.

Hence the representation of $\mathbf{A}$ with respect to $\{\mathbf{v}^1, \mathbf{v}^2, \cdots, \mathbf{v}^n\}$ is

$$\hat{\mathbf{A}} = \begin{bmatrix} \lambda_1 & 0 & 0 & \cdots & 0 \\ 0 & \lambda_2 & 0 & \cdots & 0 \\ 0 & 0 & \lambda_3 & \cdots & 0 \\ \cdot & \cdot & \cdot & & \cdot \\ \cdot & \cdot & \cdot & & \cdot \\ \cdot & \cdot & \cdot & & \cdot \\ 0 & 0 & 0 & \cdots & \lambda_n \end{bmatrix} \tag{2-44}$$

This can also be checked by using a similarity transformation. From Figure 2-5, we have

$$\mathbf{Q} = [\mathbf{v}^1 \quad \mathbf{v}^2 \quad \cdots \quad \mathbf{v}^n]$$

Since

$$\begin{aligned} \mathbf{AQ} &= \mathbf{A}[\mathbf{v}^1 \quad \mathbf{v}^2 \quad \cdots \quad \mathbf{v}^n] = [\mathbf{Av}^1 \quad \mathbf{Av}^2 \quad \cdots \quad \mathbf{Av}^n] \\ &= [\lambda_1\mathbf{v}^1 \quad \lambda_2\mathbf{v}^2 \quad \cdots \quad \lambda_n\mathbf{v}^n] = \mathbf{Q}\hat{\mathbf{A}} \end{aligned}$$

Hence we have

$$\hat{\mathbf{A}} = \mathbf{Q}^{-1}\mathbf{AQ}$$

We conclude that if the eigenvalues of a linear operator $\mathbf{A}$ that maps $(\mathbb{C}^n, \mathbb{C})$ into itself are all distinct, then by choosing the set of eigenvectors as a basis, the operator $\mathbf{A}$ has a diagonal matrix representation with the eigenvalues on the diagonal.

#### Example 2

Consider

$$\mathbf{A} = \begin{bmatrix} 1 & -1 \\ 2 & -1 \end{bmatrix}$$

The characteristic polynomial of $\mathbf{A}$ is $\lambda^2 + 1$. Hence the eigenvalues of $\mathbf{A}$ are $+i$ and $-i$. The eigenvector associated with $\lambda_1 = i$ can be obtained by solving the following homogeneous equation:

$$(\mathbf{A} - \lambda_1\mathbf{I})\mathbf{v}^1 = \begin{bmatrix} 1 - i & -1 \\ 2 & -1 - i \end{bmatrix} \begin{bmatrix} v_1{}^1 \\ v_2{}^1 \end{bmatrix} = \mathbf{0}$$

Clearly the vector $\mathbf{v}^1 = [1 \quad 1 - i]'$ is a solution. Similarly, the vector $\mathbf{v}^2 = [1 \quad 1 + i]'$ can be shown to be an eigenvector of $\lambda_2 = -i$. Hence the representation of $\mathbf{A}$ with respect to $\{\mathbf{v}^1, \mathbf{v}^2\}$ is

$$\hat{\mathbf{A}} = \begin{bmatrix} i & 0 \\ 0 & -i \end{bmatrix}$$

The reader is advised to verify this by a similarity transformation.

### Case 2: The eigenvalues of A are not all distinct

Unlike the previous case, if an operator **A** has repeated eigenvalues, it is not always possible to find a diagonal matrix representation. We shall use examples to illustrate the difficulty that may arise for matrices with repeated eigenvalues.

#### Example 3

Consider

$$\mathbf{A} = \begin{bmatrix} 1 & 0 & -1 \\ 0 & 1 & 0 \\ 0 & 0 & 2 \end{bmatrix}$$

The eigenvalues of **A** are $\lambda_1 = 1$, $\lambda_2 = 1$, and $\lambda_3 = 2$. The eigenvectors associated with $\lambda_1$ can be obtained by solving the following homogeneous equations:

$$(\mathbf{A} - \lambda_1\mathbf{I})\mathbf{v} = \begin{bmatrix} 0 & 0 & -1 \\ 0 & 0 & 0 \\ 0 & 0 & 1 \end{bmatrix} \mathbf{v} = \mathbf{0} \tag{2-45}$$

Note that the matrix $(\mathbf{A} - \lambda_1\mathbf{I})$ has rank 1; therefore, two linearly independent vector solutions can be found for (2-45) (see Corollary 2-5). Clearly, $\mathbf{v}^1 = [1 \quad 0 \quad 0]'$ and $\mathbf{v}^2 = [0 \quad 1 \quad 0]'$ are two linearly independent eigenvectors associated with $\lambda_1 = \lambda_2 = 1$. An eigenvector associated with $\lambda_3 = 2$ can be found as $\mathbf{v}^3 = [-1 \quad 0 \quad 1]'$. Since the set of vectors $\{\mathbf{v}^1, \mathbf{v}^2, \mathbf{v}^3\}$ is linearly independent, it qualifies as a basis. The representation of **A** with respect to $\{\mathbf{v}^1, \mathbf{v}^2, \mathbf{v}^3\}$ is

$$\hat{\mathbf{A}} = \begin{bmatrix} 1 & 0 & 0 \\ 0 & 1 & 0 \\ 0 & 0 & 2 \end{bmatrix}$$ ∎

In this example, although **A** has repeated eigenvalues, it can still be diagonalized. However, this is not always the case, as can be seen from the following example.

#### Example 4

Consider

$$\mathbf{A} = \begin{bmatrix} 1 & 1 & 2 \\ 0 & 1 & 3 \\ 0 & 0 & 2 \end{bmatrix} \tag{2-46}$$

The eigenvalues of **A** are $\lambda_1 = 1$, $\lambda_2 = 1$, and $\lambda_3 = 2$. The eigenvectors associated with $\lambda_1 = 1$ can be found by solving

$$(\mathbf{A} - \lambda_1\mathbf{I})\mathbf{v} = \begin{bmatrix} 0 & 1 & 2 \\ 0 & 0 & 3 \\ 0 & 0 & 1 \end{bmatrix} \mathbf{v} = \mathbf{0}$$

Since the matrix $(\mathbf{A} - \lambda_1 \mathbf{I})$ has rank 2, the null space of $(\mathbf{A} - \lambda_1 \mathbf{I})$ has dimension 1. Consequently, we can find only one linearly independent eigenvector, say $\mathbf{v}^1 = [1 \quad 0 \quad 0]'$, associated with $\lambda_1 = \lambda_2 = 1$. An eigenvector associated with $\lambda_3 = 2$ can be found as $\mathbf{v}^3 = [5 \quad 3 \quad 1]'$. Since only two eigenvectors of $\mathbf{A}$ are found, one more vector is needed to form a basis. The vector we shall use is called a *generalized eigenvector*.

### Definition 2-13

A vector $\mathbf{v}$ is said to be a *generalized eigenvector* of rank $k$ of $\mathbf{A}$ associated with $\lambda$ if and only if[7]

$$(\mathbf{A} - \lambda \mathbf{I})^k \mathbf{v} = \mathbf{0}$$

and

$$(\mathbf{A} - \lambda \mathbf{I})^{k-1} \mathbf{v} \neq \mathbf{0} \qquad \blacksquare$$

Note that if $k = 1$, Definition 2-13 reduces to $(\mathbf{A} - \lambda \mathbf{I})\mathbf{v} = \mathbf{0}$ and $\mathbf{v} \neq \mathbf{0}$, which is the definition of an eigenvector. Hence the term "generalized eigenvector" is well-justified.

Let $\mathbf{v}$ be a generalized eigenvector of rank $k$ associated with the eigenvalue $\lambda$. Define

$$\begin{aligned}
\mathbf{v}^k &\triangleq \mathbf{v} \\
\mathbf{v}^{k-1} &\triangleq (\mathbf{A} - \lambda \mathbf{I})\mathbf{v} = (\mathbf{A} - \lambda \mathbf{I})\mathbf{v}^k \\
\mathbf{v}^{k-2} &\triangleq (\mathbf{A} - \lambda \mathbf{I})^2\mathbf{v} = (\mathbf{A} - \lambda \mathbf{I})\mathbf{v}^{k-1} \qquad \text{(2-47)} \\
&\vdots
\end{aligned}$$

and

$$\mathbf{v}^1 \triangleq (\mathbf{A} - \lambda \mathbf{I})^{k-1}\mathbf{v} = (\mathbf{A} - \lambda \mathbf{I})\mathbf{v}^2$$

then for each $i$ in $1 \leq i \leq k$, $\mathbf{v}^i$ is a generalized eigenvector of rank $i$. For example,

$$(\mathbf{A} - \lambda \mathbf{I})^{k-2}\mathbf{v}^{k-2} = (\mathbf{A} - \lambda \mathbf{I})^{k-2}(\mathbf{A} - \lambda \mathbf{I})^2\mathbf{v} = (\mathbf{A} - \lambda \mathbf{I})^k\mathbf{v} = 0$$

However,

$$(\mathbf{A} - \lambda \mathbf{I})^{k-1}\mathbf{v}^{k-2} = (\mathbf{A} - \lambda \mathbf{I})^{k-1}\mathbf{v} \neq 0$$

Hence $\mathbf{v}^{k-2}$ is a generalized eigenvector of rank $k - 2$ of $\mathbf{A}$ associated with $\lambda$. We call the set of vectors $\{\mathbf{v}^1, \mathbf{v}^2, \cdots, \mathbf{v}_k\}$ a *chain* of generalized eigenvectors.

### Theorem 2-10

The set of generalized eigenvectors $\mathbf{v}^1, \mathbf{v}^2, \cdots, \mathbf{v}^k$ defined in (2-47) is linearly independent.

---

[7] $(\mathbf{A} - \lambda \mathbf{I})^k \triangleq (\mathbf{A} - \lambda \mathbf{I})(\mathbf{A} - \lambda \mathbf{I}) \cdots (\mathbf{A} - \lambda \mathbf{I})$ ($k$ terms), $(\mathbf{A} - \lambda \mathbf{I})^0 \triangleq \mathbf{I}$.

### Proof

We assume that $\mathbf{v}^1, \mathbf{v}^2, \cdots, \mathbf{v}^k$ are linearly dependent, then there exist $\alpha_1, \alpha_2, \cdots, \alpha_k$, not all zero, such that

$$\alpha_1\mathbf{v}^1 + \alpha_2\mathbf{v}^2 + \cdots + \alpha_k\mathbf{v}^k = \mathbf{0} \tag{2-48}$$

Applying $(\mathbf{A} - \lambda\mathbf{I})^{k-1}$ to (2-48) and observing

$$(\mathbf{A} - \lambda\mathbf{I})^{k-1}\mathbf{v}^i = (\mathbf{A} - \lambda\mathbf{I})^{2k-(i+1)}\mathbf{v} = \mathbf{0}$$

for $i = 1, 2, \cdots, k-1$, we obtain

$$\alpha_k(\mathbf{A} - \lambda\mathbf{I})^{k-1}\mathbf{v}^k = \mathbf{0}$$

which, with the condition $(\mathbf{A} - \lambda\mathbf{I})^{k-1}\mathbf{v} \neq \mathbf{0}$, implies $\alpha_k = 0$. Next, by applying $(\mathbf{A} - \lambda\mathbf{I})^{k-2}$, we can show that $\alpha_{k-1} = 0$. Proceeding successively, we can show that $\alpha_i = 0$, for $i = 1, 2, \cdots, k$. This contradicts the assumption. Hence the set of vectors $\mathbf{v}^1, \mathbf{v}^2, \cdots, \mathbf{v}^k$ is linearly independent. Q.E.D.

### Theorem 2-11

The generalized eigenvectors of $\mathbf{A}$ associated with different eigenvalues are linearly independent.

### Proof

Let $\mathbf{v}$ be a generalized eigenvector of rank $k$ associated with $\lambda_1$, and $\mathbf{u}$ be a generalized eigenvector of rank $l$ associated with $\lambda_2$. Define $\mathbf{v}^k = \mathbf{v}$, $\mathbf{v}^i = (\mathbf{A} - \lambda_1\mathbf{I})\mathbf{v}^{i+1} = (\mathbf{A} - \lambda_1\mathbf{I})^{k-i}\mathbf{v}$, for $i = k-1, k-2, \cdots, 1$, and define $\mathbf{u}^l = \mathbf{u}$, $\mathbf{u}^j = (\mathbf{A} - \lambda_2\mathbf{I})\mathbf{u}^{j+1} = (\mathbf{A} - \lambda_2\mathbf{I})^{l-j}\mathbf{u}$, for $j = l-1, l-2, \cdots, 1$. From Theorem 2-10, we know that $\mathbf{v}^1, \mathbf{v}^2, \cdots, \mathbf{v}^k$ are linearly independent, as are $\mathbf{u}^1, \mathbf{u}^2, \cdots, \mathbf{u}^l$. Hence what we have to show is that the set $\{\mathbf{v}^1, \mathbf{v}^2, \cdots, \mathbf{v}^k\}$ and the set $\{\mathbf{u}^1, \mathbf{u}^2, \cdots, \mathbf{u}^l\}$ are linearly independent. We prove this by contradiction. Suppose that $\mathbf{v}^i$ is linearly dependent on the set $\{\mathbf{u}^1, \mathbf{u}^2, \cdots, \mathbf{u}^l\}$, then there exist $\alpha_j$ (for $j = 1, 2, \cdots, l$) not all zero, such that

$$\mathbf{v}^i = \sum_{j=1}^{l} \alpha_j\mathbf{u}^j \tag{2-49}$$

Applying $(\mathbf{A} - \lambda_1\mathbf{I})^i$ on both sides of (2-49) and observing $(\mathbf{A} - \lambda_1\mathbf{I})^i\mathbf{v}^i = \mathbf{0}$, we obtain

$$\mathbf{0} = (\mathbf{A} - \lambda_1\mathbf{I})^i \sum_{j=1}^{l} \alpha_j\mathbf{u}^j \tag{2-50}$$

Applying next the operator $(\mathbf{A} - \lambda_2\mathbf{I})^{l-1}$ to (2-50) and observing that

$$(\mathbf{A} - \lambda_2\mathbf{I})^{l-1}(\mathbf{A} - \lambda_1\mathbf{I})^i = (\mathbf{A} - \lambda_1\mathbf{I})^i(\mathbf{A} - \lambda_2\mathbf{I})^{l-1}$$

and the fact that $(\mathbf{A} - \lambda_2\mathbf{I})^{l-1}\mathbf{u}^j = \mathbf{0}$ for $j = 1, 2, \cdots, l-1$, we obtain

$$\mathbf{0} = \alpha_l(\mathbf{A} - \lambda_1\mathbf{I})^i(\mathbf{A} - \lambda_2\mathbf{I})^{l-1}\mathbf{u}^l = \alpha_l(\mathbf{A} - \lambda_1\mathbf{I})^i\mathbf{u}^1 \tag{2-51}$$

Since $(\mathbf{A} - \lambda_2\mathbf{I})\mathbf{u}^1 = \mathbf{0}$ or $\mathbf{A}\mathbf{u}^1 = \lambda_2\mathbf{u}^1$, Equation (2-51) reduces to

$$\alpha_l(\lambda_2 - \lambda_1)^i\mathbf{u}^1 = \mathbf{0}$$

which implies that $\alpha_l = 0$ by the assumptions $\lambda_2 \neq \lambda_1$ and $\mathbf{u}^1 \neq \mathbf{0}$. Proceeding further, we can show that $\alpha_i = 0$, for $i = 1, 2, \cdots, l$. This contradicts the assumption. Hence $\mathbf{v}^i$ is linearly independent on the set $\{\mathbf{u}^1, \mathbf{u}^2, \cdots, \mathbf{u}^l\}$. Similarly we can show that $\mathbf{u}^j$ is linearly independent on the set $\{\mathbf{v}^1, \mathbf{v}^2, \cdots, \mathbf{v}^k\}$. Q.E.D.

With these two theorems, we are ready to complete Example 4.

#### Example 4 (Continued)

Consider

$$\mathbf{A} = \begin{bmatrix} 1 & 1 & 2 \\ 0 & 1 & 3 \\ 0 & 0 & 2 \end{bmatrix}$$

Its eigenvalues are $\lambda_1 = 1$, $\lambda_2 = 1$, $\lambda_3 = 2$. An eigenvector associated with $\lambda_3 = 2$ is $\mathbf{v}^3 = [5 \quad 3 \quad 1]'$. The rank of $(\mathbf{A} - \lambda_1\mathbf{I})$ is 2, and hence we can find only one eigenvector associated with $\lambda_1$. Consequently, we must use generalized eigenvectors. Let $\mathbf{v}$ be a nonzero vector such that

$$(\mathbf{A} - \lambda_1\mathbf{I})\mathbf{v} = \begin{bmatrix} 0 & 1 & 2 \\ 0 & 0 & 3 \\ 0 & 0 & 1 \end{bmatrix}\mathbf{v} \neq \mathbf{0}$$

and

$$(\mathbf{A} - \lambda_1\mathbf{I})^2\mathbf{v} = \begin{bmatrix} 0 & 1 & 2 \\ 0 & 0 & 3 \\ 0 & 0 & 1 \end{bmatrix}\begin{bmatrix} 0 & 1 & 2 \\ 0 & 0 & 3 \\ 0 & 0 & 1 \end{bmatrix}\mathbf{v} = \begin{bmatrix} 0 & 0 & 5 \\ 0 & 0 & 3 \\ 0 & 0 & 1 \end{bmatrix}\mathbf{v} = \mathbf{0}$$

Clearly, $\mathbf{v} = [0 \quad 1 \quad 0]'$ is such a vector. It is a generalized eigenvector of rank 2. Let

$$\mathbf{v}^2 \triangleq \mathbf{v} = \begin{bmatrix} 0 \\ 1 \\ 0 \end{bmatrix} \qquad \mathbf{v}^1 \triangleq (\mathbf{A} - \lambda_1\mathbf{I})\mathbf{v} = \begin{bmatrix} 0 & 1 & 2 \\ 0 & 0 & 3 \\ 0 & 0 & 1 \end{bmatrix}\begin{bmatrix} 0 \\ 1 \\ 0 \end{bmatrix} = \begin{bmatrix} 1 \\ 0 \\ 0 \end{bmatrix}$$

Theorems, 2-10 and 2-11 imply that $\mathbf{v}^1$, $\mathbf{v}^2$, and $\mathbf{v}^3$ are linearly independent. This can also be checked by computing the determinant of $[\mathbf{v}^1 \quad \mathbf{v}^2 \quad \mathbf{v}^3]$. If we use the set of vectors $\{\mathbf{v}^1, \mathbf{v}^2, \mathbf{v}^3\}$ as a basis, then the $i$th column of the new representation $\hat{\mathbf{A}}$ is the representation of $\mathbf{A}\mathbf{v}^i$ with respect to the basis $\{\mathbf{v}^1, \mathbf{v}^2, \mathbf{v}^3\}$. Since $\mathbf{A}\mathbf{v}^1 = \lambda_1\mathbf{v}^1$, $\mathbf{A}\mathbf{v}^2 = \mathbf{v}^1 + \lambda_1\mathbf{v}^2$, and $\mathbf{A}\mathbf{v}^3 = \lambda_3\mathbf{v}^3$, the representations of $\mathbf{A}\mathbf{v}^1$, $\mathbf{A}\mathbf{v}^2$, and $\mathbf{A}\mathbf{v}^3$ with respect to the basis $\{\mathbf{v}^1, \mathbf{v}^2, \mathbf{v}^3\}$ are, respectively,

$$\begin{bmatrix} \lambda_1 \\ 0 \\ 0 \end{bmatrix}, \begin{bmatrix} 1 \\ \lambda_1 \\ 0 \end{bmatrix}, \begin{bmatrix} 0 \\ 0 \\ \lambda_3 \end{bmatrix}$$

where $\lambda_1 = 1$, $\lambda_3 = 2$. Hence we have

$$\hat{\mathbf{A}} = \begin{bmatrix} 1 & 1 & 0 \\ 0 & 1 & 0 \\ 0 & 0 & 2 \end{bmatrix} \tag{2-52}$$

This can also be obtained by using the similarity transformation

$$\hat{\mathbf{A}} = \mathbf{Q}^{-1}\mathbf{A}\mathbf{Q}$$

where

$$\mathbf{Q} = [\mathbf{v}^1 \quad \mathbf{v}^2 \quad \mathbf{v}^3] = \begin{bmatrix} 1 & 0 & 5 \\ 0 & 1 & 3 \\ 0 & 0 & 1 \end{bmatrix}$$ ∎

A matrix in the form of (2-52) is said to be in the Jordan canonical form. In order to see the feature of a Jordan-canonical-form representation, we give a more general example in the following:

$$\hat{\mathbf{A}} = \left[\begin{array}{ccc|cc|c|cc} \lambda_1 & 1 & 0 & 0 & 0 & 0 & 0 & 0 \\ 0 & \lambda_1 & 1 & 0 & 0 & 0 & 0 & 0 \\ 0 & 0 & \lambda_1 & 0 & 0 & 0 & 0 & 0 \\ \hline 0 & 0 & 0 & \lambda_1 & 1 & 0 & 0 & 0 \\ 0 & 0 & 0 & 0 & \lambda_1 & 0 & 0 & 0 \\ \hline 0 & 0 & 0 & 0 & 0 & \lambda_1 & 0 & 0 \\ \hline 0 & 0 & 0 & 0 & 0 & 0 & \lambda_2 & 1 \\ 0 & 0 & 0 & 0 & 0 & 0 & 0 & \lambda_2 \end{array}\right] \triangleq \begin{bmatrix} \hat{\mathbf{A}}_{11} & \mathbf{0} & \mathbf{0} & \mathbf{0} \\ \mathbf{0} & \hat{\mathbf{A}}_{12} & \mathbf{0} & \mathbf{0} \\ \mathbf{0} & \mathbf{0} & \hat{\mathbf{A}}_{13} & \mathbf{0} \\ \mathbf{0} & \mathbf{0} & \mathbf{0} & \hat{\mathbf{A}}_{21} \end{bmatrix} \tag{2-53}$$

The eigenvalues of $\hat{\mathbf{A}}$ are $\lambda_1$ with multiplicity 6 and $\lambda_2$ with multiplicity 2. On the principal diagonal of $\hat{\mathbf{A}}$, there are four blocks, and every block is of the form

$$\begin{bmatrix} \lambda & 1 & 0 & \cdots & 0 & 0 \\ 0 & \lambda & 1 & \cdots & 0 & 0 \\ \cdot & \cdot & \cdot & & \cdot & \cdot \\ \cdot & \cdot & \cdot & & \cdot & \cdot \\ \cdot & \cdot & \cdot & & \cdot & \cdot \\ 0 & 0 & 0 & \cdots & 1 & 0 \\ 0 & 0 & 0 & \cdots & \lambda & 1 \\ 0 & 0 & 0 & \cdots & 0 & \lambda \end{bmatrix} \tag{2-54}$$

with the same eigenvalue on the main diagonal and 1's on the diagonal just above the main diagonal. A matrix of this form is called a *Jordan block* associated with $\lambda$. A matrix is said to be in the *Jordan canonical form*, or the *Jordan form*, if its principal diagonal consists of Jordan blocks and the remaining elements are zeros. The matrix $\hat{\mathbf{A}}$ in (2-53) has three Jordan blocks associated with $\lambda_1$ and one Jordan block associated with $\lambda_2$. A diagonal matrix is clearly a special case of the Jordan form: all of its Jordan blocks are of order 1.

We claim that every linear transformation which maps $(\mathbb{C}^n, \mathbb{C})$ into itself has a Jordan-form representation by a proper choice of basis. The basis vectors generally consist of eigenvectors and generalized eigenvectors. Before giving a general procedure for computing the required basis vectors, we need the following theorem.

### Theorem 2-12

Let $\mathbf{u}$ and $\mathbf{v}$ be the generalized eigenvectors of rank $k$ and $l$, respectively, associated with the same eigenvalue $\lambda$. Define $\mathbf{u}^i = (\mathbf{A} - \lambda\mathbf{I})^{k-i}\mathbf{u}$, for $i = 1, 2, \cdots, k$, and $\mathbf{v}^j = (\mathbf{A} - \lambda\mathbf{I})^{l-j}\mathbf{v}$, for $j = 1, 2, \cdots, l$. If the two vectors $\mathbf{u}^1$ and $\mathbf{v}^1$ are linearly independent, then the generalized eigenvectors $\mathbf{u}^1, \mathbf{u}^2, \cdots, \mathbf{u}^k, \mathbf{v}^1, \mathbf{v}^2, \cdots, \mathbf{v}^l$ are linearly independent. ▮

The set $\{\mathbf{u}^1, \mathbf{u}^2, \cdots, \mathbf{u}^k\}$ is a linearly independent set, so is the set $\{\mathbf{v}^1, \mathbf{v}^2, \cdots, \mathbf{v}^l\}$ (see Theorem 2-10). Hence what is left to be proved is that each $\mathbf{u}^i$ is linearly independent on the set $\{\mathbf{v}^1, \mathbf{v}^2, \cdots, \mathbf{v}^l\}$ and each $\mathbf{v}^i$ is linearly independent on the set $\{\mathbf{u}^1, \mathbf{u}^2, \cdots, \mathbf{u}^k\}$. This can be proved by the arguments used in the proof of Theorem 2-11; the proof is left as an exercise for the reader.

**A procedure for computing a Jordan-form representation.** We give in the following a procedure for computing a set of basis vectors for a linear operator $\mathbf{A}$ that maps $(\mathbb{C}^n, \mathbb{C})$ into itself such that the new representation of $\mathbf{A}$ is in the Jordan canonical form.

1. Compute the eigenvalues of $\mathbf{A}$ by solving $\det(\mathbf{A} - \lambda\mathbf{I}) = 0$. Let $\lambda_1, \lambda_2, \cdots, \lambda_m$ be the distinct eigenvalues of $\mathbf{A}$ with multiplicities $n_1, n_2, \cdots, n_m$, respectively.
2. Compute $n_1$ linearly independent generalized eigenvectors of $\mathbf{A}$ associated with $\lambda_1$ as follows: compute $(\mathbf{A} - \lambda_1\mathbf{I})^i$, for $i = 1, 2, \cdots$, until the rank of $(\mathbf{A} - \lambda_1\mathbf{I})^k$ is equal to the rank of $(\mathbf{A} - \lambda_1\mathbf{I})^{k+1}$. Find a generalized eigenvector of rank $k$, say $\mathbf{u}$. Define $\mathbf{u}^i \triangleq (\mathbf{A} - \lambda_1\mathbf{I})^{k-i}\mathbf{u}$, for $i = 1, 2, \cdots, k$. If $k = n_1$, proceed to step 3. If $k < n_1$, we try to find another linearly independent generalized eigenvector (see Theorem 2-12) with the largest possible rank; that is, try to find another generalized eigenvector with rank $k$; if this is not possible, try $k - 1$, and so forth, until $n_1$ linearly independent generalized eigenvectors are found. Note that if $\rho(\mathbf{A} - \lambda_1\mathbf{I}) = \alpha$, then there are totally $(n - \alpha)$ chains of generalized eigenvectors associated with $\lambda_1$.
3. Repeat step 2 for the eigenvalues $\lambda_2, \lambda_3, \cdots, \lambda_m$.

**4.** Let $\{\mathbf{u}^1, \mathbf{u}^2, \cdots, \mathbf{u}^k \cdots\}$ be the new basis. Observe, from (2-47), that

$$\mathbf{A}\mathbf{u}^1 = \lambda_1 \mathbf{u}^1 = [\mathbf{u}^1 \quad \mathbf{u}^2 \quad \cdots \quad \mathbf{u}^k \quad \cdots] \begin{bmatrix} \lambda_1 \\ 0 \\ 0 \\ \cdot \\ \cdot \\ \cdot \\ 0 \end{bmatrix}$$

$$\mathbf{A}\mathbf{u}^2 = \mathbf{u}^1 + \lambda_1 \mathbf{u}^2 = [\mathbf{u}^1 \quad \mathbf{u}^2 \quad \cdots \quad \mathbf{u}^k \quad \cdots] \begin{bmatrix} 1 \\ \lambda_1 \\ 0 \\ \cdot \\ \cdot \\ \cdot \\ 0 \end{bmatrix}$$

$$\vdots$$

$$\mathbf{A}\mathbf{u}^k = \mathbf{u}^{k-1} + \lambda_1 \mathbf{u}^k = [\mathbf{u}^1 \quad \mathbf{u}^2 \quad \cdots \quad \mathbf{u}^k \quad \cdots] \begin{bmatrix} 0 \\ 0 \\ \cdot \\ \cdot \\ \cdot \\ 1 \\ \lambda_1 \\ 0 \\ \cdot \\ \cdot \\ \cdot \\ 0 \end{bmatrix} \leftarrow k\text{th}$$

hence the representation of **A** with respect to this new basis is

$$\hat{\mathbf{A}} = \left[\begin{array}{cccccc|ccc} \lambda_1 & 1 & & & & & & & \\ & \lambda_1 & 1 & & & & & & \\ & & \cdot & \cdot & & & & & \\ & & & \cdot & \cdot & & & & \\ & & & & \cdot & \cdot & & & \\ & & & & & 1 & & & \\ & & & & & \lambda_1 & & & \\ \hline & (k \times k) & & & & & \square & & \\ & & & & & & & \cdot & \\ & & & & & & & & \cdot \\ & & & & & & & & \cdot \end{array}\right]$$

The representation $\hat{\mathbf{A}}$ has Jordan blocks on the diagonal and zeros elsewhere. Each chain of generalized eigenvectors generates a Jordan block whose order is equal to the length of the chain.

**5.** If so desired, we may check the result by using the similarity transformation $\hat{\mathbf{A}} = \mathbf{Q}^{-1}\mathbf{A}\mathbf{Q}$, or more easily, $\mathbf{Q}\hat{\mathbf{A}} = \mathbf{A}\mathbf{Q}$, where

$$\mathbf{Q} = [\mathbf{u}^1 \quad \mathbf{u}^2 \quad \cdots \quad \mathbf{u}^k \quad \cdots]$$

### Example 5

Transform the following matrix into the Jordan form:

$$\mathbf{A} = \begin{bmatrix} 3 & -1 & 1 & 1 & 0 & 0 \\ 1 & 1 & -1 & -1 & 0 & 0 \\ 0 & 0 & 2 & 0 & 1 & 1 \\ 0 & 0 & 0 & 2 & -1 & -1 \\ 0 & 0 & 0 & 0 & 1 & 1 \\ 0 & 0 & 0 & 0 & 1 & 1 \end{bmatrix}$$

**1.** Compute the eigenvalues of **A**.[8]

$$\det(\mathbf{A} - \lambda\mathbf{I}) = [(3-\lambda)(1-\lambda)+1](\lambda-2)^2[(1-\lambda)^2-1] = (\lambda-2)^5\lambda$$

Hence **A** has eigenvalue 2 with multiplicity 5 and eigenvalue 0 with multiplicity 1.

**2.** Compute $(\mathbf{A} - \lambda\mathbf{I})^i$, for $i = 1, 2, \cdots$, as follows:

$$(\mathbf{A} - 2\mathbf{I}) = \begin{bmatrix} 1 & -1 & 1 & 1 & 0 & 0 \\ 1 & -1 & -1 & -1 & 0 & 0 \\ 0 & 0 & 0 & 0 & 1 & 1 \\ 0 & 0 & 0 & 0 & -1 & -1 \\ 0 & 0 & 0 & 0 & -1 & 1 \\ 0 & 0 & 0 & 0 & 1 & -1 \end{bmatrix} \qquad \rho(\mathbf{A} - 2\mathbf{I}) = 4$$

$$(\mathbf{A} - 2\mathbf{I})^2 = \begin{bmatrix} 0 & 0 & 2 & 2 & 0 & 0 \\ 0 & 0 & 2 & 2 & 0 & 0 \\ 0 & 0 & 0 & 0 & 0 & 0 \\ 0 & 0 & 0 & 0 & 0 & 0 \\ 0 & 0 & 0 & 0 & 2 & -2 \\ 0 & 0 & 0 & 0 & -2 & 2 \end{bmatrix} \qquad \rho(\mathbf{A} - 2\mathbf{I})^2 = 3$$

[8] We use the fact that

$$\det \begin{bmatrix} \mathbf{A} & \mathbf{B} \\ \mathbf{0} & \mathbf{C} \end{bmatrix} = \det(\mathbf{A}) \det(\mathbf{C})$$

where **A** and **C** are square matrices, not necessarily of the same order.

$$(\mathbf{A}-2\mathbf{I})^3=\begin{bmatrix}0&0&0&0&0&0\\0&0&0&0&0&0\\0&0&0&0&0&0\\0&0&0&0&0&0\\0&0&0&0&-4&4\\0&0&0&0&4&-4\end{bmatrix}\qquad \rho(\mathbf{A}-2\mathbf{I})^3=2$$

$$(\mathbf{A}-2\mathbf{I})^4=\begin{bmatrix}0&0&0&0&0&0\\0&0&0&0&0&0\\0&0&0&0&0&0\\0&0&0&0&0&0\\0&0&0&0&8&-8\\0&0&0&0&-8&8\end{bmatrix}\qquad \rho(\mathbf{A}-2\mathbf{I})^4=2$$

The rank of $(\mathbf{A}-2\mathbf{I})^3$ is equal to the rank of $(\mathbf{A}-2\mathbf{I})^4$; hence we stop at $(\mathbf{A}-2\mathbf{I})^3$. It is easy to verify that if $\mathbf{u}=[0\ \ 0\ \ 1\ \ 0\ \ 0\ \ 0]'$, then $(\mathbf{A}-2\mathbf{I})^3\mathbf{u}=\mathbf{0}$ and $(\mathbf{A}-2\mathbf{I})^2\mathbf{u}=[1\ \ 1\ \ 0\ \ 0\ \ 0\ \ 0]'\neq\mathbf{0}$. Hence $\mathbf{u}$ is a generalized eigenvector of rank 3. Define

$$\mathbf{u}^1\triangleq(\mathbf{A}-2\mathbf{I})^2\mathbf{u}=\begin{bmatrix}2\\2\\0\\0\\0\\0\end{bmatrix}\qquad \mathbf{u}^2\triangleq(\mathbf{A}-2\mathbf{I})\mathbf{u}=\begin{bmatrix}1\\-1\\0\\0\\0\\0\end{bmatrix}$$

$$\mathbf{u}^3\triangleq\mathbf{u}=\begin{bmatrix}0\\0\\1\\0\\0\\0\end{bmatrix}$$

Since we have only three generalized eigenvectors, we have to find two more linearly independent eigenvectors. In this case it is not necessary to try to find another generalized eigenvector of rank 3. Indeed, if there is another linearly independent generalized eigenvector of rank 3, then we would have six linearly independent generalized eigenvectors associated with $\lambda_1=2$. This is impossible, since with the eigenvector associated with $\lambda_2=0$, we would have seven linearly independent vectors in a six-dimensional vector space. Hence we proceed to find a generalized eigenvector of rank 2. Let $\mathbf{v}=[0\quad 0\quad 1\quad -1\quad 1\quad 1]'$, then $(\mathbf{A}-2\mathbf{I})\mathbf{v}=[0\quad 0\quad 2\quad -2\quad 0\quad 0]'\neq\mathbf{0}$ and $(\mathbf{A}-2\mathbf{I})^2\mathbf{v}=\mathbf{0}$. Furthermore, $(\mathbf{A}-2\mathbf{I})\mathbf{v}$ is linearly independent of $\mathbf{u}^1$, hence we have another linearly independent generalized eigenvector of rank 2. Define

$$\mathbf{v}^2 \triangleq \mathbf{v} = \begin{bmatrix} 0 \\ 0 \\ 1 \\ -1 \\ 1 \\ 1 \end{bmatrix} \qquad \mathbf{v}^1 \triangleq (\mathbf{A} - 2\mathbf{I})\mathbf{v} = \begin{bmatrix} 0 \\ 0 \\ 2 \\ -2 \\ 0 \\ 0 \end{bmatrix}$$

**3.** Compute an eigenvector associated with $\lambda_2 = 0$. Let $\mathbf{w}$ be an eigenvector of $\mathbf{A}$ associated with $\lambda_2 = 0$; then

$$(\mathbf{A} - \lambda_2\mathbf{I})\mathbf{w} = \begin{bmatrix} 3 & -1 & 1 & 1 & 0 & 0 \\ 1 & 1 & -1 & -1 & 0 & 0 \\ 0 & 0 & 2 & 0 & 1 & 1 \\ 0 & 0 & 0 & 2 & -1 & -1 \\ 0 & 0 & 0 & 0 & 1 & 1 \\ 0 & 0 & 0 & 0 & 1 & 1 \end{bmatrix} \mathbf{w} = \mathbf{0}$$

Clearly, $\mathbf{w} = [0 \quad 0 \quad 0 \quad 0 \quad 1 \quad -1]'$ is a solution.

**4.** With respect to the basis $\{\mathbf{u}^1, \mathbf{u}^2, \mathbf{u}^3, \mathbf{v}^1, \mathbf{v}^2, \mathbf{w}\}$, $\mathbf{A}$ has the following Jordan-form representation:

$$\hat{\mathbf{A}} = \left[\begin{array}{ccc:cc:c} 2 & 1 & 0 & 0 & 0 & 0 \\ 0 & 2 & 1 & 0 & 0 & 0 \\ 0 & 0 & 2 & 0 & 0 & 0 \\ \hdashline 0 & 0 & 0 & 2 & 1 & 0 \\ 0 & 0 & 0 & 0 & 2 & 0 \\ \hdashline 0 & 0 & 0 & 0 & 0 & 0 \end{array}\right] \tag{2-55}$$

**5.** This may be checked by using

$$\hat{\mathbf{A}} = \mathbf{Q}^{-1}\mathbf{A}\mathbf{Q} \qquad \text{or} \qquad \mathbf{Q}\hat{\mathbf{A}} = \mathbf{A}\mathbf{Q}$$

where

$$\mathbf{Q} = [\mathbf{u}^1 \quad \mathbf{u}^2 \quad \mathbf{u}^3 \quad \mathbf{v}^1 \quad \mathbf{v}^2 \quad \mathbf{w}]$$

$$= \left[\begin{array}{c:c:c:c:c:c} 2 & 1 & 0 & 0 & 0 & 0 \\ 2 & -1 & 0 & 0 & 0 & 0 \\ 0 & 0 & 1 & 2 & 1 & 0 \\ 0 & 0 & 0 & -2 & -1 & 0 \\ 0 & 0 & 0 & 0 & 1 & 1 \\ 0 & 0 & 0 & 0 & 1 & -1 \end{array}\right]$$

∎

In this example, if we reorder the basis $\{\mathbf{u}^1, \mathbf{u}^2, \mathbf{u}^3, \mathbf{v}^1, \mathbf{v}^2, \mathbf{w}\}$ and use $\{\mathbf{w}, \mathbf{v}^2, \mathbf{v}^1, \mathbf{u}^3, \mathbf{u}^2, \mathbf{u}^1\}$ as a new basis, then the representation will be

$$\hat{\mathbf{A}} = \left[\begin{array}{c|cc|ccc} 0 & 0 & 0 & 0 & 0 & 0 \\ \hline 0 & 2 & 0 & 0 & 0 & 0 \\ 0 & 1 & 2 & 0 & 0 & 0 \\ \hline 0 & 0 & 0 & 2 & 0 & 0 \\ 0 & 0 & 0 & 1 & 2 & 0 \\ 0 & 0 & 0 & 0 & 1 & 2 \end{array}\right] \tag{2-56}$$

This is also called a Jordan-form representation. Comparing it with Equation (2-55), we see that the new Jordan block in (2-56) has 1's on the diagonal just below the main diagonal as a result of the different ordering of the basis vectors. In this book, we use mostly the Jordan block of the form in (2-54). Certainly everything discussed for this form can be modified and be applied to the form given in Equation (2-56).

A Jordan-form representation of any linear operator $\mathbf{A}$ that maps $(\mathbb{C}^n, \mathbb{C})$ into itself is unique up to the ordering of Jordan blocks. That is, the number of Jordan blocks and the order of each Jordan block are uniquely determined by $\mathbf{A}$. However, because of different orderings of basis vectors, we may have different Jordan-form representations of the same matrix.

## 2-7 Functions of a Square Matrix

In this section we shall study functions of a square matrix or a linear transformation that maps $(\mathbb{C}^n, \mathbb{C})$ into itself. We shall use the Jordan-form representation extensively, because in terms of this representation almost all properties of a function of a matrix can be visualized. We study first polynomials of a square matrix, and then define functions of a matrix in terms of polynomials of the matrix.

**Polynomials of a square matrix.** Let $\mathbf{A}$ be a square matrix that maps $(\mathbb{C}^n, \mathbb{C})$ into $(\mathbb{C}^n, \mathbb{C})$. If $k$ is a positive integer, we define

$$\mathbf{A}^k \triangleq \mathbf{A}\mathbf{A} \cdots \mathbf{A} \qquad (k \text{ terms}) \tag{2-57a}$$

and

$$\mathbf{A}^0 \triangleq \mathbf{I} \tag{2-57b}$$

where $\mathbf{I}$ is a unit matrix. Let $f(\lambda)$ be a polynomial in $\lambda$ of finite degree, then $f(\mathbf{A})$ can be defined in terms of (2-57). For example, if $f(\lambda) = \lambda^3 + 2\lambda^2 + 6$, then

$$f(\mathbf{A}) \triangleq \mathbf{A}^3 + 2\mathbf{A}^2 + 6\mathbf{I}$$

We have shown in the preceding section that every square matrix $\mathbf{A}$ that maps $(\mathbb{C}^n, \mathbb{C})$ into itself has a Jordan-form representation; or equivalently, there exists a nonsingular constant matrix $\mathbf{Q}$ such that $\mathbf{A} = \mathbf{Q}\hat{\mathbf{A}}\mathbf{Q}^{-1}$ with $\hat{\mathbf{A}}$ in the Jordan canonical form. Since

$$\mathbf{A}^k = (\mathbf{Q}\hat{\mathbf{A}}\mathbf{Q}^{-1})(\mathbf{Q}\hat{\mathbf{A}}\mathbf{Q}^{-1}) \cdots (\mathbf{Q}\hat{\mathbf{A}}\mathbf{Q}^{-1}) = \mathbf{Q}\hat{\mathbf{A}}^k\mathbf{Q}^{-1}$$

we have

$$f(\mathbf{A}) = \mathbf{Q}f(\hat{\mathbf{A}})\mathbf{Q}^{-1} \qquad \text{or} \qquad f(\hat{\mathbf{A}}) = \mathbf{Q}^{-1}f(\mathbf{A})\mathbf{Q} \tag{2-58}$$

for any polynomial $f(\lambda)$.

One of the reasons to use the Jordan-form matrix is that if

$$\mathbf{A} = \begin{bmatrix} \mathbf{A}_1 & \mathbf{0} \\ \mathbf{0} & \mathbf{A}_2 \end{bmatrix} \tag{2-59}$$

where $\mathbf{A}_1$ and $\mathbf{A}_2$ are square matrices, then

$$f(\mathbf{A}) = \begin{bmatrix} f(\mathbf{A}_1) & \mathbf{0} \\ \mathbf{0} & f(\mathbf{A}_2) \end{bmatrix} \tag{2-60}$$

This can be easily verified by observing that

$$\mathbf{A}^k = \begin{bmatrix} \mathbf{A}_1^k & \mathbf{0} \\ \mathbf{0} & \mathbf{A}_2^k \end{bmatrix}$$

### Definition 2-14

*The minimal polynomial* of a matrix $\mathbf{A}$ is the monic polynomial[9] $\psi(\lambda)$ of least degree such that $\psi(\mathbf{A}) = \mathbf{0}$. ∎

Note that the $\mathbf{0}$ in $\psi(\mathbf{A}) = \mathbf{0}$ is an $n \times n$ square matrix whose entries are all zero. A direct consequence of (2-58) is that $f(\mathbf{A}) = \mathbf{0}$ if and only if $f(\hat{\mathbf{A}}) = \mathbf{0}$. Consequently, the matrices $\mathbf{A}$ and $\hat{\mathbf{A}}$ have the same minimal polynomial; or more generally, *similar matrices have the same minimal polynomial.* Computing the minimal polynomial of a matrix is generally not a simple job; however, if the Jordan-form representation of the matrix is available, its minimal polynomial can be readily found.

Let $\lambda_1, \lambda_2, \cdots, \lambda_m$ be the distinct eigenvalues of $\mathbf{A}$ with multiplicities $n_1, n_2, \cdots, n_m$, respectively. It is the same as saying that the characteristic polynomial of $\mathbf{A}$ is

$$\Delta(\lambda) \triangleq \det(\lambda\mathbf{I} - \mathbf{A}) = \prod_{i=1}^{m} (\lambda - \lambda_i)^{n_i} \tag{2-61}$$

[9] A monic polynomial is a polynomial the coefficient of whose highest power is 1. For example, $3x + 1$ and $-x^2 + 2x + 4$ are not monic polynomials, but $x^2 - 4x + 7$ is.

Assume that a Jordan form representation of $\mathbf{A}$ is

$$\hat{\mathbf{A}} = \begin{bmatrix} \hat{\mathbf{A}}_1 & \mathbf{0} & \cdots & \mathbf{0} \\ \mathbf{0} & \hat{\mathbf{A}}_2 & \cdots & \mathbf{0} \\ \vdots & \vdots & & \vdots \\ \mathbf{0} & \mathbf{0} & \cdots & \hat{\mathbf{A}}_m \end{bmatrix} \tag{2-62}$$

where the $n_i \times n_i$ matrix $\hat{\mathbf{A}}_i$ denotes all the Jordan blocks associated with $\lambda_i$.

### Definition 2-15

The largest order of the Jordan blocks associated with $\lambda_i$ in $\mathbf{A}$ is called the *index* of $\lambda_i$ in $\mathbf{A}$. ∎

The multiplicity of $\lambda_i$ is denoted by $n_i$, the index of $\lambda_i$ is denoted by $\bar{n}_i$. For the matrix in (2-53), $n_1 = 6$, $\bar{n}_1 = 3$, $n_2 = 2$, $\bar{n}_2 = 2$; for the matrix in (2-55), $n_1 = 5$, $\bar{n}_1 = 3$; $n_2 = \bar{n}_2 = 1$. It is clear that $\bar{n}_i \leq n_i$.

### Theorem 2-13

The minimal polynomial of $\mathbf{A}$ is

$$\psi(\lambda) = \prod_{i=1}^{m} (\lambda - \lambda_i)^{\bar{n}_i}$$

where $\bar{n}_i$ is the index of $\lambda_i$ in $\mathbf{A}$.

### Proof

Since the matrices $\mathbf{A}$ and $\hat{\mathbf{A}}$ have the same minimal polynomial, it is the same as showing that $\psi(\lambda)$ is the polynomial with least degree such that $\psi(\hat{\mathbf{A}}) = \mathbf{0}$. We first show that the minimal polynomial of $\hat{\mathbf{A}}_i$ is $\psi_i(\lambda) = (\lambda - \lambda_i)^{\bar{n}_i}$. Suppose $\hat{\mathbf{A}}_i$ consists of $r$ Jordan blocks associated with $\lambda_i$. Then

$$\hat{\mathbf{A}}_i = \text{diag}\,(\hat{\mathbf{A}}_{i1} \quad \hat{\mathbf{A}}_{i2} \quad \cdots \quad \hat{\mathbf{A}}_{ir})$$

and

$$\psi_i(\hat{\mathbf{A}}_i) = \begin{bmatrix} \psi_i(\hat{\mathbf{A}}_{i1}) & \mathbf{0} & \cdots & \mathbf{0} \\ \mathbf{0} & \psi_i(\hat{\mathbf{A}}_{i2}) & \cdots & \mathbf{0} \\ \vdots & \vdots & & \vdots \\ \mathbf{0} & \mathbf{0} & \cdots & \psi_i(\hat{\mathbf{A}}_{ir}) \end{bmatrix}$$

$$= \begin{bmatrix} (\hat{\mathbf{A}}_{i1} - \lambda_i\mathbf{I})^{\bar{n}_i} & \mathbf{0} & \cdots & \mathbf{0} \\ \mathbf{0} & (\hat{\mathbf{A}}_{i1} - \lambda_i\mathbf{I})^{\bar{n}_i} & \cdots & \mathbf{0} \\ \vdots & \vdots & & \vdots \\ \mathbf{0} & \mathbf{0} & \cdots & (\hat{\mathbf{A}}_{ir} - \lambda_i\mathbf{I})^{\bar{n}_i} \end{bmatrix} \tag{2-63}$$

If the matrix $(\hat{\mathbf{A}}_{ij} - \lambda_i \mathbf{I})$ has dimension $n_{ij}$, then we have

$$\underset{(n_{ij} \times n_{ij})}{(\hat{\mathbf{A}}_{ij} - \lambda_i \mathbf{I})} = \begin{bmatrix} 0 & 1 & 0 & \cdots & 0 \\ 0 & 0 & 1 & \cdots & 0 \\ \cdot & \cdot & \cdot & & \cdot \\ \cdot & \cdot & \cdot & & \cdot \\ \cdot & \cdot & \cdot & & \cdot \\ 0 & 0 & 0 & \cdots & 1 \\ 0 & 0 & 0 & \cdots & 0 \end{bmatrix} \tag{2-64a}$$

$$(\hat{\mathbf{A}}_{ij} - \lambda_i \mathbf{I})^2 = \begin{bmatrix} 0 & 0 & 1 & 0 & \cdots & 0 \\ 0 & 0 & 0 & 1 & \cdots & 0 \\ \cdot & \cdot & \cdot & \cdot & & \cdot \\ \cdot & \cdot & \cdot & \cdot & & \cdot \\ \cdot & \cdot & \cdot & \cdot & & \cdot \\ 0 & 0 & 0 & 0 & \cdots & 0 \\ 0 & 0 & 0 & 0 & & 0 \end{bmatrix} \tag{2-64b}$$

$$(\hat{\mathbf{A}}_{ij} - \lambda_i \mathbf{I})^{n_{ij}-1} = \begin{bmatrix} 0 & 0 & 0 & \cdots & 0 & 1 \\ 0 & 0 & 0 & \cdots & 0 & 0 \\ \cdot & \cdot & \cdot & & \cdot & \cdot \\ \cdot & \cdot & \cdot & & \cdot & \cdot \\ \cdot & \cdot & \cdot & & \cdot & \cdot \\ 0 & 0 & 0 & \cdots & 0 & 0 \end{bmatrix} \tag{2-64c}$$

and

$$(\hat{\mathbf{A}}_{ij} - \lambda_i \mathbf{I})^k = 0 \qquad \text{for any integer } k \geq n_{ij} \tag{2-64d}$$

By definition, $\bar{n}_i$ is the largest order of the Jordan blocks in $\hat{\mathbf{A}}_i$ or, equivalently, $\bar{n}_i = \max (n_{ij}, j = 1, 2, \cdots, r)$. Hence $(\hat{\mathbf{A}}_{ij} - \lambda_i \mathbf{I})^{\bar{n}_i} = \mathbf{0}$ for $j = 1, 2, \cdots, r$. Consequently, $\psi_i(\hat{\mathbf{A}}_i) = \mathbf{0}$. It is easy to see from (2-63) and (2-64) that if $\psi_i(\lambda) = (\lambda - \alpha)^k$ with either $\alpha \neq \lambda_i$ or $k < \bar{n}_i$, then $\psi_i(\mathbf{A}_i) \neq \mathbf{0}$. Hence we conclude that $\psi_i = (\lambda - \lambda_i)^{\bar{n}_i}$ is the minimal polynomial of $\hat{\mathbf{A}}_i$. Now we claim that $f(\hat{\mathbf{A}}_i) = \mathbf{0}$ if and only if $f$ is divisible without remainder by $\psi_i$, denoted as $f/\psi_i$. Indeed, if $f/\psi_i$, then $f$ can be written as $f = \psi_i h$, where $h$ is the quotient polynomial, and $f(\hat{\mathbf{A}}_i) = \psi_i(\hat{\mathbf{A}}_i)h(\hat{\mathbf{A}}_i) = \mathbf{0} \cdot h(\hat{\mathbf{A}}_i) = \mathbf{0}$. If $f$ is not divisible without remainder by $\psi_i$, then $f$ can be written as $f = \psi_i h + g$ where $g$ is a polynomial of degree less than $\bar{n}_i$. Now $f(\hat{\mathbf{A}}_i) = \mathbf{0}$ implies $g(\hat{\mathbf{A}}_i) = \mathbf{0}$. This contradicts the assumption that $\psi_i$ is the minimal polynomial of $\hat{\mathbf{A}}_i$, for $g$ is a polynomial of degree less than that of $\psi_i$ and $g(\hat{\mathbf{A}}_i) = \mathbf{0}$. With these preliminaries, the theorem can be readily proved. From (2-62), we have $\psi(\hat{\mathbf{A}}) = \operatorname{diag} (\psi(\hat{\mathbf{A}}_1)\ \psi(\hat{\mathbf{A}}_2) \ \cdots \ \psi(\hat{\mathbf{A}}_m))$, since $\psi(\hat{\mathbf{A}}_i) = \mathbf{0}$ if and only if $\psi$ contains the factor $(\lambda - \lambda_i)^{\bar{n}_i}$. Hence we conclude that the minimal polynomial of $\hat{\mathbf{A}}$ and, correspondingly, of $\mathbf{A}$ is

$$\prod_{i=1}^{m} (\lambda - \lambda_i)^{\bar{n}_i} \qquad \text{Q.E.D.}$$

### Example 1

The matrices

$$\begin{bmatrix} 3 & 0 & 0 & 0 \\ 0 & 3 & 0 & 0 \\ 0 & 0 & 3 & 0 \\ 0 & 0 & 0 & 1 \end{bmatrix} \quad \begin{bmatrix} 3 & 1 & 0 & 0 \\ 0 & 3 & 0 & 0 \\ 0 & 0 & 3 & 0 \\ 0 & 0 & 0 & 1 \end{bmatrix} \quad \begin{bmatrix} 3 & 1 & 0 & 0 \\ 0 & 3 & 1 & 0 \\ 0 & 0 & 3 & 0 \\ 0 & 0 & 0 & 1 \end{bmatrix}$$

all have the same characteristic polynomial $\Delta(\lambda) = (\lambda - 3)^3(\lambda - 1)$; however, they have, respectively, $(\lambda - 3)(\lambda - 1)$, $(\lambda - 3)^2(\lambda - 1)$, and $(\lambda - 3)^3(\lambda - 1)$ as minimal polynomials. ▮

Because the characteristic polynomial is always divisible without remainder by the minimal polynomial, we have the following very important corollary of Theorem 2-13.

### Corollary 2-13 (Cayley–Hamilton theorem)

Let $\Delta(\lambda) \triangleq \det(\lambda\mathbf{I} - \mathbf{A}) \triangleq \lambda^n + \alpha_1\lambda^{n-1} + \cdots + \alpha_{n-1}\lambda + \alpha_n$ be the characteristic polynomial of $\mathbf{A}$; then

$$\Delta(\mathbf{A}) = \mathbf{A}^n + \alpha_1\mathbf{A}^{n-1} + \cdots + \alpha_{n-1}\mathbf{A} + \alpha_n\mathbf{I} = \mathbf{0}$$ ▮

The Cayley–Hamilton theorem can also be proved directly without using Theorem 2-13. See Problems 2-39 and 2-40.

The reason for introducing the concept of minimal polynomial will be seen in the following theorem.

### Theorem 2-14

Let $\lambda_1, \lambda_2, \cdots, \lambda_m$ be the distinct eigenvalues of $\mathbf{A}$ with indices $\bar{n}_1, \bar{n}_2, \cdots, \bar{n}_m$. Let $f$ and $g$ be two polynomials. Then the following statements are equivalent.

1. $f(\mathbf{A}) = g(\mathbf{A})$.
2. Either $f = h_1\psi + g$ or $g = h_2\psi + f$, where $\psi$ is the minimal polynomial of $\mathbf{A}$, and $h_1$ and $h_2$ are some polynomials.
3. $f^{(l)}(\lambda_i) = g^{(l)}(\lambda_i) \qquad \text{for } l = 0, 1, 2, \cdots, \bar{n}_i - 1; \quad i = 1, 2, \cdots, m$ **(2-65)**

### Proof

The equivalence of Statements 1 and 2 follows directly from the fact that $\psi(\mathbf{A}) = \mathbf{0}$. Statements 2 and 3 are equivalent following

$$\psi(\lambda) = \prod_{i=1}^{m} (\lambda - \lambda_i)^{\bar{n}_i}$$ Q.E.D.

In applying this theorem, the indices of the eigenvalues of **A** must be first found. Unless the Jordan-form representation of **A** is available, the indices cannot be easily obtained. Therefore Theorem 2-14 is not applicable in most of the cases. Fortunately, we have the following corollary.

### Corollary 2-14

Let the characteristic polynomial of **A** be

$$\Delta(\lambda) \triangleq \det(\lambda\mathbf{I} - \mathbf{A}) = \prod_{i=1}^{m} (\lambda - \lambda_i)^{n_i}$$

Let $f$ and $g$ be two arbitrary polynomials. If

$$f^{(l)}(\lambda_i) = g^{(l)}(\lambda_i) \qquad \text{for } l = 0, 1, 2, \cdots, n_i - 1 \quad i = 1, 2, \cdots, m \tag{2-66}$$

then $f(\mathbf{A}) = g(\mathbf{A})$. ∎

This follows immediately from Theorem 2-14 by observing that the condition (2-66) implies (2-65). The set of numbers $f^{(l)}(\lambda_i)$, for $i = 1, 2, \cdots, m$ and $l = 0, 1, 2, \cdots, n_i - 1$ (there are totally $n = \prod_{i=1}^{m} n_i$) are called *the values of $f$ on the spectrum of* **A**. Corollary 2-14 implies that any two polynomials that have the same values on the spectrum of **A** define the same matrix function. To state it in a different way: given $n$ numbers, if we can construct a polynomial which gives these numbers on the spectrum of **A**, then this polynomial defines uniquely a matrix-valued function of **A**. It is well known that given any $n$ numbers, it is possible to find a polynomial $g(\lambda)$ of degree $n - 1$ that gives these $n$ numbers at some preassigned $\lambda$. Hence if **A** is of order $n$, for any polynomial $f(\lambda)$, we can construct a polynomial of degree $n - 1$:

$$g(\lambda) = \alpha_0 + \alpha_1\lambda + \cdots + \alpha_{n-1}\lambda^{n-1}$$

such that $g(\lambda) = f(\lambda)$ on the spectrum of **A**. Hence any polynomial of **A** can be expressed as

$$f(\mathbf{A}) = g(\mathbf{A}) = \alpha_0\mathbf{I} + \alpha_1\mathbf{A} + \cdots + \alpha_{n-1}\mathbf{A}^{n-1}$$

This fact can also be deduced directly from Corollary 2-13 (Problem 2-38). This fact can be used to compute $f(\mathbf{A})$.

### Example 2

Compute $\mathbf{A}^{100}$, where

$$\mathbf{A} = \begin{bmatrix} 1 & 2 \\ 0 & 1 \end{bmatrix}$$

In other words, given $f(\lambda) = \lambda^{100}$, compute $f(\mathbf{A})$. The characteristic polynomial of $\mathbf{A}$ is $\Delta(\lambda) = \det(\lambda\mathbf{I} - \mathbf{A}) = (\lambda - 1)^2$. Let $g(\lambda)$ be a polynomial of degree $n - 1 = 1$, say

$$g(\lambda) = \alpha_0 + \alpha_1\lambda$$

Now, from Corollary 2-14, if $f(\lambda) = g(\lambda)$ on the spectrum of $\mathbf{A}$, then $f(\mathbf{A}) = g(\mathbf{A})$. On the spectrum of $\mathbf{A}$, we have

$$\begin{aligned} f(1) = g(1)&: \qquad (1)^{100} = \alpha_0 + \alpha_1 \\ f'(1) = g'(1)&: \qquad 100 \cdot (1)^{99} = \alpha_1 \end{aligned}$$

Solving these two equations, we obtain $\alpha_1 = 100$ and $\alpha_0 = -99$. Hence

$$\mathbf{A}^{100} = g(\mathbf{A}) = \alpha_0\mathbf{I} + \alpha_1\mathbf{A} = -99\begin{bmatrix} 1 & 0 \\ 0 & 1 \end{bmatrix} + 100\begin{bmatrix} 1 & 2 \\ 0 & 1 \end{bmatrix} = \begin{bmatrix} 1 & 200 \\ 0 & 1 \end{bmatrix}$$

Obviously $\mathbf{A}^{100}$ can also be obtained by multiplying $\mathbf{A}$ 100 times. ∎

## Functions of a matrix

### Definition 2-16

Let $f(\lambda)$ be a function (not necessarily a polynomial) that is defined on the spectrum of $\mathbf{A}$. If $g(\lambda)$ is a polynomial that has the same values as $f(\lambda)$ on the spectrum of $\mathbf{A}$, then the matrix-valued function $f(\mathbf{A})$ is defined as $f(\mathbf{A}) \triangleq g(\mathbf{A})$. ∎

This definition is an extension of Corollary 2-14 to include polynomials as well as functions. To be more precise, functions of a matrix should be defined by using the condition of (2-65) instead of (2-66). However, they will give the same result. The condition of (2-66), instead of that of (2-65), is used in Definition 2-16 because the characteristic polynomial is easier to obtain than the minimal polynomial.

If $\mathbf{A}$ is an $n \times n$ matrix, given the $n$ values of $f(\lambda)$ on the spectrum of $\mathbf{A}$, we can find a polynomial of degree $(n - 1)$,

$$g(\lambda) = \alpha_0 + \alpha_1\lambda + \cdots + \alpha_{n-1}\lambda^{n-1}$$

which is equal to $f(\lambda)$ on the spectrum of $\mathbf{A}$. Hence from this definition we know that every function of $\mathbf{A}$ can be expressed as

$$f(\mathbf{A}) = \alpha_0\mathbf{I} + \alpha_1\mathbf{A} + \cdots + \alpha_{n-1}\mathbf{A}^{n-1}$$

We summarize the procedure of computing a function of a matrix: Given an $n \times n$ matrix $\mathbf{A}$ and a function $f(\lambda)$,we first compute the characteristic polynomial of $\mathbf{A}$, say

$$\Delta(\lambda) = \sum_{i=1}^{m} (\lambda - \lambda_i)^{n_i}$$

Let

$$g(\lambda) = \alpha_0 + \alpha_1\lambda + \cdots + \alpha_{n-1}\lambda^{n-1}$$

where $\alpha_0, \alpha_1, \cdots, \alpha_{n-1}$ are $n$ unknown constants. Next we use the $n$ equations in (2-66) to compute these $\alpha_i$'s in terms of the values of $f$ on the spectrum of $\mathbf{A}$. Then we have $f(\mathbf{A}) = g(\mathbf{A})$.

### Example 3

Let

$$\mathbf{A}_1 = \begin{bmatrix} 0 & 0 & -2 \\ 0 & 1 & 0 \\ 1 & 0 & 3 \end{bmatrix}$$

Compute $e^{\mathbf{A}_1 t}$. Or equivalently, if $f(\lambda) = e^{\lambda t}$, what is $f(\mathbf{A}_1)$?

The characteristic polynomial of $\mathbf{A}_1$ is $(\lambda - 1)^2(\lambda - 2)$. Let $g(\lambda) = \alpha_0 + \alpha_1\lambda + \alpha_2\lambda^2$. Then

$$f(1) = g(1): \qquad e^t = \alpha_0 + \alpha_1 + \alpha_2$$
$$f'(1) = g'(1): \qquad te^t = \alpha_1 + 2\alpha_2 \text{ (note that the derivative is with respect to } \lambda \text{ not } t)$$
$$f(2) = g(2): \qquad e^{2t} = \alpha_0 + 2\alpha_1 + 4\alpha_2$$

Solving these equations, we obtain $\alpha_0 = -2te^t + e^{2t}$, $\alpha_1 = 3te^t + 2e^t - 2e^{2t}$, and $\alpha_2 = e^{2t} - e^t - te^t$. Hence,

$$e^{\mathbf{A}_1 t} = g(\mathbf{A}_1) = (-2te^t + e^{2t})\mathbf{I} + (3te^t + 2e^t - 2e^{2t})\mathbf{A}_1 + (e^{2t} - e^t - te^t)\mathbf{A}_1{}^2$$
$$= \begin{bmatrix} 2e^t - e^{2t} & 0 & 2e^t - 2e^{2t} \\ 0 & e^t & 0 \\ -e^t + e^{2t} & 0 & 2e^{2t} - e^t \end{bmatrix}$$

### Example 4

Let

$$\mathbf{A}_2 = \begin{bmatrix} 0 & 2 & -2 \\ 0 & 1 & 0 \\ 1 & -1 & 3 \end{bmatrix}$$

Its characteristic polynomial is $\Delta(\lambda) = (\lambda - 1)^2(\lambda - 2)$, which is the same as the one of $\mathbf{A}_1$ in Example 3. Hence we have the same $g(\lambda)$ as in Example 3. Consequently,

$$e^{\mathbf{A}_2 t} = g(\mathbf{A}_2) = \begin{bmatrix} 2e^t - e^{2t} & 2te^t & 2e^t - 2e^{2t} \\ 0 & e^t & 0 \\ e^{2t} - e^t & -te^t & 2e^{2t} - e^t \end{bmatrix}$$

**Example 5**

Given

$$\underset{(n \times n)}{\hat{\mathbf{A}}} = \begin{bmatrix} \lambda_1 & 1 & 0 & \cdots & 0 \\ 0 & \lambda_1 & 1 & \cdots & 0 \\ \cdot & \cdot & \cdot & & \cdot \\ \cdot & \cdot & \cdot & & \cdot \\ \cdot & \cdot & \cdot & & \cdot \\ 0 & 0 & 0 & \cdots & 1 \\ 0 & 0 & 0 & \cdots & \lambda_1 \end{bmatrix} \tag{2-67}$$

The characteristic polynomial of $\hat{\mathbf{A}}$ is $(\lambda - \lambda_1)^n$. Let the polynomial $g(\lambda)$ be of the form

$$g(\lambda) = \alpha_0 + \alpha_1(\lambda - \lambda_1) + \alpha_2(\lambda - \lambda_1)^2 + \cdots + \alpha_{n-1}(\lambda - \lambda_1)^{n-1}$$

Then the conditions in (2-66) give immediately

$$\alpha_0 = f(\lambda_1),\ \alpha_1 = f'(\lambda_1),\ \cdots,\ \alpha_{n-1} = f^{(n-1)}(\lambda_1)/(n-1)!$$

Hence,

$$f(\hat{\mathbf{A}}) = g(\hat{\mathbf{A}}) = f(\lambda_1)\mathbf{I} + \frac{f'(\lambda_1)}{1!}(\hat{\mathbf{A}} - \lambda_1\mathbf{I}) + \cdots + \frac{f^{(n-1)}(\lambda_1)}{(n-1)!}(\hat{\mathbf{A}} - \lambda_1\mathbf{I})^{n-1}$$

$$= \begin{bmatrix} f(\lambda_1) & f'(\lambda_1)/1! & f''(\lambda_1)/2! & \cdots & f^{(n-1)}(\lambda_1)/(n-1)! \\ 0 & f(\lambda_1) & f'(\lambda_1)/1! & \cdots & f^{(n-2)}(\lambda_1)/(n-2)! \\ 0 & 0 & f(\lambda_1) & \cdots & f^{(n-3)}(\lambda_1)/(n-3)! \\ \cdot & \cdot & \cdot & & \cdot \\ \cdot & \cdot & \cdot & & \cdot \\ \cdot & \cdot & \cdot & & \cdot \\ 0 & 0 & 0 & \cdots & f(\lambda_1) \end{bmatrix} \tag{2-68}$$

Here in the last step we have used (2-64).

If $f(\lambda) = e^{\lambda t}$, then

$$e^{\hat{\mathbf{A}}t} = \begin{bmatrix} e^{\lambda_1 t} & te^{\lambda_1 t} & t^2e^{\lambda_1 t}/2! & \cdots & t^{n-1}e^{\lambda_1 t}/(n-1)! \\ 0 & e^{\lambda_1 t} & te^{\lambda_1} & \cdots & t^{n-2}e^{\lambda_1 t}/(n-2)! \\ \cdot & \cdot & \cdot & & \cdot \\ \cdot & \cdot & \cdot & & \cdot \\ \cdot & \cdot & \cdot & & \cdot \\ 0 & 0 & 0 & \cdots & e^{\lambda_1 t} \end{bmatrix} \tag{2-69}$$

Note that the derivatives in (2-69) are taken with respect to $\lambda_1$, not to $t$. ∎

A function of a matrix is defined through a polynomial of the matrix; therefore, the relations that hold for polynomials can also be applied to functions of a matrix. For example, if $\mathbf{A} = \mathbf{Q}\hat{\mathbf{A}}\mathbf{Q}^{-1}$, then

$$f(\mathbf{A}) = \mathbf{Q}f(\hat{\mathbf{A}})\mathbf{Q}^{-1} \tag{2-70}$$

and if

$$\mathbf{A} = \begin{bmatrix} \mathbf{A}_1 & \mathbf{0} \\ \mathbf{0} & \mathbf{A}_2 \end{bmatrix}$$

then

$$f(\mathbf{A}) = \begin{bmatrix} f(\mathbf{A}_1) & \mathbf{0} \\ \mathbf{0} & f(\mathbf{A}_2) \end{bmatrix} \tag{2-71}$$

for any function $f$ that is defined on the spectrum of $\mathbf{A}$. Using (2-68) and (2-71), any function of a Jordan-canonical-form matrix can be obtained immediately.

### Example 6

Consider

$$\mathbf{A} = \left[\begin{array}{ccc|cc} \lambda_1 & 1 & 0 & 0 & 0 \\ 0 & \lambda_1 & 1 & 0 & 0 \\ 0 & 0 & \lambda_1 & 0 & 0 \\ \hline 0 & 0 & 0 & \lambda_2 & 1 \\ 0 & 0 & 0 & 0 & \lambda_2 \end{array}\right] \tag{2-72}$$

If $f(\lambda) = e^{\lambda t}$, then

$$f(\mathbf{A}) = e^{\mathbf{A}t} = \left[\begin{array}{ccc|cc} e^{\lambda_1 t} & te^{\lambda_1 t} & t^2e^{\lambda_1 t}/2! & 0 & 0 \\ 0 & e^{\lambda_1 t} & te^{\lambda_1 t} & 0 & 0 \\ 0 & 0 & e^{\lambda_1 t} & 0 & 0 \\ \hline 0 & 0 & 0 & e^{\lambda_2 t} & te^{\lambda_2 t} \\ 0 & 0 & 0 & 0 & e^{\lambda_2 t} \end{array}\right] \tag{2-73}$$

If $f(\lambda) = (s - \lambda)^{-1}$, where $s$ is a complex variable, then

$$f(\mathbf{A}) = (s\mathbf{I} - \mathbf{A})^{-1}$$

$$= \left[\begin{array}{ccc|cc} \frac{1}{(s-\lambda_1)} & \frac{1}{(s-\lambda_1)^2} & \frac{1}{(s-\lambda_1)^3} & 0 & 0 \\ 0 & \frac{1}{(s-\lambda_1)} & \frac{1}{(s-\lambda_1)^2} & 0 & 0 \\ 0 & 0 & \frac{1}{(s-\lambda_1)} & 0 & 0 \\ \hline 0 & 0 & 0 & \frac{1}{(s-\lambda_2)} & \frac{1}{(s-\lambda_2)^2} \\ 0 & 0 & 0 & 0 & \frac{1}{(s-\lambda_2)} \end{array}\right] \tag{2-74}$$

**Functions of a matrix defined by means of power series.** We have used a polynomial of finite degree to define a function of a matrix. We shall now use an infinite series to give an alternative expression of a function of a matrix.

### Definition 2-17

Let the power series representation of a function $f$ be

$$f(\lambda) = \sum_{i=0}^{\infty} \alpha_i \lambda^i \tag{2-75}$$

with the radius of convergence $\rho$. Then the function $f$ of a square matrix $\mathbf{A}$ is defined as

$$f(\mathbf{A}) \triangleq \sum_{i=0}^{\infty} \alpha_i \mathbf{A}^i \tag{2-76}$$

if the absolute values of all the eigenvalues of $\mathbf{A}$ are smaller than $\rho$, the radius of convergence; or the matrix $\mathbf{A}$ has the property $\mathbf{A}^k = \mathbf{0}$ for some positive integer $k$. ∎

This definition is meaningful only when the infinite series in (2-76) converges. If $\mathbf{A}^k = \mathbf{0}$ for some positive integer $k$, then (2-76) reduces to

$$f(\mathbf{A}) = \sum_{i=0}^{k-1} \alpha_i \mathbf{A}^i$$

If the absolute values of all the eigenvalues of $\mathbf{A}$ are smaller than $\rho$, it can also be shown that the infinite series converges. For a proof, see Reference [77].

Instead of proving that Definitions 2-16 and 2-17 lead to exactly the same matrix function, we shall demonstrate this by using Definition 2-17 to derive (2-68).

### Example 7

Consider the Jordan-form matrix $\hat{\mathbf{A}}$ given in (2-67). Let

$$f(\lambda) = f(\lambda_1) + f'(\lambda_1)(\lambda - \lambda_1) + \frac{f''(\lambda_1)}{2!}(\lambda - \lambda_1)^2 + \cdots$$

then

$$f(\hat{\mathbf{A}}) \triangleq f(\lambda_1)\mathbf{I} + f'(\lambda_1)(\hat{\mathbf{A}} - \lambda_1 \mathbf{I}) + \cdots + \frac{f^{(n-1)}(\lambda_1)}{(n-1)!}(\hat{\mathbf{A}} - \lambda_1 \mathbf{I})^{n-1} + \cdots \tag{2-77}$$

Since $(\hat{\mathbf{A}} - \lambda_1 \mathbf{I})^i$ is of the form of (2-64), the matrix function (2-77) reduces immediately to (2-68).

### Example 8

The exponential function

$$e^{\lambda t} = 1 + \lambda t + \frac{\lambda^2 t^2}{2!} + \cdots + \frac{\lambda^n t^n}{n!} + \cdots$$

converges for all finite $\lambda$ and $t$. Hence for any $\mathbf{A}$, we have

$$e^{\mathbf{A}t} = \sum_{k=0}^{\infty} \frac{1}{k!} t^k \mathbf{A}^k \tag{2-78}$$

∎

A remark is in order concerning the computation of $e^{\mathbf{A}t}$. If $e^{\mathbf{A}t}$ is computed by using Definition 2-16, a closed-form matrix can be obtained. However, it requires the computation of the eigenvalues of $\mathbf{A}$, or correspondingly, the roots of its characteristic polynomial. This step can be avoided if the infinite series (2-78) is used. Clearly, the disadvantage of using (2-78) is that the resulted matrix may not be in a closed form. However, since the series (2-78) converges very fast, the series is often used to compute $e^{\mathbf{A}t}$ in a digital computer.

We derive some important properties of exponential functions of matrices to close this section. Using (2-78), it can be shown that

$$e^{\mathbf{0}} = \mathbf{I}$$

$$e^{\mathbf{A}(t+s)} = e^{\mathbf{A}t}e^{\mathbf{A}s} \tag{2-79}$$

$$e^{(\mathbf{A}+\mathbf{B})t} = e^{\mathbf{A}t}e^{\mathbf{B}t} \quad \text{if and only if } \mathbf{AB} = \mathbf{BA} \tag{2-80}$$

In (2-79), if we choose $s = -t$, then from the fact that $e^{\mathbf{0}} = \mathbf{I}$, we have

$$[e^{\mathbf{A}t}]^{-1} = e^{-\mathbf{A}t} \tag{2-81}$$

By differentiation, term by term, of (2-78), we have

$$\begin{aligned} \frac{d}{dt} e^{\mathbf{A}t} &= \sum_{k=1}^{\infty} \frac{1}{(k-1)!} t^{k-1}\mathbf{A}^k = \mathbf{A}\left(\sum_{k=0}^{\infty} \frac{1}{k!} t^k \mathbf{A}^k\right) \\ &= \mathbf{A}e^{\mathbf{A}t} = e^{\mathbf{A}t}\mathbf{A} \end{aligned} \tag{2-82}$$

here we have used the fact that functions of the same matrix commute. (See Problem 2-33.)

The Laplace transform of a function $f$ defined on $[0, \infty)$ is defined as

$$\hat{f}(s) \triangleq \mathcal{L}[f(t)] \triangleq \int_0^\infty f(t)e^{-st}\,dt \tag{2-83}$$

It is easy to show that

$$\mathcal{L}\left[\frac{t^k}{k!}\right] = s^{-(k+1)}$$

By taking the Laplace transform of (2-78), we have

$$\mathcal{L}(e^{\mathbf{A}t}) = \sum_{k=0}^{\infty} s^{-(k+1)}\mathbf{A}^k = s^{-1}\sum_{k=0}^{\infty} (s^{-1}\mathbf{A})^k \tag{2-84}$$

It is well known that the infinite series

$$f(\lambda) = (1 - \lambda)^{-1} = 1 + \lambda + \lambda^2 + \cdots = \sum_{k=0}^{\infty} \lambda^k$$

converges for $|\lambda| < 1$. Now if $s$ is chosen sufficiently large, the absolute values of all the eigenvalues of $(s^{-1}\mathbf{A})$ are smaller than 1. Hence from Definition 2-17, we have

$$(\mathbf{I} - s^{-1}\mathbf{A})^{-1} = \sum_{k=0}^{\infty} (s^{-1}\mathbf{A})^k \tag{2-85}$$

Hence from (2-84) we have

$$\mathcal{L}(e^{\mathbf{A}t}) = s^{-1}(\mathbf{I} - s^{-1}\mathbf{A})^{-1} = (s\mathbf{I} - \mathbf{A})^{-1} \tag{2-86}$$

In this derivation, Equation (2-86) holds only for sufficiently large $s$. However, it can be shown by analytic continuation that Equation (2-86) does hold for all $s$ except at the eigenvalues of $\mathbf{A}$. Note that (2-86) can also be verified from (2-73) and (2-74).

## 2-8 Norms and Inner Product[10]

All the concepts introduced in this section are applicable to any linear space over the field of complex numbers or over the field of real numbers. However, for convenience in the discussion, we restrict ourself to the complex vector space $(\mathbb{C}^n, \mathbb{C})$.

The concept of the norm of a vector $\mathbf{x}$ in $(\mathbb{C}^n, \mathbb{C})$ is a generalization of the idea of length. Any real-valued function of $\mathbf{x}$, denoted by $\|\mathbf{x}\|$, can be defined as a norm if it has the properties that for any $\mathbf{x}$ in $(\mathbb{C}^n, \mathbb{C})$ and any $\alpha$ in $\mathbb{C}$

1. $\|\mathbf{x}\| \geq 0$ and $\|\mathbf{x}\| = 0$ if and only if $\mathbf{x} = \mathbf{0}$.
2. $\|\alpha\mathbf{x}\| = |\alpha| \, \|\mathbf{x}\|$.
3. $\|\mathbf{x}^1 + \mathbf{x}^2\| \leq \|\mathbf{x}^1\| + \|\mathbf{x}^2\|$.

The last inequality is called the triangular inequality.

Let $\mathbf{x} = [x_1 \quad x_2 \quad \cdots \quad x_n]'$. Then the norm of $\mathbf{x}$ can be chosen as

$$\|\mathbf{x}\|_1 \triangleq \sum_{i=1}^{n} |x_i| \tag{2-87}$$

or

$$\|\mathbf{x}\|_2 \triangleq \left(\sum_{i=1}^{n} |x_i|^2\right)^{1/2} \tag{2-88}$$

[10] The material in this section is not used until Chapter 8, and its study may be postponed until that chapter is reached.

or

$$\|\mathbf{x}\|_\infty \triangleq \max_i |x_i| \tag{2-89}$$

It is easy to verify that each of them satisfies all the properties of a norm. The norm $\|\cdot\|_2$ is called the *Euclidean norm.* In this book, the concept of norm is used mainly in the stability study; we use the fact that $\|\mathbf{x}\|$ *is finite if and only if all the components of* $\mathbf{x}$ *are finite.*

The concept of norm can be extended to linear operators that map $(\mathbb{C}^n, \mathbb{C})$ into itself, or equivalently to square matrices with complex coefficients. The norm of a matrix $\mathbf{A}$ is defined as

$$\|\mathbf{A}\| \triangleq \sup_{\mathbf{x}\neq\mathbf{0}} \frac{\|\mathbf{Ax}\|}{\|\mathbf{x}\|} = \sup_{\|\mathbf{x}\|=1} \|\mathbf{Ax}\|$$

where "sup" stands for supremum, the largest possible number of $\|\mathbf{Ax}\|$ or the least upper bound of $\|\mathbf{Ax}\|$. An immediate consequence of the definition of $\|\mathbf{A}\|$ is, for any $\mathbf{x}$ in $(\mathbb{C}^n,\mathbb{C})$,

$$\|\mathbf{Ax}\| \leq \|\mathbf{A}\|\,\|\mathbf{x}\| \tag{2-90}$$

Observe that the norm of $\mathbf{A}$ is defined through the norm of $\|\mathbf{x}\|$. Therefore, for different $\|\mathbf{x}\|$, we have different $\|\mathbf{A}\|$. For example, if $\|\mathbf{x}\|_1$ is used, $\|\mathbf{A}\|$ is equal to

$$\max_j \left(\sum_{i=1}^n |a_{ij}|\right)$$

and if $\|\mathbf{x}\|_\infty$ is used, $\|\mathbf{A}\|$ is equal to

$$\max_i \left(\sum_{j=1}^n |a_{ij}|\right)$$

where $a_{ij}$ is the $ij$th element of $\mathbf{A}$. See Figure 2-7.

The norm of a matrix has the following properties

$$\|\mathbf{A}+\mathbf{B}\| \leq \|\mathbf{A}\| + \|\mathbf{B}\| \tag{2-91}$$

$$\|\mathbf{AB}\| \leq \|\mathbf{A}\|\,\|\mathbf{B}\| \tag{2-92}$$

These inequalities can be readily verified by observing

$$\|(\mathbf{A}+\mathbf{B})\mathbf{x}\| = \|\mathbf{Ax}+\mathbf{Bx}\| \leq \|\mathbf{Ax}\| + \|\mathbf{Bx}\| \leq (\|\mathbf{A}\| + \|\mathbf{B}\|)\|\mathbf{x}\|$$

and

$$\|\mathbf{ABx}\| \leq \|\mathbf{A}\|\,\|\mathbf{Bx}\| \leq \|\mathbf{A}\|\,\|\mathbf{B}\|\,\|\mathbf{x}\|$$

for any $\mathbf{x}$.

The norm is a function of a vector. Now we shall introduce a function of two vectors, called the *scalar product* or *inner product.* The inner product of two vectors $\mathbf{x}$ and $\mathbf{y}$ in $(\mathbb{C}^n, \mathbb{C})$ is a complex number,

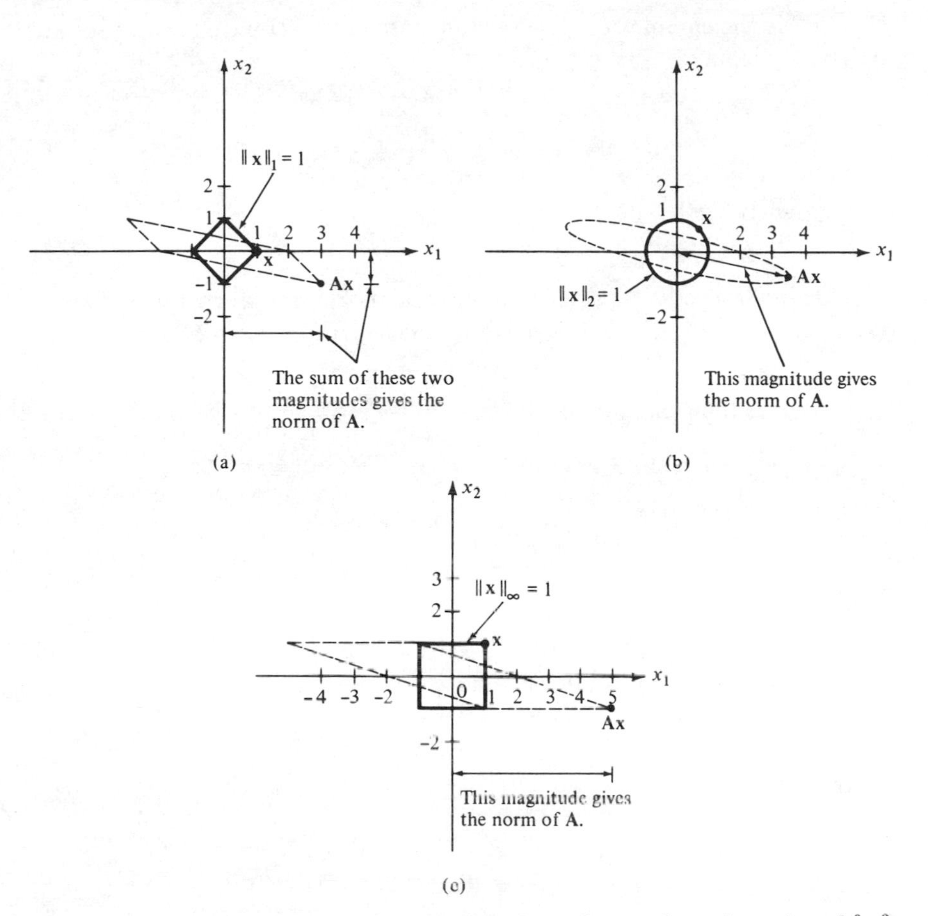


**Figure 2-7** Solid lines denote $\mathbf{x}$; broken lines denote $\mathbf{Ax}$, where $\mathbf{A} = \begin{bmatrix} 3 & 2 \\ -1 & 0 \end{bmatrix}$. (a) $\|\mathbf{A}\| = 4$. (b) $\|\mathbf{A}\| = 3.7$. (c) $\|\mathbf{A}\| = 5$.

denoted by $\langle \mathbf{x}, \mathbf{y} \rangle$, having the properties that for any $\mathbf{x}, \mathbf{y}$ in $(\mathbb{C}^n, \mathbb{C})$ and any $\alpha_1, \alpha_2$ in $\mathbb{C}$,

$$\overline{\langle \mathbf{x}, \mathbf{y} \rangle} = \langle \mathbf{y}, \mathbf{x} \rangle$$

$$\langle \alpha_1 \mathbf{x}^1 + \alpha_2 \mathbf{x}^2, \mathbf{y} \rangle = \bar{\alpha}_1 \langle \mathbf{x}^1, \mathbf{y} \rangle + \bar{\alpha}_2 \langle \mathbf{x}^2, \mathbf{y} \rangle$$

$$\langle \mathbf{x}, \mathbf{x} \rangle > 0 \qquad \text{for all } \mathbf{x} \neq \mathbf{0}$$

where the "bar" denotes the complex conjugate of a number. The first property implies that $\langle \mathbf{x}, \mathbf{x} \rangle$ is a real number. The first two properties imply that $\langle \mathbf{x}, \alpha \mathbf{y} \rangle = \alpha \langle \mathbf{x}, \mathbf{y} \rangle$.

In the complex vector space $(\mathbb{C}^n, \mathbb{C})$, the inner product is always taken to be

$$\langle \mathbf{x}, \mathbf{y} \rangle = \mathbf{x}^* \mathbf{y} = \sum_{i=1}^{n} \bar{x}_i y_i \tag{2-93}$$

where $\mathbf{x}^*$ is the complex conjugate transpose of $\mathbf{x}$. Hence, for any square matrix $\mathbf{A}$, we have

$$\langle \mathbf{x}, \mathbf{A}\mathbf{y} \rangle = \mathbf{x}^*\mathbf{A}\mathbf{y}$$

and

$$\langle \mathbf{A}^*\mathbf{x}, \mathbf{y} \rangle = (\mathbf{A}^*\mathbf{x})^*\mathbf{y} = \mathbf{x}^*\mathbf{A}\mathbf{y}$$

Consequently we have

$$\langle \mathbf{x}, \mathbf{A}\mathbf{y} \rangle = \langle \mathbf{A}^*\mathbf{x}, \mathbf{y} \rangle \tag{2-94}$$

The inner product provides a natural norm for a vector $\mathbf{x}$: $\|\mathbf{x}\| = (\langle \mathbf{x}, \mathbf{x} \rangle)^{1/2}$. In fact, this is the norm defined in Equation (2-88).

### Theorem 2-15 (Schwarz inequality)

If we define $\|\mathbf{x}\| = (\langle \mathbf{x}, \mathbf{x} \rangle)^{1/2}$, then

$$|\langle \mathbf{x}, \mathbf{y} \rangle| \leq \|\mathbf{x}\| \, \|\mathbf{y}\|$$

### Proof

The inequality is obviously true if $\mathbf{y} = \mathbf{0}$. Assume now $\mathbf{y} \neq \mathbf{0}$. From the definition of norm, we have

$$0 \leq \langle \mathbf{x} + \alpha\mathbf{y}, \mathbf{x} + \alpha\mathbf{y} \rangle = \langle \mathbf{x}, \mathbf{x} \rangle + \bar{\alpha}\langle \mathbf{y}, \mathbf{x} \rangle + \alpha\langle \mathbf{x}, \mathbf{y} \rangle + \alpha\bar{\alpha}\langle \mathbf{y}, \mathbf{y} \rangle \tag{2-95}$$

for any $\alpha$. Let $\alpha = -\langle \mathbf{y}, \mathbf{x} \rangle / \langle \mathbf{y}, \mathbf{y} \rangle$; then (2-95) becomes

$$\langle \mathbf{x}, \mathbf{x} \rangle \geq \frac{\langle \mathbf{x}, \mathbf{y} \rangle \langle \mathbf{y}, \mathbf{x} \rangle}{\langle \mathbf{y}, \mathbf{y} \rangle} = \frac{|\langle \mathbf{x}, \mathbf{y} \rangle|^2}{\langle \mathbf{y}, \mathbf{y} \rangle}$$

which gives the Schwarz inequality. Q.E.D.

## 2-9 Concluding Remarks

In this chapter we have reviewed a number of concepts and results in linear algebra which are useful in this book. The following three main topics were covered:

1. *Similarity transformation.* The basic idea of similarity transformation (change of basis vectors) and the means of carrying out the transformation are summarized in Figures 2-4 and 2-5. Similarity transformations can be carried out (a) by computing $\hat{\mathbf{A}} = \mathbf{P}\mathbf{A}\mathbf{P}^{-1} = \mathbf{Q}^{-1}\mathbf{A}\mathbf{Q}$, or more easily, (b) by using the concept of representation: the $i$th column of $\hat{\mathbf{A}}$ is the representation of $\mathbf{A}\mathbf{q}^i$ with respect to the basis $\{\mathbf{q}^1, \mathbf{q}^2, \cdots, \mathbf{q}^n\}$. The second method will be constantly employed in the remainder of this book.

**2.** *Jordan-form representation of a matrix.* This is accomplished by introducing the concepts of eigenvector and generalized eigenvector. The procedure can be programmed into a digital computer.

**3.** *Functions of a square matrix.* Three methods of computing a function of a matrix $f(\mathbf{A})$, where $\mathbf{A}$ is an $n \times n$ constant matrix, were introduced. (a) Using Definition 2-16: First compute the eigenvalues of $\mathbf{A}$, and then find a polynomial $g(\lambda)$ of degree $n - 1$ that is equal to $f(\lambda)$ on the spectrum of $\mathbf{A}$, then $f(\mathbf{A}) = g(\mathbf{A})$. (b) Using the Jordan canonical form of $\mathbf{A}$: Let $\mathbf{A} = \mathbf{Q}\hat{\mathbf{A}}\mathbf{Q}^{-1}$, then $f(\mathbf{A}) = \mathbf{Q}f(\hat{\mathbf{A}})\mathbf{Q}^{-1}$, where $\hat{\mathbf{A}}$ is in the Jordan form and $f(\hat{\mathbf{A}})$ is computed in (2-68) and (2-71). (c) Using Definition 2-17.

There were totally 15 theorems in this chapter. Many of them were introduced to establish the above three topics; for example, Theorems 2-1 and 2-2 were used to establish Figure 2-4 and 2-5; Theorems 2-9–2-12 were introduced mainly for the Jordan-form algorithm; and Theorems 2-13 and 2-14 were used to define functions of a matrix. Hence if the reader is able to carry out the foregoing computations, those theorems may be neglected. However, Theorems 2-4, 2-7, and 2-8 and the Cayley–Hamilton theorem (Corollary 2-13) will constantly be used later.

## PROBLEMS

**2-1** With the usual definition of addition and multiplication, which of the following sets forms a field?

**a.** The set of integers.
**b.** The set of rational numbers
**c.** The set of all $2 \times 2$ real matrices.
**d.** The set of polynomial of degree less than $n$ with real coefficients.

**2-2** Is it possible to define rules of addition and multiplication such that the set $\{0, 1, 2\}$ forms a field?

**2-3** Given the set $\{a, b\}$ with $a \neq b$. Define rules of addition and multiplication such that $\{a, b\}$ forms a field. What are the 0 and 1 elements in this field?

**2-4** Why is $(\mathbb{C}, \mathbb{R})$ a linear space but not $(\mathbb{R}, \mathbb{C})$?

**2-5** Let Ra denote the set of all rational functions with real coefficients. Show that (Ra, Ra) (Ra, R) are linear spaces.

**2-6** Are the following sets linearly independent in the field of real numbers?

**a.** $\begin{bmatrix} 4 \\ -9 \\ 11 \end{bmatrix}, \begin{bmatrix} 2 \\ 13 \\ 10 \end{bmatrix}, \begin{bmatrix} 2 \\ -4 \\ 1 \end{bmatrix}$

**b.** $\begin{bmatrix} 1 + i \\ 2 + 3i \end{bmatrix}, \begin{bmatrix} 10 + 2i \\ 4 - i \end{bmatrix}, \begin{bmatrix} -i \\ 3 \end{bmatrix}$

**c.** $e^{-t}, te^{-t}, e^{-2t}$

Note that these functions are vectors of the function space that consists of all piecewise continuous functions defined on $(-\infty, \infty)$.

**d.** $3x^3 + x - 10,\ -2x^2 + 1,\ x + 1$
These are vectors of the linear space $(P_3[x], \mathbb{R})$.

**e.** $\dfrac{3x^3 + 1}{4x^3 + x + 1}, \dfrac{10x^{10} + 9x^9 + 8x^8 + 1}{1}, \dfrac{1}{4x^3 + x + 1}$

These are vectors of the linear space $(\text{Ra}, \mathbb{R})$; see Problem 2-5.

**2-7** Is the set in Problem 2-6(b) linearly independent in the field of complex numbers?

**2-8** Is the set in Problem 2-6(e) linearly independent in the field of rational functions?

**2-9** What are the dimensions of the following linear spaces?

**a.** $(\mathbb{R}, \mathbb{R})$
**b.** $(\mathbb{C}, \mathbb{C})$
**c.** $(\text{Ra}, \text{Ra})$
**d.** $(\text{Ra}, \mathbb{R})$

where Ra denotes the field of rational functions with real coefficients.

**2-10** Show that the set of all $2 \times 2$ matrices with real coefficients forms a linear space over $\mathbb{R}$ with dimension 4.

**2-11** In an $n$-dimensional vector space $(\mathcal{X}, \mathcal{F})$, given the basis $\mathbf{e}^1, \mathbf{e}^2, \cdots, \mathbf{e}^n$, what is the representation of $\mathbf{e}^i$, for $i = 1, 2, \cdots, n$, with respect to the basis $\{\mathbf{e}^1, \mathbf{e}^2, \cdots, \mathbf{e}^n\}$?

**2-12** Consider Table 2-1. Suppose the representations of $\mathbf{b}$, $\bar{\mathbf{e}}^1$, $\bar{\mathbf{e}}^2$, $\mathbf{e}^1$, and $\mathbf{e}^2$ with respect to the basis $\{\mathbf{e}^1, \mathbf{e}^2\}$ are known, use Equation (2-20) to derive the representations of $\mathbf{b}$, $\bar{\mathbf{e}}^1$, $\bar{\mathbf{e}}^2$, $\mathbf{e}^1$, and $\mathbf{e}^2$ with respect to the basis $\{\bar{\mathbf{e}}^1, \bar{\mathbf{e}}^2\}$.

**2-13** Show that similar matrices have the same characteristic polynomial, and consequently, the same set of eigenvalues. (*Hint:* det (**AB**) = det **A** det **B**.)

**2-14** Find the **P** matrix in Example 3, Section 2-4, and verify $\bar{\mathbf{A}} = \mathbf{PAP}^{-1}$.

**2-15** Given

$$\mathbf{A} = \begin{bmatrix} 2 & 1 & 0 & 0 \\ 0 & 2 & 1 & 0 \\ 0 & 0 & 2 & 0 \\ 0 & 0 & 0 & 1 \end{bmatrix} \qquad \mathbf{b} = \begin{bmatrix} 0 \\ 0 \\ 1 \\ 1 \end{bmatrix} \qquad \tilde{\mathbf{b}} = \begin{bmatrix} 1 \\ 1 \\ 1 \\ 1 \end{bmatrix}$$

What are the representations of **A** with respect to the basis $\{\mathbf{b}, \mathbf{Ab}, \mathbf{A}^2\mathbf{b}, \mathbf{A}^3\mathbf{b}\}$ and the basis $\{\tilde{\mathbf{b}}, \mathbf{A}\tilde{\mathbf{b}}, \mathbf{A}^2\tilde{\mathbf{b}}, \mathbf{A}^3\tilde{\mathbf{b}}\}$ respectively. (Note that the representations are the same!)

**2-16** What are the ranks and nullities of the following matrices?

$$\mathbf{A}_1 = \begin{bmatrix} 4 & 1 & -1 \\ 3 & 2 & -3 \\ 1 & 3 & 0 \end{bmatrix} \qquad \mathbf{A}_2 = \begin{bmatrix} 0 & 1 & 0 \\ 0 & 0 & 0 \\ 0 & 0 & 1 \end{bmatrix} \qquad \mathbf{A}_3 = \begin{bmatrix} 1 & 2 & 3 & 4 & 5 \\ 2 & 3 & 4 & 1 & 2 \\ 3 & 4 & 5 & 0 & 0 \end{bmatrix}$$

**2-17** Find the bases of the range spaces and the null spaces of the matrices given in Problem 2-16.

**2-18** Consider the set of linear equations

$$\mathbf{x}(n) = \mathbf{A}^n\mathbf{x}(0) + \mathbf{A}^{n-1}\mathbf{b}u(0) + \mathbf{A}^{n-2}\mathbf{b}u(1) + \cdots + \mathbf{Ab}u(n-2) + \mathbf{b}u(n-1)$$

where **A** is an $n \times n$ constant matrix and **b** is an $n \times 1$ column vector. Given any $\mathbf{x}(n)$ and $\mathbf{x}(0)$, under what conditions on **A** and **b** will there exist $u(0), u(1), \cdots, u(n-1)$ satisfying the equation? *Hint:* Write the equation in the form

$$\mathbf{x}(n) - \mathbf{A}^n\mathbf{x}(0) = [\mathbf{b} \quad \mathbf{Ab} \quad \cdots \quad \mathbf{A}^{n-1}\mathbf{b}] \begin{bmatrix} u(n-1) \\ u(n-2) \\ \cdot \\ \cdot \\ \cdot \\ u(0) \end{bmatrix}$$

**2-19** Does there exist a solution for the following linear equations?

$$\begin{bmatrix} 3 & 3 & 0 \\ 2 & 1 & 1 \\ 1 & 2 & -1 \end{bmatrix} \begin{bmatrix} x_1 \\ x_2 \\ x_3 \end{bmatrix} = \begin{bmatrix} 6 \\ 3 \\ 3 \end{bmatrix}$$

If so, find one.

**2-20** Find the Jordan-canonical-form representations of the following matrices:

$$\mathbf{A}_1 = \begin{bmatrix} 1 & 4 & 10 \\ 0 & 2 & 0 \\ 0 & 0 & 3 \end{bmatrix} \qquad \mathbf{A}_2 = \begin{bmatrix} 0 & 1 & 0 \\ 0 & 0 & 1 \\ -2 & -4 & -3 \end{bmatrix}$$

$$\mathbf{A}_3 = \begin{bmatrix} 0 & 4 & 3 \\ 0 & -150 & -120 \\ 0 & 200 & 160 \end{bmatrix} \qquad \mathbf{A}_4 = \begin{bmatrix} 0 & 4 & 3 \\ 0 & 20 & 16 \\ 0 & -25 & -20 \end{bmatrix}$$

$$\mathbf{A}_5 = \begin{bmatrix} 7/2 & 21/2 & 14 \\ -1/2 & -3/2 & -2 \\ -1/2 & -3/2 & -2 \end{bmatrix} \qquad \mathbf{A}_6 = \begin{bmatrix} 0 & 1 & 1 & 1 & 1 \\ 0 & 0 & 1 & 1 & 1 \\ 0 & 0 & 0 & 1 & 1 \\ 0 & 0 & 0 & 0 & 1 \\ 0 & 0 & 0 & 0 & 0 \end{bmatrix}$$

**2-21** Prove Theorem 2-12.

**2-22** Let $\lambda_i$ for $i = 1, 2, \cdots, n$ be the eigenvalues of an $n \times n$ matrix $\mathbf{A}$. Show that

$$\det(\mathbf{A}) = \prod_{i=1}^{n} \lambda_i$$

**2-23** Prove that a square matrix is nonsingular if and only if there is no zero eigenvalue.

**2-24** Under what condition will $\mathbf{AB} = \mathbf{AC}$ imply $\mathbf{B} = \mathbf{C}$? ($\mathbf{A}$ is assumed to be a square matrix.)

**2-25** Show that the Vandermonde determinant

$$\begin{bmatrix} 1 & 1 & \cdots & 1 \\ \lambda_1 & \lambda_2 & \cdots & \lambda_n \\ \lambda_1^2 & \lambda_2^2 & \cdots & \lambda_n^2 \\ \cdot & \cdot & & \cdot \\ \cdot & \cdot & & \cdot \\ \cdot & \cdot & & \cdot \\ \lambda_1^{n-1} & \lambda_2^{n-1} & \cdots & \lambda_n^{n-1} \end{bmatrix}$$

is equal to $\prod_{1<i<j<n} (\lambda_i - \lambda_j)$.

**2-26** Consider the matrix

$$\mathbf{A} = \begin{bmatrix} 0 & 1 & 0 & \cdots & 0 \\ 0 & 0 & 1 & \cdots & 0 \\ \cdot & \cdot & \cdot & & \cdot \\ \cdot & \cdot & \cdot & & \cdot \\ \cdot & \cdot & \cdot & & \cdot \\ -\alpha_n & -\alpha_{n-1} & -\alpha_{n-2} & \cdots & -\alpha_1 \end{bmatrix}$$

Show that the characteristic polynomial of $\mathbf{A}$ is

$$\Delta(\lambda) \triangleq \det(\lambda\mathbf{I} - \mathbf{A}) = \lambda^n + \alpha_1\lambda^{n-1} + \alpha_2\lambda^{n-2} + \cdots + \alpha_{n-1}\lambda + \alpha_n$$

If $\lambda_1$ is an eigenvalue of $\mathbf{A}$ [that is, $\Delta(\lambda_1) = 0$], show that $[1 \quad \lambda_1 \quad \lambda_1^2 \quad \cdots \quad \lambda_1^{n-1}]'$ is an eigenvector associated with $\lambda_1$. [The matrix $\mathbf{A}$ is called the *companion matrix* of the polynomial $\Delta(\lambda)$].

**2-27**[11] Consider the matrix shown in Problem 2-26. Suppose that $\lambda_1$ is an eigenvalue of the matrix with multiplicity $k$; that is, $\Delta(\lambda)$ contains $(\lambda - \lambda_1)^k$ as a factor. Verify that the following $k$ vectors

$$\begin{bmatrix} 1 \\ \lambda_1 \\ \lambda_1^2 \\ \cdot \\ \cdot \\ \cdot \\ \lambda_1^{n-1} \end{bmatrix} \begin{bmatrix} 0 \\ 1 \\ 2\lambda_1 \\ \cdot \\ \cdot \\ \cdot \\ (n-1)\lambda_1^{n-2} \end{bmatrix} \begin{bmatrix} 0 \\ 0 \\ 1 \\ \cdot \\ \cdot \\ \cdot \\ \binom{n-1}{2}\lambda_1^{n-3} \end{bmatrix} \begin{bmatrix} 0 \\ 0 \\ 0 \\ \cdot \\ \cdot \\ \cdot \\ \binom{n-1}{3}\lambda_1^{n-4} \end{bmatrix} \cdots \begin{bmatrix} 0 \\ 0 \\ 0 \\ \cdot \\ \cdot \\ \cdot \\ \binom{n-1}{k}\lambda_1^{n-k+1} \end{bmatrix}$$

where

$$\binom{n-1}{i} \triangleq \frac{(n-1)(n-2)\cdots(n-i)}{1 \quad 2 \quad 3 \quad \cdots \quad i}$$

are generalized eigenvectors of $\mathbf{A}$ associated with $\lambda_1$.

**2-28** Find the Jordan-canonical-form representations of the following matrices:

$$\mathbf{A}_1 = \begin{bmatrix} 0 & 1 & 0 \\ 0 & 0 & 1 \\ 2 & 1 & -2 \end{bmatrix} \qquad \mathbf{A}_2 = \begin{bmatrix} 0 & 1 & 0 \\ 0 & 0 & 1 \\ -8 & -12 & -6 \end{bmatrix}$$

$$\mathbf{A}_3 = \begin{bmatrix} 0 & 1 & 0 & 0 \\ 0 & 0 & 1 & 0 \\ 0 & 0 & 0 & 1 \\ 4 & -4 & -3 & 4 \end{bmatrix}$$

[11] See Reference [6].

**2-29** Find the characteristic polynomials and the minimal polynomials of the following matrices:

$$\begin{bmatrix} \lambda_1 & 1 & 0 & 0 \\ 0 & \lambda_1 & 1 & 0 \\ 0 & 0 & \lambda_1 & 0 \\ 0 & 0 & 0 & \lambda_2 \end{bmatrix} \begin{bmatrix} \lambda_1 & 1 & 0 & 0 \\ 0 & \lambda_1 & 1 & 0 \\ 0 & 0 & \lambda_1 & 0 \\ 0 & 0 & 0 & \lambda_1 \end{bmatrix} \begin{bmatrix} \lambda_1 & 1 & 0 & 0 \\ 0 & \lambda_1 & 0 & 0 \\ 0 & 0 & \lambda_1 & 0 \\ 0 & 0 & 0 & \lambda_1 \end{bmatrix} \begin{bmatrix} \lambda_1 & 1 & 0 & 0 \\ 0 & \lambda_1 & 0 & 0 \\ 0 & 0 & \lambda_1 & 1 \\ 0 & 0 & 0 & \lambda_1 \end{bmatrix}$$

What are the multiplicities and indices?

**2-30** Show that if $\lambda_i$ is an eigenvalue of $\mathbf{A}$, then $f(\lambda_i)$ is an eigenvalue of the matrix function $f(\mathbf{A})$. (*Hint:* Show that there exists a nonzero vector $\mathbf{x}$ such that $f(\mathbf{A})\mathbf{x} = f(\lambda_i)\mathbf{x}$.)

**2-31** Given

$$\mathbf{A} = \begin{bmatrix} 1 & 1 & 0 \\ 0 & 0 & 1 \\ 0 & 0 & 1 \end{bmatrix}$$

Find $\mathbf{A}^{10}$, $\mathbf{A}^{103}$, $e^{\mathbf{A}t}$.

**2-32** Compute $e^{\mathbf{A}t}$ for the matrices

$$\begin{bmatrix} 1 & 4 & 10 \\ 0 & 2 & 0 \\ 0 & 0 & 3 \end{bmatrix} \qquad \begin{bmatrix} 0 & 4 & 3 \\ 0 & -150 & -120 \\ 0 & 200 & 160 \end{bmatrix}$$

by using Definition 2-16 and by using the Jordan-form representations.

**2-33** Show that functions of the same matrix commute; that is,

$$f(\mathbf{A})g(\mathbf{A}) = g(\mathbf{A})f(\mathbf{A})$$

Consequently, we have $\mathbf{A}e^{\mathbf{A}t} = e^{\mathbf{A}t}\mathbf{A}$.

**2-34** Let

$$\mathbf{C} = \begin{bmatrix} \lambda_1 & 0 & 0 \\ 0 & \lambda_2 & 0 \\ 0 & 0 & \lambda_3 \end{bmatrix}$$

Find a matrix $\mathbf{B}$ such that $e^{\mathbf{B}} = \mathbf{C}$. Show that if $\lambda_i = 0$ for some $i$ then the matrix $\mathbf{B}$ does not exist.

**2-35** Let

$$\mathbf{C} = \begin{bmatrix} \lambda & 1 & 0 \\ 0 & \lambda & 0 \\ 0 & 0 & \lambda \end{bmatrix}$$

Find a matrix $\mathbf{B}$ such that $e^{\mathbf{B}} = \mathbf{C}$. (*Hint:* Let $f(\lambda) = \log \lambda$ and use (2-77).)

From the Jordan-form representation of a matrix, Equation (2-71) and Problems 2-34 and 2-35, we may conclude that for any nonsingular matrix $\mathbf{C}$, there exists a matrix $\mathbf{B}$ such that $e^{\mathbf{B}} = \mathbf{C}$.

**2-36** What is the norm of the vector

$$\begin{bmatrix} 1 \\ -2 \\ 0 \end{bmatrix}?$$

Which norm do you use?

**2-37** Show that the set of all piecewise continuous complex-valued functions defined over $[0, \infty)$ forms a linear space over $\mathbb{C}$. Show that

$$\langle g, h \rangle \triangleq \int_0^\infty g^*(t)h(t)\,dt$$

qualifies as an inner product of the space, where $g$ and $h$ are two arbitrary functions of the space. What is the form of the Schwarz inequality in this space?

**2-38** Let $\mathbf{A}$ be an $n \times n$ matrix. Show by using the Cayley–Hamilton theorem that any $\mathbf{A}^k$ with $k \geq n$ can be written as a linear combination of $\{\mathbf{I}, \mathbf{A}, \cdots, \mathbf{A}^{n-1}\}$. If the degree of the minimal polynomial of $\mathbf{A}$ is known, what modification can you make?

**2-39** Define

$$(s\mathbf{I} - \mathbf{A})^{-1} \triangleq \frac{1}{\Delta(s)} [\mathbf{R}_0 s^{n-1} + \mathbf{R}_1 s^{n-2} + \cdots + \mathbf{R}_{n-2} s + \mathbf{R}_{n-1}]$$

where $\Delta(s) \triangleq \det (s\mathbf{I} - \mathbf{A}) \triangleq s^n + \alpha_1 s^{n-1} + \alpha_2 s^{n-2} + \cdots + \alpha_n$ and $\mathbf{R}_0, \mathbf{R}_1, \cdots, \mathbf{R}_{n-1}$ are constant matrices. This definition is valid because the degree in $s$ of the adjoint of $(s\mathbf{I} - \mathbf{A})$ is at most $n - 1$. Verify that

$$\begin{aligned} \mathbf{R}_0 &= \mathbf{I} \\ \mathbf{R}_1 &= \mathbf{A}\mathbf{R}_0 + \alpha_1 \mathbf{I} \\ \mathbf{R}_2 &= \mathbf{A}\mathbf{R}_1 + \alpha_2 \mathbf{I} \\ &\vdots \\ \mathbf{R}_{n-1} &= \mathbf{A}\mathbf{R}_{n-2} + \alpha_{n-1} \mathbf{I} \\ \mathbf{0} &= \mathbf{A}\mathbf{R}_{n-1} + \alpha_n \mathbf{I} \end{aligned}$$

(*Hint:* Compute $\Delta(s)\mathbf{I} = (s\mathbf{I} - \mathbf{A})(\mathbf{R}_0 s^{n-1} + \mathbf{R}_1 s^{n-2} + \cdots + \mathbf{R}_{n-1})$.)

**2-40** Prove the Cayley–Hamilton theorem. (*Hint:* Use Problem 2-39 and eliminate $\mathbf{R}_{n-1}, \mathbf{R}_{n-2}, \cdots$ from $\mathbf{0} = \mathbf{A}\mathbf{R}_{n-1} + \alpha_n \mathbf{I}$.)

CHAPTER

# 3

# Mathematical Descriptions of Systems

## 3-1 Introduction

The very first step in the analytical study of a system is to set up mathematical equations to describe the system. Because of different analytical methods used, or because of different questions asked, we may often set up different mathematical equations to describe the same system. For example, in network analysis, if we are interested in only the terminal properties, we may use the impedance or transfer function to describe the network; if we want to know the current and voltage of each branch of the network, then loop analysis or node analysis has to be used to find a set of differential equations to describe the network. The transfer function that describes only the terminal property of a system may be called the *external* or *input–output description* of the system. The set of differential equations that describes the internal as well as the terminal behavior of a system may be called the *internal* or *state-variable description* of the system.

In this chapter we shall introduce the input–output description and the state-variable description of systems from a very general setting.

They will be developed from the concepts of linearity, relaxedness, time-invariance, and causality. Therefore they will be applicable to any system, be it an electrical, a mechanical, or a chemical system, provided the system has the aforementioned properties.

The class of systems studied in this book is assumed to have some input terminals and output terminals. The inputs, or the causes, or the excitations $\mathbf{u}$ are applied at the input terminals; the outputs, or the effects, or the responses $\mathbf{y}$ are measurable at the output terminals. In Section 3-2 we show that if the input $\mathbf{u}$ and the output $\mathbf{y}$ of a system satisfy the linearity property, then they can be related by an equation of the form

$$\mathbf{y}(t) = \int_{-\infty}^{\infty} \mathbf{G}(t, \tau)\mathbf{u}(\tau)\, d\tau \tag{3-1a}$$

If the input and output have, in addition, the causality property, then (3-1a) can be reduced to

$$\mathbf{y}(t) = \int_{-\infty}^{t} \mathbf{G}(t, \tau)\mathbf{u}(\tau)\, d\tau \tag{3-1b}$$

which can be further reduced to

$$\mathbf{y}(t) = \int_{t_0}^{t} \mathbf{G}(t, \tau)\mathbf{u}(\tau)\, d\tau \tag{3-1c}$$

if the system is relaxed at $t_0$. Equation (3-1) describes the relation between the input and output of a system and is called the *input–output description* or the *external description* of the system. We also introduce in Section 3-2 the concepts of time-invariance and the transfer function. In Section 3-3 the concept of state is introduced. The set of equations

$$\dot{\mathbf{x}}(t) = \mathbf{A}(t)\mathbf{x}(t) + \mathbf{B}(t)\mathbf{u}(t) \tag{3-2a}$$
$$\mathbf{y}(t) = \mathbf{C}(t)\mathbf{x}(t) + \mathbf{D}(t)\mathbf{u}(t) \tag{3-2b}$$

that relates the input $\mathbf{u}$, the output $\mathbf{y}$, and the state $\mathbf{x}$ is then introduced. The set of two equations of the form (3-2) is called a *dynamical equation*. If it is used to describe a system, it is called the *dynamical-equation description* or *state-variable description* of the system. We give in Section 3-4 many examples to illustrate the procedure of setting up these two mathematical descriptions. Comparisons between the input–output description and the dynamical-equation description are given in Section 3-5. We study in Section 3-6 the mathematical descriptions of parallel, tandem, and feedback connections of two systems. Finally, the discrete-time versions of Equations (3-1) and (3-2) are introduced in the last section.

We are concerned with descriptions of systems that are models of actual physical systems; hence all the variables and functions in this chapter are assumed to be real-valued. Before proceeding, we classify a system as a single-variable or a multivariable system according to the following definition.

Definition 3-1

A system is said to be a *single-variable system* if and only if it has only one input terminal and only one output terminal. A system is said to be a *multivariable system* if and only if it has more than one input terminal or more than one output terminal.

The references for this chapter are [24], [27], [31], [53], [60], [68], [70], [73], [92], [97], [109], and [116]. The main objective of Sections 3-2 and 3-3 is to introduce the concepts of linearity, causality, time-invariance, and the state, and to illustrate their importance in developing the linear equations. They are not introduced very rigorously. For a more rigorous exposition, see References [60], [68], [109], and [116].

## 3-2 The Input–Output Description

The input–output description of a system gives a mathematical relation between the input and output of the system. In developing this description, the knowledge of the internal structure of a system may be assumed

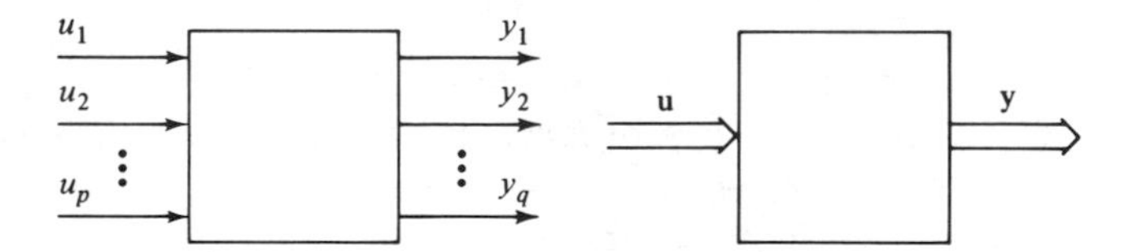


**Figure 3-1** A system with $p$ input terminals and $q$ output terminals.

to be unavailable to us; the only access to the system is by means of the input terminals and the output terminals. Under this assumption, a system may be considered as a "black box," as shown in Figure 3-1. Clearly what we can do to a black box is to apply all kinds of inputs and measure their corresponding outputs, and then try to abstract key properties of the system from these input–output pairs.

We digress at this point to introduce some notations. The system shown in Figure 3-1 is assumed to have $p$ input terminals and $q$ output terminals. The inputs are denoted by $u_1, u_2, \cdots, u_p$, or by a $p \times 1$ column vector $\mathbf{u} = [u_1 \quad u_2 \quad \cdots \quad u_p]'$. The outputs or responses are denoted by $y_1, y_2, \cdots, y_q$, or by a $q \times 1$ column vector $\mathbf{y} = [y_1 \quad y_2 \quad \cdots \quad y_q]'$. The time interval in which the inputs and outputs will be defined is from $-\infty$ to $+\infty$. We use $\mathbf{u}$ or $\mathbf{u}(\cdot)$ to denote a vector function defined over $(-\infty, \infty)$; $\mathbf{u}(t)$ is used to denote the value of $\mathbf{u}$ at time $t$. If the function $\mathbf{u}$ is defined only over $[t_0, t_1)$, we write $\mathbf{u}_{[t_0,t_1)}$.

If the output at time $t_1$ of a system depends only on the input applied at time $t_1$, the system is called an *instantaneous* or *zero-memory system*. A network that consists of only resistors is such a system. Most systems of interest, however, have memory; that is, the output at time $t_1$ depends not only on the input applied at $t_1$, but also on the input applied before and/or after $t_1$. Hence, if an input $\mathbf{u}_{[t_1,\infty)}$ is applied to a system, unless we know the input applied before $t_1$, the output $\mathbf{y}_{[t_1,\infty)}$ is generally not uniquely determinable. In fact, for different inputs applied before $t_1$, we will obtain different output $\mathbf{y}_{[t_1,\infty)}$ although the same input $\mathbf{u}_{[t_1,\infty)}$ is applied. It is clear that such an input–output pair, which lacks a unique relation, is of no use in determining the key properties of the system. Hence, in developing the input–output description, before an input is applied the system must be assumed to be *relaxed* or *at rest*, and that *the output is excited solely and uniquely by the input applied thereafter.* If the concept of energy is applicable to a system, the system is said to be relaxed at time $t_1$ if no energy is stored in the system at that instant. As in the engineering literature, we shall assume that every system is relaxed at time $-\infty$. Consequently if an input $\mathbf{u}_{(-\infty,\infty)}$ is applied at $t = -\infty$, the corresponding output will be excited solely and uniquely by $\mathbf{u}$. Hence, under the relaxedness assumption, it is legitimate to write

$$\mathbf{y} = H\mathbf{u} \tag{3-3}$$

where $H$ is some operator or function that specifies uniquely the output $\mathbf{y}$ in terms of the input $\mathbf{u}$ of the system. In this book, we shall call a system that is initially relaxed at $-\infty$ an *initially relaxed system*—or a *relaxed system*, for short. Note that Equation (3-3) is applicable only to a relaxed system. In this section, whenever we talk about the input–output pairs of a system, we mean only those input–output pairs that can be related by Equation (3-3).

**Linearity.** We introduce next the concept of linearity. This concept is exactly the same as the linear operators introduced in the preceding chapter.

Definition 3-2

A relaxed system is said to be *linear* if and only if

$$H(\alpha_1\mathbf{u}^1 + \alpha_2\mathbf{u}^2) = \alpha_1 H\mathbf{u}^1 + \alpha_2 H\mathbf{u}^2 \tag{3-4}$$

for any inputs $\mathbf{u}^1$ and $\mathbf{u}^2$, and for any real numbers $\alpha_1$ and $\alpha_2$. Otherwise the relaxed system is said to be *nonlinear*. ∎

In engineering literature, the condition of Equation (3-4) is often written as

$$H(\mathbf{u}^1 + \mathbf{u}^2) = H\mathbf{u}^1 + H\mathbf{u}^2 \tag{3-5}$$

$$H(\alpha\mathbf{u}^1) = \alpha H\mathbf{u}^1 \tag{3-6}$$

for any $\mathbf{u}^1$, $\mathbf{u}^2$ and any real number $\alpha$. It is easy to verify that the condition given in (3-4) and the set of conditions in (3-5) and (3-6) are equivalent. The relationship in (3-5) is called the property of *additivity*, and the relationship in (3-6) is called the property of *homogeneity*. If a relaxed system has these two properties, the system is said to satisfy the *principle of superposition*. The reader may wonder whether or not there is redundancy in (3-5) and (3-6). Generally, the property of homogeneity does not imply the property of additivity, as can be seen from the following example.

### Example 1

Consider a single-variable system whose input and output are related by

$$
\begin{aligned}
y(t) &= \frac{u^2(t)}{u(t-1)} \qquad && \text{if } u(t-1) \neq 0 \\
&= 0 && \text{if } u(t-1) = 0
\end{aligned}
$$

for all $t$. It is easy to verify that the input–output pair satisfies the property of homogeneity but not the property of additivity. ▮

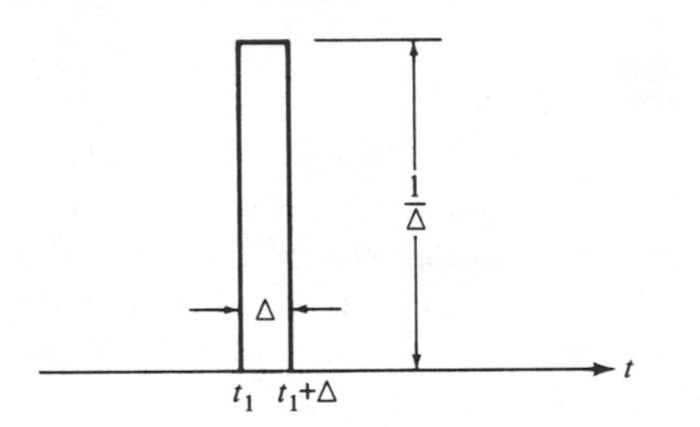


**Figure 3-2** A pulse function $\delta_\Delta(t - t_1)$.

The property of additivity, however, almost implies the property of homogeneity. To be precise, the condition $H(\mathbf{u}^1 + \mathbf{u}^2) = H\mathbf{u}^1 + H\mathbf{u}^2$ for any $\mathbf{u}^1$ and $\mathbf{u}^2$ implies that $H(\alpha\mathbf{u}^1) = \alpha H\mathbf{u}^1$, for any rational number $\alpha$ (see Problem 3-9). Since any real number can be approximated as closely as desired by a rational number, if a relaxed system has the continuity property that $\mathbf{u}^n \to \mathbf{u}$ implies $H\mathbf{u}^n \to H\mathbf{u}$, then the property of additivity implies the property of homogeneity.

We shall develop in the following a mathematical description for a linear relaxed system. Before proceeding we need the concept of the delta function or impulse function. We proceed intuitively because a detailed exposition would lead us too far astray.[1] First let $\delta_\Delta(t - t_1)$ be the pulse function defined in Figure 3-2; that is,

$$
\delta_\Delta(t - t_1) = \begin{cases} 0 & \text{for } t < t_1 \\ \dfrac{1}{\Delta} & \text{for } t_1 \leq t < t_1 + \Delta \\ 0 & \text{for } t \geq t_1 + \Delta \end{cases}
$$

[1] For a rigorous development of the subsequent material, the theory of distributions is needed; see Reference [96].

Note that $\delta_\Delta(t - t_1)$ has unit area for all $\Delta$. As $\Delta$ approaches zero, the limiting "function,"

$$\delta(t - t_1) \triangleq \lim_{\Delta \to 0} \delta_\Delta(t - t_1)$$

is called the *impulse function* or the *Dirac delta function* or simply *δ-function.* Thus the delta function $\delta(t - t_1)$ has the properties that

$$\int_{-\infty}^{\infty} \delta(t - t_1)\, dt = \int_{t_1-\epsilon}^{t_1+\epsilon} \delta(t - t_1)\, dt = 1$$

for any positive $\epsilon$, and that

$$\int_{-\infty}^{\infty} f(t)\delta(t - t_1)\, dt = f(t_1) \tag{3-7}$$

for any function $f$ that is continuous at $t_1$.

With the concept of impulse function, we are ready to develop a mathematical description for linear relaxed systems. We discuss first single-variable systems, the result can then be easily extended to multivariable systems. Consider a relaxed single-variable system whose input and output are related by

$$y = Hu$$

As shown in Figure 3-3, every piecewise continuous input can be approxi-

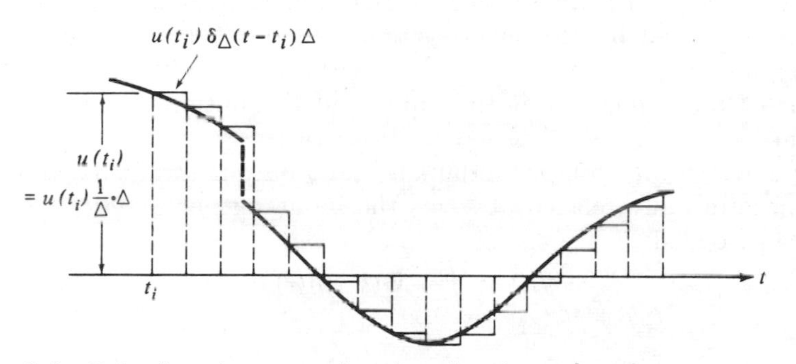


**Figure 3-3** Pulse-function approximation of an input function.

mated by a series of pulse functions. Since every pulse function can be described by $u(t_i)\delta_\Delta(t - t_i)\Delta$, we can write the input function as

$$u \doteqdot \sum_i u(t_i)\delta_\Delta(t - t_i)\Delta$$

If the input–output pairs of the relaxed system satisfy the linearity property, then we have

$$y = Hu \doteqdot \sum_i (H\delta_\Delta(t - t_i))u(t_i)\Delta \tag{3-8}$$

Now as $\Delta$ tends to zero, the approximation tends to an exact equality, the summation becomes an integration and the pulse function $\delta_\Delta(t - t_i)$

tends toward a $\delta$-function. Consequently, as $\Delta \to 0$, Equation (3-8) becomes

$$y = \int_{-\infty}^{\infty} (H\delta(t - \tau))u(\tau)\, d\tau \tag{3-9}$$

Now if $H\delta(t - \tau)$ is known for all $\tau$, then for any input, the output can be computed from (3-9). The physical meaning of $H\delta(t - \tau)$ is that it is the output of the *relaxed* system due to an impulse function input applied at time $\tau$. Define

$$H\delta(t - \tau) = g(\cdot, \tau) \tag{3-10}$$

Note that $g$ is a function of two variables—the second variable denoting the time at which the $\delta$-function is applied and the first variable denoting the time at which the output is observed. Since $g(\cdot, \tau)$ is the response of an impulse function, it is called the *impulse response* of the system. Using (3-10), we can write the output at time $t$ as

$$y(t) = \int_{-\infty}^{\infty} g(t, \tau)u(\tau)\, d\tau \tag{3-11}$$

Hence if $g(\cdot, \tau)$ for all $\tau$ is known then for any input $u$, the output can be computed from (3-11). In other words, a linear relaxed system is completely described by the superposition integral (3-11), where $g(\cdot, \tau)$ is the impulse response of the system, and theoretically it can be obtained by direct measurements at the input and the output terminals of the system.

If a system has $p$ input terminals and $q$ output terminals, and if the system is initially relaxed at $-\infty$, the input–output description (3-11) can be extended to

$$\mathbf{y}(t) = \int_{-\infty}^{\infty} \mathbf{G}(t, \tau)\mathbf{u}(\tau)\, d\tau \tag{3-12}$$

where

$$\mathbf{G}(t, \tau) = \begin{bmatrix} g_{11}(t, \tau) & g_{12}(t, \tau) & \cdots & g_{1p}(t, \tau) \\ g_{21}(t, \tau) & g_{22}(t, \tau) & \cdots & g_{2p}(t, \tau) \\ \vdots & \vdots & & \vdots \\ g_{q1}(t, \tau) & g_{q2}(t, \tau) & \cdots & g_{qp}(t, \tau) \end{bmatrix}$$

and $g_{ij}(t, \tau)$ is the response at time $t$ at the $i$th output terminal due to an impulse function applied at time $\tau$ at the $j$th input terminal, the inputs at other terminals being identically zero. Equivalently, *$g_{ij}$ is the impulse response between the $j$th input terminal and the $i$th output terminal.* Hence $\mathbf{G}$ is called the *impulse-response matrix* of the system.

We shall assume that the impulse responses $\mathbf{G}(t, \tau)$ studied in this book may contain a series of delta functions and that, excluding these delta functions, the remainder of $\mathbf{G}(t, \tau)$ is piecewise continuous in $t$ and $\tau$ for $t > \tau$. Under this assumption, if the input is a piecewise continuous function, so is the output. Hence, a linear relaxed system may be considered as a linear operator that maps the infinite-dimensional function space consisting of all piecewise continuous functions defined over $(-\infty, \infty)$ into another infinite-dimensional function space (see Example 3 in Section 2-3).

Although the input and output of a linear relaxed system can be related by an equation of the form of Equation (3-12), the equation is not readily applicable because the integration required is from $-\infty$ to $\infty$ and because there is no way of checking whether or not a system is initially relaxed at $-\infty$. These difficulties will be removed in the subsequent development.

**Causality.** A system is said to be *causal* or *nonanticipatory* if the output of the system at time $t$ does not depend on the input applied after time $t$; it depends only on the input applied before and at time $t$. In short, the past affects the future, but not conversely. Hence, if a relaxed system is causal, its input and output relation can be written as

$$\mathbf{y}(t) = H\mathbf{u}_{(-\infty,t]} \tag{3-13}$$

for all $t$ in $(-\infty, \infty)$. If a system is not causal, it is said to be *noncausal* or *anticipatory*. The output of a noncausal system depends not only on the past input but also on the future value of the input. This implies that a noncausal system is able to *predict* the input that will be applied in the future. For real physical systems, this is impossible. Hence *causality is an intrinsic property of every physical system*. Since in this book systems are models of physical systems, hence all the systems we study are assumed to be causal.

If a relaxed system is linear and causal, what can we say about the impulse response matrix, $\mathbf{G}(t, \tau)$, of the system? Every element of $\mathbf{G}(t, \tau)$ is, by definition, the output due to a $\delta$-function input applied at time $\tau$. Now if a relaxed system is causal, the output is identically zero before any input is applied. Hence for a linear, causal, relaxed system, we have

$$\mathbf{G}(t, \tau) = \mathbf{0} \qquad \text{for all } \tau \text{ and all } t < \tau \tag{3-14}$$

Consequently, the input–output description of a linear, causal, relaxed system becomes

$$\mathbf{y}(t) = \int_{-\infty}^{t} \mathbf{G}(t, \tau)\mathbf{u}(\tau)\, d\tau \tag{3-15}$$

**Relaxedness.** Recall that the equation $\mathbf{y} = H\mathbf{u}$ of a system holds only when the system is relaxed at $-\infty$ or, equivalently, only when the output $\mathbf{y}$ is excited solely and uniquely by $\mathbf{u}_{(-\infty,\infty)}$. We may apply this concept to an arbitrary $t_0$.

### Definition 3-3

A system is said to be *relaxed* at time $t_0$ if and only if the output $\mathbf{y}_{[t_0,\infty)}$ is solely and uniquely excited by $\mathbf{u}_{[t_0,\infty)}$. ∎

This concept can be easily understood if we consider an *RLC* network. If all the voltages across capacitors and all the currents through inductors are zero at time $t_0$, then the network is relaxed at $t_0$. If a network is not relaxed at $t_0$ and if an input $\mathbf{u}_{[t_0,\infty)}$ is applied, then part of the response will be excited by the initial conditions; and for different initial conditions, different responses will be excited by the same input $\mathbf{u}_{[t_0,\infty)}$. If a system is known to be relaxed at $t_0$, then its input–output relation can be written as

$$\mathbf{y}_{[t_0,\infty)} = H\mathbf{u}_{[t_0,\infty)} \tag{3-16}$$

It is clear that if a system is relaxed at $-\infty$, and if $\mathbf{u}_{(-\infty,t_0)} \equiv \mathbf{0}$, then the system is still relaxed at $t_0$. For the class of systems whose inputs and outputs are linearly related, it is easy to show that *the necessary and sufficient condtion for a system to be relaxed at $t_0$ is that $\mathbf{y}(t) = H\mathbf{u}_{(-\infty,t_0)} = \mathbf{0}$ for all $t \geq t_0$.* In words, if the net effect of $\mathbf{u}_{(-\infty,t_0)}$ on the output after $t_0$ is identically zero, the system is relaxed at $t_0$.

### Example 2

A unit-time-delay system is a device whose output is equal to the input delayed one unit of time; that is, $y(t) = u(t-1)$ for all $t$. The system is relaxed at $t_0$ if $u_{[t_0-1,t_0)} \equiv 0$, although $u_{(-\infty,t_0-1)} \not\equiv 0$. ∎

Now if a system whose inputs and outputs are linearly related is known to be relaxed at $t_0$, then the input–output description reduces to

$$\mathbf{y}(t) = \int_{t_0}^{\infty} \mathbf{G}(t,\tau)\mathbf{u}(\tau)\,d\tau$$

Hence, if a system is described by

$$\mathbf{y}(t) = \int_{t_0}^{t} \mathbf{G}(t,\tau)\mathbf{u}(\tau)\,d\tau \tag{3-17}$$

we know that the input–output pairs of the system satisfy the linearity property and that the system is causal and is relaxed at $t_0$.

A legitimate question may be raised at this point: Given a system at time $t_0$, how do we know that the system is relaxed? For a system whose inputs and outputs are linearly related, this can be determined without knowing the previous history of the system, by the use of the following theorem.

### Theorem 3-1

A system that is describable by

$$\mathbf{y}(t) = \int_{-\infty}^{\infty} \mathbf{G}(t,\tau)\mathbf{u}(\tau)\,d\tau$$

is relaxed at $t_0$ if and only if $\mathbf{u}_{[t_0,\infty)} \equiv \mathbf{0}$ implies $\mathbf{y}_{[t_0,\infty)} \equiv \mathbf{0}$.

### Proof

*Necessity:* If a system is relaxed at $t_0$, the output $\mathbf{y}(t)$ for $t \geq t_0$ is given by

$$\int_{t_0}^{\infty} \mathbf{G}(t, \tau)\mathbf{u}(\tau)\, d\tau$$

Hence, if $\mathbf{u}_{[t_0,\infty)} \equiv \mathbf{0}$, then $\mathbf{y}_{[t_0,\infty)} \equiv \mathbf{0}$. *Sufficiency:* We show that if $\mathbf{u}_{[t_0,\infty)} \equiv \mathbf{0}$ implies $\mathbf{y}_{[t_0,\infty)} \equiv \mathbf{0}$, then the system is relaxed at $t_0$. Since

$$\mathbf{y}(t) = \int_{-\infty}^{\infty} \mathbf{G}(t, \tau)\mathbf{u}(\tau)\, d\tau = \int_{-\infty}^{t_0} \mathbf{G}(t, \tau)\mathbf{u}(\tau)\, d\tau + \int_{t_0}^{\infty} \mathbf{G}(t, \tau)\mathbf{u}(\tau)\, d\tau$$

the assumptions $\mathbf{u}_{[t_0,\infty)} \equiv \mathbf{0}$, $\mathbf{y}_{[t_0,\infty)} \equiv \mathbf{0}$ imply that

$$\int_{-\infty}^{t_0} \mathbf{G}(t, \tau)\mathbf{u}(\tau)\, d\tau = \mathbf{0} \qquad \text{for all } t \geq t_0$$

In words, the net effect of $\mathbf{u}_{(-\infty,t_0)}$ on the output $\mathbf{y}(t)$ for $t \geq t_0$ is zero, and hence the system is relaxed at $t_0$. Q.E.D.

An implication of Theorem 3-1 is that given a system at time $t_0$, if the system is known to be describable by

$$\int_{-\infty}^{\infty} \mathbf{G}(t, \tau)\mathbf{u}(\tau)\, d\tau$$

the relaxedness of the system can be determined from the behavior of the system after $t_0$ without knowing the previous history of the system. Certainly it is impractical or impossible to observe the output from time $t_0$ to infinity; fortunately, for a large class of systems, it is not necessary to do so.

### Corollary 3-1

If the impulse-response matrix $\mathbf{G}(t, \tau)$ of a system can be decomposed into $\mathbf{G}(t, \tau) = \mathbf{M}(t)\mathbf{N}(\tau)$, and if every element of $\mathbf{M}$ is analytic (see Appendix A) on $(-\infty, \infty)$, then the system is relaxed at $t_0$ if and only if for some fixed positive $\epsilon$, $\mathbf{u}_{[t_0,t_0+\epsilon)} \equiv \mathbf{0}$ implies $\mathbf{y}_{[t_0,t_0+\epsilon)} \equiv \mathbf{0}$.

### Proof

If $\mathbf{u}_{[t_0,\infty)} \equiv \mathbf{0}$, the output $\mathbf{y}(t)$ of the system is given by

$$\mathbf{y}(t) = \int_{-\infty}^{t_0} \mathbf{G}(t, \tau)\mathbf{u}(\tau)\, d\tau = \mathbf{M}(t) \int_{-\infty}^{t_0} \mathbf{N}(\tau)\mathbf{u}(\tau)\, d\tau \qquad \text{for } t \geq t_0$$

Since

$$\int_{-\infty}^{t_0} \mathbf{N}(\tau)\mathbf{u}(\tau)\, d\tau$$

is a constant vector, the analyticity assumption of $\mathbf{M}$ implies that $\mathbf{y}(t)$ is analytic on $[t_0, \infty)$. Consequently, if $\mathbf{y}_{[t_0,t_0+\epsilon)} \equiv \mathbf{0}$ and if $\mathbf{u}_{[t_0,\infty)} \equiv \mathbf{0}$, then $\mathbf{y}_{[t_0,\infty)} \equiv \mathbf{0}$ (see Appendix A), and the corollary follows from Theorem 3-1.

Q.E.D.

This is an important result. For any system that satisfies the conditions of Corollary 3-1, its relaxedness can be easily determined by observing the output over any nonzero interval of time, say 10 seconds. If the output is zero in this interval, then the system is relaxed at that moment. It will be shown in the next chapter that the class of systems that are describable by rational transfer-function matrices or linear time-invariant ordinary differential equations satisfies the conditions of Corollary 3-1. Hence Corollary 3-1 is widely applicable.

We give an example to illustrate that Theorem 3-1 does not hold for systems whose inputs and outputs are not linearly related.

### Example 3

Consider the system shown in Figure 3-4, the input is a voltage source, the output is the voltage across the nonlinear capacitor. If the electric

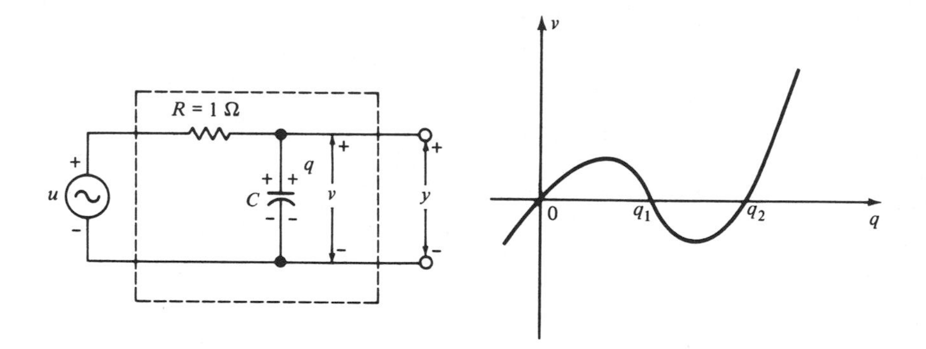


**Figure 3-4** A nonlinear network. (a) The network. (b) The characteristic of the nonlinear capacitor $C$.

charge stored in the capacitor at time $t_0$ is either 0, $q_1$, or $q_2$, the output will be identically zero if no voltage is applied at the input. However, the system is not necessarily relaxed at $t_0$, because if an input is applied, we may obtain different outputs depending on which initial charge the capacitor has. ∎

**Time-invariance.** If the characteristics of a system do not change with time, then the system is said to be *time-invariant*, *fixed*, or *stationary*. In order to define it precisely, we need the concept of a shifting operator $Q_\alpha$. The effect of the shifting operator $Q_\alpha$ is illustrated in Figure 3-5. The output of $Q_\alpha$ is equal to the input delayed by $\alpha$ seconds. Mathematically it is defined as $\bar{u} \triangleq Q_\alpha u$ if and only if $\bar{u}(t) = u(t - \alpha)$ or $\bar{u}(t + \alpha) = u(t)$ for all $t$.

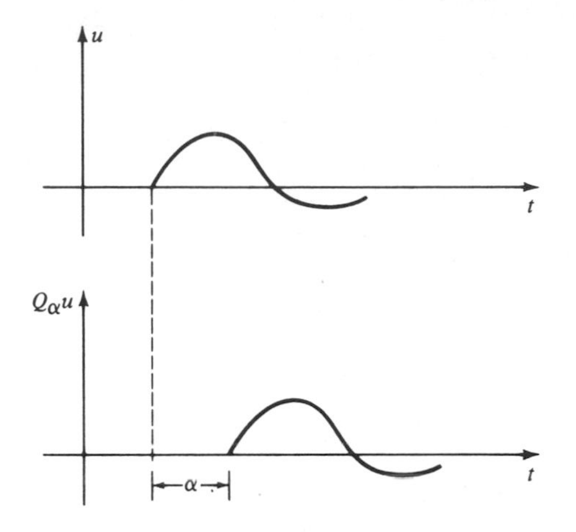


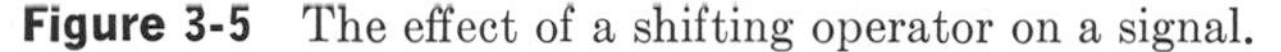

**Figure 3-5** The effect of a shifting operator on a signal.

Definition 3-4

A relaxed system is said to be *time-invariant* (or *stationary* or *fixed*) if and only if

$$HQ_\alpha \mathbf{u} = Q_\alpha H \mathbf{u} \tag{3-18}$$

for any input $\mathbf{u}$ and any real number $\alpha$. Otherwise the relaxed system is said to be *time-varying*. ∎

The relation $HQ_\alpha \mathbf{u} = Q_\alpha H \mathbf{u}$ can also be written as $HQ_\alpha \mathbf{u} = Q_\alpha \mathbf{y}$, which implies that if an input is shifted by $\alpha$ seconds, the waveform of the output remains the same except for a shift by $\alpha$ seconds. In other words, no matter at what time an input is applied to a time-invariant relaxed system, the waveform of the output is always the same.

If a linear relaxed system is known to be time-invariant, what condition does that impose on its impulse response? The impulse response $g(\cdot, \tau)$ is the output due to a $\delta$-function input applied at time $\tau$; that is, $g(\cdot, \tau) = H\delta(t - \tau)$. If the system is time-invariant, then we have

$$\begin{aligned} Q_\alpha g(\cdot, \tau) &= Q_\alpha H\delta(t - \tau) = HQ_\alpha \delta(t - \tau) \\ &= H\delta(t - (\tau + \alpha)) = g(\cdot, \tau + \alpha) \end{aligned}$$

Now by the definition of $Q_\alpha$, the equation $Q_\alpha g(\cdot, \tau) = g(\cdot, \tau + \alpha)$ implies $g(t, \tau) = g(t + \alpha, \tau + \alpha)$, which holds for any $t$, $\tau$, and $\alpha$. By choosing $\alpha = -\tau$, we have $g(t, \tau) = g(t - \tau, 0)$ for all $t$, $\tau$. Hence *the impulse response $g(t, \tau)$ of a linear, time-invariant, relaxed system depends only on the difference of $t$ and $\tau$.* Extending this to the multivariable case, we obtain[2]

$$\mathbf{G}(t, \tau) = \mathbf{G}(t - \tau, 0) = \mathbf{G}(t - \tau)$$

[2] Note that $\mathbf{G}(t, \tau)$ and $\mathbf{G}(t - \tau)$ are two different functions. However, for convenience, the same symbol $\mathbf{G}$ is used.

for all $t$ and $\tau$. Hence, if a system is described by

$$\mathbf{y}(t) = \int_{t_0}^{t} \mathbf{G}(t - \tau)\mathbf{u}(\tau)\, d\tau \tag{3-19}$$

we know that its input–output pairs satisfy the linearity, causality, and time-invariance properties; furthermore, the system is relaxed at time $t_0$. In the time-invariant case, the initial time $t_0$ is always chosen, without loss of generality, to be 0 and the time interval of interest is $[0, \infty)$. Note that $t_0 = 0$ is the instant we start to consider the system or to apply the input $\mathbf{u}$. If $t_0 = 0$, Equation (3-19) becomes

$$\mathbf{y}(t) = \int_{0}^{t} \mathbf{G}(t - \tau)\mathbf{u}(\tau)\, d\tau = \int_{0}^{t} \mathbf{G}(\tau)\mathbf{u}(t - \tau)\, d\tau \tag{3-20}$$

The second equality of (3-20) can be easily verified by changing the variables. The integration in (3-20) is called the *convolution integral.* Since $\mathbf{G}(t - \tau) = \mathbf{G}(t, \tau)$ represents the responses at time $t$ due to $\delta$-function inputs applied at time $\tau$, $\mathbf{G}(t)$ represents the responses at time $t$ due to $\delta$-function inputs applied at $\tau = 0$. If the system is causal, then $\mathbf{G}(t) = \mathbf{0}$ for $t < 0$.

**Transfer-function matrix.** In the study of the class of systems that are describable by convolution integrals, it is of great advantage to use the Laplace transform, because it will change a convolution integral in the time domain into an algebraic equation in the frequency domain. Let $\hat{\mathbf{y}}(s)$ be *the Laplace transform* of $\mathbf{y}$; that is,[3]

$$\hat{\mathbf{y}}(s) \triangleq \mathcal{L}(\mathbf{y}) = \int_{0}^{\infty} \mathbf{y}(t)e^{-st}\, dt$$

Since $\mathbf{G}(t - \tau) = \mathbf{0}$ for $\tau > t$, the upper limit of the integration in (3-20) can be set at $\infty$; hence, from (3-20), we have

$$\begin{aligned}
\hat{\mathbf{y}}(s) &= \int_{0}^{\infty} \left( \int_{0}^{\infty} \mathbf{G}(t - \tau)\mathbf{u}(\tau)\, d\tau \right) e^{-st}\, dt \\
&= \int_{0}^{\infty} \left( \int_{0}^{\infty} \mathbf{G}(t - \tau)e^{-s(t-\tau)}\, dt \right) \mathbf{u}(\tau)e^{-s\tau}\, d\tau \\
&= \int_{0}^{\infty} \mathbf{G}(v)e^{-sv}\, dv \int_{0}^{\infty} \mathbf{u}(\tau)e^{-s\tau}\, d\tau \\
&\triangleq \hat{\mathbf{G}}(s)\hat{\mathbf{u}}(s)
\end{aligned} \tag{3-21}$$

Here we have changed the order of integration, changed the variables, and used the fact that $\mathbf{G}(t) = \mathbf{0}$ for $t < 0$. As defined in (3-21), $\hat{\mathbf{G}}(s)$ is the Laplace transform of the impulse-response matrix; that is,

$$\hat{\mathbf{G}}(s) = \int_{0}^{\infty} \mathbf{G}(t)e^{-st}\, dt$$

[3] If $\mathbf{y}$ contains delta functions at $t = 0$, the lower limit of the integration should start from $0-$ to include the delta functions in the transform.

and is called the *transfer-function matrix* of the system. For single-variable systems, $\hat{\mathbf{G}}(s)$ reduces to a scalar and is called the *transfer function.* Hence the *transfer function is the Laplace transform of the impulse response;* it can also be defined, following (3-21), as

$$\hat{g}(s) = \left.\frac{\mathcal{L}[y(t)]}{\mathcal{L}[u(t)]}\right|_{\substack{\text{the system is} \\ \text{relaxed at } t\,=\,0}} = \left.\frac{\hat{y}(s)}{\hat{u}(s)}\right|_{\substack{\text{relaxed} \\ \text{at } t=0}} \tag{3-22}$$

where the circumflex (^) over a variable denotes the Laplace transform of the same variable; for example,

$$\hat{\mathbf{x}}(s) \triangleq \mathcal{L}[\mathbf{x}] \triangleq \int_0^\infty \mathbf{x}(t)e^{-st}\,dt$$

We see that the familiar transfer functions are the input–output descriptions of systems. It is important to note that this input–output description is obtained under the relaxedness assumption of a system; hence, if the system is not relaxed at $t = 0$, the transfer function cannot be directly applied (see Problem 3-36). Thus *whenever a transfer function is used, the system is always implicity assumed to be relaxed at $t = 0$.*

A transfer function is not necessarily a rational function of $s$. For example, the impulse response $g(t)$ of the unit-time-delay system introduced in Example 2 is $\delta(t - 1)$ and its transfer function is $e^{-s}$, which is not a rational function of $s$. However, the transfer functions we shall encounter in this book are mostly rational functions of $s$. Furthermore, the degree of the denominator of each transfer function is at least equal to that of its numerator. For rational transfer functions, we may define poles and zeros.

**Definition 3-5**

A number $\lambda$ (real or complex) is said to be a *pole* of a rational function $\hat{g}(s)$ if and only if $|\hat{g}(\lambda)| = \infty$. A number $\lambda$ is said to be a *zero* of $\hat{g}(s)$ if and only if $\hat{g}(\lambda) = 0$. A number is said to be a pole of a rational matrix $\hat{\mathbf{G}}(s)$ if and only if it is a pole of at least one element of $\hat{\mathbf{G}}(s)$. ▮

We shall assume throughout this book that every rational function is irreducible—that is, there is no common factor (except a constant) between the numerator and denominator. Under this assumption, it is clear that every root of the denominator of $\hat{g}(s)$ is a pole of $\hat{g}(s)$; every root of the numerator of $\hat{g}(s)$ is a zero of $\hat{g}(s)$. Without this irreducibility assumption, a root of the denominator of $\hat{g}(s)$ may not be a pole of $\hat{g}(s)$; for example, $-1$ is not a pole of

$$\hat{g}(s) = \frac{s + 1}{s^2 + 3s + 2}$$

although it is a root of $(s^2 + 3s + 2)$.

## 3-3 The State-Variable Description

**The concept of state.** The input–output description of a system is applicable only when the system is initially relaxed. If a system is not initially relaxed, say at time $t_0$, then the equation $\mathbf{y}_{[t_0,\infty)} = H\mathbf{u}_{[t_0,\infty)}$ does not hold, because in this case the output $\mathbf{y}_{[t_0,\infty)}$ is not determined by the input $\mathbf{u}_{[t_0,\infty)}$ alone; it depends also on the initial conditions at $t_0$. Hence, in order to determine the output $\mathbf{y}_{[t_0,\infty)}$ uniquely, in addition to the input $\mathbf{u}_{[t_0,\infty)}$ we need a set of initial conditions at $t_0$. This set of intial conditions is called the *state*. Hence the state at $t_0$ is the information that, together with the input $u_{[t_0,\infty)}$, determines *uniquely* the output $y_{[t_0,\infty)}$. For example, in Newtonian mechanics, if an external force (input) is applied to a particle (system) at time $t_0$, the motion (output) of the particle for $t \geq t_0$ is not uniquely determinable unless the position and velocity at time $t_0$ are also known. How the particle actually attained the position and velocity at $t_0$ is immaterial in determining the motion of the particle after $t_0$. Hence the set of two numbers, the position and velocity at time $t_0$, is qualified to be called the state of the system at time $t_0$. Note that the set of two numbers, the position and the momentum at $t_0$, is also qualified.

### Definition 3-6

The *state* of a system at time $t_0$ is the amount of information at $t_0$ that, together with $\mathbf{u}_{[t_0,\infty)}$, determines uniquely the behavior of the system for all $t \geq t_0$. ∎

This concept will be further illustrated in the following examples.

### Example 1

Consider the network shown in Figure 3-6. It is well known that if the initial current through the inductor and the initial voltage across the

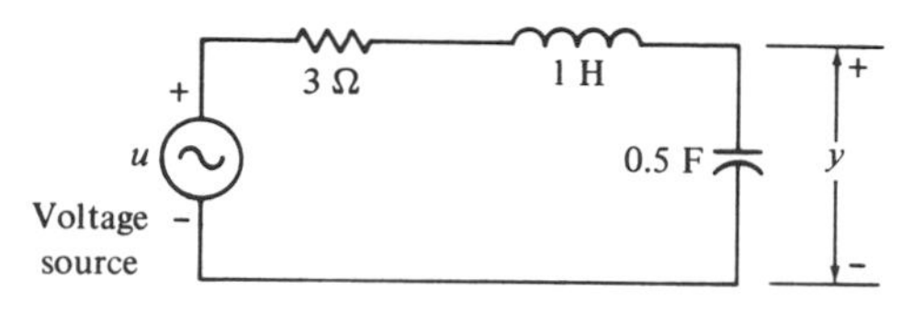


**Figure 3-6** A network.

capacitor are known, then for any driving voltage the behavior of the network can be determined uniquely. Hence the inductor current and the capacitor voltage qualify as the state of the network.

Example 2

We consider again the network in Example 1. The transfer function from $u$ to $y$ of the network can be easily found as

$$\hat{g}(s) = \frac{2}{(s+1)(s+2)} = \frac{2}{s+1} - \frac{2}{s+2}$$

Hence the impulse response of the network is

$$g(t) = 2e^{-t} - 2e^{-2t} \tag{3-23}$$

which is the inverse Laplace transform of the transfer function $\hat{g}(s)$. Now we apply an input $\mathbf{u}_{[t_0,\infty)}$ to the network. If the network is relaxed at $t_0$, the output is given by

$$y(t) = \int_{t_0}^{t} g(t-\tau)u(\tau)\,d\tau \qquad \text{for } t \geq t_0$$

If the network is not relaxed at $t_0$, the output must be computed from

$$y(t) = \int_{-\infty}^{t} g(t-\tau)u(\tau)\,d\tau = \int_{-\infty}^{t_0} g(t-\tau)u(\tau)\,d\tau + \int_{t_0}^{t} g(t-\tau)u(\tau)\,d\tau \qquad \text{for } t \geq t_0 \tag{3-24}$$

because the input that had been applied before $t_0$ may still have some effect on the output after $t_0$ through the energy stored in the capacitor and inductor. We consider now the effect on $y_{[t_0,\infty)}$ due to the unknown input $u_{(-\infty,t_0)}$. From (3-23) we have

$$\begin{aligned}\int_{-\infty}^{t_0} g(t-\tau)u(\tau)\,d\tau &= 2e^{-t}\int_{-\infty}^{t_0} e^{\tau}u(\tau)\,d\tau - 2e^{-2t}\int_{-\infty}^{t_0} e^{2\tau}u(\tau)\,d\tau \\ &\triangleq 2e^{-t}c_1 - 2e^{-2t}c_2\end{aligned} \tag{3-25}$$

for $t \geq t_0$, where

$$c_1 \triangleq \int_{-\infty}^{t_0} e^{\tau}u(\tau)\,d\tau \qquad \text{and} \qquad c_2 \triangleq \int_{-\infty}^{t_0} e^{2\tau}u(\tau)\,d\tau$$

note that $c_1$ and $c_2$ are independent of $t$. Hence if $c_1$ and $c_2$ are known, the output after $t \geq t_0$ excited by the unknown input $u_{(-\infty,t_0)}$ is completely determinable. From (3-24) and (3-25), we have

$$y(t_0) = 2e^{-t_0}c_1 - 2e^{-2t_0}c_2 \tag{3-26}$$

Taking the derivative with respect to $t$, Equation (3-24) yields[4]

$$\dot{y}(t) = -2e^{-t}c_1 + 4e^{-2t}c_2 + g(0)\mathrm{u}(t) + \int_{t_0}^{t} \frac{\partial}{\partial t} g(t-\tau)u(\tau)\,d\tau$$

---

[4] $\dfrac{d}{dt}\displaystyle\int_{t_0}^{t} g(t-\tau)u(\tau)\,d\tau = g(t-\tau)u(\tau)\Big|_{\tau=t} + \int_{t_0}^{t} \frac{\partial}{\partial t} g(t-\tau)u(\tau)\,d\tau$

which, together with $g(0) = 0$, implies that

$$\dot{y}(t_0) = -2e^{-t_0}c_1 + 4e^{-2t_0}c_2 \tag{3-27}$$

Solving for $c_1$ and $c_2$ from (3-26) and (3-27), we obtain

$$\begin{aligned} c_1 &= 0.5e^{t_0}(2y(t_0) + \dot{y}(t_0)) \\ c_2 &= 0.5e^{2t_0}(y(t_0) + \dot{y}(t_0)) \end{aligned}$$

Hence if the network is not relaxed at $t_0$, the output $y(t)$ is given by

$$y(t) = (2y(t_0) + \dot{y}(t_0))e^{-(t-t_0)} - (y(t_0) + \dot{y}(t_0))e^{-2(t-t_0)} + \int_{t_0}^{t} g(t-\tau)u(\tau)\,d\tau \qquad \text{for } t \geq t_0$$

We see that if $y(t_0)$ and $\dot{y}(t_0)$ are known, the output after $t \geq t_0$ can be uniquely determined even if the network is not relaxed at $t_0$. Hence the set of numbers $y(t_0)$ and $\dot{y}(t_0)$ qualifies as the state of the network at $t_0$. Clearly, the set $\{c_1, c_2\}$ also qualifies as the state of the network.

## Example 3

Consider the network shown in Figure 3-7. It is clear that if all the capacitor voltages are known, then the behavior of the network is uniquely

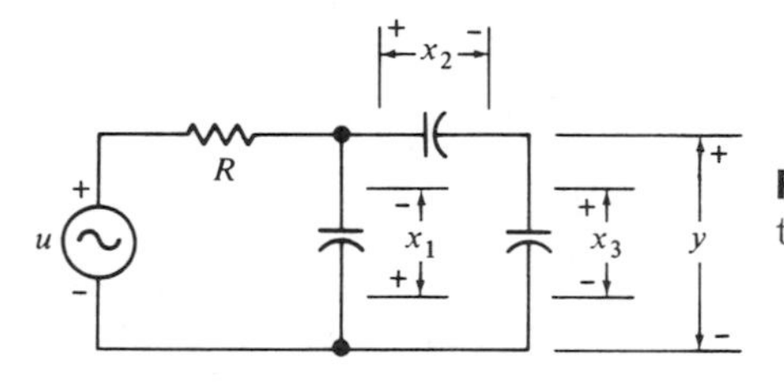


**Figure 3-7** A network with a loop that consists of capacitors only.

determinable for any applied input. Hence the set of capacitor voltages $x_1$, $x_2$, and $x_3$ qualifies as the state of the network. Let us examine the network more carefully. If we apply the Kirchhoff voltage law to the loop that consists of three capacitors, we have $x_1(t) + x_2(t) + x_3(t) = 0$ for all $t$. It implies that if any two of $x_1$, $x_2$, and $x_3$ are known, the third one is also known. Consequently, if any two of the capacitor voltages are known, the behavior of the network is uniquely determinable for any input applied thereafter. In other words, any two of the three capacitor voltages qualify as the state. If all the three capacitor voltages are chosen as the state, then there is a redundancy. In choosing the state of a system, it is desirable to choose a state that consists of the least number of variables. How to pick the state with the least number of variables for general *RLC* networks will be studied in the next section.

### Example 4

A unit-time-delay system is a device whose output $y(t)$ is equal to $u(t-1)$ for all $t$. For this system, in order to determine $y_{[t_0,\infty)}$ uniquely from $u_{[t_0,\infty)}$, we need the information $u_{[t_0-1,t_0)}$. Hence the information $u_{[t_0-1,t_0)}$ is qualified to be called the state of the system at time $t_0$. ∎

From these examples, we may have the following observations concerning the state of a system. First, the choice of the state is not unique. For the network shown in Figure 3-6, the state may be chosen as the inductor current and the capacitor voltage, or chosen as $y(t_0)$ and $\dot{y}(t_0)$ or $c_1$ and $c_2$. For the network shown in Figure 3-7, any two of the three capacitor voltages can be chosen as the state. Different analyses often lead to different choices of state. Second, the state chosen in Example 1 is associated with physical quantities, whereas in Example 2 the state is introduced for mathematical necessity. Hence, the state of a system is an auxiliary quantity that may or may not be easily interpretable in physical terms. Finally the state at each instant may consist of only finite set of numbers, as in Examples 1 and 3, or consist of infinite set of numbers, as in Example 4. Note that there is an infinite number of points between $[t_0-1, t_0)$; hence the state of this example consists of an infinite set of numbers.

In this book we study only the class of systems whose states may be chosen to consist of a finite number of variables. The state of a system can then be represented by a finite-dimensional column vector $\mathbf{x}$, called the *state vector*. The components of $\mathbf{x}$ are called *state variables*. The linear space in which the state vector ranges is denoted by $\Sigma$. Since state variables are usually real-valued, and since we consider only cases in which states consist of a finite number of variables, the state spaces we encounter in this book usually are the familiar finite-dimensional real vector space $(\mathbb{R}^n, \mathbb{R})$.

**Dynamical equations.** In addition to the input and output of a system we have now the state of the system. The state at time $t_0$ is, by definition, the required information at $t_0$ that, together with input $\mathbf{u}_{[t_0,\infty)}$, determines uniquely the behavior (output and state) of the system for all $t \geq t_0$. *The set of equations that describes the unique relations between the input, output, and state is called a dynamical equation.* In this book, we study only the dynamical equations of the form

$$\dot{\mathbf{x}}(t) = \mathbf{h}(\mathbf{x}(t), \mathbf{u}(t), t) \qquad \text{(state equation)} \qquad \textbf{(3-28a)}$$
$$\mathbf{y}(t) = \mathbf{g}(\mathbf{x}(t), \mathbf{u}(t), t) \qquad \text{(output equation)} \qquad \textbf{(3-28b)}$$

or, more explicitly,

$$\left\{\begin{aligned}\dot{x}_1(t) &= h_1(x_1(t), x_2(t), \cdots, x_n(t), u_1(t), u_2(t), \cdots, u_p(t), t)\\ \dot{x}_2(t) &= h_2(x_1(t), x_2(t), \cdots, x_n(t), u_1(t), u_2(t), \cdots, u_p(t), t)\\ &\cdot\\ &\cdot\\ &\cdot\\ \dot{x}_n(t) &= h_n(x_1(t), x_2(t), \cdots, x_n(t), u_1(t), u_2(t), \cdots, u_p(t), t)\end{aligned}\right. \tag{3-29a}$$

$$\left\{\begin{aligned}y_1(t) &= g_1(x_1(t), x_2(t), \cdots, x_n(t), u_1(t), u_2(t), \cdots, u_p(t), t)\\ y_2(t) &= g_2(x_1(t), x_2(t), \cdots, x_n(t), u_1(t), u_2(t), \cdots, u_p(t), t)\\ &\cdot\\ &\cdot\\ &\cdot\\ y_q(t) &= g_q(x_1(t), x_2(t), \cdots, x_n(t), u_1(t), u_2(t), \cdots, u_p(t), t)\end{aligned}\right. \tag{3-29b}$$

where $\mathbf{x} = [x_1 \quad x_2 \quad \cdots \quad x_n]'$ is the state, $\mathbf{y} = [y_1 \quad y_2 \quad \cdots \quad y_q]'$ is the output, and $\mathbf{u} = [u_1 \quad u_2 \quad \cdots \quad u_p]'$ is the input. The input $\mathbf{u}$, the output $\mathbf{y}$, and the state $\mathbf{x}$ are real-valued vector functions of $t$ defined over $(-\infty, \infty)$. In order for (3-28) to qualify as a dynamical equation, we must assume that for any initial state $\mathbf{x}(t_0)$ and any given $\mathbf{u}$, Equation (3-28) has a unique solution. A sufficient condition for (3-28) to have a unique solution for a given $\mathbf{u}$ and a given initial state is that $h_i$ and $\partial h_i/\partial x_j$ are continuous functions of $t$ for $i, j = 1, 2, \cdots, n$; see References [24], [77], and [92]. If a unique solution exists in (3-28), it can be shown that the solution can be solved in terms of $\mathbf{x}(t_0)$ and $\mathbf{u}_{[t_0,t)}$. Hence $\mathbf{x}$ serves as the state at $t_0$, as expected. Equation (3-28a) governs the behavior of the state and is called a *state equation*. Equation (3-28b) gives the output and is called an *output equation*. Note that there is no loss of generality in writing the output equation in the form of (3-28b), because by the definition of the state, the knowledge of $\mathbf{x}(t)$ and $\mathbf{u}(t)$ suffices to determine $\mathbf{y}(t)$.

The state space of (3-28) is an $n$-dimensional real vector space, hence the set of equation (3-28) is called an $n$-dimensional dynamical equation.

If $\mathbf{h}$ and $\mathbf{g}$ in (3-28) are linear functions of $\mathbf{x}$ and $\mathbf{u}$, then (3-28) is called a *linear dynamical equation*. It is easy to show that if $\mathbf{h}$ and $\mathbf{g}$ are linear functions of $\mathbf{x}$ and $\mathbf{u}$, then they are of the form

$$\mathbf{h}(\mathbf{x}(t), \mathbf{u}(t), t) = \mathbf{A}(t)\mathbf{x}(t) + \mathbf{B}(t)\mathbf{u}(t), \quad \mathbf{g}(\mathbf{x}(t), \mathbf{u}(t), t) = \mathbf{C}(t)\mathbf{x}(t) + \mathbf{D}(t)\mathbf{u}(t)$$

where $\mathbf{A}$, $\mathbf{B}$, $\mathbf{C}$, and $\mathbf{D}$ are, respectively, $n \times n$, $n \times p$, $q \times n$, and $q \times p$ matrices (Problem 3-33). Hence an $n$-dimensional *linear* dynamical equation is of the form

$$E: \quad \dot{\mathbf{x}} = \mathbf{A}(t)\mathbf{x} + \mathbf{B}(t)\mathbf{u} \quad \text{(state equation)} \tag{3-30a}$$

$$\mathbf{y} = \mathbf{C}(t)\mathbf{x} + \mathbf{D}(t)\mathbf{u} \quad \text{(output equation)} \tag{3-30b}$$

A sufficient condition for (3-30) to have a unique solution is that every entry of $\mathbf{A}(\cdot)$ is a continuous function of $t$ defined over $(-\infty, \infty)$. For convenience, the entries of $\mathbf{B}$, $\mathbf{C}$, and $\mathbf{D}$ are also assumed to be continuous in $(-\infty, \infty)$. Since the values of $\mathbf{A}$, $\mathbf{B}$, $\mathbf{C}$, and $\mathbf{D}$ change with time, the dynamical equation $E$ in (3-30) is more suggestively called a *linear time-varying dynamical equation.* If the matrices $\mathbf{A}$, $\mathbf{B}$, $\mathbf{C}$, and $\mathbf{D}$ are independent of $t$, then the equation is called a *linear time-invariant dynamical equation and is denoted by FE (fixed equation).* Hence an $n$-dimensional, linear, time-invariant dynamical equation is of the form

$$FE: \quad \dot{\mathbf{x}} = \mathbf{A}\mathbf{x} + \mathbf{B}\mathbf{u} \tag{3-31a}$$

$$\mathbf{y} = \mathbf{C}\mathbf{x} + \mathbf{D}\mathbf{u} \tag{3-31b}$$

where $\mathbf{A}$, $\mathbf{B}$, $\mathbf{C}$, and $\mathbf{D}$ are, respectively, $n \times n$, $n \times p$, $q \times n$, and $q \times p$ real constant matrices. In the time-invariant case, the characteristics of the equation do not change with time; hence there is no loss of generality in choosing the initial time to be 0. The time interval of interest then becomes $[0, \infty)$.

The state space $\Sigma$ of $E$ or $FE$ is an $n$-dimensional real vector space $(\mathbb{R}^n, \mathbb{R})$. Hence we can think of the $n \times n$ matrix $\mathbf{A}$ as a linear operator which maps $\Sigma$ into $\Sigma$. As mentioned in the preceding chapter, it is very convenient to introduce the set of the orthonormal vectors $\{\mathbf{n}^1, \mathbf{n}^2, \cdots, \mathbf{n}^n\}$, where $\mathbf{n}^i$ is an $n \times 1$ column vector with 1 at its $i$th component, and zero elsewhere, as the basis of the state space. In doing so, we may also think of the matrix $\mathbf{A}$ as representing a linear operator with respect to this orthonormal basis. Hence, *unless otherwise stated, the basis of the state space of E or FE is assumed to be the set of the orthonormal vectors*

$$\{\mathbf{n}^1, \mathbf{n}^2, \cdots, \mathbf{n}^n\}$$

In the study of linear, time-invariant dynamical equations, we may also apply the Laplace transform. Taking the Laplace transform of $FE$ and assuming $\mathbf{x}(0) = \mathbf{x}^0$, we obtain

$$s\hat{\mathbf{x}}(s) - \mathbf{x}^0 = \mathbf{A}\hat{\mathbf{x}}(s) + \mathbf{B}\hat{\mathbf{u}}(s) \tag{3-32a}$$

$$\hat{\mathbf{y}}(s) = \mathbf{C}\hat{\mathbf{x}}(s) + \mathbf{D}\hat{\mathbf{u}}(s) \tag{3-32b}$$

where the circumflex over a variable denotes the Laplace transform of the same variable; for example,

$$\hat{\mathbf{x}}(s) = \int_0^\infty \mathbf{x}(t)e^{-st}\,dt$$

From (3-32), we have

$$\hat{\mathbf{x}}(s) = (s\mathbf{I} - \mathbf{A})^{-1}\mathbf{x}^0 + (s\mathbf{I} - \mathbf{A})^{-1}\mathbf{B}\hat{\mathbf{u}}(s) \tag{3-33}$$

$$\hat{\mathbf{y}}(s) = \mathbf{C}(s\mathbf{I} - \mathbf{A})^{-1}\mathbf{x}^0 + \mathbf{C}(s\mathbf{I} - \mathbf{A})^{-1}\mathbf{B}\hat{\mathbf{u}}(s) + \mathbf{D}\hat{\mathbf{u}}(s) \tag{3-34}$$

We see that Equations (3-33) and (3-34) are algebraic equations. If $\mathbf{x}^0$ and $\mathbf{u}$ are known, $\hat{\mathbf{x}}(s)$ and $\hat{\mathbf{y}}(s)$ can be computed from (3-33) and (3-34). Note that the determinant of $(s\mathbf{I} - \mathbf{A})$ is different from zero (the zero of the field of rational functions of $s$)[5]; hence, the inverse of the matrix $(s\mathbf{I} - \mathbf{A})$ always exists. If the initial state $\mathbf{x}^0$ is $\mathbf{0}$—that is, the system is relaxed at $t = 0$—then (3-34) reduces to

$$\hat{\mathbf{y}}(s) = [\mathbf{C}(s\mathbf{I} - \mathbf{A})^{-1}\mathbf{B} + \mathbf{D}]\hat{\mathbf{u}}(s) \triangleq \hat{\mathbf{G}}(s)\hat{\mathbf{u}}(s) \tag{3-35}$$

where $\hat{\mathbf{G}}(s) \triangleq [\mathbf{C}(s\mathbf{I} - \mathbf{A})^{-1}\mathbf{B} + \mathbf{D}]$ is called the *transfer-function matrix of FE.*

In this book, a system that is described by a linear dynamical equation is called a *linear system;* a system that is describable by a linear, time-invariant dynamical equation is called a *linear time-invariant* system. As will be seen later, if a system has the input–output description as well as the state-variable description, then the transfer functions developed in (3-21) and (3-35) will be the same.

**Analog computer simulation of linear dynamical equations.** As will be illustrated in the next section, systems that have a finite number of state variables can always be described by finite-dimensional dynamical equations. We show in this subsection that every finite-dimensional linear dynamical equation can be readily simulated in an analog computer by interconnecting integrators, summers, and amplifiers (or attenuators). The integrator,[6] summer, and amplifier are three basic components of an analog computer, and their functions are illustrated in Figure 3-8. We

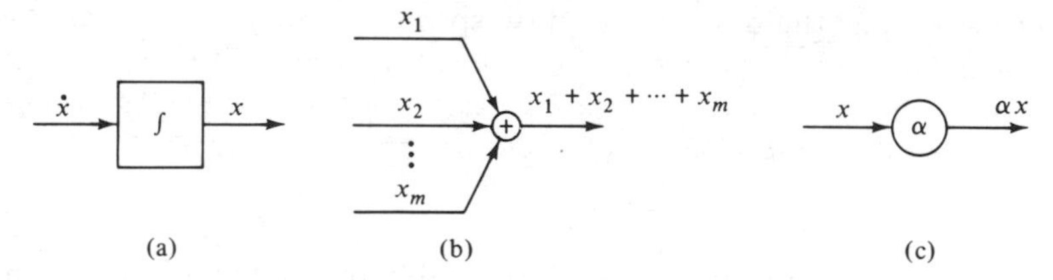


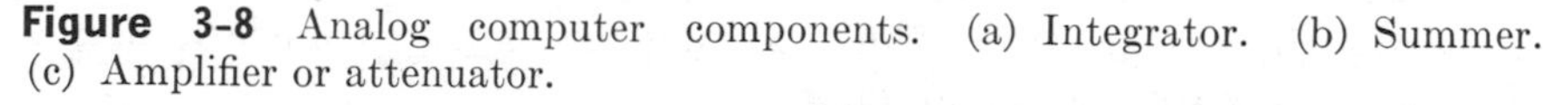

**Figure 3-8** Analog computer components. (a) Integrator. (b) Summer. (c) Amplifier or attenuator.

give in Figure 3-9 a block diagram of analog computer connections of the following two-dimensional dynamical equation:

$$\begin{bmatrix} \dot{x}_1(t) \\ \dot{x}_2(t) \end{bmatrix} = \begin{bmatrix} a_{11}(t) & a_{12}(t) \\ a_{21}(t) & a_{22}(t) \end{bmatrix} \begin{bmatrix} x_1(t) \\ x_2(t) \end{bmatrix} + \begin{bmatrix} b_{11}(t) & b_{12}(t) \\ b_{21}(t) & b_{22}(t) \end{bmatrix} \begin{bmatrix} u_1(t) \\ u_2(t) \end{bmatrix}$$

$$\begin{bmatrix} y_1(t) \\ y_2(t) \end{bmatrix} = \begin{bmatrix} c_{11}(t) & c_{12}(t) \\ c_{21}(t) & c_{22}(t) \end{bmatrix} \begin{bmatrix} x_1(t) \\ x_2(t) \end{bmatrix} + \begin{bmatrix} d_{11}(t) & d_{12}(t) \\ d_{21}(t) & d_{22}(t) \end{bmatrix} \begin{bmatrix} u_1(t) \\ u_2(t) \end{bmatrix}$$

Note that for a two-dimensional dynamical equation, we need two integrators. *The output of every integrator is a state variable.* We see that

[5] That is, the determinant of $(s\mathbf{I} - \mathbf{A})$ is not identically equal to zero. Note that $\det(s\mathbf{I} - \mathbf{A}) = 0$ for some $s$ is permitted.

[6] In practice, pure differentiators are not used for the reason that they will amplify the noise. On the other hand, integrators will smooth or suppress the noise.

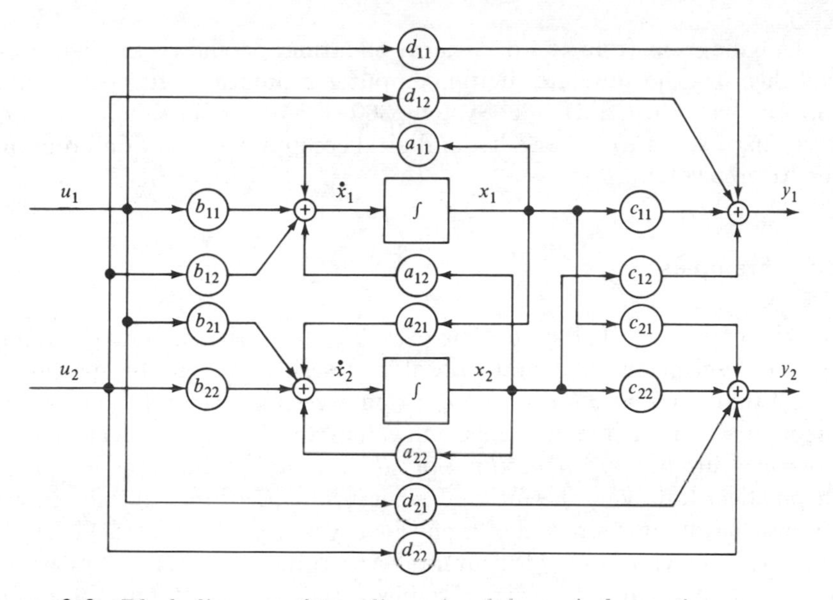


**Figure 3-9** Block diagram of two-dimensional dynamical equation.

even for a two-dimensional dynamical equation, the wiring of the block diagram is complicated; hence, for the general case, we usually use a matrix block diagram. The matrix block diagram of the dynamical equation $E$ is shown in Figure 3-10. If $E$ is $n$-dimensional, the integration block in Figure 3-10 consists of $n$ integrators. The matrix $\mathbf{D}$ represents the direct transmission part from the input $\mathbf{u}$ to the output

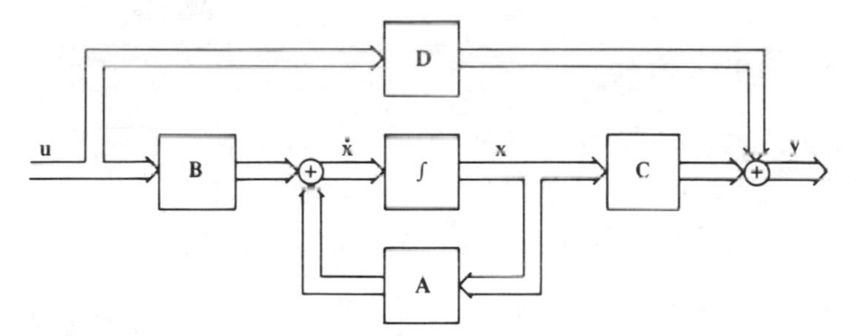


**Figure 3-10** Matrix block diagram of the dynamical equation $E$.

$\mathbf{y}$. If the matrix $\mathbf{A}$ is a zero matrix, then there is no "feedback" in the block diagram.

Since every dynamical equation can be simulated in an analog computer, if the dynamical-equation description of a system is found, then the system can be readily simulated on an analog computer. We also note that the dynamical equation $E$ can readily be programmed in a digital computer by use of one of the standard subroutines for integrating differential equations. Or more conveniently, we may first form a block diagram from the dynamical equation, as shown in Figure 3-9, and then use the standard digital computer simulation programs such as the IBM 1130 CSMP or System/360 CSMP (Continuous System Modeling Program). For a list of digital computer simulation programs, see Reference [26].

## 3-4 Examples

In this section we shall give some examples to illustrate how the input–output descriptions and state-variable descriptions can be developed. Recall that the input–output description gives only the relation between the inputs and outputs of a relaxed system and it can be developed, say from measurements, *without knowing the internal structure of the system.* In practice, however, the internal structure of a system is known to us because a system is a model of a physical system and the model is chosen or decided by us. Now if the internal structure of a system is known and if every component of the system is linear, then the input–output description of the system can be obtained directly from analysis, as will be seen in the following examples.

### Example 1

Consider the mechanical system shown in Figure 3-11(a). It is a model of some physical system. Let $k_2$ be the spring constant and let $k_1$ be

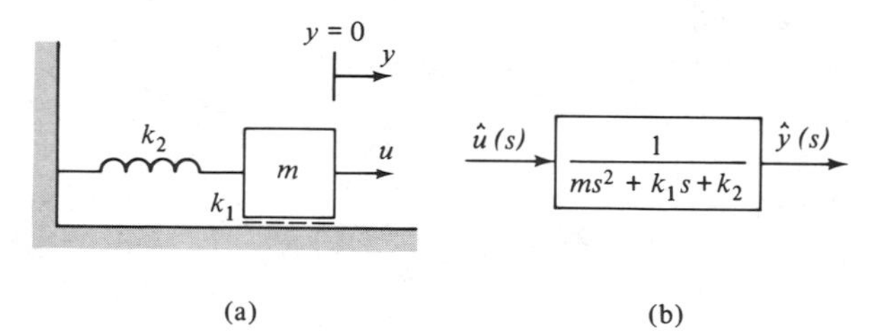


**Figure 3-11** A mechanical system.

the damping coefficient between the mass $m$ and the floor. We shall now derive the relation between the displacement $y$ (output) and the external force $u$ (input). Newton's law, $ma = F$, gives

$$m\frac{d^2y}{dt^2} = u - k_1\frac{dy}{dt} - k_2y$$

Taking the Laplace transform and assuming the zero initial condition, we obtain

$$\hat{u}(s) = (ms^2 + k_1s + k_2)\hat{y}(s)$$

Hence the input–output description in the frequency domain of the system is

$$\hat{y}(s) = \frac{1}{ms^2 + k_1 s + k_2}\hat{u}(s)$$

If $m = 1$, $k_1 = 3$, $k_2 = 2$, then the impulse response of the system is

$$g(t) = \mathcal{L}^{-1}\left[\frac{1}{s^2 + 3s + 2}\right] = \mathcal{L}^{-1}\left[\frac{1}{(s+1)} - \frac{1}{(s+2)}\right] = e^{-t} - e^{-2t}$$

and the input and output are related by

$$y(t) = \int_0^t g(t - \tau)u(\tau)\, d\tau$$

We next derive the dynamical-equation description of the system. Let the displacement and the velocity of the mass $m$ be the state variables; that is, $x_1 = y$, $x_2 = \dot{y}$. Then we have

$$\begin{aligned} \dot{x}_1 &= x_2 \\ u &= m\dot{x}_2 + k_1 x_2 + k_2 x_1 \end{aligned}$$

or

$$\begin{bmatrix} \dot{x}_1 \\ \dot{x}_2 \end{bmatrix} = \begin{bmatrix} 0 & 1 \\ -k_2/m & -k_1/m \end{bmatrix}\begin{bmatrix} x_1 \\ x_2 \end{bmatrix} + \begin{bmatrix} 0 \\ 1/m \end{bmatrix} u$$

$$y = [1 \quad 0]\begin{bmatrix} x_1 \\ x_2 \end{bmatrix}$$

This is a state-variable description of the system. By choosing a different set of state variables, we may obtain a different state-variable description of the system.

## Example 2

An armature-controlled dc motor can be modeled as shown in Figure 3-12. $R_a$ and $L_a$ represent the resistive and inductive components of

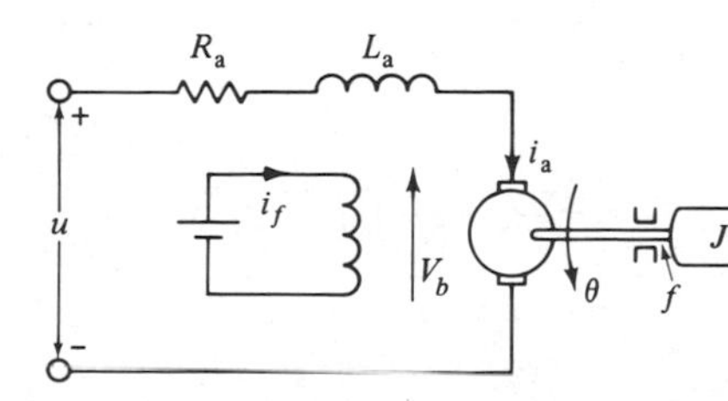


**Figure 3-12** An armature-controlled dc motor.

the armature circuit. The field current $i_f$ is constant. The motor is shown driving a load with an inertia $J$ and damping $f$. The input voltage $u$ is applied to the armature terminals. The desired information (output) is the angular displacement $\theta$ of the shaft.

The torque developed by the motor is a function of the flux developed by the field current $i_f$ and the armature current $i_a$. Since $i_f$ is assumed to be constant, the developed torque $T_m$, under the assumption that there is no saturation, is given by

$$T_m = K_m i_a$$

where $K_m$ is a motor constant. The developed torque is used to drive the load and overcome the damping. Hence,

$$K_m i_a = J \frac{d^2\theta}{dt^2} + f \frac{d\theta}{dt} \tag{3-36}$$

The armature current $i_a$ is related to the input voltage by

$$u(t) = R_a i_a + L_a \frac{di_a}{dt} + V_b(t) \tag{3-37}$$

where $V_b(t)$ is the back-electromotive-force voltage, which is proportional to the motor speed. Therefore we have

$$V_b(t) = K_b \frac{d\theta}{dt} \tag{3-38}$$

Taking the Laplace transform of Equations (3-36)–(3-38) and assuming zero initial conditions, the transfer function of the motor can be found as

$$\hat{g}(s) = \frac{\hat{\theta}(s)}{\hat{u}(s)} = \frac{K_m}{s[(R_a + L_a s)(Js + f) + K_b K_m]}$$

The impulse response of the system is the inverse Laplace transform of $\hat{g}(s)$.

To derive the state-variable description of the system, we choose the current through the inductor, the angular displacement, and angular velocity of the motor as the state variables; that is,

$$x_1 = i_a \qquad x_2 = \theta \qquad x_3 = \frac{d}{dt}\theta = \dot{\theta}$$

Then we have

$$\begin{aligned} u &= R_a x_1 + L_a \dot{x}_1 + K_b x_3 \\ \dot{x}_2 &= x_3 \\ K_m x_1 &= J\dot{x}_3 + f x_3 \end{aligned}$$

Hence the state-variable description of the motor is

$$\begin{bmatrix} \dot{x}_1 \\ \dot{x}_2 \\ \dot{x}_3 \end{bmatrix} = \begin{bmatrix} -R_a/L_a & 0 & -K_b/L_a \\ 0 & 0 & 1 \\ K_m/J & 0 & -f/J \end{bmatrix} \mathbf{x} + \begin{bmatrix} 1/L_a \\ 0 \\ 0 \end{bmatrix} u$$

$$\theta = [0 \quad 1 \quad 0]\mathbf{x}$$

Example 3

Consider the single-loop feedback system shown in Figure 3-13. The open-loop system $P$ is described by

$$y(t) = \int_0^t g(t-\tau)u(\tau)\,d\tau$$

The element $\phi$ is a memoryless element; that is, its output $v(t_0)$ depends

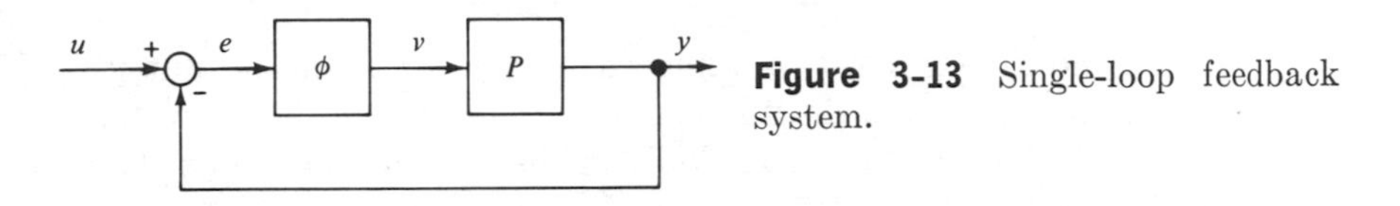


**Figure 3-13** Single-loop feedback system.

only on the instantaneous value of the input $e$ at time $t_0$, and is denoted by $v(t_0) = \phi(e(t_0))$. What is the input–output description of the feedback system?

**Case 1.** $v(t) = \phi(e(t)) = ke(t)$, where $k$ is a constant; that is, the memoryless element is a constant gain.

Since $P$ and $\phi$ are time-invariant, we may apply the Laplace transform to compute the transfer function $\hat{h}(s)$ of the feedback system. It is clear that $\hat{v}(s) = k(\hat{u}(s) - \hat{y}(s))$ and $\hat{y}(s) = \hat{g}(s)\hat{u}(s)$; hence

$$\hat{h}(s) = \frac{\hat{y}(s)}{\hat{u}(s)} = \frac{k\hat{g}(s)}{1 + k\hat{g}(s)} \tag{3-39}$$

In the time domain, (3-39) can be written as[7]

$$\begin{aligned} h(t) &= kg(t) - k\int_0^t h(t-\tau)g(\tau)\,d\tau && \text{for } t \geq 0 \\ &= 0 && \text{for } t < 0 \end{aligned} \tag{3-40}$$

Having obtained either $\hat{h}(s)$ or $h(t)$ we may use either $\hat{y}(s) = \hat{h}(s)\hat{u}(s)$ or

$$y(t) = \int_0^t h(t-\tau)u(\tau)\,d\tau$$

to describe the system.

**Case 2.** $v(t) = \phi(e(t)) = k(t)e(t)$, in which the memoryless element is a time-varying gain.

This system is linear but time-varying, hence the Laplace transform is not a convenient tool in the study. We redraw Figure 3-13 as in Fig. 3-14 with the variables chosen as shown. Since

$$v(t) = k(t)e(t) = k(t)[u(t) - y(t)]$$

[7] $\mathcal{L}^{-1}[\hat{h}(s)\hat{g}(s)] = \int_0^t g(t-\tau)h(\tau)\,d\tau = \int_0^t h(t-\tau)g(\tau)\,d\tau.$

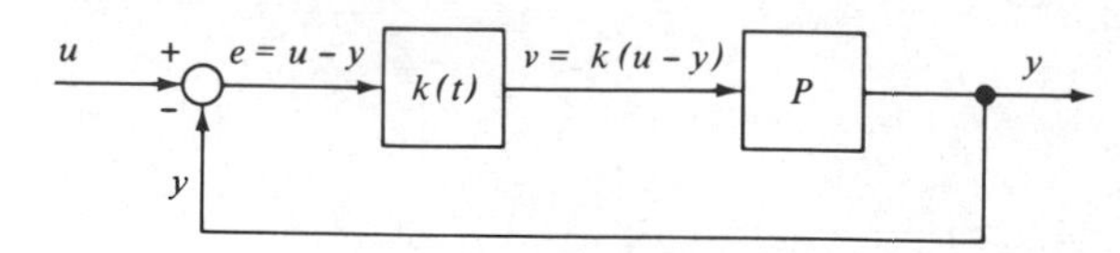

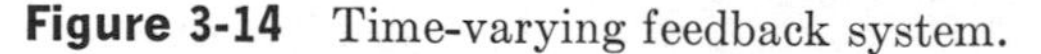

**Figure 3-14** Time-varying feedback system.

for all $t$, if the system is relaxed at $t_0$, then we have

$$y(t) = \int_{t_0}^{t} g(t-\tau)v(\tau)\,d\tau = \int_{t_0}^{t} g(t-\tau)k(\tau)[u(\tau) - y(\tau)]\,d\tau \quad t \geq t_0$$

In this equation, if $u(t)$ is replaced by $\delta(t - t_0)$, the output $y(t)$ will be, by definition, the impulse response of the feedback system. Let $h(t, t_0)$ be the impulse response of the feedback system. Then we have

$$h(t, t_0) = \int_{t_0}^{t} g(t-\tau)k(\tau)[\delta(\tau - t_0) - h(\tau, t_0)]\,d\tau$$

which can be simplified as

$$h(t, t_0) = g(t - t_0)k(t_0) - \int_{t_0}^{t} g(t-\tau)k(\tau)h(\tau, t_0)\,d\tau \qquad t \geq t_0 \qquad \textbf{(3-41)}$$

That $h(t, t_0) = 0$ for $t < t_0$ follows from the causality of the system. Note that if $k(\cdot)$ is independent of $t$ and if $t_0 = 0$, then (3-41) reduces to (3-40).

**Case 3.** $v(t) = \phi(e(t))$, a nonlinear memoryless element (for example, a model of a saturation).

Since $\phi$ is a nonlinear element, the input and the output of the feedback system are not linearly related; consequently, the impulse response of the feedback system is of no use in describing the system. In this case, the input $u$ and the output $y$ can only be described by the following implicit equation:

$$y(t) = \int_{t_0}^{t} g(t-\tau)\phi(u(\tau) - y(\tau))\,d\tau$$

This equation can be solved by iteration.

## Example 4

Consider the single-loop feedback system shown in Figure 3-13. It is assumed that $P$ is describable by the linear, time-invariant, dynamical equation

$$\begin{aligned}\dot{\mathbf{x}} &= \mathbf{Ax} + \mathbf{b}v \\ y &= \mathbf{cx}\end{aligned}$$

and the input and output of $\phi$ is related by either $v(t) = k(t)e(t)$ or $v(t) = \phi(e(t))$. What is the dynamical equation of the entire feedback system?

Since $e(t) = u(t) - y(t) = u(t) - \mathbf{c}\mathbf{x}(t)$, if $v(t) = k(t)e(t)$, then

$$\begin{cases} \dot{\mathbf{x}} = \mathbf{A}\mathbf{x} + \mathbf{b}k(t)(u(t) - y(t)) = (\mathbf{A} - k(t)\mathbf{b}\mathbf{c})\mathbf{x} + k(t)\mathbf{b}u(t) \\ y = \mathbf{c}\mathbf{x} \end{cases}$$

which is the dynamical-equation description of the feedback system. If $v(t) = \phi(e(t))$, then

$$\dot{\mathbf{x}} = \mathbf{A}\mathbf{x} + \mathbf{b}\phi(u(t) - \mathbf{c}\mathbf{x}(t))$$
$$y = \mathbf{c}\mathbf{x}$$

This is a nonlinear dynamical equation that describes the feedback system.

**Dynamical equations for *RLC* networks.** We introduce in this subsection a systematic procedure for assigning state variables and writing dynamical equations for general lumped linear $RLC$ networks which may contain independent voltage and current sources. It is well known that if all the inductor currents and the capacitor voltages of an $RLC$ network are known, then the behavior of the network is uniquely determinable for any input. However, it is not necessary to choose all the inductor currents and capacitor voltages as state variables. This can be seen from the simple circuits shown in Figure 3-15. If we assign all the capacitor

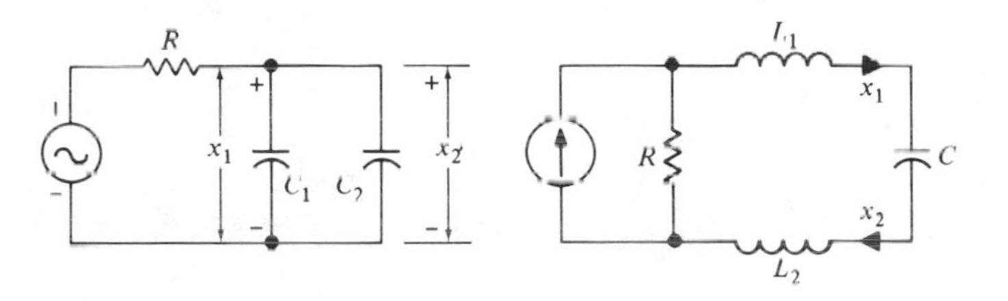


**Figure 3-15** Circuits with a loop which consists of capacitors only or a cutset which consists of inductors only.

voltages and inductor currents as state variables as shown, then we see that $x_1(t) = x_2(t)$ for all $t$. Clearly there is a redundancy here. Hence *the state of an RLC network can be chosen to consist of only independent capacitor voltages and independent inductor currents.*

Before proceeding, we review briefly the concepts of tree, link, and cutset of a network. (We consider only connected networks.) A *tree* of a network is defined as any connected graph (connection of branches) containing all the nodes of the network and not containing any loop. Every branch in a given tree is called a *tree branch.* Every branch not in the tree is called a *link.* A *cutset* of a connected network is any minimal set of branches such that the removal of all the branches in this set causes the remaining network to be unconnected. With respect to any fixed tree, every link and some tree branches form a unique loop called a *fundamental loop;* every tree branch with some links form a unique cutset called a *fundamental cutset.* Hence every fundamental loop includes only one link and every fundamental cutset includes only one tree branch. With these concepts, we are ready to give a systematic procedure for developing a dynamical-equation description of any *RLC* network that may contain independent voltage sources or current sources:[8]

**1.** Choose a tree called a *normal tree.* The branches of the normal tree are chosen in the order of voltage sources, capacitors, resistors, inductors, and current sources. Hence, a normal tree consists of all the voltage sources, the maximal number of permissible capacitors (those that do not form a loop), the resistors, and finally the minimal number of inductors. Usually it does not contain any current source.
**2.** Assign the charges or voltages of *the capacitors in the normal tree* and the flux or current of *the inductors in the links* as state variables. The voltages or charges of the capacitors in the links and the flux or current of the inductors in the normal tree need *not* be chosen as state variables.
**3.** Express the branch variables (branch voltage and current) of all the resistors, the capacitors in the links, and the inductors in the normal tree in terms of the state variables by applying the Kirchhoff voltage or current law to the fundamental loops or cutsets of these branches.
**4.** Apply the Kirchhoff voltage or current laws to the fundamental loop or cutset of every branch that is assigned as a state variable.

Example 5

Consider the linear network shown in Figure 3-16. The normal tree is chosen as shown (heavy lines); it consists of the voltage source, two capacitors, and one resistor. The voltages of the capacitors in the normal tree and the current of the inductor in the link are chosen as state variables. Next we express the variables of resistors ① and ② in terms of the state variables. By applying the Kirchhoff voltage law (KVL) to the fundamental loop of branch ①, the voltage across ① is found as $(u_1 - x_1)$; hence its current is $(u_1 - x_1)$. By applying the Kirchhoff current law (KCL) to the fundamental cutset of ②, we have immediately that the current through resistor ② is $x_3$. Consequently the voltage across resistor ② is $x_3$. Now the characteristics of every branch are expressed in terms

[8] A network with a loop that consists of only voltage sources and capacitors or with a cutset that consists of only current sources and inductors is excluded, because in this case its dynamical-equation description cannot be of the form in (3-30).

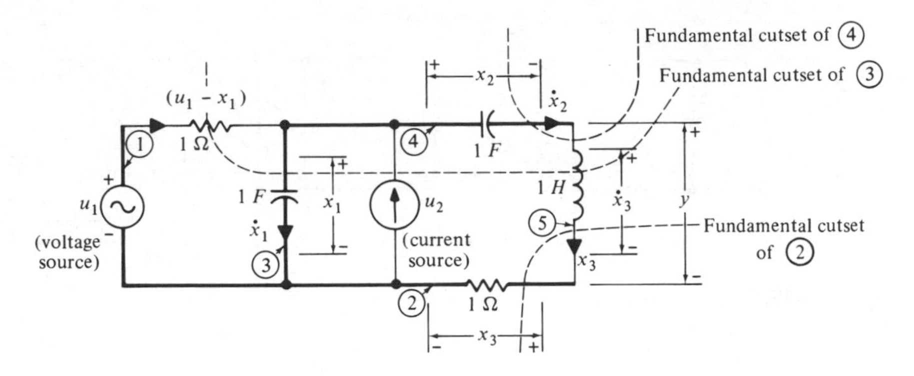


**Figure 3-16** A network with voltage and current sources.

of the state variables as shown. If we apply the KCL to the fundamental cutset of branch ③, we have

$$(u_1 - x_1) - \dot{x}_1 + u_2 - x_3 = 0$$

The application of the KCL to the fundamental cutset of ④ yields

$$\dot{x}_2 = x_3$$

The application of the KVL to the fundamental loop of ⑤ yields

$$\dot{x}_3 + x_3 - x_1 + x_2 = 0$$

These three equations can be rearranged in matrix form as

$$\begin{bmatrix} \dot{x}_1 \\ \dot{x}_2 \\ \dot{x}_3 \end{bmatrix} = \begin{bmatrix} -1 & 0 & -1 \\ 0 & 0 & 1 \\ 1 & -1 & -1 \end{bmatrix} \mathbf{x} + \begin{bmatrix} 1 & 1 \\ 0 & 0 \\ 0 & 0 \end{bmatrix} \begin{bmatrix} u_1 \\ u_2 \end{bmatrix}$$

The output equation can be easily found as

$$y = \dot{x}_3 = [1 \quad -1 \quad -1]\mathbf{x}$$

### Example 6

Find the input–output description of the network shown in Figure 3-16, or equivalently, the transfer-function matrix of the network. Since there are two inputs, one output, the transfer-function matrix is a $1 \times 2$ matrix. Let $\hat{\mathbf{G}}(s) = [\hat{g}_{11}(s) \quad \hat{g}_{12}(s)]$, where $\hat{g}_{11}$ is the transfer function from $u_1$ to $y$ with $u_2 = 0$, and $\hat{g}_{12}(s)$ is the transfer function from $u_2$ to $y$ with $u_1 = 0$. With $u_2 = 0$, the network reduces to the one in Figure 3-17(a). By loop analysis, we have

$$\left(1 + \frac{1}{s}\right)\hat{I}_1(s) - \frac{1}{s}\hat{I}_2(s) = \hat{u}_1(s)$$

$$-\frac{1}{s}\hat{I}_1(s) + \left(1 + s + \frac{2}{s}\right)\hat{I}_2(s) = 0$$

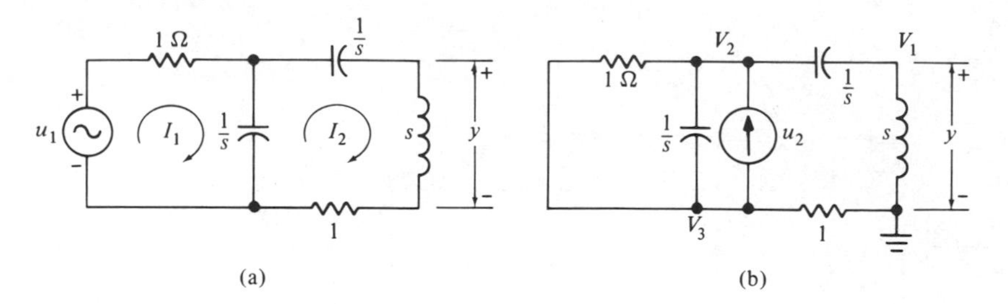


**Figure 3-17** Two reduced networks.

Solving for $\hat{I}_2(s)$, we obtain

$$\hat{I}_2(s) = \frac{\frac{1}{s}\hat{u}_1(s)}{\left(1+\frac{1}{s}\right)\left(1+s+\frac{2}{s}\right)-\frac{1}{s^2}}$$

Hence

$$\hat{g}_{11}(s) = \left.\frac{\hat{y}(s)}{\hat{u}_1(s)}\right|_{\substack{u_2=0\\ \text{initially}\\ \text{relaxed}}} = \frac{s\hat{I}_2(s)}{\hat{u}_1(s)} = \frac{s^2}{s^3+2s^2+3s+1}$$

If $u_1 = 0$, the network in Figure 3-16 reduces to the one in Figure 3-17(b). By node analysis,

$$\begin{aligned} \left(\frac{1}{s}+s\right)\hat{V}_1(s) - s\hat{V}_2(s) &= 0 \\ -s\hat{V}_1(s) + (1+2s)\hat{V}_2(s) - (1+s)\hat{V}_3(s) &= \hat{u}_2(s) \\ -(1+s)\hat{V}_2(s) + (2+s)\hat{V}_3(s) &= -\hat{u}_2(s) \end{aligned}$$

Solving for $\hat{V}_1(s)$, we obtain

$$\hat{V}_1(s) = \frac{s^2}{s^3+2s^2+3s+1}\hat{u}_2(s)$$

Hence

$$\hat{g}_{12}(s) = \left.\frac{\hat{V}_1(s)}{\hat{u}_2(s)}\right|_{\substack{u_1=0\\ \text{initially}\\ \text{relaxed}}} = \frac{s^2}{s^3+2s^2+3s+1}$$

Consequently, the input–output description in the frequency domain of the network is given by

$$\hat{y}(s) = \begin{bmatrix} \dfrac{s^2}{s^3+2s^2+3s+1} & \dfrac{s^2}{s^3+2s^2+3s+1} \end{bmatrix} \begin{bmatrix} \hat{u}_1(s) \\ \hat{u}_2(s) \end{bmatrix} \tag{3-42}$$

### Example 7

Consider the network shown in Figure 3-18. The capacitor branches are labeled 1, 2, and 3; the resistors are 4, 5, and 6; the voltage source

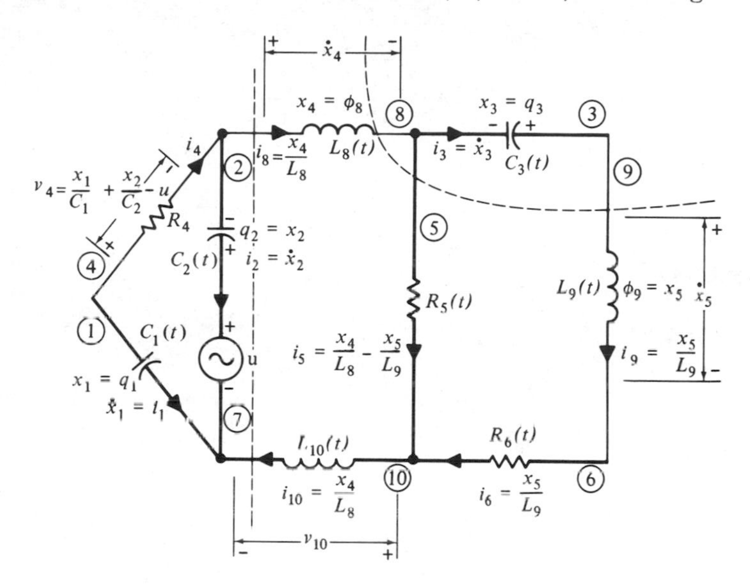


**Figure 3-18** A network with linear, time-varying elements.

is 7; and the inductors are 8, 9, and 10. It is assumed that all the elements are linear but time-varying. The characteristics of branches are given by $q_j(t) = C_j(t)v_j(t)$, for $j = 1, 2, 3$; $v_k(t) = R_k(t)i_k(t)$, for $k = 4, 5, 6$; $\phi_l(t) = L_l(t)i_l(t)$, for $l = 8, 9, 10$. The $i_k$'s and $v_k$'s are branch currents and branch voltages, the $q_j$'s are capacitor charges, and the $\phi_l$'s are inductor fluxes. The capacitances $C_j(\cdot)$, the resistances $R_k(\cdot)$, and the inductances $L_l(\cdot)$ are assumed to be continuous and positive for all $t$.

First we form a normal tree, which consists of the branches $\{1, 2, 7, 10, 5, 3, 6\}$ (heavy lines). Note that the normal tree contains one inductor. The branches $\{4, 8, 9\}$ are links. Next we choose the charges of the capacitors in the normal tree ($q_1$, $q_2$, and $q_3$) and the fluxes of the inductors in the links ($\phi_8$, $\phi_9$) as the state variables. Note that the inductor flux $\phi_{10}$ in the normal tree is *not* a state variable. Let $x_1 = q_1$, $x_2 = q_2$, $x_3 = q_3$, $x_4 = \phi_8$, and $x_5 = \phi_9$. Observe that $i_j = \dot{x}_j$, $v_j = x_j/C_j$, for $j = 1, 2, 3$; $v_l = \dot{x}_l$, $i_l = x_l/L_l$, for $l = 8, 9$. Applying the KVL and KCL to the fundamental loops and fundamental cutsets of branches ④, ⑤, ⑥, and ⑩, we have

$$v_4 = \frac{x_1}{C_1} + \frac{x_2}{C_2} - u \qquad i_{10} = \frac{x_4}{L_8} \qquad i_6 = \frac{x_5}{L_9} \qquad i_5 = \frac{x_4}{L_8} - \frac{x_5}{L_9}$$

Now the variables of every branch are expressed in terms of state variables, as shown in Figure 3-18. The application of the KCL to the fundamental cutsets of $C_1$, $C_2$, and $C_3$ yields, respectively,

$$\dot{x}_1 = -i_4 = -\frac{x_1}{R_4C_1} - \frac{x_2}{R_4C_2} + \frac{u}{R_4}$$

$$\dot{x}_2 = -i_4 + i_8 = -\frac{x_1}{R_4C_1} - \frac{x_2}{R_4C_2} + \frac{u}{R_4} + \frac{x_4}{L_8}$$

$$\dot{x}_3 = -i_9 = -\frac{x_5}{L_9}$$

The application of the KVL to the fundamental loops of $L_8$ and $L_9$ yields, respectively,

$$\dot{x}_4 + \frac{R_5}{L_8}x_4 - \frac{R_5}{L_9}x_5 + \frac{d}{dt}\left(\frac{L_{10}}{L_8}x_4\right) - u + \frac{x_2}{C_2} = 0$$

$$\dot{x}_5 + \frac{R_6}{L_9}x_5 - \frac{R_5}{L_8}x_4 + \frac{R_5}{L_9}x_5 - \frac{x_3}{C_3} = 0$$

With some manipulation, these equations can be written as

$$\begin{bmatrix} \dot{x}_1 \\ \dot{x}_2 \\ \dot{x}_3 \\ \dot{x}_4 \\ \dot{x}_5 \end{bmatrix} = \begin{bmatrix} -\frac{1}{R_4C_1} & -\frac{1}{R_4C_2} & 0 & 0 & 0 \\ -\frac{1}{R_4C_1} & -\frac{1}{R_4C_2} & 0 & \frac{1}{L_8} & 0 \\ 0 & 0 & 0 & 0 & -\frac{1}{L_9} \\ 0 & \frac{Y_1}{C_2} & 0 & Y_1Y_2 & \frac{Y_1R_5}{L_9} \\ 0 & 0 & \frac{1}{C_3} & \frac{R_5}{L_8} & -\frac{R_5+R_6}{L_9} \end{bmatrix} \mathbf{x} + \begin{bmatrix} \frac{1}{R_4} \\ \frac{1}{R_4} \\ 0 \\ 1 \\ 0 \end{bmatrix} u$$

where

$$Y_1 = -\left(1 + \frac{L_{10}}{L_8}\right)^{-1} \quad \text{and} \quad Y_2 = \frac{R_5L_8 + \dot{L}_{10}L_8 - L_{10}\dot{L}_8}{L_8{}^2}$$

This is the state equation of the network. If the output is specified, the output equation can be easily found.

## 3-5 Comparisons of the Input–Output Description and the State-Variable Description

We compare in this section the input–output description and the state-variable description of systems.

1. The input–output description of a system covers only the relationship between the input and the output under the assumption

that the system is initially relaxed. Hence if the system is not initially relaxed, the description is not applicable. A more serious problem is that it does not reveal what will happen if the system is not initially relaxed nor does it reveal the behavior inside the system. For example, consider the networks shown in Figure 3-19. The networks are assumed to have a capacitor with $-1$ farad. If the initial voltage of the capacitor is zero, the current will distribute equally among the two branches. Hence the transfer function of the network in Figure 3-19(a) is 0.5; the transfer function of the network in Figure 3-19(b) is 1. Because of the

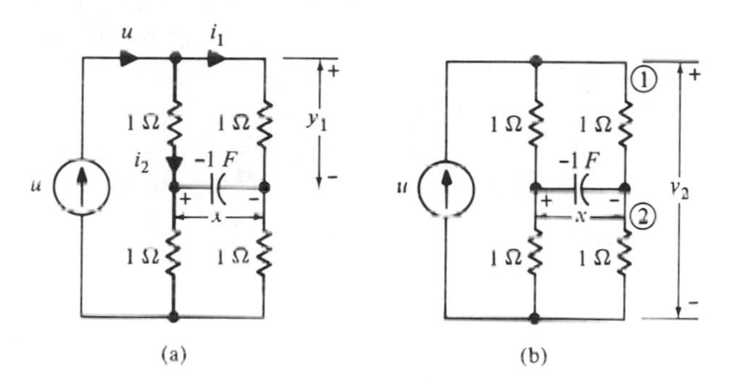


**Figure 3-19** Active networks.

negative capacitance, if the initial voltage across the capacitor is different from zero, then the voltage $y_1$ in Figure 3-19(a) will increase with time, whereas the voltage $y_2$ in Figure 3-19(b) remains equal to $u(t)$ for all $t$. It is clear that the network in Figure 3-19(a) is not satisfactory, because the output will increase without bound if the initial condition is different from zero. Although the output of the network in Figure 3-19(b) behaves well, the network is still not satisfactory, because the voltages in branches 1 and 2 will increase with time (in different polarity), and the network will eventually burn out. Hence the networks in Figure 3-19 can never function properly. If the internal structure of the network is not known, this fact cannot be detected from its transfer function. Consequently, the input–output description is sometimes not sufficient to describe a system completely.

The dynamical-equation descriptions of the networks in Figure 3-19 (see Problem 3-18) can be found, respectively, as

$$\begin{cases} \dot{x} = x \\ y = 0.5x + 0.5u \end{cases}$$

$$\begin{cases} \dot{x} = x \\ y = u \end{cases}$$

A dynamical equation describes not only the relationship between the input and output, but also the behavior inside a system under any initial condition; hence a dynamical equation describes a system more completely.

**2.** For extremely complicated linear systems, it might be very involved to find the dynamical-equation descriptions. In these cases, it might be easier to find the input–output descriptions by direct measurements. We apply at each input terminal a very sharp and narrow pulse, the responses of the output terminals give us immediately the impulse-response matrix of the system. In practice, we may have difficulty in generating a very sharp and narrow pulse, or ideally a $\delta$-function; however, this can be avoided by using a unit step function. A *unit step function* $\delta_1(t - t_0)$ is defined as

$$\delta_1(t - t_0) = \begin{cases} 1 & \text{for } t \geq t_0 \\ 0 & \text{for } t < t_0 \end{cases} \tag{3-43}$$

The response of a linear causal system due to a unit step function input is given by

$$g_1(t, t_0) = \int_{t_0}^{t} g(t, \tau)\, d\tau \tag{3-44}$$

where $g_1(t, t_0)$ is called the *step response* (due to a unit step function applied at $t_0$). Differentiating (3-44) with respect to $t_0$, we obtain

$$g(t, t_0) = -\frac{\partial}{\partial t_0} g_1(t, t_0) \tag{3-45}$$

For the time-invariant case, (3-45) reduces to

$$g(t) = \frac{d}{dt} g_1(t) \tag{3-46}$$

Thus the impulse response can be obtained from the step response by using (3-45) or (3-46).

For linear time-invariant systems, we may measure either transfer functions or impulse responses. The transfer function of a system can be measured easily and accurately by employing signal generators and sinusoidal measurement equipments (see Corollary 8-3). If the transfer function is measured, the impulse response can be obtained by taking the inverse Laplace transform.

**3.** In classical control theory, analysis and synthesis are all performed on transfer functions; for example, designs of feedback systems can be easily carried out by employing the root-locus technique or the Bode plot. In modern control theory, designs are carried out on dynamical equations. Although the analytical solution in a dynamical equation may seem simple, the computation of numerical solutions is very involved. In general, a digital computer has to be used. However, modern control theory solves problems that cannot be handled by classical theory.

**4.** The dynamical equations we study in this book are restricted to finite-dimensional ones; therefore, they are applicable only to lumped systems. The input–output descriptions are, however, applicable to lumped as well as distributed systems. In the study of nonlinear systems, depending on the approach taken, either description can be used. For example, in the study of stability, we use the input–output description if the functional analysis approach is used; see References [28], [95], and [117]. If Lyapunov's second method is employed, we have to use the dynamical-equation description.

**5.** If the dynamical-equation description of a system is available, the system can be readily simulated on an analog or a digital computer.

From the foregoing discussion, we see that the input output and the state-variable descriptions have their own merits. In order to carry out a design efficiently, a designer should make himself familiar with these two mathematical descriptions. In this book, these two descriptions will be developed equally and their relationships will be explored.

## *3-5 Mathematical Descriptions of Composite Systems [9]

**Time-varying case.** In engineering it is often useful to view a system as consisting of two or more subsystems; it is then called a composite system. Whether or not a system is a composite system depends on what we consider as building blocks. For example, an analog computer can be looked upon as a system, or as a composite system that consists of operational amplifiers, function generators, potentiometers, and others as subsystems. There are many forms of composite systems; however, mostly they are built from the following three basic connections: the parallel, the tandem, and the feedback connections, as shown in Figure 3-20. In this section we shall study the input–output and state-variable descriptions of these three basic composite systems.

We study first the input–output description of composite systems. Consider two multivariable systems $S_i$, which are described by

$$\mathbf{y}_i(t) = \int_{-\infty}^{t} \mathbf{G}_i(t, \tau)\mathbf{u}_i(\tau)\, d\tau \qquad i = 1, 2 \tag{3-47}$$

where $\mathbf{u}_i$ and $\mathbf{y}_i$ are the input and the output,[10] $\mathbf{G}_i$ is the impulse-response matrix of the system $S_i$. Let $\mathbf{u}$, $\mathbf{y}$, and $\mathbf{G}$ be, respectively, the input, the

[9] The material in this section is not used until Chapter 9; thus its study may be postponed.

[10] In the study of composite systems, we use vectors with subscripts, instead of superscripts as used elsewhere, to denote different vectors. In any case, every notation is explained wherever it is introduced, hence no confusion should arise.

output, and the impulse-response matrix of a composite system. We see from Figure 3-20 that in the parallel connection, we have $\mathbf{u}_1 = \mathbf{u}_2 = \mathbf{u}$,

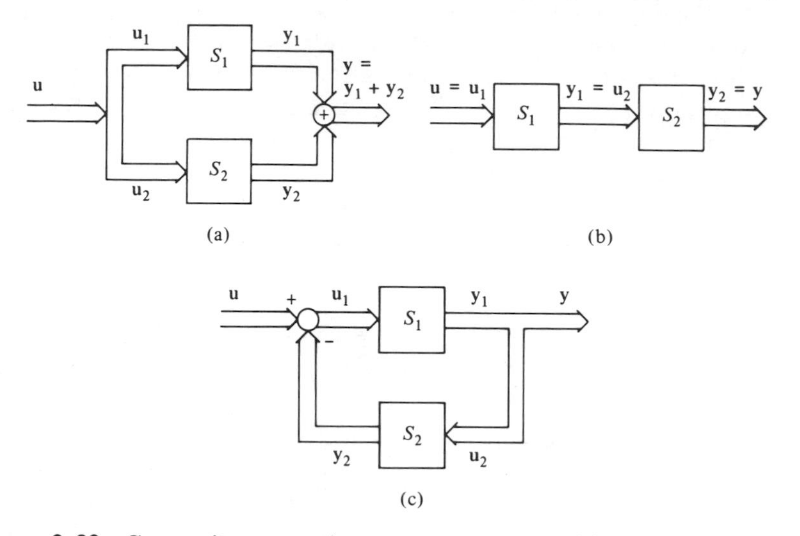


**Figure 3-20** Composite connections of two systems. (a) Parallel connection. (b) Tandem connection. (c) Feedback connection.

$\mathbf{y} = \mathbf{y}_1 + \mathbf{y}_2$; in the tandem connection, we have $\mathbf{u} = \mathbf{u}_1$, $\mathbf{y}_1 = \mathbf{u}_2$, $\mathbf{y}_2 = \mathbf{y}$; in the feedback connection, we have $\mathbf{u}_1 = \mathbf{u} - \mathbf{y}_2$, $\mathbf{y} = \mathbf{y}_1$. Here we have implicitly assumed that the systems $S_1$ and $S_2$ have compatible numbers of input and output; otherwise, the systems cannot be properly connected. It is also assumed that there is no loading effect in the connection; that is, the impulse response matrices $\mathbf{G}_1$, $\mathbf{G}_2$ remain unchanged after connection. It is easy to show that the impulse response matrix of the parallel connection of $S_1$ and $S_2$ shown in Figure 3-20(*a*) is

$$\mathbf{G}(t, \tau) = \mathbf{G}_1(t, \tau) + \mathbf{G}_2(t, \tau) \tag{3-48}$$

For the tandem connection shown in Figure 3-20(b), we have

$$\mathbf{G}(t, \tau) = \int_\tau^t \mathbf{G}_2(t, v)\mathbf{G}_1(v, \tau)\, dv \tag{3-49}$$

We prove (3-49) for the single-variable case. The impulse response $g(t, \tau)$ is, by definition, the response at $y_2$ due to a $\delta$-function applied at time $\tau$ at $u_1$. The response at $y_1$ due to this $\delta$-function is $g_1(t, \tau)$. The output of $S_2$ due to the input $g_1(t, \tau)$ is

$$\int_\tau^t g_2(t, v)g_1(v, \tau)\, dv$$

Hence,

$$g(t, \tau) = \int_\tau^t g_2(t, v)g_1(v, \tau)\, dv$$

The same procedure can be used to prove the multivariable case.

For the feedback connection shown in Figure 3-20(c), the impulse-response matrix is the solution of the integral equation

$$\mathbf{G}(t, \tau) = \mathbf{G}_1(t, \tau) - \int_\tau^t \mathbf{G}_1(t, v) \int_\tau^v \mathbf{G}_2(v, s)\mathbf{G}(s, \tau)\, ds\, dv \tag{3-50}$$

where $\mathbf{G}_1$ and $\mathbf{G}_2$ are known and $\mathbf{G}$ is unknown. This equation can be verified by using the idea of proving Equation (3-41) (see Problem 3-24). There is a general iterative method for solving Equation (3-50), but it is rather involved.

Now we study the state-variable descriptions of composite systems. Let the system $S_1$ and $S_2$ in Figure 3-20 be described by

$$E^i: \quad \dot{\mathbf{x}}_i = \mathbf{A}_i(t)\mathbf{x}_i + \mathbf{B}_i(t)\mathbf{u}_i \qquad i = 1, 2 \tag{3-51a}$$

$$\mathbf{y}_i = \mathbf{C}_i(t)\mathbf{x}_i + \mathbf{D}_i(t)\mathbf{u}_i \tag{3-51b}$$

where $\mathbf{x}_i$ is the state, $\mathbf{u}_i$ is the input, and $\mathbf{y}_i$ is the output; $\mathbf{A}_i$, $\mathbf{B}_i$, $\mathbf{C}_i$ and $\mathbf{D}_i$ are matrices of compatible order whose entries are continuous function of $t$ defined over $(-\infty, \infty)$. The state space of $S_i$ is denoted by $\Sigma_i$.

Let us introduce the concept of the direct sum of two linear spaces. The linear space $\Sigma$ is the direct sum of two linear spaces $\Sigma_1$ and $\Sigma_2$, written as $\Sigma = \Sigma_1 \oplus \Sigma_2$, if every vector in $\Sigma$ is of the form $[\mathbf{x}_1' \ \ \mathbf{x}_2']'$ where $\mathbf{x}_1$ is a vector in $\Sigma_1$ and $\mathbf{x}_2$ is a vector in $\Sigma_2$. The dimension of $\Sigma$ is the sum of those of $\Sigma_1$ and $\Sigma_2$.

It is clear that the composite vector

$$\begin{bmatrix} \mathbf{x}_1 \\ \mathbf{x}_2 \end{bmatrix}$$

qualifies as the state of any composite connection of $S_1$ and $S_2$; its state space is the direct sum of the state spaces of $S_1$ and $S_2$, $\Sigma_1 \oplus \Sigma_2$. For the parallel connection, we have $\mathbf{u}_1 = \mathbf{u}_2 = \mathbf{u}$, $\mathbf{y} = \mathbf{y}_1 + \mathbf{y}_2$; hence its dynamical equation is

$$\begin{bmatrix} \dot{\mathbf{x}}_1 \\ \dot{\mathbf{x}}_2 \end{bmatrix} = \begin{bmatrix} \mathbf{A}_1(t) & 0 \\ 0 & \mathbf{A}_2(t) \end{bmatrix} \begin{bmatrix} \mathbf{x}_1 \\ \mathbf{x}_2 \end{bmatrix} + \begin{bmatrix} \mathbf{B}_1(t) \\ \mathbf{B}_2(t) \end{bmatrix} \mathbf{u} \tag{3-52a}$$

$$\mathbf{y} = [\mathbf{C}_1(t) \quad \mathbf{C}_2(t)] \begin{bmatrix} \mathbf{x}_1 \\ \mathbf{x}_2 \end{bmatrix} + (\mathbf{D}_1(t) + \mathbf{D}_2(t))\mathbf{u} \tag{3-52b}$$

The dynamical equation of the tandem connection of $S_1$ and $S_2$ is given by

$$\begin{bmatrix} \dot{\mathbf{x}}_1 \\ \dot{\mathbf{x}}_2 \end{bmatrix} = \begin{bmatrix} \mathbf{A}_1(t) & \mathbf{0} \\ \mathbf{B}_2(t)\mathbf{C}_1(t) & \mathbf{A}_2(t) \end{bmatrix} \begin{bmatrix} \mathbf{x}_1 \\ \mathbf{x}_2 \end{bmatrix} + \begin{bmatrix} \mathbf{B}_1(t) \\ \mathbf{B}_2(t)\mathbf{D}_1(t) \end{bmatrix} \mathbf{u} \tag{3-53a}$$

$$\mathbf{y} = [\mathbf{D}_2(t)\mathbf{C}_1(t) \quad \mathbf{C}_2(t)] \begin{bmatrix} \mathbf{x}_1 \\ \mathbf{x}_2 \end{bmatrix} + \mathbf{D}_2(t)\mathbf{D}_1(t)\mathbf{u} \tag{3-53b}$$

which can be easily obtained by observing $\mathbf{u}_1 = \mathbf{u}$, $\mathbf{y}_1 = \mathbf{u}_2$, $\mathbf{y} = \mathbf{y}_2$.

For the feedback connection shown in Figure 3-20(c), its dynamical-equation description is

$$\begin{bmatrix} \dot{\mathbf{x}}_1 \\ \dot{\mathbf{x}}_2 \end{bmatrix} = \begin{bmatrix} \mathbf{A}_1(t) - \mathbf{B}_1(t)\mathbf{D}_2(t)\mathbf{Y}_1(t)\mathbf{C}_1(t) & -\mathbf{B}_1(t)\mathbf{C}_2(t) + \mathbf{B}_1(t)\mathbf{D}_2(t)\mathbf{Y}_1(t)\mathbf{D}_1(t)\mathbf{C}_2(t) \\ \mathbf{B}_2(t)\mathbf{Y}_1(t)\mathbf{C}_1(t) & \mathbf{A}_2(t) - \mathbf{B}_2(t)\mathbf{Y}_1(t)\mathbf{D}_1(t)\mathbf{C}_2(t) \end{bmatrix} \begin{bmatrix} \mathbf{x}_1 \\ \mathbf{x}_2 \end{bmatrix} + \begin{bmatrix} \mathbf{B}_1(t) - \mathbf{B}_1(t)\mathbf{D}_2(t)\mathbf{Y}_1(t)\mathbf{D}_1(t) \\ \mathbf{B}_2(t)\mathbf{Y}_1(t)\mathbf{D}_1(t) \end{bmatrix} \mathbf{u} \tag{3-54a}$$

$$\mathbf{y} = [\mathbf{Y}_1(t)\mathbf{C}_1(t) \quad -\mathbf{Y}_1(t)\mathbf{D}_1(t)\mathbf{C}_2(t)] \begin{bmatrix} \mathbf{x}_1 \\ \mathbf{x}_2 \end{bmatrix} + \mathbf{Y}_1(t)\mathbf{D}_1(t)\mathbf{u} \tag{3-54b}$$

where $\mathbf{Y}_1(t) = (\mathbf{I} + \mathbf{D}_1(t)\mathbf{D}_2(t))^{-1}$. It is clear that in order for (3-54) to be defined, we must assume that the inverse of the matrix $(\mathbf{I} + \mathbf{D}_1(t)\mathbf{D}_2(t))$ exists, or equivalently, $\det(\mathbf{I} + \mathbf{D}_1(t)\mathbf{D}_2(t)) \neq 0$ for all $t$. The dynamical equation (3-54) can be easily verified by observing $\mathbf{u}_1 = \mathbf{u} - \mathbf{y}_2$, $\mathbf{y}_1 = \mathbf{u}_2$, $\mathbf{y} = \mathbf{y}_1$. (See Problem 3-28.)

**Time-invariant case.** All the results in the preceding subsection can be applied to the time-invariant case without any modification. We shall now discuss the transfer-function matrices of composite systems. Let $\hat{\mathbf{G}}_1(s)$ and $\hat{\mathbf{G}}_2(s)$ be the rational transfer-function matrices of $S_1$ and $S_2$, respectively; then the transfer-function matrix of the parallel connection of $S_1$ and $S_2$ is $\hat{\mathbf{G}}_1(s) + \hat{\mathbf{G}}_2(s)$. The transfer-function matrix of the tandem connection of $S_1$ followed by $S_2$ is $\hat{\mathbf{G}}_2(s)\hat{\mathbf{G}}_1(s)$. Note that the order of $\hat{\mathbf{G}}_2(s)\hat{\mathbf{G}}_1(s)$ cannot be reversed. In order to discuss the transfer-function matrix of the feedback connection, we need some preliminary results. Let $\hat{\mathbf{G}}_1$ and $\hat{\mathbf{G}}_2$ be, respectively, $q_1 \times p_1$ and $q_2 \times p_2$ rational function matrices. In the feedback connection, we have $p_1 = q_2$ and $q_1 = p_2$.

Theorem 3-2

$$\det(\mathbf{I} + \hat{\mathbf{G}}_2(s)\hat{\mathbf{G}}_1(s)) = \det(\mathbf{I} + \hat{\mathbf{G}}_1(s)\hat{\mathbf{G}}_2(s)) \tag{3-55}$$

∎

Observe that the matrix on the right-hand side of (3-55) is a $q_1 \times q_1$ matrix, while the matrix on the left-hand side is a $p_1 \times p_1$ matrix. **I** is a unit matrix of compatible order. The elements of these matrices are rational functions of $s$, and since the rational functions form a field, standard results in matrix theory can be applied.

Proof of theorem 3-2

It is well known that $\det(\mathbf{NQ}) = \det(\mathbf{N})\det(\mathbf{Q})$. Hence we have

$$\det(\mathbf{NQP}) = \det(\mathbf{N})\det(\mathbf{Q})\det(\mathbf{P}) = \det(\mathbf{PQN}) \tag{3-56}$$

where **N**, **Q** and **P** are any square matrices of the same order. Let us choose

$$\mathbf{N} = \begin{bmatrix} \mathbf{I} & \mathbf{0} \\ -\hat{\mathbf{G}}_1(s) & \mathbf{I} \end{bmatrix} \qquad \mathbf{Q} = \begin{bmatrix} \mathbf{I} & -\hat{\mathbf{G}}_2(s) \\ \hat{\mathbf{G}}_1(s) & \mathbf{I} \end{bmatrix} \qquad \mathbf{P} = \begin{bmatrix} \mathbf{I} & \hat{\mathbf{G}}_2(s) \\ \mathbf{0} & \mathbf{I} \end{bmatrix}$$

They are square matrices of order $(q_1 + p_1)$. It is easy to verify that

$$\mathbf{NQP} = \begin{bmatrix} \mathbf{I} & \mathbf{0} \\ \mathbf{0} & \mathbf{I} + \hat{\mathbf{G}}_1(s)\hat{\mathbf{G}}_2(s) \end{bmatrix}$$

and

$$\mathbf{PQN} = \begin{bmatrix} \mathbf{I} + \hat{\mathbf{G}}_2(s)\hat{\mathbf{G}}_1(s) & \mathbf{0} \\ \mathbf{0} & \mathbf{I} \end{bmatrix}$$

hence

$$\det(\mathbf{NQP}) = \det(\mathbf{I} + \hat{\mathbf{G}}_1(s)\hat{\mathbf{G}}_2(s))$$
$$\det(\mathbf{PQN}) = \det(\mathbf{I} + \hat{\mathbf{G}}_2(s)\hat{\mathbf{G}}_1(s))$$

and the theorem follows immediately from (3-56). Q.E.D.

Theorem 3-3

If $\det(\mathbf{I} + \hat{\mathbf{G}}_1(s)\hat{\mathbf{G}}_2(s)) \neq 0$, then

$$\hat{\mathbf{G}}_1(s)(\mathbf{I} + \hat{\mathbf{G}}_2(s)\hat{\mathbf{G}}_1(s))^{-1} = (\mathbf{I} + \hat{\mathbf{G}}_1(s)\hat{\mathbf{G}}_2(s))^{-1}\hat{\mathbf{G}}_1(s)$$

Proof

Note that the zero in $\det(\mathbf{I} + \hat{\mathbf{G}}_1(s)\hat{\mathbf{G}}_2(s)) \neq 0$ is the zero element in the field of rational functions. Hence it can be written more suggestively as $\det(\mathbf{I} + \hat{\mathbf{G}}_1(s)\hat{\mathbf{G}}_2(s)) \neq 0$ for some $s$. The condition $\det(\mathbf{I} + \hat{\mathbf{G}}_1(s)\hat{\mathbf{G}}_2(s)) \neq 0$ implies that the inverse of the matrix $(\mathbf{I} + \hat{\mathbf{G}}_1(s)\hat{\mathbf{G}}_2(s))$ exists. From Theorem 3-2, we have

$$\det(\mathbf{I} + \hat{\mathbf{G}}_1(s)\hat{\mathbf{G}}_2(s)) = \det(\mathbf{I} + \hat{\mathbf{G}}_2(s)\hat{\mathbf{G}}_1(s)) \neq 0$$

Hence, both $(\mathbf{I} + \hat{\mathbf{G}}_1(s)\hat{\mathbf{G}}_2(s))^{-1}$ and $(\mathbf{I} + \hat{\mathbf{G}}_2(s)\hat{\mathbf{G}}_1(s))^{-1}$ exist. Consider the identity

$$\hat{\mathbf{G}}_1(s)(\mathbf{I} + \hat{\mathbf{G}}_2(s)\hat{\mathbf{G}}_1(s))(\mathbf{I} + \hat{\mathbf{G}}_2(s)\hat{\mathbf{G}}_1(s))^{-1} = \hat{\mathbf{G}}_1(s)$$

which can be written as

$$(\mathbf{I} + \hat{\mathbf{G}}_1(s)\hat{\mathbf{G}}_2(s))\hat{\mathbf{G}}_1(s)(\mathbf{I} + \hat{\mathbf{G}}_2(s)\hat{\mathbf{G}}_1(s))^{-1} = \hat{\mathbf{G}}_1(s) \tag{3-57}$$

Premultiplying $(\mathbf{I} + \hat{\mathbf{G}}_1(s)\hat{\mathbf{G}}_2(s))^{-1}$ on both sides of (3-57), we obtain the desired equality. Q.E.D.

Knowing Theorem 3-3, we are ready to investigate the transfer-function matrix of the feedback connection of $S_1$ and $S_2$.

### Corollary 3-3

Consider the feedback system shown in Figure 3-20(c). Let $\hat{\mathbf{G}}_1(s)$ and $\hat{\mathbf{G}}_2(s)$ be the transfer-function matrices of $S_1$ and $S_2$, respectively. If $\det(\mathbf{I} + \hat{\mathbf{G}}_1(s)\hat{\mathbf{G}}_2(s)) \neq 0$, then the transfer-function matrix of the feedback system is given by

$$\hat{\mathbf{G}}(s) = \hat{\mathbf{G}}_1(s)(\mathbf{I} + \hat{\mathbf{G}}_2(s)\hat{\mathbf{G}}_1(s))^{-1} = (\mathbf{I} + \hat{\mathbf{G}}_1(s)\hat{\mathbf{G}}_2(s))^{-1}\hat{\mathbf{G}}_1(s)$$

### Proof

From Figure 3-20(c), we have $\hat{\mathbf{G}}_1(s)(\hat{\mathbf{u}}(s) - \hat{\mathbf{G}}_2(s)\hat{\mathbf{y}}(s)) = \hat{\mathbf{y}}(s)$, or

$$(\mathbf{I} + \hat{\mathbf{G}}_1(s)\hat{\mathbf{G}}_2(s))\hat{\mathbf{y}}(s) = \hat{\mathbf{G}}_1(s)\hat{\mathbf{u}}(s) \tag{3-58}$$

which, together with Theorem 3-3, implies this corollary. Q.E.D.

Note that the condition $\det(\mathbf{I} + \hat{\mathbf{G}}_1(s)\hat{\mathbf{G}}_2(s)) \neq 0$ is essential for a feedback system to be defined. Without this condition, a feedback system may become meaningless in the sense that for certain inputs, there are no outputs satisfying Equation (3-58).

### Example 1

Consider a feedback system with

$$\hat{\mathbf{G}}_1(s) = \begin{bmatrix} \dfrac{-s}{s+1} & \dfrac{1}{s+2} \\ \dfrac{1}{s+1} & \dfrac{-s-1}{s+2} \end{bmatrix} \qquad \hat{\mathbf{G}}_2(s) = \begin{bmatrix} 1 & 0 \\ 0 & 1 \end{bmatrix}$$

It is easy to verify that $\det(\mathbf{I} + \hat{\mathbf{G}}_1(s)\hat{\mathbf{G}}_2(s)) = 0$. Let us choose

$$\hat{\mathbf{u}}(s) = \begin{bmatrix} \dfrac{1}{s+2} \\ \dfrac{1}{(s+1)^2} \end{bmatrix}$$

Then (3-58) becomes

$$\begin{bmatrix} \dfrac{1}{s+1} & \dfrac{1}{s+2} \\ \dfrac{1}{s+1} & \dfrac{1}{s+2} \end{bmatrix} \hat{\mathbf{y}}(s) = \begin{bmatrix} \dfrac{1 - s(s+1)}{(s+1)^2(s+2)} \\ 0 \end{bmatrix} \tag{3-59}$$

Obviously there is no $\hat{\mathbf{y}}(s)$ satisfying (3-59). In matrix theory, Equation (3-59) is said to be inconsistent. Hence in a feedback connection of two systems, certain caution should be taken. ∎

Recall from (3-54) that in order for the existence of the dynamical-equation description of a feedback system, we need the condition det (**I** +

$\mathbf{D}_1\mathbf{D}_2) \neq 0$ where $\mathbf{D}_i$ is the direct transmission part of the dynamical equation of $S_i$. It can be verified that if $\det(\mathbf{I} + \mathbf{D}_1\mathbf{D}_2) \neq 0$, then $\det(\mathbf{I} + \hat{\mathbf{G}}_1(s)\hat{\mathbf{G}}_2(s)) \neq 0$. However, the converse is not true; that is, the condition $\det(\mathbf{I} + \hat{\mathbf{G}}_1(s)\hat{\mathbf{G}}_2(s)) \neq 0$ may not imply $\det(\mathbf{I} + \mathbf{D}_1\mathbf{D}_2) \neq 0$. Hence a feedback system may have the transfer-function matrix description without having the state-variable description.

## *3-7 Discrete-Time Systems

The inputs and outputs of the systems we studied in the previous sections are defined for all $t$ in $(-\infty, \infty)$ for the time-varying case or in $[0, \infty)$ for the time-invariant case. They are called continuous-time systems. In this section we shall study a different class of systems, called discrete-time systems. The inputs and outputs of discrete-time systems are defined only at discrete instants of time. For example, a digital computer reads and prints out data that are the values of variables at discrete instants of time, hence it is a discrete-time system. A continuous-time system can also be modeled as a discrete-time system if its responses are of interest or measurable only at certain instants of time.

For convenience, the discrete instants of time at which the input and the output appear will be assumed to be equally spaced and will be denoted by the set of integers ranging from $-\infty$ to $\infty$. The input and the output are sequences and will be denoted by $\{u(n)\}$ and $\{y(n)\}$, where $n$ stands for an integer. In the following, we discuss only single-variable discrete-time systems. As in the continuous-time system, a discrete-time system that is initially relaxed is called a *relaxed discrete-time system*. If the inputs and outputs of a relaxed discrete-time system satisfy the linearity property, then they can be related by

$$y(n) = \sum_{-\infty}^{\infty} g(n, m)u(m) \tag{3-60}$$

where $g(n, m)$ is called the *weighting sequence* and is the response of the system due to the application of the input

$$\begin{aligned} u(k) &= 1 \qquad k = m \\ &= 0 \qquad k \neq m \end{aligned}$$

provided the system is relaxed at time $m-$. If the discrete-time system is causal—that is, the output does not depend on the future values of the input—then $g(n, m) = 0$ for $n < m$. Consequently, if a discrete-time system is causal and is relaxed at $n_0$, then (3-60) reduces to

$$y(n) = \sum_{m=n_0}^{n} g(n, m)u(m) \tag{3-61}$$

If a linear causal relaxed discrete-time system is time-invariant, then we have $g(n, m) = g(n - m)$ for all $n \geq m$. In this case, the initial time is chosen to be $n_0 = 0$ and the set of time of interest is the set of positive integers. Hence for a linear time-invariant causal relaxed discrete-time system, we have

$$y(n) = \sum_{m=0}^{n} g(n - m)u(m) \qquad n = 0, 1, 2, \cdots \tag{3-62}$$

Comparing with continuous-time systems, we see that in discrete-time systems, we use summation instead of integration; otherwise all the concepts are the same. In the continuous-time case, if we apply the Laplace transform, a convolution integral can be transformed into an algebraic equation. We have the same situation here, the transformation which will be used is called the $z$-transform.

### Definition 3-7

The *z-transform* of a sequence

$$\{u(n)\}_{n=0}^{\infty}$$

is defined to be the function of the complex variable $z$ given by

$$\hat{u}(z) \triangleq \sum_{n=0}^{\infty} u(n)z^{-n}$$

### Example 1

If $u(n) = 1$ for $n = 0, 1, 2, \cdots$. Then

$$\hat{u}(z) = \sum_{n=0}^{\infty} z^{-n} = \frac{1}{1 - z^{-1}} = \frac{z}{z - 1}$$

### Example 2

If $u(n) = e^{-2n}$ for $n = 0, 1, 2, \cdots$. Then

$$u(z) = \sum_{n=0}^{\infty} e^{-2n}z^{-n} = \frac{1}{1 - e^{-2}z^{-1}} = \frac{z}{z - e^{-2}}$$ ∎

Now we shall apply the $z$-transform to (3-62). Since for a causal, relaxed system, we have $g(n - m) = 0$ for $n < m$; hence we may also write (3-62) as

$$y(n) = \sum_{m=0}^{\infty} g(n - m)u(m)$$

Consequently, we have

$$\hat{\mathbf{y}}(z) = \sum_{n=0}^{\infty} y(n)z^{-n} = \sum_{n=0}^{\infty}\sum_{m=0}^{\infty} g(n-m)u(m)z^{-(n-m)}z^{-m}$$
$$= \sum_{m=0}^{\infty}\sum_{n=0}^{\infty} g(n-m)z^{-(n-m)}u(m)z^{-m}$$
$$= \sum_{n=0}^{\infty} g(n)z^{-n} \sum_{m=0}^{\infty} u(m)z^{-m} = \hat{g}(z)\hat{u}(z) \tag{3-63}$$

Here we have changed the order of summations and used the fact that $g(n - m) = 0$ for $n < m$. The function $\hat{g}(z)$ is the $z$-transform of the weighting sequence $\{g(n)\}_{n=0}^{\infty}$ and is called the *z-transfer function* or *sampled transfer function.*

The extension of the input–output description of single-variable discrete-time systems to the multivariable case is straightforward and its discussion is omitted. We introduce now the discrete-time dynamical equation. *A linear, time varying, discrete-time dynamical equation* is defined as

$$DE: \quad \begin{aligned} \mathbf{x}(n+1) &= \mathbf{A}(n)\mathbf{x}(n) + \mathbf{B}(n)\mathbf{u}(n) \\ \mathbf{y}(n) &= \mathbf{C}(n)\mathbf{x}(n) + \mathbf{D}(n)\mathbf{u}(n) \end{aligned}$$

where $\mathbf{x}$ is the state vector, $\mathbf{u}$ the input, and $\mathbf{y}$ the output. Note that a discrete-time dynamical equation is a set of first order *difference* equations instead of a set of first-order *differential* equations, as in the continuous-time case.

If $\mathbf{A}(n)$, $\mathbf{B}(n)$, $\mathbf{C}(n)$, and $\mathbf{D}(n)$ are independent of $n$, then $DE$ reduces to

$$DFE: \quad \mathbf{x}(n+1) = \mathbf{A}\mathbf{x}(n) + \mathbf{B}\mathbf{u}(n) \tag{3-64a}$$
$$\mathbf{y}(n) = \mathbf{C}\mathbf{x}(n) + \mathbf{D}\mathbf{u}(n) \tag{3-64b}$$

which is called a *linear, time-invariant, discrete-time dynamical equation.* Let $\hat{\mathbf{x}}(z)$ be the $z$-transform of

$$\{\mathbf{x}(n)\}_{n=0}^{\infty}$$

That is,

$$\hat{\mathbf{x}}(z) \triangleq \mathcal{Z}[x(n)] \triangleq \sum_{n=0}^{\infty} x(n)z^{-n}$$

Then

$$\mathcal{Z}[\mathbf{x}(n+1)] = \sum_{n=0}^{\infty} \mathbf{x}(n+1)z^{-n} = z\sum_{n=0}^{\infty} \mathbf{x}(n+1)z^{-(n+1)}$$
$$= z\left[\sum_{n=-1}^{\infty} \mathbf{x}(n+1)z^{-(n+1)} - \mathbf{x}(0)\right] = z[\hat{\mathbf{x}}(z) - \mathbf{x}(0)]$$

Hence the application of the $z$-transform to (3-64) yields

$$z\hat{\mathbf{x}}(z) - z\mathbf{x}^0 = \mathbf{A}\hat{\mathbf{x}}(z) + \mathbf{B}\hat{\mathbf{u}}(z) \tag{3-65a}$$
$$\hat{\mathbf{y}}(z) = \mathbf{C}\hat{\mathbf{x}}(z) + \mathbf{D}\hat{\mathbf{u}}(z) \tag{3-65b}$$

where $\mathbf{x}(0) = \mathbf{x}^0$. Equation (3-65) can be arranged as

$$\hat{\mathbf{x}}(z) = (z\mathbf{I} - \mathbf{A})^{-1}z\mathbf{x}^0 + (z\mathbf{I} - \mathbf{A})^{-1}\mathbf{B}\hat{\mathbf{u}}(z) \tag{3-66}$$
$$\hat{\mathbf{y}}(z) = \mathbf{C}(z\mathbf{I} - \mathbf{A})^{-1}z\mathbf{x}^0 + \mathbf{C}(z\mathbf{I} - \mathbf{A})^{-1}\mathbf{B}\hat{\mathbf{u}}(z) + \mathbf{D}\hat{\mathbf{u}}(z) \tag{3-67}$$

They are algebraic equations. The manipulation of these equations is exactly the same as (3-33) and (3-34). If $\mathbf{x}(0) = \mathbf{0}$, then (3-67) reduces to

$$\hat{\mathbf{y}}(z) = [\mathbf{C}(z\mathbf{I} - \mathbf{A})^{-1}\mathbf{B} + \mathbf{D}]\hat{\mathbf{u}}(z) \triangleq \hat{\mathbf{G}}(z)\hat{\mathbf{u}}(z)$$

where

$$\hat{\mathbf{G}}(z) = \mathbf{C}(z\mathbf{I} - \mathbf{A})^{-1}\mathbf{B} + \mathbf{D} \tag{3-68}$$

is called the *sampled transfer-function matrix* of (3-64).

In this book, discrete-time systems will not be further discussed, because most of the concepts and results in the continuous-time case can be applied to the discrete-time case with only slight modifications. Therefore, those results that are applicable to the discrete-time case will be stated in the problems instead of discussed in the text.

## 3-8 Concluding Remarks

In this chapter we have developed systematically the input–output description and the state-variable description of linear systems. Although the input–output description can be obtained by analysis, it can also be developed from the measurements at the input and output terminals without knowing the internal structure of the system. For linear, time-invariant systems, the transfer-function description can also be used. Whenever transfer functions (including impedances and admittances in network theory) are used, the systems are implicitly assumed to be relaxed at $t = 0$.

The condition for a set of first-order differential equations $\dot{\mathbf{x}} = \mathbf{f}(\mathbf{x}, \mathbf{u}, t)$, $\mathbf{y} = \mathbf{g}(\mathbf{x}, \mathbf{u}, t)$—in particular, $\dot{\mathbf{x}} = \mathbf{A}(t)\mathbf{x} + \mathbf{B}(t)\mathbf{u}$, $\mathbf{y} = \mathbf{C}(t)\mathbf{x} + \mathbf{D}(t)\mathbf{u}$—to be qualified as a dynamical equation is that for any initial state $\mathbf{x}^0$ and any input $\mathbf{u}_{[t_0,\infty)}$, there is a *unique* solution $\mathbf{y}_{[t_0,\infty)}$ satisfying the equations. This uniqueness condition is essential in the study of solutions of a dynamical equation. The dynamical equations studied in this book are of the form $\dot{\mathbf{x}} = \mathbf{A}(t)\mathbf{x} + \mathbf{B}(t)\mathbf{u}$, $\mathbf{y} = \mathbf{C}(t)\mathbf{x} + \mathbf{D}(t)\mathbf{u}$. They can be extended to include derivatives of $\mathbf{u}$ in the output equation, such as

$$\mathbf{y}(t) = \mathbf{C}(t)\mathbf{x}(t) + \mathbf{D}(t)\mathbf{u} + \mathbf{D}_1(t)\dot{\mathbf{u}}(t) + \mathbf{D}_2(t)\ddot{\mathbf{u}}(t) + \cdots$$

However, these extensions are of limited interest in practice.

One may wonder why we study a set of first-order differential equations instead of studying high-order differential equations. The reasons are as follows: (1) every high-order differential equation can be written as a set of first-order differential equations; (2) the notations used to describe first-order equations are compact and very simple; and (3) first-order differential equations can be readily simulated on an analog or a digital computer.

Both the input–output description and the state-variable description studied in this chapter are useful in practice. Which description should be used depends on the problem, on the data available, and on the question asked.

## PROBLEMS

**3-1** Consider a linear relaxed system that has the input–output pair shown in Figure P3-1.

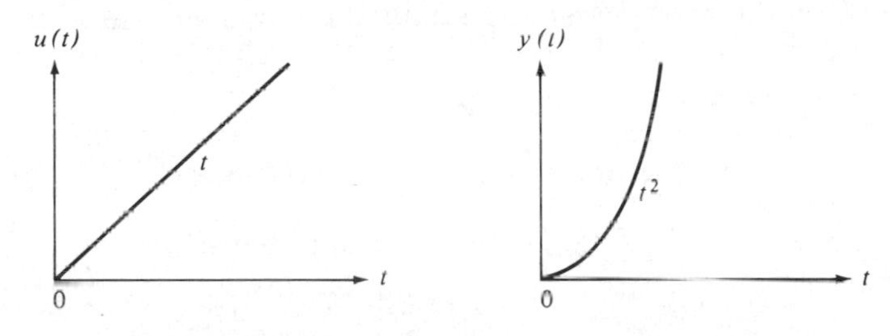


**Figure P3-1**

**a.** If we know that the relaxed system is time-invariant, is the system causal? What is its impulse response? What is its transfer function?
**b.** If we know that the relaxed system is time-varying, do we have enough information to conclude that the system is causal? What is its impulse response?

**3-2** We may define the impulse response of a nonlinear relaxed system $g(\cdot, \tau)$ as the response due to a $\delta$-function input applied at time $\tau$. Is this impulse response useful in describing the system? Why?

**3-3** The impulse response of a linear relaxed system is found to be $g(t, \tau) = e^{-|t-\tau|}$ for all $t$ and $\tau$. Is this system causal? Is it time-invariant?

**3-4** The impulse response of an ideal low-pass filter is given by

$$g(t) = 2\omega \frac{\sin 2\omega(t - t_0)}{2\omega(t - t_0)} \qquad \text{for all } t$$

where $\omega$ and $t_0$ are constants. Is the ideal low-pass filter causal? Is it possible to build an ideal low-pass filter in the real world?

**3-5** Consider a relaxed system whose input and output are related by

$$\mathbf{y}(t) = (P_\alpha \mathbf{u})(t) \triangleq \mathbf{u}(t) \qquad \text{for } t \leq \alpha$$
$$\triangleq \mathbf{0} \qquad \text{for } t > \alpha$$

for any $\mathbf{u}$, where $\alpha$ is a fixed constant. In words, this system chops off the input after time $\alpha$, and is called a *truncation operator*. Is this system linear? Is it time-invariant? Is it causal?

**3-6** Let $P_\alpha$ be the truncation operator defined in Problem 3-5. Show that for any causal relaxed system $\mathbf{y} = H\mathbf{u}$, we have

$$P_\alpha \mathbf{y} = P_\alpha H \mathbf{u} = P_\alpha H P_\alpha \mathbf{u}$$

This fact is often used in stability studies of nonlinear feedback systems.

**3-7** In Problem 3-6, is it true that

$$(P_\alpha \mathbf{y})(t) = (P_\alpha H \mathbf{u})(t) = (H P_\alpha \mathbf{u})(t)$$

for all $t$ in $(-\infty, \infty)$? If not, find the interval of time in which the equation holds.

**3-8** Show that for linear relaxed systems, if $\mathbf{u} \equiv \mathbf{0}$, then $\mathbf{y} \equiv \mathbf{0}$.

**3-9** Show that if $H(\mathbf{u}^1 + \mathbf{u}^2) = H\mathbf{u}^1 + H\mathbf{u}^2$ for any $\mathbf{u}^1$, $\mathbf{u}^2$, then $H\alpha\mathbf{u} = \alpha H\mathbf{u}$ for any rational number $\alpha$ and any $\mathbf{u}$.

**3-10** Show that for a fixed $\alpha$, the shifting operator $O_\alpha$ defined in Figure 3-5 is a linear time-invariant system. What is its impulse response? What is its transfer function?

**3-11** The transfer function of a single-variable, linear, time-invariant, causal, relaxed system can be found from one input–output pair of the system. [See Equation (3-22).] For a $p$-input, $q$-output, linear, time-invariant, causal, relaxed system, what is the minimal number of input–output pairs required to determine its transfer function matrix? *Hint:* See Theorem 2-2.

**3-12** Let $g(t, \tau) = g(t + \alpha, \tau + \alpha)$ for all $t$, $\tau$, and $\alpha$. Define $x = t + \tau$, $y = t - \tau$, then $g(t, \tau) = g\left(\frac{x+y}{2}, \frac{x-y}{2}\right)$. Show that $\frac{\partial g(t, \tau)}{\partial x} = 0$.

(From this fact we may conclude that if $g(t, \tau) = g(t + \alpha, \tau + \alpha)$ for all $t$, $\tau$, and $\alpha$, then $g(t, \tau)$ depends only on $t - \tau$.)

**3-13** Consider a relaxed system that is described by

$$y(t) = \int_0^t g(t - \tau)u(\tau)\, d\tau$$

If the impulse response $g$ is given by Figure P3-13(a), what is the output due to the input shown in Figure P3-13(b)? (Use graphical method.)

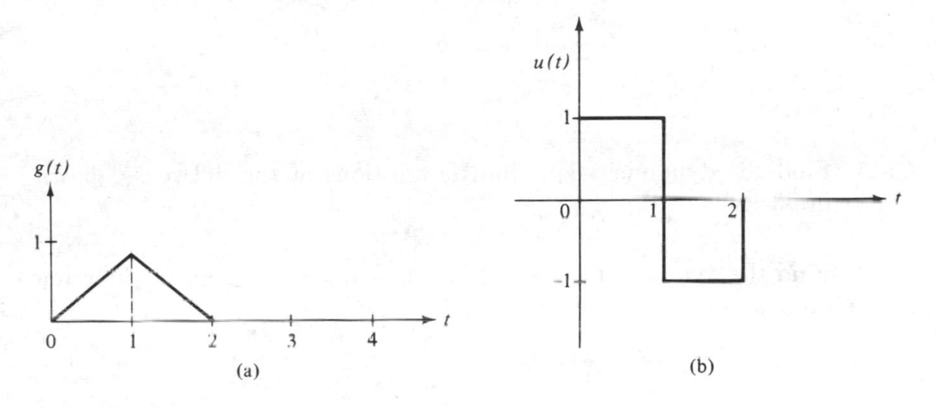

**Figure P3-13**

**3-14** The input $u$ and the output $y$ of a system is described by

$$\dddot{y} + 2\ddot{y} + 3y = 2\dot{u} + u$$

What is the transfer function of the system?

**3-15** Consider a multivariable system that is describable by $\hat{\mathbf{y}}(s) = \hat{\mathbf{G}}(s)\hat{\mathbf{u}}(s)$. Show that the $ij$th element of $\hat{\mathbf{G}}(s)$ can be defined as

$$\hat{g}_{ij}(s) = \left.\frac{\mathcal{L}[y_i(t)]}{\mathcal{L}[u_j(t)]}\right|_{\substack{\text{Initially relaxed} \\ \text{and } u_k = 0 \text{ for } k \neq j}}$$

where $y_i$ is the $i$th component of $\mathbf{y}$ and $u_j$ is the $j$th component of $\mathbf{u}$.

**3-16** Consider a multivariable system whose inputs and outputs are described by

$$\begin{aligned} N_{11}(p)y_1(t) + N_{12}(p)y_2(t) &= D_{11}(p)u_1(t) + D_{12}(p)u_2(t) \\ N_{21}(p)y_1(t) + N_{22}(p)y_2(t) &= D_{21}(p)u_1(t) + D_{22}(p)u_2(t) \end{aligned}$$

where the $N_{ij}$'s and $D_{ij}$'s are polynomials of $p \triangleq d/dt$. What is the transfer function matrix of the system?

**3-17** Find the dynamical-equation descriptions of the systems shown in Figure P3-17.

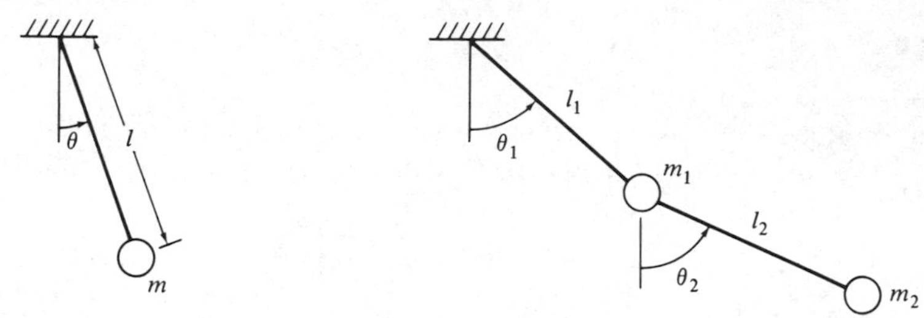


**Figure P3-17**

**3-18** Find the dynamical-equation descriptions of the networks shown in Figure 3-19.

**3-19** Find the dynamical-equation description and the transfer-function matrix of the network in Figure P3-19.

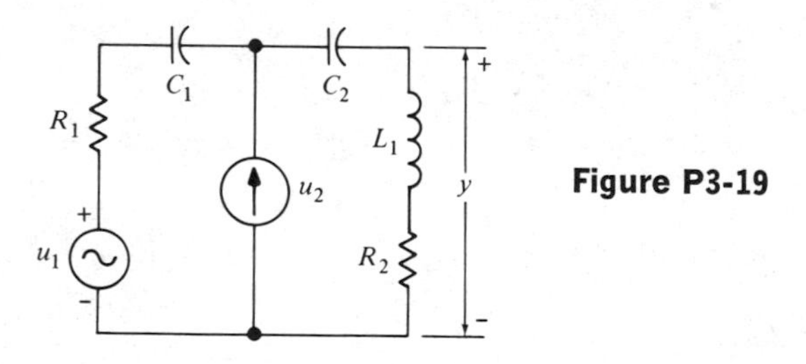


**Figure P3-19**

**3-20** Find the dynamical-equation description of the network in Figure P3-20.

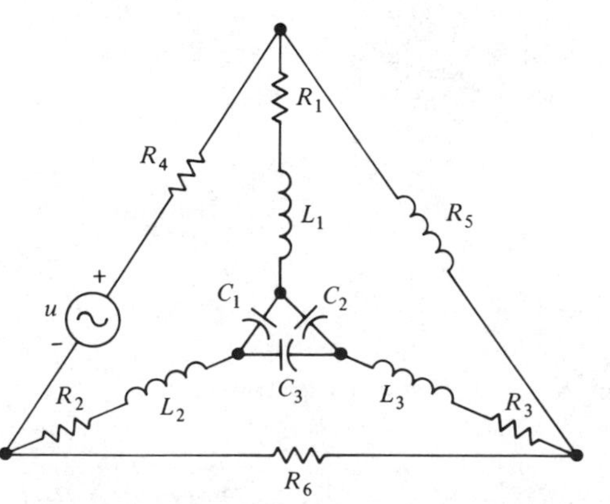


**Figure P3-20**

**3-21** Show that the output of a linear causal relaxed system due to the input $u(t)\delta_1(t - t_0)$ is given by

$$y(t) = u(t_0)g_1(t, t_0) + \int_{t_0}^{t} \dot{u}(\tau)g_1(t, \tau)\, d\tau \qquad \text{for } t \geq t_0$$

where $g_1(t, t_0)$ is the step response and $\delta_1$ is the step function defined in (3-43) and (3-44). *Hint:* This can be proved either by decomposing $u$ into a sum of step functions or by using (3-45).

**3-22** Draw block diagrams of computer simulations for the systems in Problems 3-17 and 3-19.

**3-23** Find the transfer function and the dynamical-equation description of the network in Figure P3-23. Do you think the transfer function is a good description of this system?

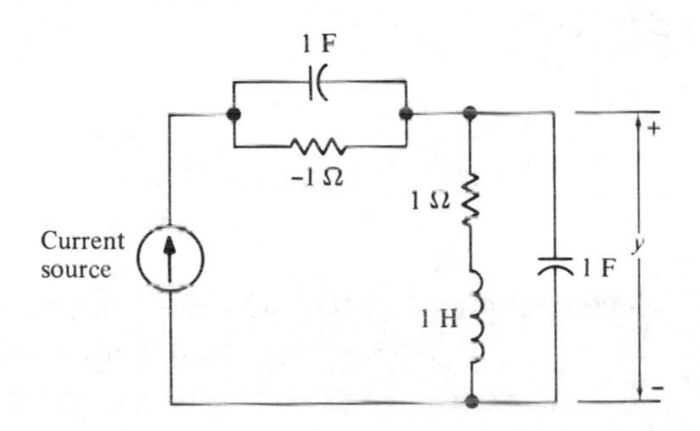


**Figure P3-23**

**3-24** Show that single-variable, linear, time-invariant, causal, relaxed systems commute in the sense that the order of the tandem connection of two systems is immaterial. Is this true for the time-varying systems?

**3-25** The impulse and step responses of a single-variable, linear time-invariant, causal, relaxed system are, by definition, given by $g = H\delta(t)$ and $g_1 = H\delta_1(t)$, where $\delta$ is a delta function and $\delta_1$ is a step function. It can be shown that $\delta(t) = \frac{d}{dt}\delta_1(t)$. Verify (3-46) by using the property given in Problem 3-24.

**3-26** Verify that the impulse-response matrix of a feedback system is given by Equation (3-50).

**3-27** Verify the identity

$$\mathbf{I} - \mathbf{D}_2(\mathbf{I} + \mathbf{D}_1\mathbf{D}_2)^{-1}\mathbf{D}_1 = (\mathbf{I} + \mathbf{D}_2\mathbf{D}_1)^{-1}$$

where $\mathbf{D}_1$ is a $q_1 \times p_1$ constant matrix and $\mathbf{D}_2$ is a $p_1 \times q_1$ constant matrix.

**3-28** Verify (3-54) and show that (3-54a) can be reduced to

$$\begin{bmatrix} \dot{\mathbf{x}}_1 \\ \dot{\mathbf{x}}_2 \end{bmatrix} = \begin{bmatrix} \mathbf{A}_1(t) - \mathbf{B}_1(t)\mathbf{D}_2(t)\mathbf{Y}_1(t)\mathbf{C}_1(t) & -\mathbf{B}_1(t)\mathbf{Y}_2(t)\mathbf{C}_2(t) \\ \mathbf{B}_2(t)\mathbf{Y}_1(t)\mathbf{C}_1(t) & \mathbf{A}_2(t) - \mathbf{B}_2\mathbf{Y}_1(t)\mathbf{D}_1(t)\mathbf{C}_2(t) \end{bmatrix} \begin{bmatrix} \mathbf{x}_1 \\ \mathbf{x}_2 \end{bmatrix} + \begin{bmatrix} \mathbf{B}_1(t)\mathbf{Y}_2(t) \\ \mathbf{B}_2(t)\mathbf{Y}_1(t)\mathbf{D}_1(t) \end{bmatrix} \mathbf{u}$$

where $\mathbf{Y}_2 = (\mathbf{I} + \mathbf{D}_2(t)\mathbf{D}_1(t))^{-1}$.

**3-29** Show that

$$\det \left( \mathbf{I} + \begin{bmatrix} a_1 \\ a_2 \\ \cdot \\ \cdot \\ \cdot \\ a_n \end{bmatrix} [b_1 \quad b_2 \quad \cdots \quad b_n] \right) = 1 + \sum_{i=1}^{n} a_i b_i$$

Note that the matrix in the left-hand side is an $n \times n$ matrix.

**3-30** Find the transfer-function matrix of the feedback system shown in Figure 3-20(c), where the transfer-function matrices of $S_1$ and $S_2$ are, respectively,

$$\hat{\mathbf{G}}_1 = \begin{bmatrix} \dfrac{1}{s+1} & \dfrac{1}{s+2} \\ 0 & \dfrac{s+1}{s+2} \end{bmatrix} \qquad \hat{\mathbf{G}}_2 = \begin{bmatrix} \dfrac{1}{s+3} & \dfrac{1}{s+4} \\ \dfrac{1}{s+1} & 0 \end{bmatrix}$$

**3-31** Find the dynamical-equation description of the feedback system in Figure 3-20(c), where $S_1$ and $S_2$ are, respectively, described by

$$E^1: \quad \begin{bmatrix} \dot{x}_{11} \\ \dot{x}_{12} \end{bmatrix} = \begin{bmatrix} -2 & 1 \\ 0 & -1 \end{bmatrix} \begin{bmatrix} x_{11} \\ x_{12} \end{bmatrix} + \begin{bmatrix} 4 & 1 \\ -1 & 2 \end{bmatrix} \mathbf{u}_1$$

$$\mathbf{y}_1 = [0 \quad 1]\mathbf{x}_1 + [1 \quad -1]\mathbf{u}_1$$

and

$$E^2: \quad \begin{bmatrix} \dot{x}_{21} \\ \dot{x}_{22} \end{bmatrix} = \begin{bmatrix} 2 \\ 1 \end{bmatrix} u_2$$

$$\mathbf{y}_2 = \begin{bmatrix} 2 & 0 \\ 1 & -1 \end{bmatrix} \mathbf{x}_2$$

Draw a block diagram of a computer simulation of this feedback system.

**3-32** Find the transfer function of the feedback system shown in Figure P3-32.

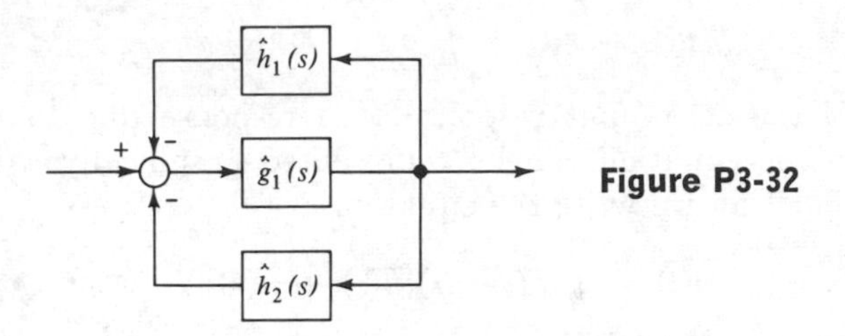


Figure P3-32

**3-33** A function $\mathbf{h}(\mathbf{x}(t), \mathbf{u}(t))$ is said to be a linear function of $\mathbf{x}(t)$ and $\mathbf{u}(t)$ if and only if

$$\alpha_1\mathbf{h}(\mathbf{x}^1(t), \mathbf{u}^1(t)) + \alpha_2\mathbf{h}(\mathbf{x}^2(t), \mathbf{u}^2(t)) = \mathbf{h}(\alpha_1\mathbf{x}^1(t) + \alpha_2\mathbf{x}^2(t), \alpha_1\mathbf{u}^1(t) + \alpha_2\mathbf{u}^2(t))$$

for any real numbers $\alpha_1$, $\alpha_2$, any $\mathbf{x}^1(t)$, $\mathbf{x}^2(t)$, and any $\mathbf{u}^1(t)$, $\mathbf{u}^2(t)$. Show that $\mathbf{h}(\mathbf{x}(t), \mathbf{u}(t))$ is a linear function of $\mathbf{x}(t)$ and $\mathbf{u}(t)$, if and only if $\mathbf{h}$ is of the form

$$\mathbf{h}(\mathbf{x}(t), \mathbf{u}(t)) = \mathbf{A}(t)\mathbf{x}(t) + \mathbf{B}(t)\mathbf{u}(t)$$

for some $\mathbf{A}$ and $\mathbf{B}$. If $\mathbf{h}$, $\mathbf{x}$, and $\mathbf{u}$ are square matrix functions instead of vector functions, does the assertion hold? If not, what modification do you need?

**3-34** The causality of a relaxed system may also be defined as follows: A relaxed system is causal if and only if $\mathbf{u}^1(t) = \mathbf{u}^2(t)$ for all $t \leq t_0$ implies $(H\mathbf{u}^1)(t) = (H\mathbf{u}^2)(t)$ for all $t \leq t_0$. Show that this definition implies that $\mathbf{y}(t) = H\mathbf{u}_{(-\infty,t]}$, and vice versa.

**3-35** Find the impedance (the transfer function from $u$ to $i$) of the network in Figure P3-35. If the initial conditions of the inductor and the capacitor are zero and if an input voltage $u(t) = e^{-t}$ is applied, what are $i(t)$, $i_1(t)$ and $i_2(t)$? Note that $i_1(t)$ and $i_2(t)$ contain some exponential functions that do not appear in $i(t)$. How do you explain this?

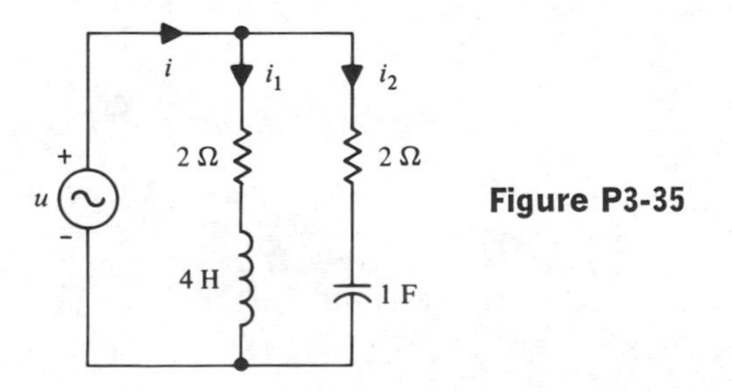


Figure P3-35

**3-36** The input–output description of a linear time-invariant system describes only the zero-state response (the response under the assumption that the initial state is zero). Hence if the initial state of the system is not zero, the total response may be assumed to be

$$y(t) = z(t) + \int_{t_0}^{t} g(t - \tau)u(\tau)\, d\tau$$

where $z(t)$ is the zero-input response (the response due to the initial condition). The equation can be represented graphically as shown in Figure P3-36. Can we write the equation as

$$y(t) = \int_{t_0}^{t} g(t - \tau)[u(\tau) + \tilde{u}(\tau)]\, d\tau$$

on the assumption that $z(t)$ is a response to some input? (See Problem 5-33.)

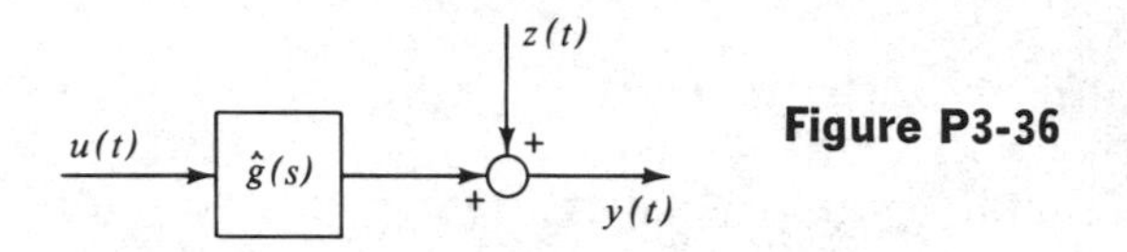


**Figure P3-36**

**3-37** Find the dynamical-equation and transfer-function descriptions of the systems shown in Figure P3-37.

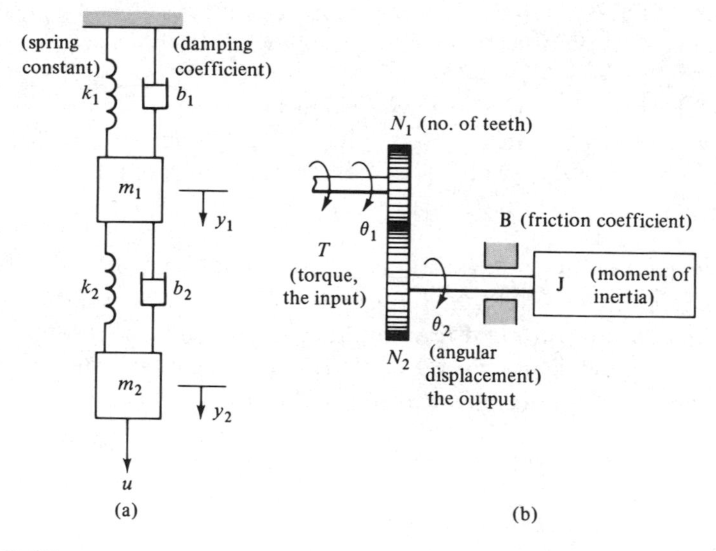


**Figure P3-37**

CHAPTER

# 4

# Linear Dynamical Equations and Impulse-Response Matrices

## 4-1 Introduction

In the preceding chapter we introduced the input–output description and the state-variable description of linear systems. We have also illustrated, by many examples, how these descriptions can be set up. The next step in the analysis obviously is to find the solutions of these equations. In the present chapter, this and other related problems will be studied.

If the impulse-response matrix $\mathbf{G}(t, \tau)$ of a system is known, then for any input the output $\mathbf{y}$ can be obtained from

$$\mathbf{y}(t) = \int_{t_0}^{t} \mathbf{G}(t, \tau)\mathbf{u}(\tau)\, d\tau$$

by direct computation or by a graphical method. In the time-invariant case, the equation $\hat{\mathbf{y}}(s) = \hat{\mathbf{G}}(s)\hat{\mathbf{u}}(s)$ can also be used. Unless the transfer-function matrix $\hat{\mathbf{G}}(s)$ is a rational matrix, generally it is easier to compute $\mathbf{y}$ directly in the time domain than from the frequency domain. It is elementary to compute $\mathbf{y}$ from the input–output description, and hence it will not be discussed further.

Solutions of linear dynamical equations are studied in Section 4-2. Solutions are stated in terms of the state transition matrix $\mathbf{\Phi}(t, \tau)$, which is the unique solution of

$$\frac{\partial}{\partial t}\mathbf{\Phi}(t, \tau) = \mathbf{A}(t)\mathbf{\Phi}(t, \tau) \qquad \mathbf{\Phi}(\tau, \tau) = \mathbf{I}$$

In the time-invariant case, we have $\mathbf{\Phi}(t, \tau) = e^{\mathbf{A}(t-\tau)}$. Various methods for the computation of $e^{\mathbf{A}t}$ and $(s\mathbf{I} - \mathbf{A})^{-1}$ are discussed. In Section 4-3 the concept of equivalent dynamical equations is introduced. Equivalent dynamical equations are obtained by changing the basis of the state space. We also show that every linear dynamical equation with a periodic matrix $\mathbf{A}$ has an equivalent linear dynamical equation that has a constant $\mathbf{A}$ matrix. This is the so-called *theory of Floquet.* In the last section, the relation between linear dynamical equations and impulse-response matrices are studied. The necessary and sufficient condition for an impulse-response matrix to be realizable by a linear dynamical equation is established. We also show that every proper rational matrix has a linear, time-invariant, dynamical-equation realization.

The references for this chapter are [24], [31], [60], [68], [77], [109], [114], and [116].

## 4-2 Solutions of a Dynamical Equation

**Time-varying case.** Consider the $n$-dimensional linear time-varying dynamical equation

$$E: \quad \begin{aligned} \dot{\mathbf{x}}(t) &= \mathbf{A}(t)\mathbf{x}(t) + \mathbf{B}(t)\mathbf{u}(t) \qquad \text{(state equation)} \qquad (4\text{-}1a)\\ \mathbf{y}(t) &= \mathbf{C}(t)\mathbf{x}(t) + \mathbf{D}(t)\mathbf{u}(t) \qquad \text{(output equation)} \qquad (4\text{-}1b)\end{aligned}$$

where $\mathbf{A}(\cdot)$, $\mathbf{B}(\cdot)$, $\mathbf{C}(\cdot)$, and $\mathbf{D}(\cdot)$ are $n \times n$, $n \times p$, $q \times n$, and $q \times p$ matrices whose entries are real-valued continuous functions of $t$ defined over $(-\infty, \infty)$. Since $\mathbf{A}(\cdot)$ is assumed to be continuous, for any initial state $\mathbf{x}(t_0)$ and any $\mathbf{u}$, there exists a unique solution in the dynamical equation $E$. This fact will be used frequently in the following development. Before studying the entire dynamical equation $E$, we study first the solutions of the homogeneous part of $E$; namely,

$$\dot{\mathbf{x}} = \mathbf{A}(t)\mathbf{x} \qquad (4\text{-}2$$

### Solutions of $\dot{\mathbf{x}} = \mathbf{A}(t)\mathbf{x}$

**Theorem 4-1**

The set of all solutions of $\dot{\mathbf{x}}(t) = \mathbf{A}(t)\mathbf{x}(t)$ forms an $n$-dimensional vector space over the field of real numbers.

Proof

Let $\psi^1$ and $\psi^2$ be two arbitrary solutions of (4-2). Then $\alpha_1\psi^1 + \alpha_2\psi^2$ is also a solution of (4-2) for any real $\alpha_1$ and $\alpha_2$. We prove this by direct verification:

$$\frac{d}{dt}(\alpha_1\psi^1 + \alpha_2\psi^2) = \alpha_1\frac{d}{dt}\psi^1 + \alpha_2\frac{d}{dt}\psi^2 = \alpha_1\mathbf{A}(t)\psi^1 + \alpha_2\mathbf{A}(t)\psi^2$$
$$= \mathbf{A}(t)(\alpha_1\psi^1 + \alpha_2\psi^2)$$

Hence the set of solutions form a linear space over $\mathbb{R}$. It is called the *solution space* of (4-2). We next show that the solution space has dimension $n$. Let $\mathbf{e}^1, \mathbf{e}^2, \cdots, \mathbf{e}^n$ be any linearly independent vectors in $(\mathbb{R}^n, \mathbb{R})$ and $\psi^i$ be the solutions of (4-2) with the initial condition $\psi^i(t_0) = \mathbf{e}^i$, for $i = 1, 2, \cdots, n$. If we show that $\psi^i$, for $i = 1, 2, \cdots, n$ are linearly independent and that every solution of (4-2) can be written as a linear combination of $\psi^i$, for $i = 1, 2, \cdots, n$, then the assertion is proved. We prove by contradiction the fact that the $\psi^i$'s are linearly independent. Suppose that $\psi^i$, for $i = 1, 2, \cdots, n$, are linearly dependent; then, by definition, there exists a nonzero $n \times 1$ real vector $\boldsymbol{\alpha}$ such that

$$[\psi^1 \quad \psi^2 \quad \cdots \quad \psi^n]\boldsymbol{\alpha} = \mathbf{0} \tag{4-3}$$

Note that the $\mathbf{0}$ in the right-hand side of (4-3) is the zero vector of the solution space; therefore, it is more informative to write (4-3) as

$$[\psi^1(t) \quad \psi^2(t) \quad \cdots \quad \psi^n(t)]\boldsymbol{\alpha} = \mathbf{0} \qquad \text{for all } t \text{ in } (-\infty, \infty)$$

In particular, we have

$$[\psi^1(t_0) \quad \psi^2(t_0) \quad \cdots \quad \psi^n(t_0)]\boldsymbol{\alpha} = [\mathbf{e}^1 \quad \mathbf{e}^2 \quad \cdots \quad \mathbf{e}^n]\boldsymbol{\alpha} = \mathbf{0}$$

which implies that $\mathbf{e}^i$, for $i = 1, 2, \cdots, n$, are linearly dependent. This contradicts the hypothesis; hence $\psi^i$, for $i = 1, 2, \cdots, n$ are linearly independent over $(-\infty, \infty)$.

Let $\psi$ be any solution of (4-2) and let $\psi(t_0) = \mathbf{e}$. Since $\mathbf{e}^1, \mathbf{e}^2, \cdots, \mathbf{e}^n$ are $n$ linearly independent vectors in the $n$-dimensional vector space $(\mathbb{R}^n, \mathbb{R})$, $\mathbf{e}$ can be written as a unique linear combination of $\mathbf{e}^i$, for $i = 1, 2, \cdots, n$—for example, as

$$\mathbf{e} = \sum_{i=1}^{n} \alpha_i \mathbf{e}^i$$

It is clear that

$$\sum_{i=1}^{n} \alpha_i \psi^i(\cdot)$$

is a solution of (4-2) with the initial condition

$$\sum_{i=1}^{n} \alpha_i \psi^i(t_0) = \mathbf{e}$$

Hence, from the uniqueness of the solution, we conclude that

$$\boldsymbol{\psi}(\cdot) = \sum_{i=1}^{n} \alpha_i \boldsymbol{\psi}^i(\cdot)$$

This completes the proof that the solutions of (4-2) form an $n$-dimensional vector space. Q.E.D.

### Definition 4-1

An $n \times n$ matrix function $\boldsymbol{\Psi}$ is said to be a *fundamental matrix* of $\dot{\mathbf{x}} = \mathbf{A}(t)\mathbf{x}$ if and only if the $n$ columns of $\boldsymbol{\Psi}$ consist of $n$ linearly independent solutions of $\dot{\mathbf{x}} = \mathbf{A}(t)\mathbf{x}$.

### Example 1

Consider the dynamical equation

$$\dot{\mathbf{x}} = \begin{bmatrix} 0 & 0 \\ t & 0 \end{bmatrix} \mathbf{x}$$

It actually consists of two equations: $\dot{x}_1 = 0$, $\dot{x}_2 = tx_1$. The solutions of these two systems are $x_1(t) = x_1(t_0)$ and $x_2(t) = 0.5t^2x_1(t_0) + x_2(t_0)$, which are obtained by first solving for $x_1(t)$ and then substituting $x_1(t)$ into $\dot{x}_2 = tx_1$. Now two linearly independent solutions $\boldsymbol{\psi}^1 = [0 \quad 1]'$ and $\boldsymbol{\psi}^2 = [2 \quad t^2]'$ can be easily obtained by setting $x_1(t_0) = 0$, $x_2(t_0) = 1$ and $x_1(t_0) = 2$, $x_2(t_0) = 0$. Hence, the matrix

$$\begin{bmatrix} 0 & 2 \\ 1 & t^2 \end{bmatrix}$$

is a fundamental matrix. ∎

Each column of $\boldsymbol{\Psi}$, by definition, satisfies the differential equation $\dot{\mathbf{x}} = \mathbf{A}(t)\mathbf{x}$; hence it is evident that $\boldsymbol{\Psi}$ satisfies the matrix equation

$$\dot{\boldsymbol{\Psi}} = \mathbf{A}(t)\boldsymbol{\Psi} \tag{4-4}$$

with $\boldsymbol{\Psi}(t_0) = \mathbf{E}$, where $\mathbf{E}$ is some nonsingular real constant matrix. Conversely, if a matrix $\mathbf{M}$ satisfies (4-4) and if $\mathbf{M}(t)$ is nonsingular for some $t$, then from the proof of Theorem 4-1 we know that all the columns of $\mathbf{M}$ are linearly independent. Hence the matrix function $\mathbf{M}$ qualifies as a fundamental matrix. Thus we conclude that *a matrix function* $\boldsymbol{\Psi}$ *is a fundamental matrix of* $\dot{\mathbf{x}} = \mathbf{A}(t)\mathbf{x}$ *if and only if* $\boldsymbol{\Psi}$ *satisfies* (4-4) *and* $\boldsymbol{\Psi}(t)$ *is nonsingular for some* $t$.

An important property of a fundamental matrix $\boldsymbol{\Psi}(\cdot)$ is that the inverse of $\boldsymbol{\Psi}(t)$ exists for each $t$ in $(-\infty, \infty)$. This follows from the following theorem.

### Theorem 4-2

Every fundamental matrix $\boldsymbol{\Psi}$ is nonsingular for all $t$ in $(-\infty, \infty)$.

### Proof

Before we prove the theorem, we need the following fact: If $\boldsymbol{\psi}(\cdot)$ is a solution of $\dot{\mathbf{x}} = \mathbf{A}(t)\mathbf{x}$ and if $\boldsymbol{\psi}(t_0) = \mathbf{0}$ for some $t_0$, then the solution $\boldsymbol{\psi}(\cdot)$ is identically zero; that is, $\boldsymbol{\psi}(\cdot) \equiv \mathbf{0}$. It is obvious that $\boldsymbol{\psi}(\cdot) \equiv \mathbf{0}$ is a solution of $\dot{\mathbf{x}} = \mathbf{A}(t)\mathbf{x}$ with $\boldsymbol{\psi}(t_0) = \mathbf{0}$. Again, from the uniqueness of the solution, we conclude that $\boldsymbol{\psi}(\cdot) \equiv \mathbf{0}$ is the only solution with $\boldsymbol{\psi}(t_0) = \mathbf{0}$.

We shall now prove the theorem; we prove it by contradiction. Suppose that $\det \boldsymbol{\Psi}(t_0) = \det [\boldsymbol{\psi}^1(t_0) \quad \boldsymbol{\psi}^2(t_0) \quad \cdots \quad \boldsymbol{\psi}^n(t_0)] = 0$ for some $t_0$. Then the set of $n$ constant column vectors $\boldsymbol{\psi}^1(t_0)$, $\boldsymbol{\psi}^2(t_0)$, $\cdots$, $\boldsymbol{\psi}^n(t_0)$ is linearly dependent in $(\mathbb{R}^n, \mathbb{R})$. It follows that there exist real $\alpha_i$, for $i = 1, 2, \cdots, n$, not all zero, such that

$$\sum_{i=1}^{n} \alpha_i \boldsymbol{\psi}^i(t_0) = \mathbf{0}$$

which, together with the fact that

$$\sum_{i=1}^{n} \alpha_i \boldsymbol{\psi}^i(\cdot)$$

is a solution of $\dot{\mathbf{x}} = \mathbf{A}(t)\mathbf{x}$, implies that

$$\sum_{i=1}^{n} \alpha_i \boldsymbol{\psi}^i(\cdot) \equiv \mathbf{0}$$

This contradicts the assumption that $\boldsymbol{\psi}^i(\cdot)$, for $i = 1, 2, \cdots, n$, are linearly independent. Hence we conclude that $\det \boldsymbol{\Psi}(t) \neq 0$ for all $t$ in $(-\infty, \infty)$. Q.E.D.

### Definition 4-2

Let $\boldsymbol{\Psi}(\cdot)$ be *any* fundamental matrix of $\dot{\mathbf{x}} = \mathbf{A}(t)\mathbf{x}$. Then

$$\boldsymbol{\Phi}(t, t_0) \triangleq \boldsymbol{\Psi}(t)\boldsymbol{\Psi}^{-1}(t_0) \qquad \text{for all } t, t_0 \text{ in } (-\infty, \infty)$$

is said to be *the state transition matrix* of $\dot{\mathbf{x}} = \mathbf{A}(t)\mathbf{x}$. ▮

The physical meaning of $\boldsymbol{\Phi}(t, t_0)$ will be seen later. Since $\boldsymbol{\Psi}(t)$ is nonsingular for all $t$, its inverse is well-defined for each $t$. From the definition we have immediately the following very important properties of the state transition matrix:

$$\boldsymbol{\Phi}(t, t) = \mathbf{I} \tag{4-5}$$

$$\boldsymbol{\Phi}^{-1}(t, t_0) = \boldsymbol{\Psi}(t_0)\boldsymbol{\Psi}^{-1}(t) = \boldsymbol{\Phi}(t_0, t) \tag{4-6}$$

$$\boldsymbol{\Phi}(t_2, t_0) = \boldsymbol{\Phi}(t_2, t_1)\boldsymbol{\Phi}(t_1, t_0) \tag{4-7}$$

for any $t$, $t_0$, $t_1$, and $t_2$ in $(-\infty, \infty)$.

Note that $\mathbf{\Phi}(t, t_0)$ is uniquely determined by $\mathbf{A}(t)$ and is independent of the particular $\mathbf{\Psi}$ chosen. Let $\mathbf{\Psi}^1$ and $\mathbf{\Psi}^2$ be two different fundamental matrices of $\dot{\mathbf{x}} = \mathbf{A}(t)\mathbf{x}$. Since the columns of $\mathbf{\Psi}^1$, as well as the columns of $\mathbf{\Psi}^2$, qualify as basis vectors, we have shown in Section 2-3 that there exists a nonsingular real constant matrix $\mathbf{P}$ such that $\mathbf{\Psi}^2 = \mathbf{\Psi}^1\mathbf{P}$. In fact, the $i$th column of $\mathbf{P}$ is the representation of the $i$th column of $\mathbf{\Psi}^2$ with respect to the basis that consists of the columns of $\mathbf{\Psi}^1(\cdot)$. By definition, we have

$$\begin{aligned}\mathbf{\Phi}(t, t_0) &= \mathbf{\Psi}^2(t)(\mathbf{\Psi}^2(t_0))^{-1} = \mathbf{\Psi}^1(t)\mathbf{P}\mathbf{P}^{-1}(\mathbf{\Psi}^1(t_0))^{-1} \\ &= \mathbf{\Psi}^1(t)(\mathbf{\Psi}^1(t_0))^{-1}\end{aligned}$$

which shows the uniqueness of $\mathbf{\Phi}(t, t_0)$. From Equation (4-4), it is evident that $\mathbf{\Phi}(t, t_0)$ is the unique solution of the matrix equation

$$\frac{\partial}{\partial t}\mathbf{\Phi}(t, t_0) = \mathbf{A}(t)\mathbf{\Phi}(t, t_0) \tag{4-8}$$

with the initial condition $\mathbf{\Phi}(t_0, t_0) = \mathbf{I}$.

Remarks are in order concerning the solutions of (4-4) and (4-8). If $\mathbf{A}(t)$ is a continuous function of $t$, then $\mathbf{\Phi}(t, t_0)$ and $\mathbf{\Psi}(t)$ are continuously differentiable[1] in $t$. More generally, if $\mathbf{A}(t)$ is $n$ times continuously differentiable in $t$, then $\mathbf{\Phi}(t, t_0)$ and $\mathbf{\Psi}(t)$ are $(n + 1)$ times continuously differentiable in $t$; see References [24] and [77]. If $\mathbf{A}(t)$ and $\int_{t_0}^{t}\mathbf{A}(\tau)\,d\tau$ commute for all $t$, then the unique solution of (4-8) is given by

$$\mathbf{\Phi}(t, t_0) = \exp\left[\int_{t_0}^{t}\mathbf{A}(\tau)\,d\tau\right] \tag{4-9}$$

(See Problem 4-30). It is clear that if $\mathbf{A}(t)$ is a diagonal matrix or a constant matrix, then we have

$$\mathbf{A}(t)\left(\int_{t_0}^{t}\mathbf{A}(\tau)\,d\tau\right) = \left(\int_{t_0}^{t}\mathbf{A}(\tau)\,d\tau\right)\mathbf{A}(t)$$

However, most of the time-varying matrices do not have this commutativity property and (4-9) does not hold. In this case, there is no simple relation between $\mathbf{A}(t)$ and $\mathbf{\Phi}(t, t_0)$, and solving for $\mathbf{\Phi}$ is generally a very difficult task.

From the concept of state transition matrix, the solution of $\dot{\mathbf{x}} = \mathbf{A}(t)\mathbf{x}$ follows immediately. To be more informative, we use $\boldsymbol{\phi}(t; t_0, \mathbf{x}^0, \mathbf{0})$ to denote the solution of $\dot{\mathbf{x}} = \mathbf{A}(t)\mathbf{x}$ at time $t$ due to the initial condition $\mathbf{x}(t_0) = \mathbf{x}^0$. The fourth argument of $\boldsymbol{\phi}$ denotes the fact that $\mathbf{u} \equiv \mathbf{0}$. The solution of $\dot{\mathbf{x}} = \mathbf{A}(t)\mathbf{x}$ with $\mathbf{x}(t_0) = \mathbf{x}^0$ is given by

$$\mathbf{x}(t) \triangleq \boldsymbol{\phi}(t; t_0, \mathbf{x}^0, \mathbf{0}) = \mathbf{\Phi}(t, t_0)\mathbf{x}^0$$

[1] A function is said to be continuously differentiable if its first derivative exists and is continuous.

which can be verified by direct substitution. The physical meaning of the state transition matrix $\mathbf{\Phi}(t, t_0)$ is now clear. It governs the motion of the state vector in the time interval in which the input is identically zero. $\mathbf{\Phi}(t, t_0)$ is a linear transformation that maps the state $\mathbf{x}^0$ at $t_0$ into the state $\mathbf{x}$ at time $t$.

### Solutions of the dynamical equation *E*

We use $\boldsymbol{\phi}(t; t_0, \mathbf{x}^0, \mathbf{u})$ to denote the state resulted at time $t$ due to the initial state $\mathbf{x}(t_0) = \mathbf{x}^0$ and the application of the input $\mathbf{u}$.

### Theorem 4-3

The solution of the state equation

$$\dot{\mathbf{x}} = \mathbf{A}(t)\mathbf{x} + \mathbf{B}(t)\mathbf{u} \qquad \mathbf{x}(t_0) = \mathbf{x}^0$$

is given by

$$\mathbf{x}(t) \triangleq \boldsymbol{\phi}(t; t_0, \mathbf{x}^0, \mathbf{u}) = \mathbf{\Phi}(t, t_0)\mathbf{x}^0 + \int_{t_0}^{t} \mathbf{\Phi}(t, \tau)\mathbf{B}(\tau)\mathbf{u}(\tau)\, d\tau \tag{4-10}$$

$$= \mathbf{\Phi}(t, t_0)\left[\mathbf{x}^0 + \int_{t_0}^{t} \mathbf{\Phi}(t_0, \tau)\mathbf{B}(\tau)\mathbf{u}(\tau)\, d\tau\right] \tag{4-11}$$

where $\mathbf{\Phi}(t, \tau)$ is the state transition matrix of $\dot{\mathbf{x}} = \mathbf{A}(t)\mathbf{x}$; or, correspondingly, the unique solution of

$$\frac{\partial}{\partial t}\mathbf{\Phi}(t, \tau) = \mathbf{A}(t)\mathbf{\Phi}(t, \tau), \quad \mathbf{\Phi}(\tau, \tau) = \mathbf{I} \qquad \blacksquare$$

### Proof

Equation (4-11) is obtained from (4-10) by using $\mathbf{\Phi}(t, \tau) = \mathbf{\Phi}(t, t_0)\mathbf{\Phi}(t_0, \tau)$. We show that (4-10) is a solution by direct substitution[2]:

$$\begin{aligned}
\frac{d}{dt}\mathbf{x}(t) &= \frac{\partial}{\partial t}\mathbf{\Phi}(t, t_0)\mathbf{x}^0 + \frac{\partial}{\partial t}\int_{t_0}^{t} \mathbf{\Phi}(t, \tau)\mathbf{B}(\tau)\mathbf{u}(\tau)\, d\tau \\
&= \mathbf{A}(t)\mathbf{\Phi}(t, t_0)\mathbf{x}^0 + \mathbf{\Phi}(t, t)\mathbf{B}(t)\mathbf{u}(t) + \int_{t_0}^{t} \frac{\partial}{\partial t}\mathbf{\Phi}(t, \tau)\mathbf{B}(\tau)\mathbf{u}(\tau)\, d\tau \\
&= \mathbf{A}(t)\left[\mathbf{\Phi}(t, t_0)\mathbf{x}^0 + \int_{t_0}^{t} \mathbf{\Phi}(t, \tau)\mathbf{B}(\tau)\mathbf{u}(\tau)\, d\tau\right] + \mathbf{B}(t)\mathbf{u}(t) \\
&= \mathbf{A}(t)\mathbf{x}(t) + \mathbf{B}(t)\mathbf{u}(t)
\end{aligned}$$

Q.E.D.

We consider again the solution given by Equation (4-10). If $\mathbf{u} \equiv \mathbf{0}$, then (4-10) reduces to

$$\boldsymbol{\phi}(t; t_0, \mathbf{x}^0, \mathbf{0}) = \mathbf{\Phi}(t, t_0)\mathbf{x}^0 \tag{4-12}$$

---

[2] $\displaystyle \frac{\partial}{\partial t}\int_{t_0}^{t} f(t, \tau)\, d\tau = f(t, \tau)\Big|_{\tau = t} + \int_{t_0}^{t} \frac{\partial}{\partial t} f(t, \tau)\, d\tau$

If $\mathbf{x}^0 = \mathbf{0}$, Equation (4-10) reduces to

$$\boldsymbol{\phi}(t; t_0, \mathbf{0}, \mathbf{u}) = \int_{t_0}^{t} \boldsymbol{\Phi}(t, \tau)\mathbf{B}(\tau)\mathbf{u}(\tau)\, d\tau \tag{4-13}$$

For obvious reasons, $\boldsymbol{\phi}(t; t_0, \mathbf{x}^0, \mathbf{0})$ is called the *zero-input response,* and $\boldsymbol{\phi}(t; t_0, \mathbf{0}, \mathbf{u})$ is called the *zero-state response* of the state equation. It is clear that $\boldsymbol{\phi}(t; t_0, \mathbf{x}^0, \mathbf{0})$ and $\boldsymbol{\phi}(t; t_0, \mathbf{0}, \mathbf{u})$ are linear functions of $\mathbf{x}^0$ and $\mathbf{u}$, respectively. Using (4-12) and (4-13), the solution given by Equation (4-10) can be written as

$$\boldsymbol{\phi}(t; t_0, \mathbf{x}^0, \mathbf{u}) = \boldsymbol{\phi}(t; t_0, \mathbf{x}^0, \mathbf{0}) + \boldsymbol{\phi}(t; t_0, \mathbf{0}, \mathbf{u}) \tag{4-14}$$

This is a very important property; it says that *the response of a linear state equation can always be decomposed into the zero-state response and the zero-input response.*

Note that Equation (4-13) can be derived directly from the fact that it is a linear function of $\mathbf{u}$. The procedure is exactly the same as the one in deriving

$$\int_{t_0}^{t} \mathbf{G}(t, \tau)\mathbf{u}(\tau)\, d\tau$$

in Section 3-2. The response $\boldsymbol{\phi}(t; t_0, \mathbf{0}, \mathbf{u})$ is, by definition, the solution of $\dot{\mathbf{x}} = \mathbf{A}(t)\mathbf{x} + \mathbf{B}(t)\mathbf{u}$ with $\mathbf{0}$ as the initial state. If we cut the input $\mathbf{u}$ into small pulses, say

$$\mathbf{u} = \sum_i \mathbf{u}_{[t_i, t_i+\Delta)}$$

then we have

$$\boldsymbol{\phi}(t; t_0, \mathbf{0}, \mathbf{u}_{[t_i, t_i+\Delta)}) \doteqdot \boldsymbol{\Phi}(t, t_i + \Delta)\mathbf{B}(t_i)\mathbf{u}(t_i)\Delta \tag{4-15}$$

where we have used the fact that if $\Delta$ is very small, the solution of $\dot{\mathbf{x}} = \mathbf{A}(t)\mathbf{x} + \mathbf{B}(t)\mathbf{u}$ due to the input $\mathbf{u}_{[t_i, t_i+\Delta)}$ with $\mathbf{0}$ as the initial state is approximately equal to $\mathbf{B}(t_i)\mathbf{u}(t_i)\Delta$. The input $\mathbf{u}_{[t_i, t_i+\Delta)}$ outside the time interval $[t_i, t_i + \Delta)$ is identically zero; hence, the response between $t_i + \Delta$ and $t$ is governed by $\boldsymbol{\Phi}(t, t_i + \Delta)$. Summing up (4-15) for all $i$ and taking the limit $\Delta \rightarrow 0$, we immediately obtain the equation

$$\boldsymbol{\phi}(t; t_0, \mathbf{0}, \mathbf{u}) = \int_{t_0}^{t} \boldsymbol{\Phi}(t, \tau)\mathbf{B}(\tau)\mathbf{u}(\tau)\, d\tau$$

We give now the solution of the entire dynamical equation $E$.

### Corollary 4.3

The solution of the dynamical equation $E$ in (4-1) is given by

$$\begin{aligned}\mathbf{y}(t) &= \mathbf{C}(t)\boldsymbol{\Phi}(t, t_0)\mathbf{x}^0 + \mathbf{C}(t)\int_{t_0}^{t} \boldsymbol{\Phi}(t, \tau)\mathbf{B}(\tau)\mathbf{u}(\tau)\, d\tau + \mathbf{D}(t)\mathbf{u}(t) \\ &= \mathbf{C}(t)\boldsymbol{\Phi}(t, t_0)\left[\mathbf{x}^0 + \int_{t_0}^{t} \boldsymbol{\Phi}(t_0, \tau)\mathbf{B}(\tau)\mathbf{u}(\tau)\, d\tau\right] + \mathbf{D}(t)\mathbf{u}(t)\end{aligned} \tag{4-16}$$

∎

By substituting (4-10) and (4-11) into (4-1b), we immediately obtain Corollary 4-3. The output $\mathbf{y}$ can also be decomposed into the zero-state response and the zero-input response. If the dynamical equation is in the zero state, Equation (4-16) becomes

$$\begin{aligned} \mathbf{y}(t) &= \int_{t_0}^{t} [\mathbf{C}(t)\mathbf{\Phi}(t, \tau)\mathbf{B}(\tau) + \mathbf{D}(t)\delta(t - \tau)]\mathbf{u}(\tau)\, d\tau \\ &\triangleq \int_{t_0}^{t} \mathbf{G}(t, \tau)\mathbf{u}(\tau)\, d\tau \end{aligned} \tag{4-17}$$

The matrix function

$$\mathbf{G}(t, \tau) \triangleq \mathbf{C}(t)\mathbf{\Phi}(t, \tau)\mathbf{B}(\tau) + \mathbf{D}(t)\delta(t - \tau) \tag{4-18}$$

is called the *impulse-response matrix* of the dynamical equation $E$. It governs the input–output relation of $E$ if $E$ is initially in the zero state.

We see from (4-16) that if the state transition matrix $\mathbf{\Phi}(t, \tau)$ is known, then the solution of the dynamical equation can be computed. Recall that $\mathbf{\Phi}(t, \tau)$ is the unique solution of

$$\frac{\partial}{\partial t}\mathbf{\Phi}(t, \tau) = \mathbf{A}(t)\mathbf{\Phi}(t, \tau), \quad \mathbf{\Phi}(\tau, \tau) = \mathbf{I}$$

Unfortunately, there is in general no simple relation between $\mathbf{\Phi}(t, \tau)$ and $\mathbf{A}(t)$, and except for very simple cases, state transition matrices $\mathbf{\Phi}(t, \tau)$ cannot be easily found. Therefore, Equations (4-10) and (4-16) are used mainly in the theoretical study of linear system theory. If we are required to find the solution of a dynamical equation due to a given $\mathbf{x}^0$ and $\mathbf{u}$, it is not necessary to compute $\mathbf{\Phi}(t, \tau)$. The solution can be easily obtained by direct integration, using a digital computer.

**Time-invariant case.** In this subsection, we study the linear, time-invariant (fixed) dynamical equation

$$FE: \quad \dot{\mathbf{x}} = \mathbf{A}\mathbf{x} + \mathbf{B}\mathbf{u} \tag{4-19a}$$

$$\mathbf{y} = \mathbf{C}\mathbf{x} + \mathbf{D}\mathbf{u} \tag{4-19b}$$

where $\mathbf{A}$, $\mathbf{B}$, $\mathbf{C}$, and $\mathbf{D}$ are $n \times n$, $n \times p$, $q \times n$, and $q \times p$ real constant matrices, respectively. Since the equation $FE$ is a special case of the linear, time-varying, dynamical equation $E$, all the results derived in the preceding subsection can be applied here. We have shown in (2-82) that

$$\frac{d}{dt}e^{\mathbf{A}t} = \mathbf{A}e^{\mathbf{A}t}$$

and since $e^{\mathbf{A}t}$ is nonsingular at $t = 0$, hence $e^{\mathbf{A}t}$ is a fundamental matrix of $\dot{\mathbf{x}} = \mathbf{A}\mathbf{x}$. In fact, $e^{\mathbf{A}t}$ is nonsingular for all $t$ and $(e^{\mathbf{A}t})^{-1} = e^{-\mathbf{A}t}$; therefore, the state transition matrix of $\dot{\mathbf{x}} = \mathbf{A}\mathbf{x}$ is, by the use of (2-81) and (2-79),

$$\mathbf{\Phi}(t, t_0) = e^{\mathbf{A}t}(e^{\mathbf{A}t_0})^{-1} = e^{\mathbf{A}(t-t_0)} = \mathbf{\Phi}(t - t_0)$$

It follows from (4-10) that the solution of (4-19a) is

$$\boldsymbol{\phi}(t; t_0, \mathbf{x}^0, \mathbf{u}) = e^{\mathbf{A}(t-t_0)}\mathbf{x}^0 + \int_{t_0}^{t} e^{\mathbf{A}(t-\tau)}\mathbf{B}\mathbf{u}(\tau)\, d\tau \tag{4-20}$$

If $t_0 = 0$, as is usually assumed in the time-invariant equation, then we have the following theorem.

### Theorem 4-4

The solution of the linear, time-invariant dynamical equation $FE$ given in (4-19) is

$$\mathbf{x}(t) \triangleq \boldsymbol{\phi}(t; 0, \mathbf{x}^0, \mathbf{u}) = e^{\mathbf{A}t}\mathbf{x}^0 + \int_0^t e^{\mathbf{A}(t-\tau)}\mathbf{B}\mathbf{u}(\tau)\, d\tau \tag{4-21}$$

and

$$\mathbf{y}(t) = \mathbf{C}e^{\mathbf{A}t}\mathbf{x}^0 + \mathbf{C}e^{\mathbf{A}t}\int_0^t e^{-\mathbf{A}\tau}\mathbf{B}\mathbf{u}(\tau)\, d\tau + \mathbf{D}\mathbf{u}(t) \tag{4-22}$$

∎

The impulse response matrix of $FE$ is

$$\mathbf{G}(t, \tau) = \mathbf{G}(t - \tau) = \mathbf{C}e^{\mathbf{A}(t-\tau)}\mathbf{B} + \mathbf{D}\delta(t - \tau)$$

or, as more commonly written,

$$\mathbf{G}(t) = \mathbf{C}e^{\mathbf{A}t}\mathbf{B} + \mathbf{D}\delta(t) \tag{4-23}$$

The solution of a linear, time-invariant dynamical equation can also be computed in the frequency domain. Taking the Laplace transform of (4-21) and (4-22), and using $\mathcal{L}[e^{\mathbf{A}t}] = (s\mathbf{I} - \mathbf{A})^{-1}$ [see (2-86)], we obtain

$$\hat{\mathbf{x}}(s) = (s\mathbf{I} - \mathbf{A})^{-1}\mathbf{x}(0) + (s\mathbf{I} - \mathbf{A})^{-1}\mathbf{B}\hat{\mathbf{u}}(s) \tag{4-24}$$

and

$$\hat{\mathbf{y}}(s) = \mathbf{C}(s\mathbf{I} - \mathbf{A})^{-1}\mathbf{x}(0) + \mathbf{C}(s\mathbf{I} - \mathbf{A})^{-1}\mathbf{B}\hat{\mathbf{u}}(s) + \mathbf{D}\hat{\mathbf{u}}(s) \tag{4-25}$$

where the circumflex denotes the Laplace transform of a variable. These equations have been derived in Section 3-3 directly from the dynamical equation. As is defined there, the rational-function matrix

$$\hat{\mathbf{G}}(s) = \mathbf{C}(s\mathbf{I} - \mathbf{A})^{-1}\mathbf{B} + \mathbf{D} \tag{4-26}$$

is called the *transfer-function matrix* of the dynamical equation $FE$. It is the Laplace transform of the impulse-response matrix given in (4-23). The transfer-function matrix governs the zero-state response of the equation $FE$.

We give now some remarks concerning the computation of $e^{\mathbf{A}t}$. We have introduced in Section 2-7 three methods of computing functions of a matrix. We may apply them to compute $e^{\mathbf{A}t}$: (1) Using Definition 2-16: First, compute the eigenvalues of $\mathbf{A}$; next, find a polynomial $g(\lambda)$ of degree $n - 1$ that is equal to $e^{\lambda t}$ on the spectrum of $\mathbf{A}$, then $e^{\mathbf{A}t} = g(\mathbf{A})$. (2) Using the Jordan canonical form of $\mathbf{A}$: Let $\mathbf{A} = \mathbf{Q}\hat{\mathbf{A}}\mathbf{Q}^{-1}$; then $e^{\mathbf{A}t} =$

$\mathbf{Q}e^{\hat{\mathbf{A}}t}\mathbf{Q}^{-1}$ where $\hat{\mathbf{A}}$ is of the Jordan form. $e^{\hat{\mathbf{A}}t}$ can be obtained immediately by the use of (2-69). (3) Using the infinite series $e^{\mathbf{A}t} = \sum_{k=0}^{\infty} t^k\mathbf{A}^k/k!$: This series will not, generally, give a closed-form solution and is mainly used in digital computer computation. We introduce one more method of computing $e^{\mathbf{A}t}$. Since $\mathcal{L}[e^{\mathbf{A}t}] = (s\mathbf{I} - \mathbf{A})^{-1}$, we have

$$e^{\mathbf{A}t} = \mathcal{L}^{-1}(s\mathbf{I} - \mathbf{A})^{-1} \tag{4-27}$$

Hence, to compute $e^{\mathbf{A}t}$, we first invert the matrix $(s\mathbf{I} - \mathbf{A})$ and then take the inverse Laplace transform of each element of $(s\mathbf{I} - \mathbf{A})^{-1}$. Computing the inverse of a matrix is generally not an easy job. However, if a matrix is triangular[3] or with order less than 4, its inverse can be easily computed. Note that the inverse of a triangular matrix is again a triangular matrix.

Note that $(s\mathbf{I} - \mathbf{A})^{-1}$ is a function of the matrix $\mathbf{A}$; therefore, again we have many methods to compute it: (1) Taking the inverse of $(s\mathbf{I} - \mathbf{A})$. (2) Using Definition 2-16. (3) Using $(s\mathbf{I} - \mathbf{A})^{-1} = \mathbf{Q}(s\mathbf{I} - \hat{\mathbf{A}})^{-1}\mathbf{Q}^{-1}$ and (2-74). (4) Using Definition 2-17. (5) Taking the Laplace transform of $e^{\mathbf{A}t}$. In addition, there is an iterative scheme to compute $(s\mathbf{I} - \mathbf{A})^{-1}$ (Problem 4-15).

### Example 2

We use methods 1 and 2 to compute $(s\mathbf{I} - \mathbf{A})^{-1}$, where

$$\mathbf{A} = \begin{bmatrix} 0 & -1 \\ 1 & -2 \end{bmatrix}$$

**1.** $$(s\mathbf{I} - \mathbf{A})^{-1} = \begin{bmatrix} s & 1 \\ -1 & s+2 \end{bmatrix}^{-1} = \frac{1}{s^2 + 2s + 1}\begin{bmatrix} s+2 & -1 \\ 1 & s \end{bmatrix} = \begin{bmatrix} \dfrac{s+2}{(s+1)^2} & \dfrac{-1}{(s+1)^2} \\ \dfrac{1}{(s+1)^2} & \dfrac{s}{(s+1)^2} \end{bmatrix}$$

**2.** The eigenvalues of $\mathbf{A}$ are $-1, -1$. Let $g(\lambda) = \alpha_0 + \alpha_1\lambda$. If $f(\lambda) \triangleq (s - \lambda)^{-1} = g(\lambda)$ on the spectrum of $\mathbf{A}$, then

$$\begin{aligned} f(-1) &= g(-1): \qquad (s+1)^{-1} = \alpha_0 - \alpha_1 \\ f'(-1) &= g'(-1): \qquad (s+1)^{-2} = \alpha_1 \end{aligned}$$

Hence

$$g(\lambda) = [(s+1)^{-1} + (s+1)^{-2}] + (s+1)^{-2}\lambda$$

[3] A square matrix is said to be triangular if all the elements below or above the main diagonal are zero.

and

$$(s\mathbf{I} - \mathbf{A})^{-1} = g(\mathbf{A}) = [(s+1)^{-1} + (s+1)^{-2}]\mathbf{I} + (s+1)^{-2}\mathbf{A}$$
$$= \begin{bmatrix} \dfrac{s+2}{(s+1)^2} & \dfrac{-1}{(s+1)^2} \\ \dfrac{1}{(s+1)^2} & \dfrac{s}{(s+1)^2} \end{bmatrix}$$

Example 3

Consider the state equation

$$\begin{bmatrix} \dot{x}_1 \\ \dot{x}_2 \end{bmatrix} = \begin{bmatrix} 0 & -1 \\ 1 & -2 \end{bmatrix} \begin{bmatrix} x_1 \\ x_2 \end{bmatrix} + \begin{bmatrix} 0 \\ 1 \end{bmatrix} u$$

The solution is given by

$$\mathbf{x}(t) = e^{\mathbf{A}t}\mathbf{x}(0) + \int_0^t e^{\mathbf{A}(t-\tau)}\mathbf{B}u(\tau)\, d\tau$$

The matrix function $e^{\mathbf{A}t}$ can be obtained by taking the inverse Laplace transform of $(s\mathbf{I} - \mathbf{A})^{-1}$, which is computed in Example 2. Hence

$$e^{\mathbf{A}t} = \mathcal{L}^{-1} \begin{bmatrix} \dfrac{s+2}{(s+1)^2} & \dfrac{-1}{(s+1)^2} \\ \dfrac{1}{(s+1)^2} & \dfrac{s}{(s+1)^2} \end{bmatrix} = \begin{bmatrix} (1+t)e^{-t} & -te^{-t} \\ te^{-t} & (1-t)e^{-t} \end{bmatrix}$$

and

$$\mathbf{x}(t) = \begin{bmatrix} (1+t)e^{-t} & -te^{-t} \\ te^{-t} & (1-t)e^{-t} \end{bmatrix} \mathbf{x}(0) + \int_0^t \begin{bmatrix} -(t-\tau)e^{-(t-\tau)} \\ [1-(t-\tau)]e^{-(t-\tau)} \end{bmatrix} u(\tau)\, d\tau$$

The solutions with $u \equiv 0$ are plotted in Figure 4-1 for the initial states $\mathbf{x}(0) = [6 \quad 8]'$ and $\mathbf{x}(0) = [8 \quad 5]'$. Note that the slope at each point of the trajectories in Figure 4-1 is equal to $\mathbf{A}\mathbf{x}$. ∎

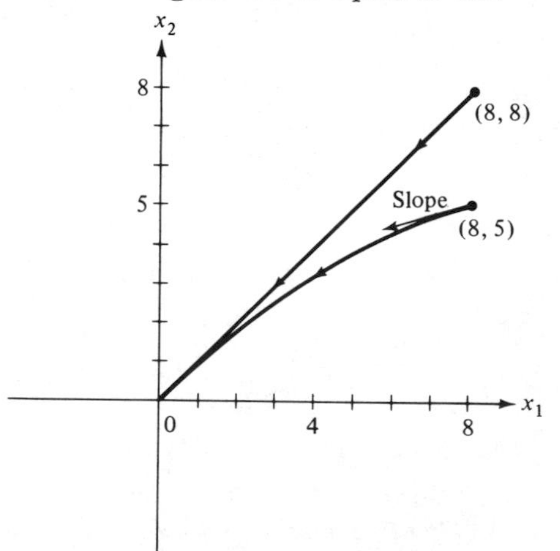


**Figure 4-1** Trajectories.

If the matrix $\mathbf{A}$ has $m$ distinct eigenvalues $\lambda_i$ with index $\bar{n}_i$, for $i = 1, 2, \cdots, m$ (see Definition 2-15), we claim that every element of $e^{\mathbf{A}t}$ is a linear combination of the factors $t^k e^{\lambda_i t}$, for $k = 0, 1, \cdots, \bar{n}_i - 1$; $i = 1, 2, \cdots, m$. Let $\hat{\mathbf{A}}$ be a Jordan-form representation of $\mathbf{A}$ and $\mathbf{A} = \mathbf{Q}\hat{\mathbf{A}}\mathbf{Q}^{-1}$. Then $e^{\mathbf{A}t} = \mathbf{Q}e^{\hat{\mathbf{A}}t}\mathbf{Q}^{-1}$. From (2-69) we know that every element of $e^{\hat{\mathbf{A}}t}$ is of the form $t^k e^{\lambda_i t}$, for $k = 0, 1, \cdots, \bar{n}_i - 1$; $i = 1, 2, \cdots, m$. Hence every element of $e^{\mathbf{A}t}$ is a linear combination of these factors. Since the function $t^k e^{\lambda_i t}$ is an analytic function over $(-\infty, \infty)$, we conclude that $e^{\mathbf{A}t}$ is *analytic over* $(-\infty, \infty)$. We call $t^k e^{\lambda_i t}$ a *mode* of the dynamical equation $FE$.

We introduce an interesting result to conclude this section.

#### Theorem 4-5

Consider the single-variable linear time-invariant dynamical equation

$$FE_1: \quad \begin{aligned} \dot{\mathbf{x}} &= \mathbf{A}\mathbf{x} + \mathbf{b}u \\ y &= \mathbf{c}\mathbf{x} + du \end{aligned}$$

where $\mathbf{A}$, $\mathbf{b}$, and $\mathbf{c}$ are $n \times n$, $n \times 1$ and $1 \times n$ real constant matrices, and $d$ is a real number. If the input $u$ is of the form $e^{\lambda_1 t}$, where $\lambda_1$ may be a real, imaginary, or complex number, and if $\lambda_1$ is not an eigenvalue of $\mathbf{A}$, then there exists an initial state in $FE_1$ such that the output $y$ will be immediately of the form $e^{\lambda_1 t}$ without containing any transient. ∎

We need the following identity to prove this theorem:

$$(s\mathbf{I} - \mathbf{A})^{-1}(s - \lambda_1)^{-1} = (\lambda_1\mathbf{I} - \mathbf{A})^{-1}(s - \lambda_1)^{-1} + (s\mathbf{I} - \mathbf{A})^{-1}(\mathbf{A} - \lambda_1\mathbf{I})^{-1} \quad \text{(4-28)}$$

for any $\lambda_1$ that is not an eigenvalue of $\mathbf{A}$. Note the similarity of this identity to the partial fraction expansion

$$\frac{1}{(s - \lambda)(s - \lambda_1)} = \frac{1}{(\lambda_1 - \lambda)} \cdot \frac{1}{s - \lambda_1} + \frac{1}{s - \lambda} \cdot \frac{1}{\lambda - \lambda_1}$$

The identity (4-28) can be proved by premultiplying $(s\mathbf{I} - \mathbf{A})$ and postmultiplying $(\mathbf{A} - \lambda_1\mathbf{I})$ by both sides of (4-28). (Prove!)

#### Proof of Theorem 4-5

Taking the Laplace transform of $FE_1$, we obtain

$$\hat{\mathbf{x}}(s) = (s\mathbf{I} - \mathbf{A})^{-1}\mathbf{x}(0) + (s\mathbf{I} - \mathbf{A})^{-1}\mathbf{b}\hat{u}(s) \quad \text{(4-29)}$$

If $u(t) = e^{\lambda_1 t}$, then $\hat{u}(s) = (s - \lambda_1)^{-1}$. Using (4-28), Equation (4-29) becomes

$$\hat{\mathbf{x}}(s) = (s\mathbf{I} - \mathbf{A})^{-1}[\mathbf{x}(0) + (\mathbf{A} - \lambda_1\mathbf{I})^{-1}\mathbf{b}] + (\lambda_1\mathbf{I} - \mathbf{A})^{-1}\mathbf{b}(s - \lambda_1)^{-1}$$

If we choose $\mathbf{x}(0) = -(\mathbf{A} - \lambda_1\mathbf{I})^{-1}\mathbf{b}$, then the output of $FE_1$ is

$$\hat{y}(s) = \mathbf{c}(\lambda_1\mathbf{I} - \mathbf{A})^{-1}\mathbf{b}(s - \lambda_1)^{-1} + d(s - \lambda_1)^{-1}$$

or

$$y(t) = [\mathbf{c}(\lambda_1\mathbf{I} - \mathbf{A})^{-1}\mathbf{b} + d]e^{\lambda_1 t} = \hat{g}(\lambda_1)e^{\lambda_1 t} \tag{4-30}$$

where we have used (4-26). Thus, if we choose the initial state $\mathbf{x}(0) = -(\mathbf{A} - \lambda_1\mathbf{I})^{-1}\mathbf{b}$, then the output $y(t)$ will not contain any transient. Q.E.D.

In this theorem, the assumption that $\lambda_1$ is not an eigenvalue of $\mathbf{A}$ is essential. Otherwise the theorem does not hold. This theorem is very useful in establishing that the impedance of a linear, time-invariant, lumped, passive network is a positive real function. See Reference [31]. We may also give a physical interpretation of the zeros of a transfer function by the use of this theorem. If $\lambda_1$ is a zero of $\hat{g}(s)$, that is, $\hat{g}(\lambda_1) = 0$, then by a proper choice of the initial state, the output of the system can be made identically zero even if the input $e^{\lambda_1 t}$ is applied.

## 4-3 Equivalent Dynamical Equations

We introduce in this section the concept of equivalent dynamical equations. This concept in the time-invariant case is identical to the one of change of basis introduced in Figure 2-5. Hence we study first the time-invariant case, and then the time-varying case.

**Time-invariant case.** Consider the linear time-invariant (fixed) dynamical equation

$$FE: \quad \dot{\mathbf{x}} = \mathbf{A}\mathbf{x} + \mathbf{B}\mathbf{u} \tag{4-31a}$$
$$\mathbf{y} = \mathbf{C}\mathbf{x} + \mathbf{D}\mathbf{u} \tag{4-31b}$$

where $\mathbf{A}$, $\mathbf{B}$, $\mathbf{C}$, and $\mathbf{D}$ are, respectively, $n \times n$, $n \times p$, $q \times n$, and $q \times p$ real constant matrices; $\mathbf{u}$ is the $p \times 1$ input vector, $\mathbf{y}$ is the $q \times 1$ output vector, and $\mathbf{x}$ is the $n \times 1$ state vector. The state space $\Sigma$ of the dynamical equation is an $n$-dimensional real vector space and the matrix $\mathbf{A}$ maps $\Sigma$ into itself. We have agreed in Section 3-3 to choose the orthonormal vectors $\{\mathbf{n}^1, \mathbf{n}^2, \cdots, \mathbf{n}^n\}$ as the basis vectors of the state space $\Sigma$, where $\mathbf{n}^i$ is an $n \times 1$ vector with 1 at its $i$th component and zeros elsewhere. We study now the effect of changing the basis of the state space. The dynamical equations that result from changing the basis of the state space are called equivalent dynamical equations. In order to allow a broader class of equivalent dynamical equations, we shall in this section extend the field of real numbers to the field of complex numbers and consider the state space as an $n$-dimensional complex vector space. This generalization is needed in order for the dynamical equation $FE$ to have an equivalent Jordan-form dynamical equation.

**Definition 4-3**

Let $\mathbf{P}$ be an $n \times n$ nonsingular matrix with coefficients in the field of complex numbers $\mathbb{C}$, and let $\bar{\mathbf{x}} = \mathbf{Px}$. Then the dynamical equation

$$\overline{FE}: \quad \dot{\bar{\mathbf{x}}} = \bar{\mathbf{A}}\bar{\mathbf{x}} + \bar{\mathbf{B}}\mathbf{u} \tag{4-32a}$$

$$\mathbf{y} = \bar{\mathbf{C}}\bar{\mathbf{x}} + \bar{\mathbf{D}}\mathbf{u} \tag{4-32b}$$

where

$$\bar{\mathbf{A}} = \mathbf{PAP}^{-1} \qquad \bar{\mathbf{B}} = \mathbf{PB} \qquad \bar{\mathbf{C}} = \mathbf{CP}^{-1} \qquad \mathbf{D} = \bar{\mathbf{D}} \tag{4-33}$$

is said to be *equivalent* to the dynamical equation $FE$ in (4-31), and $\mathbf{P}$ is said to be an *equivalence transformation.* ∎

The dynamical equation $\overline{FE}$ in (4-32) is obtained from (4-31) by the substitution of $\bar{\mathbf{x}} = \mathbf{Px}$. In this substitution we have changed the basis vectors of the state space from the orthonormal vectors to the columns of $\mathbf{P}^{-1}$. (See Figure 2-5). Observe that the matrices $\mathbf{A}$ and $\bar{\mathbf{A}}$ are similar; they are different representations of the same operator. If we let $\mathbf{Q} = \mathbf{P}^{-1} = [\mathbf{q}^1 \ \ \mathbf{q}^2 \ \ \cdots \ \ \mathbf{q}^n]$, then the $i$th column of $\bar{\mathbf{A}}$ is the representation of $\mathbf{Aq}^i$ with respect to the basis $\{\mathbf{q}^1, \mathbf{q}^2, \cdots, \mathbf{q}^n\}$. From the equation $\bar{\mathbf{B}} = \mathbf{PB}$ or $\mathbf{B} = \mathbf{P}^{-1}\bar{\mathbf{B}} = [\mathbf{q}^1 \ \ \mathbf{q}^2 \ \ \cdots \ \ \mathbf{q}^n]\bar{\mathbf{B}}$, we see that the $i$th column of $\bar{\mathbf{B}}$ is the representation of the $i$th column of $\mathbf{B}$ with respect to the basis $\{\mathbf{q}^1, \mathbf{q}^2, \cdots, \mathbf{q}^n\}$. The matrix $\bar{\mathbf{C}}$ is to be computed from $\mathbf{CP}^{-1}$. The matrix $\mathbf{D}$ is the direct transmission part between the input and the output. Since it has nothing to do with the state space, it is not affected by any equivalence transformation.

We explore now the physical meaning of equivalent dynamical equations. Recall that the state of a system is an auxiliary quantity introduced to give a unique relation between the input and the output when the system is not initially relaxed. The choice of the state is not unique; different methods of analysis often lead to different choices of the state.

**Example 1**

Consider the network shown in Figure 4-2. If the current passing through the inductor $x_1$ and the voltage across the capacitor $x_2$ are chosen

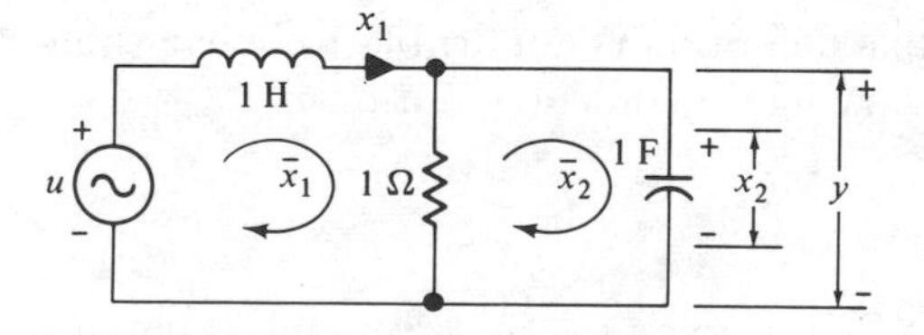


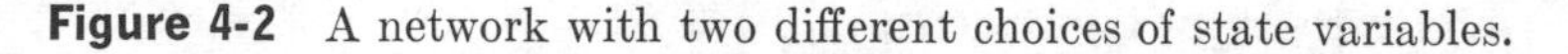

**Figure 4-2** A network with two different choices of state variables.

as the state variables, then the dynamical equation description of the network is

$$\begin{bmatrix} \dot{x}_1 \\ \dot{x}_2 \end{bmatrix} = \begin{bmatrix} 0 & -1 \\ 1 & -1 \end{bmatrix} \begin{bmatrix} x_1 \\ x_2 \end{bmatrix} + \begin{bmatrix} 1 \\ 0 \end{bmatrix} u \tag{4-34a}$$

$$y = [0 \quad 1] \begin{bmatrix} x_1 \\ x_2 \end{bmatrix} \tag{4-34b}$$

If, instead, the loop currents $\bar{x}_1$ and $\bar{x}_2$ are chosen as the state variables, then the dynamical equation is

$$\begin{bmatrix} \dot{\bar{x}}_1 \\ \dot{\bar{x}}_2 \end{bmatrix} = \begin{bmatrix} -1 & 1 \\ -1 & 0 \end{bmatrix} \begin{bmatrix} \bar{x}_1 \\ \bar{x}_2 \end{bmatrix} + \begin{bmatrix} 1 \\ 1 \end{bmatrix} u \tag{4-35a}$$

$$y = [1 \quad -1] \begin{bmatrix} \bar{x}_1 \\ \bar{x}_2 \end{bmatrix} \tag{4-35b}$$

The dynamical equations (4-34) and (4-35) have the same dimension and describe the same system. Hence they are equivalent. The equivalence transformation between these two equations can be found from Figure 4-2. It is clear that $x_1 = \bar{x}_1$. Since $x_2$ is equal to the voltage across the 1-ohm resistor, we have $x_2 = (\bar{x}_1 - \bar{x}_2)$. Thus

$$\begin{bmatrix} x_1 \\ x_2 \end{bmatrix} = \begin{bmatrix} 1 & 0 \\ 1 & -1 \end{bmatrix} \begin{bmatrix} \bar{x}_1 \\ \bar{x}_2 \end{bmatrix}$$

or

$$\begin{bmatrix} \bar{x}_1 \\ \bar{x}_2 \end{bmatrix} = \begin{bmatrix} 1 & 0 \\ 1 & -1 \end{bmatrix}^{-1} \begin{bmatrix} x_1 \\ x_2 \end{bmatrix} = \begin{bmatrix} 1 & 0 \\ 1 & -1 \end{bmatrix} \begin{bmatrix} x_1 \\ x_2 \end{bmatrix} \tag{4-36}$$

It is easy to verify that the dynamical equations (4-34) and (4-35) are indeed related by the equivalence transformation (4-36). ▮

### Definition 4-4

Two linear dynamical equations are said to be *zero-state equivalent* if and only if they have the same impulse-response matrix. Two linear dynamical equations are said to be *zero-input equivalent* if and only if for any initial state in one equation, there exists a state in the other equation, and vice versa, such that the two equations have the same zero-input response. ▮

Note that this definition is applicable to linear time-invariant as well as linear time-varying dynamical equations.

### Theorem 4-6

Two equivalent linear time-invariant dynamical equations are zero-state equivalent and zero-input equivalent.

### Proof

The impulse-response matrix of $FE$ is

$$\mathbf{G}(t) = \mathbf{C}e^{\mathbf{A}t}\mathbf{B} + \mathbf{D}\delta(t)$$

The impulse-response matrix of $F\bar{E}$ is

$$\bar{\mathbf{G}}(t) = \bar{\mathbf{C}}e^{\bar{\mathbf{A}}t}\bar{\mathbf{B}} + \bar{\mathbf{D}}\delta(t)$$

If the equations $FE$ and $F\bar{E}$ are equivalent, we have $\bar{\mathbf{A}} = \mathbf{PAP}^{-1}$, $\bar{\mathbf{B}} = \mathbf{PB}$, $\bar{\mathbf{C}} = \mathbf{CP}^{-1}$, and $\mathbf{D} = \bar{\mathbf{D}}$. Consequently we have

$$e^{\bar{\mathbf{A}}t} = \mathbf{P}e^{\mathbf{A}t}\mathbf{P}^{-1}$$

and

$$\bar{\mathbf{C}}e^{\bar{\mathbf{A}}t}\bar{\mathbf{B}} + \bar{\mathbf{D}}\delta(t) = \mathbf{CP}^{-1}\mathbf{P}e^{\mathbf{A}t}\mathbf{P}^{-1}\mathbf{PB} + \mathbf{D}\delta(t) = \mathbf{C}e^{\mathbf{A}t}\mathbf{B} + \mathbf{D}\delta(t)$$

Hence, two equivalent dynamical equations are zero-state equivalent.

The zero-input response of $FE$ is

$$\mathbf{y}(t) = \mathbf{C}e^{\mathbf{A}(t-t_0)}\mathbf{x}(t_0)$$

The zero-input response of $F\bar{E}$ is

$$\bar{\mathbf{y}}(t) = \bar{\mathbf{C}}e^{\bar{\mathbf{A}}(t-t_0)}\bar{\mathbf{x}}(t_0) = \mathbf{C}e^{\bar{\mathbf{A}}(t-t_0)}\mathbf{P}^{-1}\bar{\mathbf{x}}(t_0)$$

Hence, for any $\mathbf{x}(t_0)$, if we choose $\bar{\mathbf{x}}(t_0) = \mathbf{Px}(t_0)$, then $FE$ and $F\bar{E}$ have the same zero-input response. Q.E.D.

We note that although equivalence implies zero-state equivalence and zero-input equivalence, the converse is not true. That is, *two linear dynamical equations can be zero-state equivalent and zero-input equivalent without being equivalent.* Furthermore, two linear equations can be zero-state equivalent without being zero-input equivalent.

### Example 2

Consider the two networks shown in Figure 4-3. If the initial state in the capacitor is zero, the two networks have the same input–output pairs; their impulse responses are all equal to $\delta(t)$, the Dirac $\delta$-function. Therefore they are zero-state equivalent or, more precisely, their dynamical-equation descriptions are zero-state equivalent. Because of the symmetry of the network, for any initial voltage in the capacitor the zero-input responses of these two networks are all identically zero. Hence they are zero-input equivalent.

A necessary condition for two dynamical equations to be equivalent is that they have the same dimension. The dynamical-equation description of the network in Figure 4-3(a) has dimension 0; the dynamical-equation description of the network in Figure 4-3(b) has dimension 1. Hence they are not equivalent. ▮

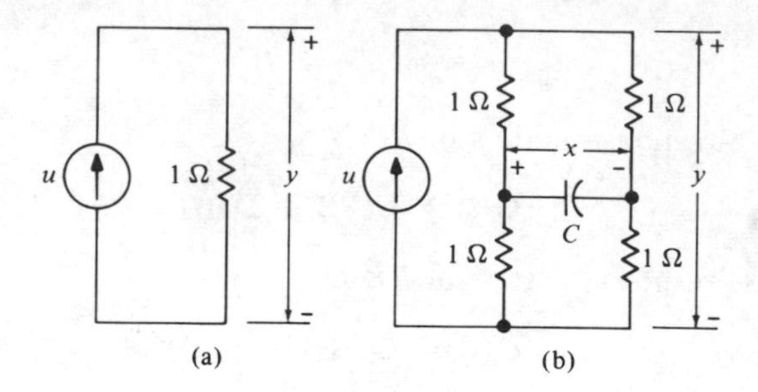


**Figure 4-3** Two networks whose dynamical equations are zero-state equivalent and zero-input equivalent without being equivalent.

Example 3

The two networks in Figure 4-4 or, more precisely, their dynamical-equation descriptions, are zero-state equivalent but not zero-input equivalent. Indeed, if there is a nonzero initial condition in the capacitor, the zero-input response of Figure 4-4(b) is nonzero, whereas the zero-input response of Figure 4-4(a) is identically zero. ∎

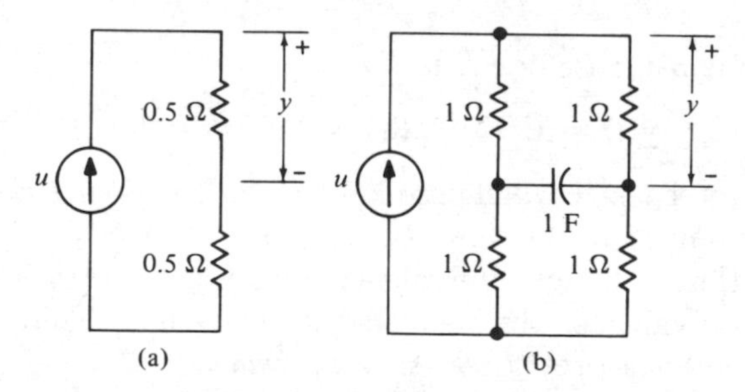


**Figure 4-4** Two networks whose dynamical equations are zero-state equivalent.

If the matrix $\mathbf{A}$ in a dynamical equation is in the Jordan form, then the equation is said to be a Jordan-form dynamical equation. We have shown in Section 2-6 that every operator that maps $(\mathbb{C}^n, \mathbb{C})$ into itself has a Jordan-form matrix representation. Hence *every linear, time-invariant dynamical equation has an equivalent Jordan-form-dynamical equation.*

Example 4

Consider the dynamical equation

$$FE: \quad \begin{bmatrix} \dot{x}_1 \\ \dot{x}_2 \\ \dot{x}_3 \end{bmatrix} = \begin{bmatrix} 1 & 1 & 2 \\ 0 & 1 & 3 \\ 0 & 0 & 2 \end{bmatrix} \begin{bmatrix} x_1 \\ x_2 \\ x_3 \end{bmatrix} + \begin{bmatrix} 1 & 0 \\ 3 & 2 \\ -1 & -2 \end{bmatrix} \begin{bmatrix} u_1 \\ u_2 \end{bmatrix} \tag{4-37a}$$

$$\begin{bmatrix} y_1 \\ y_2 \end{bmatrix} = \begin{bmatrix} 1 & 0 & 0 \\ 1 & 2 & 2 \end{bmatrix} \begin{bmatrix} x_1 \\ x_2 \\ x_3 \end{bmatrix} \tag{4-37b}$$

It is easy to verify that if we choose

$$\mathbf{P} = \mathbf{Q}^{-1} = \begin{bmatrix} 1 & 0 & 5 \\ 0 & 1 & 3 \\ 0 & 0 & 1 \end{bmatrix}^{-1} = \begin{bmatrix} 1 & 0 & -5 \\ 0 & 1 & -3 \\ 0 & 0 & 1 \end{bmatrix}$$

then the new **A** matrix will be in the Jordan form. The method for finding the matrix **P** was discussed in Section 2-6. If we substitute $\bar{\mathbf{x}} = \mathbf{P}\mathbf{x}$ into (4-37), we obtain immediately the following equivalent Jordan-form dynamical equation:

$$\begin{bmatrix} \dot{\bar{x}}_1 \\ \dot{\bar{x}}_2 \\ \dot{\bar{x}}_3 \end{bmatrix} = \left[\begin{array}{cc|c} 1 & 1 & 0 \\ 0 & 1 & 0 \\ \hline 0 & 0 & 2 \end{array}\right] \begin{bmatrix} \bar{x}_1 \\ \bar{x}_2 \\ \bar{x}_3 \end{bmatrix} + \begin{bmatrix} 6 & 10 \\ 6 & 8 \\ -1 & -2 \end{bmatrix} \begin{bmatrix} u_1 \\ u_2 \end{bmatrix}$$

$$\begin{bmatrix} y_1 \\ y_2 \end{bmatrix} = \begin{bmatrix} 1 & 0 & 5 \\ 1 & 2 & 13 \end{bmatrix} \begin{bmatrix} \bar{x}_1 \\ \bar{x}_2 \\ \bar{x}_3 \end{bmatrix}$$

***Time-varying case.** In this subsection we study equivalent linear time varying dynamical equations. This is an extension of the time-invariant case. All the discussion and interpretation in the preceding subsection applies here as well. The only conceptual difference is that the basis vectors in the time-invariant case are fixed (independent of time), whereas the basis vectors in the time-varying case may change with time.

Consider the linear time-varying dynamical equation

$$E: \quad \dot{\mathbf{x}} = \mathbf{A}(t)\mathbf{x} + \mathbf{B}(t)\mathbf{u} \qquad \textbf{(4-38a)}$$
$$\mathbf{y} = \mathbf{C}(t)\mathbf{x} + \mathbf{D}(t)\mathbf{u} \qquad \textbf{(4-38b)}$$

where **A**, **B**, **C**, and **D** are $n \times n$, $n \times p$, $q \times n$, and $q \times p$ matrices whose entries are real-valued continuous functions of $t$. As in the time-invariant case, we extend the field of real numbers to the field of complex numbers.

**Definition 4-5**[4]

Let $\mathbf{P}(\cdot)$ be an $n \times n$ complex-valued matrix defined over $(-\infty, \infty)$. It is assumed that $\mathbf{P}(t)$ is nonsingular for all $t$ and continuously differentiable in $t$. Let $\bar{\mathbf{x}} = \mathbf{P}(t)\mathbf{x}$. Then the dynamical equation

$$\bar{E}: \quad \dot{\bar{\mathbf{x}}} = \bar{\mathbf{A}}(t)\bar{\mathbf{x}} + \bar{\mathbf{B}}(t)\mathbf{u} \qquad \textbf{(4-39a)}$$
$$\mathbf{y} = \bar{\mathbf{C}}(t)\bar{\mathbf{x}} + \bar{\mathbf{D}}(t)\mathbf{u} \qquad \textbf{(4-39b)}$$

where

$$\bar{\mathbf{A}}(t) = (\mathbf{P}(t)\mathbf{A}(t) + \dot{\mathbf{P}}(t))\mathbf{P}^{-1}(t) \qquad \textbf{(4-40a)}$$
$$\bar{\mathbf{B}}(t) = \mathbf{P}(t)\mathbf{B}(t) \qquad \textbf{(4-40b)}$$
$$\bar{\mathbf{C}}(t) = \mathbf{C}(t)\mathbf{P}^{-1}(t) \qquad \textbf{(4-40c)}$$
$$\bar{\mathbf{D}}(t) = \mathbf{D}(t) \qquad \textbf{(4-40d)}$$

[4] This definition reduces to Definition 4-3 if **P** is independent of time.

is said to be *equivalent*[5] to the dynamical equation $E$ in (4-38), and $P(\cdot)$ is said to be an *equivalence transformation.* ∎

The dynamical equation $\bar{E}$ in Equation (4-39) is obtained from (4-38) by the substitution of $\bar{\mathbf{x}} = \mathbf{P}(t)\mathbf{x}$ and $\dot{\bar{\mathbf{x}}} = \dot{\mathbf{P}}(t)\mathbf{x} + \mathbf{P}(t)\dot{\mathbf{x}}$. Let $\mathbf{\Psi}$ be a fundamental matrix of $E$. Then we claim that

$$\bar{\mathbf{\Psi}}(t) \triangleq \mathbf{P}(t)\mathbf{\Psi}(t) \tag{4-41}$$

is a fundamental matrix of $\bar{E}$. The matrix $\mathbf{\Psi}$ is, by assumption, a fundamental matrix of $E$; hence $\dot{\mathbf{\Psi}}(t) = \mathbf{A}(t)\mathbf{\Psi}(t)$, and $\mathbf{\Psi}(t)$ is nonsingular for all $t$. Consequently, the matrix $\mathbf{P}(t)\mathbf{\Psi}(t)$ is nonsingular for all $t$ (see Theorem 2-7). Now we show that $\mathbf{P}(t)\mathbf{\Psi}(t)$ satisfies the matrix equation $\dot{\bar{\mathbf{x}}} = \bar{\mathbf{A}}(t)\bar{\mathbf{x}}$. Indeed,

$$\begin{aligned}\frac{d}{dt}(\mathbf{P}(t)\mathbf{\Psi}(t)) &= \dot{\mathbf{P}}(t)\mathbf{\Psi}(t) + \mathbf{P}(t)\dot{\mathbf{\Psi}}(t)\\ &= (\dot{\mathbf{P}}(t) + \mathbf{P}(t)\mathbf{A}(t))\mathbf{P}^{-1}(t)\mathbf{P}(t)\mathbf{\Psi}(t)\\ &= \bar{\mathbf{A}}(t)(\mathbf{P}(t)\mathbf{\Psi}(t))\end{aligned}$$

Hence $\mathbf{P}(t)\mathbf{\Psi}(t)$ is a fundamental matrix of $\dot{\bar{\mathbf{x}}} = \bar{\mathbf{A}}\bar{\mathbf{x}}$. With this fact, it is easy to show the following theorem.

### Theorem 4-7

Two equivalent linear time-varying dynamical equations are zero-state equivalent and zero-input equivalent. ∎

The proof of this theorem is left as an exercise. Zero-state equivalence and zero-input equivalence were defined in Definition 4-4. As in the time-invariant case, two linear time-varying dynamical equations can be zero-state equivalent and zero-input equivalent without being equivalent; two linear time-varying equations can be zero-state equivalent without being zero-input equivalent.

### Linear dynamical equation with periodic A(·)

Consider the linear time-varying dynamical equation

$$E: \quad \dot{\mathbf{x}}(t) = \mathbf{A}(t)\mathbf{x}(t) + \mathbf{B}(t)\mathbf{u}(t) \tag{4-42a}$$
$$\mathbf{y}(t) = \mathbf{C}(t)\mathbf{x}(t) + \mathbf{D}(t)\mathbf{u}(t) \tag{4-42b}$$

[5] It is also called *algebraically equivalent*. In Definition 4-5, if the matrix $\mathbf{P}(\cdot)$ has, in addition, the properties $\|\mathbf{P}(t)\| \leq k_1$, $\|\mathbf{P}^{-1}(t)\| \leq k_2$ for all $t$, where $k_1$ and $k_2$ are fixed constants and $\|\cdot\|$ is a norm, then $\mathbf{P}(\cdot)$ is called a *Lyapunov transformation*, and the dynamical equations $E$ and $\bar{E}$ are said to be *topologically equivalent*. In general, (algebraic) equivalence does not preserve the stability properties of an equation, but topological equivalence does; see Problems 8-19 and 8-20. In the time-invariant case, the matrix $\mathbf{P}$ is a constant matrix; therefore, there is no difference between (algebraic) equivalence and topological equivalence.

where $\mathbf{A}$, $\mathbf{B}$, $\mathbf{C}$, and $\mathbf{D}$ are matrices whose entries are continuous functions of $t$ defined over $(-\infty, \infty)$. We assume, in addition, that

$$\mathbf{A}(t + T) = \mathbf{A}(t) \tag{4-43}$$

for all $t$ and for some positive constant $T$. This means that every element of $\mathbf{A}(t)$ is a periodic function with the same period $T$. We shall show that, under this condition, the equation $E$ has an equivalent dynamical equation with a constant $\mathbf{A}$ matrix.

Let $\mathbf{\Psi}(t)$ be a fundamental matrix of $\dot{\mathbf{x}} = \mathbf{A}(t)\mathbf{x}$, then $\mathbf{\Psi}(t + T)$ is also a fundamental matrix of $\dot{\mathbf{x}} = \mathbf{A}(t)\mathbf{x}$. This is checked by direct verification:

$$\dot{\mathbf{\Psi}}(t + T) = \mathbf{A}(t + T)\mathbf{\Psi}(t + T) = \mathbf{A}(t)\mathbf{\Psi}(t + T)$$

The matrix function $\mathbf{\Psi}(t)$ is nonsingular for all $t$, consequently so is $\mathbf{\Psi}(t + T)$. Columns of $\mathbf{\Psi}(t)$ and columns of $\mathbf{\Psi}(t + T)$ form two sets of basis vectors in the solution space; hence there exists a nonsingular constant matrix $\mathbf{Q}$ (see Section 2-3) such that

$$\mathbf{\Psi}(t + T) = \mathbf{\Psi}(t)\mathbf{Q} \tag{4-44}$$

For the nonsingular matrix $\mathbf{Q}$ there exists a constant matrix $\bar{\mathbf{A}}$ such that $e^{\bar{\mathbf{A}}T} = \mathbf{Q}$ (Problem 2-35). Hence (4-44) can be written as

$$\mathbf{\Psi}(t + T) = \mathbf{\Psi}(t)e^{\bar{\mathbf{A}}T} \tag{4-45}$$

Define

$$\mathbf{P}(t) \triangleq e^{\bar{\mathbf{A}}t}\mathbf{\Psi}^{-1}(t) \tag{4-46}$$

We show that $\mathbf{P}(\cdot)$ is a periodic function with period $T$:

$$\mathbf{P}(t + T) = e^{\bar{\mathbf{A}}(t+T)}\mathbf{\Psi}^{-1}(t + T) = e^{\bar{\mathbf{A}}t}e^{\bar{\mathbf{A}}T}e^{-\bar{\mathbf{A}}T}\mathbf{\Psi}^{-1}(t) = \mathbf{P}(t)$$

### Theorem 4-8

Assume that the matrix $\mathbf{A}$ in the dynamical equation $E$ in (4-42) is periodic with period $T$. Let $\mathbf{P}$ be defined as in (4-46). Then the dynamical equation $E$ in (4-42) and the dynamical equation

$$\bar{E}: \quad \begin{aligned} \dot{\bar{\mathbf{x}}}(t) &= \bar{\mathbf{A}}\bar{\mathbf{x}}(t) + \mathbf{P}(t)\mathbf{B}(t)\mathbf{u}(t) \\ \bar{\mathbf{y}}(t) &= \mathbf{C}(t)\mathbf{P}^{-1}(t)\bar{\mathbf{x}}(t) + \mathbf{D}(t)\mathbf{u}(t) \end{aligned}$$

where $\bar{\mathbf{A}}$ is a constant matrix, are equivalent.

### Proof

First we show that the matrix $\mathbf{P}$ qualifies as an equivalence transformation. That $\mathbf{P}(t)$ is nonsingular for all $t$ follows from the fact that $e^{\bar{\mathbf{A}}t}$ and $\mathbf{\Psi}^{-1}(t)$ are nonsingular for all $t$. Both $e^{\bar{\mathbf{A}}t}$ and $\mathbf{\Psi}^{-1}(t)$ are continuously differentiable; therefore, so is $\mathbf{P}(t)$. Consequently, $\mathbf{P}$ qualifies as an equivalence transformation.

Let $\bar{\mathbf{x}}(t) = \mathbf{P}(t)\mathbf{x}$; then we obtain the following equivalent dynamical equation of $E$:

$$\begin{aligned}\dot{\bar{\mathbf{x}}} &= (\mathbf{P}(t)\mathbf{A}(t) + \dot{\mathbf{P}}(t))\mathbf{P}^{-1}(t)\bar{\mathbf{x}} + \mathbf{P}(t)\mathbf{B}(t)\mathbf{u}(t)\\ \dot{\bar{\mathbf{y}}} &= \mathbf{C}(t)\mathbf{P}^{-1}(t)\bar{\mathbf{x}} + \mathbf{D}(t)\mathbf{u}(t)\end{aligned}$$

Clearly, if $(\mathbf{P}(t)\mathbf{A}(t) + \dot{\mathbf{P}}(t))\mathbf{P}^{-1}(t) = \bar{\mathbf{A}}$, then the theorem is proved. This is checked by direct verification:

$$\begin{aligned}&(\mathbf{P}(t)\mathbf{A}(t) + \dot{\mathbf{P}}(t))\mathbf{P}^{-1}(t) = (e^{\bar{\mathbf{A}}t}\mathbf{\Psi}^{-1}(t)\mathbf{A}(t) + \bar{\mathbf{A}}e^{\bar{\mathbf{A}}t}\mathbf{\Psi}^{-1}(t)\\ &\quad - e^{\bar{\mathbf{A}}t}\mathbf{\Psi}^{-1}(t)\mathbf{A}(t))\mathbf{P}^{-1}(t) = \bar{\mathbf{A}}\end{aligned}$$

Q.E.D.

The homogeneous part of this theorem is the so-called *theory of Floquet.* It states that if $\dot{\mathbf{x}} = \mathbf{A}(t)\mathbf{x}$ and if $\mathbf{A}(t+T) = \mathbf{A}(t)$ for all $t$, then its fundamental matrix is of the form $\mathbf{P}^{-1}(t)e^{\bar{\mathbf{A}}t}$ where $\mathbf{P}^{-1}(t)$ is a periodic function. Furthermore, $\dot{\mathbf{x}} = \mathbf{A}(t)\mathbf{x}$ is equivalent to $\dot{\bar{\mathbf{x}}} = \bar{\mathbf{A}}\bar{\mathbf{x}}$.

### Example 5

Consider the one-dimensional dynamical equation

$$\begin{aligned}\dot{x} &= (\sin t)x + u\\ y &= 2x\end{aligned}$$

where $\sin t$ is a periodic function with a period of $2\pi$. It is clear that $\psi(t) = e^{-\cos t}$ is a fundamental matrix. Since $\psi(t+2\pi) = e^{-\cos(t+2\pi)} = e^{-\cos t}$, the $Q$ in (4-44) is equal to 1. It follows that $\bar{A} = 0$, and $\mathrm{P}(t) = e^{\cos t}$. Hence the dynamical equation

$$\begin{aligned}\dot{\bar{x}} &= e^{\cos t}u(t)\\ \bar{y} &= 2e^{-\cos t}\bar{x}\end{aligned}$$

is equivalent to the original equation.

## 4.4 Impulse-Response Matrices and Dynamical Equations

**Time-varying case.** In this section we shall study the relation between the impulse-response matrix and the dynamical equation. Let the input–output description of a system with $p$ input terminals and $q$ output terminals be

$$\mathbf{y}(t) = \int_{t_0}^{t} \mathbf{G}(t, \tau)\mathbf{u}(\tau)\, d\tau \qquad \textbf{(4-47)}$$

where $\mathbf{y}$ is the $q \times 1$ output vector, $\mathbf{u}$ is the $p \times 1$ input vector, and $\mathbf{G}$ is the $q \times p$ impulse-response matrix of the system. We have implicitly assumed in (4-47) that the system is initially relaxed at $t_0$. The $ij$th ($i$th row, $j$th column) element of $\mathbf{G}(\cdot, \tau)$ is the response at the $i$th output terminal due to a $\delta$-function input applied at time $\tau$ at the $j$th input terminal.

Suppose now that the internal structure of the same system is accessible, and that analysis of this system leads to a dynamical equation of the form

$$E: \quad \begin{aligned} \dot{\mathbf{x}}(t) &= \mathbf{A}(t)\mathbf{x}(t) + \mathbf{B}(t)\mathbf{u}(t) \qquad \text{(4-48a)} \\ \mathbf{y}(t) &= \mathbf{C}(t)\mathbf{x}(t) + \mathbf{D}(t)\mathbf{u}(t) \qquad \text{(4-48b)} \end{aligned}$$

where $\mathbf{x}$ is the $n \times 1$ state vector of the system, and $\mathbf{A}$, $\mathbf{B}$, $\mathbf{C}$, and $\mathbf{D}$ are $n \times n$, $n \times p$, $q \times n$, and $q \times p$ matrices whose entries are continuous functions of $t$ defined over $(-\infty, \infty)$. Since Equations (4-47) and (4-48) are two different descriptions of the same system, they should give the same input–output pairs if the system is initially relaxed. The solution of the dynamical equation $E$ with $\mathbf{x}(t_0) = \mathbf{0}$ is given by

$$\begin{aligned} \mathbf{y}(t) &= \mathbf{C}(t) \int_{t_0}^{t} \mathbf{\Phi}(t, \tau)\mathbf{B}(\tau)\mathbf{u}(\tau)\, d\tau + \mathbf{D}(t)\mathbf{u}(t) \\ &= \int_{t_0}^{t} [\mathbf{C}(t)\mathbf{\Phi}(t, \tau)\mathbf{B}(\tau) + \mathbf{D}(t)\delta(t - \tau)]\mathbf{u}(\tau)\, d\tau \qquad \text{(4-49)} \end{aligned}$$

where $\mathbf{\Phi}(t, \tau)$ is the state transition matrix of $\dot{\mathbf{x}} = \mathbf{A}(t)\mathbf{x}$. By comparing (4-47) and (4-49), we immediately obtain

$$\begin{aligned} \mathbf{G}(t, \tau) &= \mathbf{C}(t)\mathbf{\Phi}(t, \tau)\mathbf{B}(\tau) + \mathbf{D}(t)\delta(t - \tau) \qquad && \text{for } t \geq \tau \\ &= \mathbf{0} && \text{for } t < \tau \qquad \text{(4-50)} \end{aligned}$$

That $\mathbf{G}(t, \tau) = \mathbf{0}$ for $t < \tau$ follows from the causality assumption which is implicitly embodded in writing (4-47); that is, the integration is stopped at $t$.

If the state-variable description of a system is available, the input–output description of the system can be easily obtained from (4-50). The converse problem—to find the state-variable description from the input–output description of a system—is much more complicated, however. It actually consists of two problems: (1) Is it possible at all to obtain the state-variable description from the impulse-response matrix of a system? (2) If yes, how do we obtain the state-variable description from the impulse-response matrix? We shall study the first problem in the remainder of this section. The second problem will be studied in Chapter 6.

Consider a system with the impulse-response matrix $\mathbf{G}(t, \tau)$. If there exists a linear, finite-dimensional dynamical equation $E$ that has $\mathbf{G}(t, \tau)$ as its impulse-response matrix, then $\mathbf{G}(t, \tau)$ is said to be *realizable.* We call the dynamical equation $E$, or more specifically, the matrices $\{\mathbf{A}, \mathbf{B}, \mathbf{C}, \mathbf{D}\}$, a *realization* of $\mathbf{G}(t, \tau)$. The terminology "realization" is justified by the fact that by using the dynamical equation, we can build an operational amplifier circuit that will generate $\mathbf{G}(t, \tau)$. Note that the state of a dynamical-equation realization of the impulse-response matrix of a system is purely an auxiliary variable and it may not have any physical meaning. Note also that *the dynamical-equation realization gives only the same zero-*

*state response of the system.* If the dynamical equation is not in the zero-state, its response may not have any relation to the system.

If the realization of an impulse response $\mathbf{G}(t, \tau)$ is restricted to a finite-dimensional, linear dynamical equation of the form (4-48), it is conceivable that not every $\mathbf{G}(t, \tau)$ is realizable. For example, there is no linear equation of the form (4-48) that will generate the impulse response of a unit-time-delay system or the impulse response $1/(t - \tau)$. We give in the following the necessary and sufficient condition for $\mathbf{G}(t, \tau)$ to be realizable.

### Theorem 4-9

A $q \times p$ impulse-response matrix $\mathbf{G}(t, \tau)$ is realizable by a finite-dimensional, linear dynamical equation of the form (4-48) if and only if $\mathbf{G}(t, \tau)$ can be decomposed into

$$\mathbf{G}(t, \tau) = \mathbf{D}(t)\delta(t - \tau) + \mathbf{M}(t)\mathbf{N}(\tau) \qquad \text{for all } t \geq \tau \tag{4-51}$$

where $\mathbf{D}$ is a $q \times p$ matrix, and $\mathbf{M}$ and $\mathbf{N}$ are, respectively, $q \times n$ and $n \times p$ continuous matrices of $t$.

### Proof

*Necessity:* Suppose the dynamical equation

$$E: \qquad \begin{aligned} \dot{\mathbf{x}} &= \mathbf{A}(t)\mathbf{x} + \mathbf{B}(t)\mathbf{u} \\ \mathbf{y} &= \mathbf{C}(t)\mathbf{x} + \mathbf{D}(t)\mathbf{u} \end{aligned}$$

is a realization of $\mathbf{G}(t, \tau)$; then

$$\begin{aligned} \mathbf{G}(t, \tau) &= \mathbf{D}(t)\delta(t - \tau) + \mathbf{C}(t)\mathbf{\Phi}(t - \tau)\mathbf{B}(\tau) \\ &= \mathbf{D}(t)\delta(t - \tau) + \mathbf{C}(t)\mathbf{\Psi}(t)\mathbf{\Psi}^{-1}(\tau)\mathbf{B}(\tau) \end{aligned}$$

where $\mathbf{\Psi}$ is a fundamental matrix of $\dot{\mathbf{x}} = \mathbf{A}(t)\mathbf{x}$. The proof is completed by identifying

$$\mathbf{M}(t) = \mathbf{C}(t)\mathbf{\Psi}(t) \qquad \text{and} \qquad \mathbf{N}(t) = \mathbf{\Psi}^{-1}(t)\mathbf{B}(t)$$

*Sufficiency:* Let

$$\mathbf{G}(t, \tau) = \mathbf{D}(t)\delta(t - \tau) + \mathbf{M}(t)\mathbf{N}(\tau)$$

where $\mathbf{M}$ and $\mathbf{N}$ are $q \times n$ and $n \times p$ continuous matrices, respectively. Then the following $n$-dimensional dynamical equation

$$\bar{E}: \qquad \dot{\mathbf{x}}(t) = \mathbf{N}(t)\mathbf{u}(t) \tag{4-52a}$$

$$\mathbf{y}(t) = \mathbf{M}(t)\mathbf{x}(t) + \mathbf{D}(t)\mathbf{u}(t) \tag{4-52b}$$

is a realization of $\mathbf{G}(t, \tau)$. Indeed, the state transition matrix of $\bar{E}$ is an $n \times n$ identity matrix; hence

$$\mathbf{G}(t, \tau) = \mathbf{M}(t)\mathbf{I}\mathbf{N}(\tau) + \mathbf{D}(t)\delta(t - \tau) \qquad \text{Q.E.D.}$$

It is of interest to note that the dynamical equation $\bar{E}$ can be simulated without using feedback, as shown in Figure 4-5.

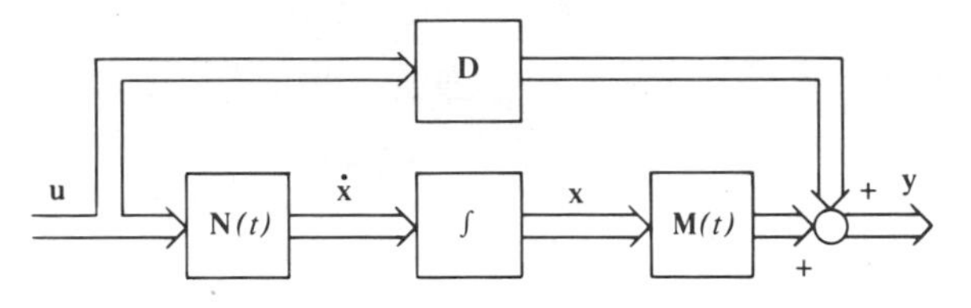


**Figure 4-5** Matrix block diagram of the dynamical equation $\bar{E}$ in (4-52).

Example 1

Consider $g(t, \tau) = g(t - \tau) = (t - \tau)e^{\lambda(t-\tau)}$. It is easy to verify that

$$g(t - \tau) = (t - \tau)e^{\lambda(t-\tau)} = [e^{\lambda t} \quad te^{\lambda t}]\begin{bmatrix} -\tau e^{-\lambda\tau} \\ e^{-\lambda\tau} \end{bmatrix}$$

Hence the dynamical equation

$$E: \qquad \begin{bmatrix} \dot{x}_1 \\ \dot{x}_2 \end{bmatrix} = \begin{bmatrix} -te^{-\lambda t} \\ e^{-\lambda t} \end{bmatrix} u(t)$$

$$y(t) = [e^{\lambda t} \quad te^{\lambda t}]\begin{bmatrix} x_1 \\ x_2 \end{bmatrix}$$

is a realization of $g(t, \tau)$. ∎

All the equivalent dynamical equations have the same impulse-response matrix; hence, if we find a realization of $\mathbf{G}(t, \tau)$, we may obtain different realizations of $\mathbf{G}(t, \tau)$ by applying equivalence transformations. Note that an impulse-response matrix may have different dimensional realizations; for example, the networks in Figure 4-3 are two different dimensional realizations of $g(t, \tau) = \delta(t - \tau)$.

**Time-invariant case.** We shall first apply Theorem 4-9 to the time-invariant case and see what can be established. For the time-invariant case, we have $\mathbf{G}(t, \tau) = \mathbf{G}(t - \tau)$. Consequently, Theorem 4-9 can be read as: An impulse response $\mathbf{G}(t - \tau)$ is realizable by a finite-dimensional (time-varying) dynamical equation if and only if there exist continuous matrices $\mathbf{M}$ and $\mathbf{N}$ such that

$$\mathbf{G}(t - \tau) = \mathbf{M}(t)\mathbf{N}(\tau) + \mathbf{D}(t)\delta(t - \tau) \qquad \text{for all } t \geq \tau$$

There are two objections to using this theorem. First, the condition is stated in terms of $\mathbf{G}(t - \tau)$ instead of $\mathbf{G}(t)$. Second, the condition is given for $\mathbf{G}(t - \tau)$ to be realizable by a linear *time-varying* dynamical equation. What is more desirable is to have conditions on $\mathbf{G}(t)$ under which $\mathbf{G}(t)$ has a linear *time-invariant* dynamical-equation realization. Since, in the time-invariant case, we may as well study a system in the frequency domain, we shall first derive the condition of realization in terms of transfer-function matrix, and then state it in terms of $\mathbf{G}(t)$.

Consider a system with the input–output description

$$\mathbf{y}(t) = \int_0^t \mathbf{G}(t-\tau)\mathbf{u}(\tau)\, d\tau$$

or, in the frequency domain,

$$\hat{\mathbf{y}}(s) = \hat{\mathbf{G}}(s)\hat{\mathbf{u}}(s) \tag{4-53}$$

where $\mathbf{G}(\cdot)$ is the impulse-response matrix and $\hat{\mathbf{G}}(s)$ is the transfer-function matrix of the system. Suppose now a state-variable description of the system is found to be

$$FE: \qquad \dot{\mathbf{x}} = \mathbf{A}\mathbf{x} + \mathbf{B}\mathbf{u} \tag{4-54a}$$

$$\mathbf{y} = \mathbf{C}\mathbf{x} + \mathbf{D}\mathbf{u} \tag{4-54b}$$

By taking the Laplace transform and assuming the zero initial state, we obtain

$$\hat{\mathbf{y}}(s) = [\mathbf{C}(s\mathbf{I} - \mathbf{A})^{-1}\mathbf{B} + \mathbf{D}]\hat{\mathbf{u}}(s) \tag{4-55}$$

Since (4-53) and (4-55) describe the same system, we have

$$\begin{aligned}\hat{\mathbf{G}}(s) &= \mathbf{C}(s\mathbf{I} - \mathbf{A})^{-1}\mathbf{B} + \mathbf{D} \\ &= \frac{1}{\det(s\mathbf{I} - \mathbf{A})}\mathbf{C}[\operatorname{Adj}(s\mathbf{I} - \mathbf{A})]\mathbf{B} + \mathbf{D}\end{aligned} \tag{4-56}$$

The determinant of $(s\mathbf{I} - \mathbf{A})$ is a polynomial of degree $n$ in $s$. Every entry of the adjoint of $(s\mathbf{I} - \mathbf{A})$ is a polynomial of degree equal to or less than $n - 1$. Hence every entry of $\mathbf{C}(s\mathbf{I} - \mathbf{A})^{-1}\mathbf{B}$ is a rational function of $s$ and the degree of its denominator is at least one degree higher than that of its numerator. Now $\mathbf{D}$ is a constant matrix, hence every entry of $\mathbf{C}(s\mathbf{I} - \mathbf{A})^{-1}\mathbf{B} + \mathbf{D}$ is again a rational function of $s$ and the degree of its denominator is equal to or greater than that of its numerator. Note that

$$\hat{\mathbf{G}}(\infty) = \mathbf{D} \tag{4-57}$$

**Definition 4-6**

A rational-function matrix $\hat{\mathbf{G}}(s)$ is said to be *proper* if $\hat{\mathbf{G}}(\infty)$ is a constant matrix. $\hat{\mathbf{G}}(s)$ is said to be *strictly proper* if $\hat{\mathbf{G}}(\infty) = \mathbf{0}$.

**Example 2**

Consider the rational-function matrices

$$\hat{\mathbf{G}}_1(s) = \begin{bmatrix} \dfrac{1}{s+1} & \dfrac{s^2}{s^3+1} \\ \dfrac{s-1}{s^2+1} & \dfrac{1}{s} \end{bmatrix} \qquad \hat{\mathbf{G}}_2(s) = \begin{bmatrix} \dfrac{s-1}{s+1} & \dfrac{s^2}{s^3+1} \\ \dfrac{s-1}{s^2+1} & \dfrac{1}{s} \end{bmatrix}$$

$$\hat{\mathbf{G}}_3(s) = \begin{bmatrix} \dfrac{s-1}{s+1} & \dfrac{s^2}{s^3+1} \\ \dfrac{s-1}{s^2+1} & s \end{bmatrix}$$

$\hat{\mathbf{G}}_1$ is a strictly proper matrix. $\hat{\mathbf{G}}_2$ is a proper matrix. $\hat{\mathbf{G}}_3(s)$ is not a proper matrix. ∎

By Definition 4-6, the rational matrix $\mathbf{C}(s\mathbf{I} - \mathbf{A})^{-1}\mathbf{B}$ is strictly proper and the rational-function matrix $\mathbf{C}(s\mathbf{I} - \mathbf{A})^{-1}\mathbf{B} + \mathbf{D}$ is proper.

### Theorem 4-10

A transfer-function matrix $\hat{\mathbf{G}}(s)$ is realizable by a finite-dimensional, linear, time-invariant dynamical equation if and only if $\hat{\mathbf{G}}(s)$ is a proper rational matrix.

### Proof

If $\hat{\mathbf{G}}(s)$ is realizable by a finite-dimensional, linear, time-invariant dynamical equation, then from (4-56), we know that $\hat{\mathbf{G}}(s)$ is a proper rational matrix. Before proving that every proper rational matrix $\hat{\mathbf{G}}(s)$ is realizable, we first prove that every scalar $(1 \times 1)$ proper rational function is realizable. The most general form of a proper rational function is

$$\hat{g}(s) = d + \frac{\beta_1 s^{n-1} + \cdots + \beta_{n-1}s + \beta_n}{s^n + \alpha_1 s^{n-1} + \cdots + \alpha_{n-1}s + \alpha_n} \tag{4-58}$$

We claim that the $n$-dimensional, linear, time-invariant dynamical equation

$$\begin{bmatrix} \dot{x}_1 \\ \dot{x}_2 \\ \cdot \\ \cdot \\ \cdot \\ \dot{x}_{n-1} \\ \dot{x}_n \end{bmatrix} = \begin{bmatrix} 0 & 1 & 0 & \cdots & 0 & 0 \\ 0 & 0 & 1 & \cdots & 0 & 0 \\ \cdot & \cdot & \cdot & & \cdot & \cdot \\ \cdot & \cdot & \cdot & & \cdot & \cdot \\ \cdot & \cdot & \cdot & & \cdot & \cdot \\ 0 & 0 & 0 & \cdots & 0 & 1 \\ -\alpha_n & -\alpha_{n-1} & -\alpha_{n-2} & \cdots & -\alpha_2 & -\alpha_1 \end{bmatrix} \begin{bmatrix} x_1 \\ x_2 \\ \cdot \\ \cdot \\ \cdot \\ x_{n-1} \\ x_n \end{bmatrix} + \begin{bmatrix} 0 \\ 0 \\ \cdot \\ \cdot \\ \cdot \\ 0 \\ 1 \end{bmatrix} u \tag{4-59a}$$

$$y = [\beta_n \quad \beta_{n-1} \quad \beta_{n-2} \quad \cdots \quad \beta_2 \quad \beta_1]\mathbf{x} + d\,u \tag{4-59b}$$

is a realization of $\hat{g}(s)$. What we have to show is that the transfer function of (4-59) is $\hat{g}(s)$. We shall demonstrate this by using Mason's formula for a signal-flow graph.[6,7] Let us choose $x_1, \dot{x}_1, x_2, \dot{x}_2, \cdots, x_n, \dot{x}_n$

[6] Mason's gain formula for signal-flow graph is as follows: The transfer function of a signal-flow graph is

$$\frac{\Sigma g_i \Delta_i}{\Delta}$$

where $\Delta = 1 - (\Sigma$ all individual loop gains$) + (\Sigma$ all possible gain products of two nontouching loops$)$; $g_i$ = gain of the $i$th forward path; and $\Delta_i$ = the part of $\Delta$ not touching the $i$th forward path. Two loops or two parts of a signal-flow graph are said to be nontouching if they do not have any common node.

[7] This can also be proved by computing algebraically the transfer function of (4-59). This is done in Chapter 6.

as nodes; then the signal-flow graph of (4-59) is of the form shown in Figure 4-6. There are $n$ loops with loop gain $-\alpha_1/s$, $-\alpha_2/s^2$, $\cdots$,

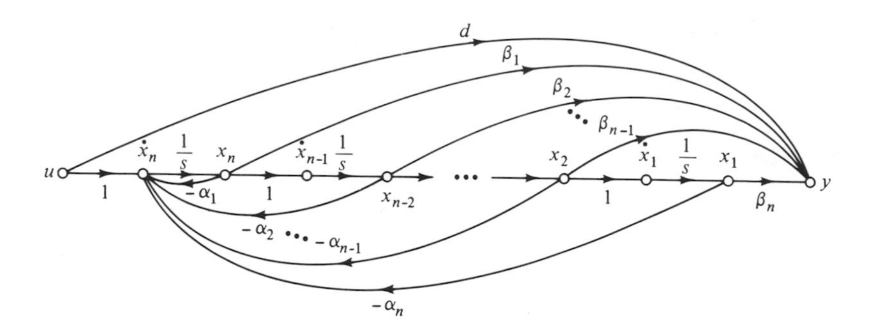


**Figure 4-6** Signal-flow graph of the dynamical equation in (4-59).

$-\alpha_n/s^n$; and there are, except the direct transmission path $d$, $n$ forward paths with gains $\beta_1/s$, $\beta_2/s^2$, $\cdots$, $\beta_n/s^n$. Since all the forward paths and loops have common nodes, we have

$$\Delta = 1 + \frac{\alpha_1}{s} + \frac{\alpha_2}{s^2} + \cdots + \frac{\alpha_n}{s^n}$$

and $\Delta_i = 1$, for $i = 1, 2, \cdots, n$. Hence the transfer function of (4-59) from the input $u$ to the output $y$ is:

$$d + \frac{\dfrac{\beta_1}{s} + \dfrac{\beta_2}{s^2} + \cdots + \dfrac{\beta_n}{s^n}}{1 + \dfrac{\alpha_1}{s} + \dfrac{\alpha_2}{s^2} + \cdots + \dfrac{\alpha_n}{s^n}} = d + \frac{\beta_1 s^{n-1} + \cdots + \beta_n}{s^n + \alpha_1 s^{n-1} + \cdots + \alpha_n} = \hat{g}(s)$$

This proves the assertion that every scalar proper rational function is realizable.

We are now ready to show that every proper rational matrix is realizable. In order to avoid cumbersome notations, we assume that $\hat{\mathbf{G}}(s)$ is a $2 \times 2$ matrix. Let

$$\hat{\mathbf{G}}(s) = \begin{bmatrix} \hat{g}_{11}(s) & \hat{g}_{12}(s) \\ \hat{g}_{21}(s) & \hat{g}_{22}(s) \end{bmatrix} \tag{4-60}$$

and let

$$\begin{aligned} \dot{\mathbf{x}}_{ij} &= \mathbf{A}_{ij}\mathbf{x}_{ij} + \mathbf{b}_{ij}u_j \\ y_{ij} &= \mathbf{c}_{ij}\mathbf{x}_{ij} + d_{ij}u_j \end{aligned}$$

be a realization of $\hat{g}_{ij}$, for $i, j = 1, 2$; that is, $\hat{g}_{ij}(s) = \mathbf{c}_{ij}(s\mathbf{I} - \mathbf{A}_{ij})^{-1}\mathbf{b}_{ij} + d_{ij}$. Note that the $\mathbf{b}_{ij}$'s are column vectors and the $\mathbf{c}_{ij}$'s are row vectors.

Then the composite dynamical equation

$$\begin{bmatrix} \dot{\mathbf{x}}_{11} \\ \dot{\mathbf{x}}_{12} \\ \dot{\mathbf{x}}_{21} \\ \dot{\mathbf{x}}_{22} \end{bmatrix} = \begin{bmatrix} \mathbf{A}_{11} & \mathbf{0} & \mathbf{0} & \mathbf{0} \\ \mathbf{0} & \mathbf{A}_{12} & \mathbf{0} & \mathbf{0} \\ \mathbf{0} & \mathbf{0} & \mathbf{A}_{21} & \mathbf{0} \\ \mathbf{0} & \mathbf{0} & \mathbf{0} & \mathbf{A}_{22} \end{bmatrix} \begin{bmatrix} \mathbf{x}_{11} \\ \mathbf{x}_{12} \\ \mathbf{x}_{21} \\ \mathbf{x}_{22} \end{bmatrix} + \begin{bmatrix} \mathbf{b}_{11} & \mathbf{0} \\ \mathbf{0} & \mathbf{b}_{12} \\ \mathbf{b}_{21} & \mathbf{0} \\ \mathbf{0} & \mathbf{b}_{22} \end{bmatrix} \begin{bmatrix} u_1 \\ u_2 \end{bmatrix} \tag{4-61a}$$

$$\begin{bmatrix} y_1 \\ y_2 \end{bmatrix} = \begin{bmatrix} \mathbf{c}_{11} & \mathbf{c}_{12} & \mathbf{0} & \mathbf{0} \\ \mathbf{0} & \mathbf{0} & \mathbf{c}_{21} & \mathbf{c}_{22} \end{bmatrix} \begin{bmatrix} \mathbf{x}_{11} \\ \mathbf{x}_{12} \\ \mathbf{x}_{21} \\ \mathbf{x}_{22} \end{bmatrix} + \begin{bmatrix} d_{11} & d_{12} \\ d_{21} & d_{22} \end{bmatrix} \begin{bmatrix} u_1 \\ u_2 \end{bmatrix} \tag{4-61b}$$

is a realization of $\hat{\mathbf{G}}(s)$. Indeed, the transfer-function matrix of (4-61) is

$$\begin{bmatrix} \mathbf{c}_{11} & \mathbf{c}_{12} & \mathbf{0} & \mathbf{0} \\ \mathbf{0} & \mathbf{0} & \mathbf{c}_{21} & \mathbf{c}_{22} \end{bmatrix} \cdot \begin{bmatrix} (s\mathbf{I} - \mathbf{A}_{11})^{-1} & \mathbf{0} & \mathbf{0} & \mathbf{0} \\ \mathbf{0} & (s\mathbf{I} - \mathbf{A}_{12})^{-1} & \mathbf{0} & \mathbf{0} \\ \mathbf{0} & \mathbf{0} & (s\mathbf{I} - \mathbf{A}_{21})^{-1} & \mathbf{0} \\ \mathbf{0} & \mathbf{0} & \mathbf{0} & (s\mathbf{I} - \mathbf{A}_{22})^{-1} \end{bmatrix} \cdot \begin{bmatrix} \mathbf{b}_{11} & \mathbf{0} \\ \mathbf{0} & \mathbf{b}_{12} \\ \mathbf{b}_{21} & \mathbf{0} \\ \mathbf{0} & \mathbf{b}_{22} \end{bmatrix} + \begin{bmatrix} d_{11} & d_{12} \\ d_{21} & d_{22} \end{bmatrix}$$

$$= \begin{bmatrix} \mathbf{c}_{11}(s\mathbf{I} - \mathbf{A}_{11})^{-1}\mathbf{b}_{11} + d_{11} & \mathbf{c}_{12}(s\mathbf{I} - \mathbf{A}_{12})^{-1}\mathbf{b}_{12} + d_{12} \\ \mathbf{c}_{21}(s\mathbf{I} - \mathbf{A}_{21})^{-1}\mathbf{b}_{21} + d_{21} & \mathbf{c}_{22}(s\mathbf{I} - \mathbf{A}_{22})^{-1}\mathbf{b}_{22} + d_{22} \end{bmatrix} = \hat{\mathbf{G}}(s) \tag{4-62}$$

Thus every proper rational matrix is realizable by a finite-dimensional, linear, time-invariant dynamical equation. Q.E.D.

Given a proper rational matrix $\hat{\mathbf{G}}(s)$, a finite-dimensional, linear, time-invariant dynamical equation realization can be obtained by using (4-59) and (4-61). However, this realization is in general not satisfactory because its dimension is much larger than the minimum required. How to find a realization of a proper rational matrix with the minimal dimension will be studied in Chapter 6.

The condition of Theorem 4-10 is stated in terms of transfer-function matrices. We may translate it into the time domain as follows:

### Corollary 4-10

An impulse-response matrix $\mathbf{G}(t)$ is realizable by a finite-dimensional, linear, time-invariant dynamical equation if and only if every entry of $\mathbf{G}(t)$ is a linear combination of terms of the form $t^k e^{\lambda_i t}$ (for $k = 0, 1, 2, \cdots$ and $i = 1, 2, \cdots$) and possibly contains a $\delta$-function at $t = 0$. ∎

The impulse-response matrix $\mathbf{G}(t)$ is the inverse Laplace transform of the transfer-function matrix $\hat{\mathbf{G}}(s)$. If a proper rational function is of the form

$$\hat{g}_{ij}(s) = d + \frac{N(s)}{(s - \lambda_1)^{k_1}(s - \lambda_2)^{k_2} \cdots}$$

then its inverse Laplace transform $g_{ij}(t)$ is a linear combination of the terms

$$\delta(t),\ e^{\lambda_1 t},\ te^{\lambda_1 t},\ \cdots,\ t^{k_1-1}e^{\lambda_1 t};\ e^{\lambda_2 t},\ te^{\lambda_2 t},\ \cdots,\ t^{k_2-1}e^{\lambda_2 t},\ \cdots$$

Hence, Corollary 4-10 follows directly from Theorem 4-10.

Since a realizable $\mathbf{G}(t)$ can be decomposed into $\mathbf{G}(t - \tau) = \mathbf{M}(t)\mathbf{N}(\tau)$, and since entries of $\mathbf{G}(t)$ are linear combinations of $t^k e^{\lambda_i t}$, for $k = 0, 1, \cdots$ and $i = 1, 2, \cdots$, the entries of $\mathbf{M}(t)$ and $\mathbf{N}(t)$ must be linear combinations of $t^k e^{\lambda_i t}$. Consequently, the matrices $\mathbf{M}(t)$ and $\mathbf{N}(t)$ are analytic functions of $t$ on the whole real line. As a consequence of this fact, we have the following theorem.

#### Theorem 4-11

A system that has a proper rational-function matrix description is relaxed at $t_0$ if and only if $\mathbf{u}_{[t_0, t_0+\epsilon]} = \mathbf{0}$ implies $\mathbf{y}_{[t_0, t_0+\epsilon]} = \mathbf{0}$ for some positive real $\epsilon$. ∎

This is a restatement of Corollary 3-1. Hence, for the class of systems that have rational transfer-function matrices—or equivalently, are describable by linear, time-invariant dynamical equations—if the output is identically zero in an interval (no matter how small), then the system is relaxed at the end of that interval.

## 4-5 Concluding Remarks

The solutions of linear dynamical equations were studied in this chapter. The solution hinges on the state transition matrix $\mathbf{\Phi}(t, \tau)$, which has the properties $\mathbf{\Phi}(t, t) = \mathbf{I}$, $\mathbf{\Phi}^{-1}(t, \tau) = \mathbf{\Phi}(\tau, t)$, and $\mathbf{\Phi}(t, \tau)\mathbf{\Phi}(\tau, t_0) = \mathbf{\Phi}(t, t_0)$. For the time-varying case, $\mathbf{\Phi}(t, \tau)$ is very difficult to compute; for the time-invariant case, $\mathbf{\Phi}(t, \tau)$ is equal to $e^{\mathbf{A}(t-\tau)}$, which can be computed by using the methods introduced in Section 2-7. In both cases, if only a specific solution is of interest, we may bypass $\mathbf{\Phi}(t, \tau)$ and compute the solution on a digital computer by direct integration.

Different analyses often lead to different dynamical-equation descriptions of a system. Mathematically, it means that dynamical equations depend on the basis chosen for the state space. However, the input–output description has nothing to do with the basis; no matter what analysis is used, it always leads to the same input–output description.

This is an advantage of the input-output description over the state-variable description.

Every proper rational function has been shown to be realizable by a finite-dimensional, linear, time-invariant dynamical equation. This corresponds to the synthesis problem in network theory. A realization of a proper rational matrix was also constructed. However the realization is not satisfactory, because generally it is possible to construct a lesser-dimensional realization. This will be discussed in Chapter 6 after the introduction of controllability and observability.

## PROBLEMS

**4-1** Find the fundamental matrices and the state transition matrices of the following homogeneous equations:

$$\begin{bmatrix} \dot{x}_1(t) \\ \dot{x}_2(t) \end{bmatrix} = \begin{bmatrix} 0 & 1 \\ 0 & t \end{bmatrix} \begin{bmatrix} x_1(t) \\ x_2(t) \end{bmatrix}$$

and

$$\begin{bmatrix} \dot{x}_1(t) \\ \dot{x}_2(t) \end{bmatrix} = \begin{bmatrix} 1 & e^{2t} \\ 0 & -1 \end{bmatrix} \begin{bmatrix} x_1(t) \\ x_2(t) \end{bmatrix}$$

**4-2** Show that $\dfrac{\partial}{\partial \tau} \mathbf{\Phi}(t, \tau) = -\mathbf{\Phi}(t, \tau)\mathbf{A}(\tau)$.

**4-3** Find the solution of

$$\dot{\mathbf{x}} = \begin{bmatrix} 0 & 1 & 0 \\ 0 & 0 & 1 \\ -2 & -4 & -3 \end{bmatrix} \mathbf{x} + \begin{bmatrix} 1 & 0 \\ 0 & 1 \\ -1 & 1 \end{bmatrix} \mathbf{u}$$

$$\mathbf{y} = \begin{bmatrix} 0 & 1 & -1 \\ 1 & 2 & 1 \end{bmatrix} \mathbf{x}$$

with the initial state

$$\mathbf{x}(0) = \begin{bmatrix} 1 \\ 0 \\ 0 \end{bmatrix}$$

Verify for this problem that $e^{\mathbf{A}(t-\tau)} = e^{\mathbf{A}t}e^{-\mathbf{A}\tau} = e^{-\mathbf{A}\tau}e^{\mathbf{A}t}$.

**4-4** Let

$$\mathbf{A} = \begin{bmatrix} 1 & 0 & 1 & 1 \\ 0 & 1 & 0 & 0 \\ 0 & 0 & 1 & -1 \\ 0 & 0 & 0 & 1 \end{bmatrix}$$

Find $e^{\mathbf{A}t}$ by using the formula $\mathcal{L}[e^{\mathbf{A}t}] = (s\mathbf{I} - \mathbf{A})^{-1}$.

**4-5** If $\mathbf{T}^{-1}(t)$ exists and is differentiable for all $t$, show that

$$\frac{d}{dt}[\mathbf{T}^{-1}(t)] = -\mathbf{T}^{-1}(t)\left[\frac{d}{dt}\mathbf{T}(t)\right]\mathbf{T}^{-1}(t)$$

**4-6** From $\mathbf{\Phi}(t, \tau)$, show how to compute $\mathbf{A}(t)$.

**4-7** Given

$$\mathbf{A}(t) = \begin{bmatrix} a_{11}(t) & a_{12}(t) \\ a_{21}(t) & a_{22}(t) \end{bmatrix}$$

show that

$$\det \mathbf{\Phi}(t, t_0) = \exp\left[\int_{t_0}^{t} (a_{11}(\tau) + a_{22}(\tau))\, d\tau\right]$$

where $\frac{\partial}{\partial t}\mathbf{\Phi}(t, t_0) = \mathbf{A}(t)\mathbf{\Phi}(t, t_0)$ and $\mathbf{\Phi}(t_0, t_0) = \mathbf{I}$. *Hint:* Show that

$$\frac{\partial}{\partial t}\det(\mathbf{\Phi}(t, t_0)) = (a_{11}(t) + a_{22}(t))\det(\mathbf{\Phi}(t, t_0))$$

**4-8** Prove Theorem 4-7.

**4-9** Given $\dot{\mathbf{x}}(t) = \mathbf{A}(t)\mathbf{x}$. The equation $\dot{\mathbf{z}} = -\mathbf{A}^*\mathbf{z}$ is called the *adjoint equation* of $\dot{\mathbf{x}} = \mathbf{A}(t)\mathbf{x}$, where $\mathbf{A}^*$ is the complex conjugate transpose of $\mathbf{A}$. Let $\mathbf{\Phi}(t, t_0)$ and $\mathbf{\Phi}_1(t, t_0)$ be the state transition matrices of $\dot{\mathbf{x}} = \mathbf{A}(t)\mathbf{x}$ and $\dot{\mathbf{z}} = -\mathbf{A}^*\mathbf{z}$, respectively. Verify that

$$\mathbf{\Phi}_1(t, t_0)\mathbf{\Phi}^*(t, t_0) = \mathbf{\Phi}^*(t, t_0)\mathbf{\Phi}_1(t, t_0) = \mathbf{\Phi}_1^*(t, t_0)\mathbf{\Phi}(t, t_0) = \mathbf{I}$$

**4-10** Verify that $\mathbf{B}(t) = \mathbf{\Phi}(t, t_0)\mathbf{B}_0\mathbf{\Phi}^*(t, t_0)$ is the solution of

$$\frac{d}{dt}\mathbf{B}(t) = \mathbf{A}(t)\mathbf{B}(t) + \mathbf{B}(t)\mathbf{A}^*(t) \qquad \mathbf{B}(t_0) = \mathbf{B}_0$$

where $\mathbf{\Phi}(t, t_0)$ is the state transition matrix of $\dot{\mathbf{x}} = \mathbf{A}(t)\mathbf{x}$.

**4-11** Verify that $\mathbf{X}(t) = e^{\mathbf{A}t}\mathbf{C}e^{\mathbf{B}t}$ is the solution of

$$\frac{d}{dt}\mathbf{X} = \mathbf{AX} + \mathbf{XB} \qquad \mathbf{X}(0) = \mathbf{C}$$

**4-12** Every element of $\mathbf{\Phi}(t, t_0)$ can be interpreted as the impulse response of some input–output pair. What is the input and the output of the $ij$th element of $\mathbf{\Phi}(t, t_0)$?

**4-13** Let

$$\mathbf{\Phi}(t, t_0) = \begin{bmatrix} \mathbf{\Phi}_{11}(t, t_0) & \mathbf{\Phi}_{12}(t, t_0) \\ \mathbf{\Phi}_{21}(t, t_0) & \mathbf{\Phi}_{22}(t, t_0) \end{bmatrix}$$

be the state transition matrix of

$$\dot{\mathbf{x}} = \begin{bmatrix} \mathbf{A}_{11} & \mathbf{A}_{12} \\ \mathbf{0} & \mathbf{A}_{22} \end{bmatrix} \mathbf{x}$$

Show that $\mathbf{\Phi}_{21}(t, t_0) = \mathbf{0}$ for all $t$, $t_0$ and $\frac{\partial}{\partial t} \mathbf{\Phi}_{ii}(t, t_0) = A_{ii}\mathbf{\Phi}(t, t_0)$ for $i = 1, 2$.

**4-14** Given an $n$-dimensional realization $\{\mathbf{A}, \mathbf{B}, \mathbf{C}\}$ of $\mathbf{G}(t, \tau)$. Find the equivalence transformation that will transform $\{\mathbf{A}, \mathbf{B}, \mathbf{C}\}$ into $\{\mathbf{0}, \bar{\mathbf{B}}, \bar{\mathbf{C}}\}$.

**4-15** Let

$$(s\mathbf{I} - \mathbf{A})^{-1} \triangleq \frac{1}{\Delta(s)} [\mathbf{R}_0 s^{n-1} + \mathbf{R}_1 s^{n-2} + \cdots + \mathbf{R}_{n-2} s + \mathbf{R}_{n-1}]$$

where

$$\Delta(s) \triangleq \det (s\mathbf{I} - \mathbf{A}) = s^n + \alpha_1 s^{n-1} + \alpha_2 s^{n-2} + \cdots + \alpha_n$$

is the characteristic polynomial of $\mathbf{A}$. Verify that

$$\begin{aligned} \mathbf{R}_0 &= \mathbf{I} \\ \mathbf{R}_1 &= \mathbf{A} + \alpha_1 \mathbf{I} = \mathbf{A}\mathbf{R}_0 + \alpha_1 \mathbf{I} \\ \mathbf{R}_2 &= \mathbf{A}^2 + \alpha_1 \mathbf{A} + \alpha_2 \mathbf{I} = \mathbf{A}\mathbf{R}_1 + \alpha_2 \mathbf{I} \\ &\ \ \vdots \\ \mathbf{R}_{n-1} &= \mathbf{A}^{n-1} + \alpha_1 \mathbf{A}^{n-2} + \cdots + \alpha_{n-1}\mathbf{I} = \mathbf{A}\mathbf{R}_{n-2} + \alpha_{n-1}\mathbf{I} \end{aligned}$$

These equations can be used to compute the coefficients $\alpha_i$ of the characteristic polynomial: $\alpha_1 = -\text{tr}(\mathbf{A})$, $\alpha_2 = -\frac{1}{2}\text{tr}(\mathbf{R}_1\mathbf{A})$, $\ldots$, $\alpha_{n-1} = -\frac{1}{(n-1)}\text{tr}(\mathbf{R}_{n-2}\mathbf{A})$, and $\alpha_n = -\frac{1}{n}\text{tr}(\mathbf{R}_{n-1}\mathbf{A})$. The trace of a matrix, tr $(\mathbf{M})$, is the sum of all the diagonal elements of $\mathbf{M}$. The proof of this statement can be found, for example, in References [39] and [116].

**4-16** Use a signal-flow graph to show that the transfer function of the following single-variable, linear, time-invariant dynamical equation

$$\dot{\mathbf{x}} = \begin{bmatrix} 0 & 0 & 0 & \cdots & 0 & -\alpha_n \\ 1 & 0 & 0 & \cdots & 0 & -\alpha_{n-1} \\ 0 & 1 & 0 & \cdots & 0 & -\alpha_{n-2} \\ \cdot & \cdot & \cdot & & \cdot & \cdot \\ \cdot & \cdot & \cdot & & \cdot & \cdot \\ \cdot & \cdot & \cdot & & \cdot & \cdot \\ 0 & 0 & 0 & \cdots & 1 & -\alpha_1 \end{bmatrix} \mathbf{x} + \begin{bmatrix} \beta_n \\ \beta_{n-1} \\ \beta_{n-2} \\ \cdot \\ \cdot \\ \cdot \\ \beta_1 \end{bmatrix} u$$

$$y = [0 \ \ 0 \ \ 0 \ \ \cdots \ \ 0 \ \ 1]\mathbf{x} + du$$

is

$$\hat{g}(s) = d + \frac{\beta_1 s^{n-1} + \cdots + \beta_n}{s^n + \alpha_1 s^{n-1} + \cdots + \alpha_{n-1}s + \alpha_n}$$

**4-17** Consider the network shown in Figure P4-17. Find the initial inductor current and the initial capacitor voltage such that for the input

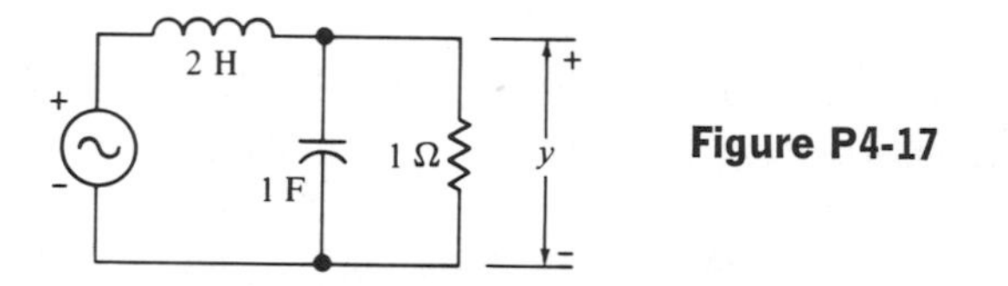


**Figure P4-17**

$u(t) = e^{-4t}$. The output $y(t)$ will be immediately of the form $e^{-4t}$ without containing any transient.

**4-18** Find an equivalent time-invariant dynamical equation of

$$\dot{x} = (\cos t \sin t)x$$

**4-19** Find an equivalent Jordan-canonical-form dynamical equation of

$$\begin{bmatrix} \dot{x}_1 \\ \dot{x}_2 \\ \dot{x}_3 \end{bmatrix} = \begin{bmatrix} 0 & 4 & 3 \\ 0 & 20 & 16 \\ 0 & -25 & -20 \end{bmatrix} \begin{bmatrix} x_1 \\ x_2 \\ x_3 \end{bmatrix} + \begin{bmatrix} -1 \\ 3 \\ 0 \end{bmatrix} u$$

$$y = [-1 \quad 3 \quad 0] \begin{bmatrix} x_1 \\ x_2 \\ x_3 \end{bmatrix} + 4u$$

**4-20** Find a time-varying dynamical equation realization and a time-invariant dynamical realization of the impulse response $g(t) = t^2 e^{\lambda t}$.

**4-21** Realize the proper rational matrix

$$\begin{bmatrix} \dfrac{2+s}{s+1} & \dfrac{1}{s+3} \\ \dfrac{5}{s+1} & \dfrac{5s+1}{s+2} \end{bmatrix}$$

and draw an analog computer simulation block diagram for it.

**4-22** Find a dynamical-equation realization of $g(t, \tau) = \sin t\, e^{-(t-\tau)} \cos \tau$. Is it possible to find a linear, time-invariant dynamical-equation realization for it?

**4-23** Let $FE$ and $F\bar{E}$ be two linear, time-invariant dynamical equations (not necessarily of the same dimension). Show that $FE$ and $F\bar{E}$ are zero-state equivalent if and only if

$$\mathbf{CA}^k\mathbf{B} = \bar{\mathbf{C}}\bar{\mathbf{A}}^k\bar{\mathbf{B}} \qquad k = 0, 1, 2, \cdots$$

and

$$\mathbf{D} = \bar{\mathbf{D}}.$$

**4-24** Consider $\mathbf{x}(n + 1) = \mathbf{A}(n)\mathbf{x}(n)$. Define

$$\mathbf{\Phi}(n, m) \triangleq \mathbf{A}(n - 1)\mathbf{A}(n - 2)\mathbf{A}(n - 3) \cdots \mathbf{A}(m) \qquad \text{for } n > m$$
$$\mathbf{\Phi}(m, m) \triangleq \mathbf{I}$$

Show that, given the initial state $\mathbf{x}(m) = \mathbf{x}^0$, the state at time $n$ is given by $\mathbf{x}(n) = \mathbf{\Phi}(n, m)\mathbf{x}^0$. If $\mathbf{A}$ is independent of $n$, what is $\mathbf{\Phi}(n, m)$?

**4-25** For continuous-time dynamical equations, the state transition matrix $\mathbf{\Phi}(t, \tau)$ is defined for all $t, \tau$. However, in discrete-time dynamical equation, $\mathbf{\Phi}(n, m)$ is defined only for $n \geq m$. What condition do we need on $\mathbf{A}(n)$ in order for $\mathbf{\Phi}(n, m)$ to be defined for $n < m$?

**4-26** Show that the solution of $\mathbf{x}(n + 1) = \mathbf{A}(n)\mathbf{x}(n) + \mathbf{B}(n)\mathbf{u}(n)$ is given by

$$\mathbf{x}(n) = \mathbf{\Phi}(n, m)\mathbf{x}(m) + \sum_{l=m}^{n-1} \mathbf{\Phi}(n, l + 1)\mathbf{B}(l)\mathbf{u}(l)$$

[This can be easily verified by considering $\mathbf{B}(l)\mathbf{u}(l)$ as an initial state at time $(l + 1)$.]

**4-27** Find an equivalent discrete-time Jordan-canonical-form dynamical equation for

$$\begin{bmatrix} x_1(n+1) \\ x_2(n+1) \\ x_3(n+1) \end{bmatrix} = \begin{bmatrix} 0 & 4 & 3 \\ 0 & 20 & 16 \\ 0 & -25 & -20 \end{bmatrix} \begin{bmatrix} x_1(n) \\ x_2(n) \\ x_3(n) \end{bmatrix} + \begin{bmatrix} -1 \\ 3 \\ 0 \end{bmatrix} u(n)$$

$$y(n) = [-1 \quad 3 \quad 0] \begin{bmatrix} x_1(n) \\ x_2(n) \\ x_3(n) \end{bmatrix} + 4u(n)$$

**4-28** Consider the $n$-dimensional, linear, time-invariant dynamical equation

$$FE: \quad \begin{aligned} \dot{\mathbf{x}} &= \mathbf{Ax} + \mathbf{Bu} \\ \mathbf{y} &= \mathbf{Cx} + \mathbf{Du} \end{aligned}$$

If the input $\mathbf{u}(t)$ is piecewise constant on intervals of constant length $T > 0$ as illustrated in Figure P4-28, show that the behavior of the system

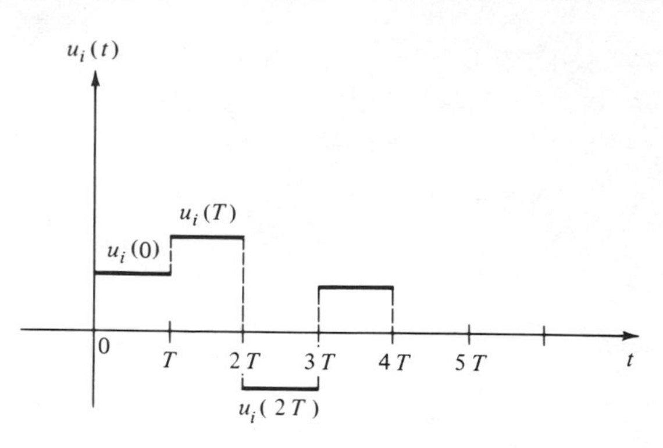


**Figure P4-28**

at discrete instants of time $0, T, 2T, 3T, \cdots$ is governed by the following $n$-dimensional, linear, time-invariant discrete-time equation:

$$\mathbf{x}(n+1) = \bar{\mathbf{A}}\mathbf{x}(n) + \bar{\mathbf{B}}\mathbf{u}(n)$$
$$\mathbf{y}(n) = \bar{\mathbf{C}}\mathbf{x}(n) + \bar{\mathbf{D}}\mathbf{u}(n)$$

where

$$\bar{\mathbf{A}} = e^{\mathbf{A}T} \qquad \bar{\mathbf{C}} = \mathbf{C}$$
$$\bar{\mathbf{B}} = \int_0^T e^{\mathbf{A}\tau}\mathbf{B}\,d\tau \qquad \bar{\mathbf{D}} = \mathbf{D}$$

(See Appendix C.)

**4-29** Let $\mathbf{A}(t) \triangleq (a_{ij}(t))$; then, by definition,

$$\frac{d}{dt}\mathbf{A}(t) = \left(\frac{d}{dt}a_{ij}(t)\right)$$

Verify that

$$\frac{d}{dt}(\mathbf{A}(t)\mathbf{B}(t)) = \dot{\mathbf{A}}(t)\mathbf{B}(t) + \mathbf{A}(t)\dot{\mathbf{B}}(t)$$

Verify also that

$$\frac{d}{dt}[\mathbf{A}(t)]^2 \triangleq \frac{d}{dt}(\mathbf{A}(t)\mathbf{A}(t)) = 2\dot{\mathbf{A}}(t)\mathbf{A}(t)$$

if and only if $\dot{\mathbf{A}}(t)$ and $\mathbf{A}(t)$ commute; that is, $\dot{\mathbf{A}}(t)\mathbf{A}(t) = \mathbf{A}(t)\dot{\mathbf{A}}(t)$.

**4-30** Show that if $\int_{t_0}^t \mathbf{A}(\tau)\,d\tau$ and $\mathbf{A}(t)$ commute for all $t$, then the unique solution of

$$\frac{\partial}{\partial t}\mathbf{\Phi}(t, t_0) = \mathbf{A}(t)\mathbf{\Phi}(t, t_0) \qquad \mathbf{\Phi}(t_0, t_0) = \mathbf{I}$$

is

$$\mathbf{\Phi}(t, t_0) = \exp \int_{t_0}^t \mathbf{A}(\tau)\,d\tau$$

*Hint:* Use Problem 4-29 and Equation (2-78).

**4-31** Extend Theorem 4-5 to the multivariable case.

CHAPTER

# 5

# Controllability and Observability of Linear Dynamical Equations

## 5-1 Introduction

System analyses generally consist of two parts: quantitative and qualitative. In the quantitative study we are interested in the exact response of the system to certain input and initial conditions, as we studied in the preceding chapter. In the qualitative study, we are interested in the general properties of a system. In this chapter we shall introduce two qualitative properties of linear dynamical equations: controllability and observability. We shall first give the reader some rough ideas of these two concepts by using the network shown in Figure 5-1. The

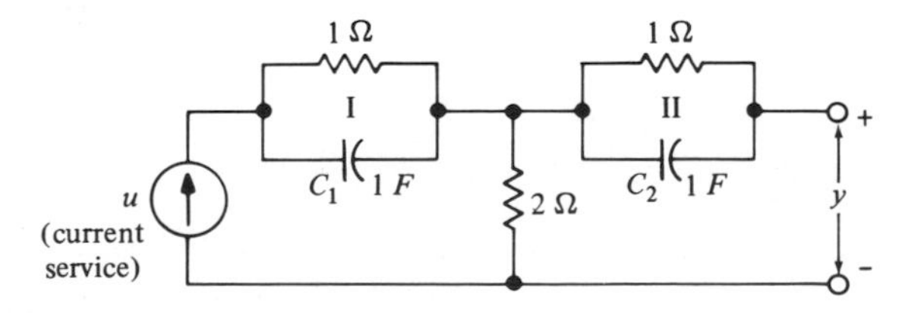


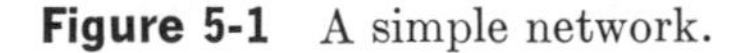
**Figure 5-1** A simple network.

input $u$ of the network is a current source. It is clear that if the initial voltage in the capacitor $C_2$ in loop II is zero, no matter what input $u$ is applied, the mode $e^{-t}$ in II can never be excited. Hence the mode $e^{-t}$ in II is said to be not controllable by the input $u$. On the other hand, the mode $e^{-t}$ in loop I can be excited by the application of the input $u$; hence the mode in I is controllable by the input $u$. Although the mode $e^{-t}$ in I can be excited by the input, its presence can never be detected from the output terminal $y$. Hence it is said to be not observable from the output $y$. On the other hand, the presence of the mode $e^{-t}$ in II can be detected from the output $y$; hence the mode is said to be observable. This illustration, though not very accurate, may convey the ideas of the concepts of controllability and observability.

The concepts of controllability and observability are very important in the study of control and filtering problems. As an example, consider the platform system shown in Figure 5-2. The system consists of one

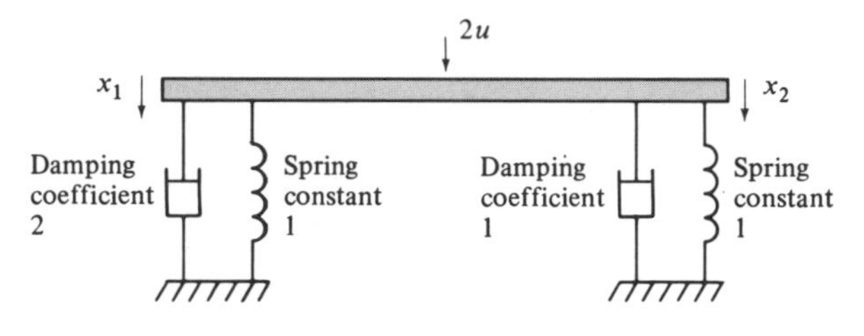


**Figure 5-2** A platform system.

platform; both ends of the platform are supported on the ground by means of springs and dashpots. The mass of the platform is, for simplicity, assumed to be zero; hence the movements of the two spring systems are independent. If the initial displacements of both ends of the platform are different from zero, the platform will start to vibrate. If no force is applied, it will take an infinite time for the platform to come back to rest. Now we may ask: For any initial displacements, is it possible to apply a force to bring the platform to rest in a *finite* time? In order to answer this question, the concept of controllability is needed.

This chapter is organized as follows. In Section 5-2 the required mathematical background is introduced. Three theorems that give the conditions for linear independence of a set of vector functions are presented. All the results in controllability and observability follow almost directly from these three theorems. The concept of controllability is introduced in Section 5-3. Necessary and sufficient conditions for linear time-varying dynamical equations and linear time-invariant dynamical equations to be controllable are derived. The concept of observability is introduced in Section 5-4. It is dual to the concept of controllability, and hence its discussion is rather brief. Duality theorem is introduced there. The controllability and observability of linear time-invariant Jordan-form dynamical equations are studied in Section 5-5. In terms of Jordan-form equations, controllability and observability can be checked almost by inspection. In Section 5-6 we study the dynamical equations which are uncontrollable and/or unobservable. The canonical decomposition theorem is introduced. A consequence of this theorem is that the transfer-function matrix of a dynamical equation depends solely on the part of equation that is controllable and observable. In the last section, the concepts of output controllability and output function controllability are introduced. It is shown that they are properties of the input–output description of a system.

Although elements of the matrices **A**, **B**, **C**, and **D** are all real-valued, for mathematical convenience they are considered as elements of the field of complex numbers. Consequently, the state space of an $n$-dimensional dynamical equation will be taken as an $n$-dimensional complex vector space $(\mathbb{C}^n, \mathbb{C})$.

The references for this chapter are [2], [8], [11], [13], [14], [20], [21], [48], [55], [56], [60], [61], [69], [71], [98], [103], and [105]–[107].

The reader who is interested in only the time-invariant case may skip Theorems 5-2, 5-5, and 5-6.

## 5-2 Linear Independence of Time Functions

The concept of linear independence of a set of vectors of a linear space was introduced in Section 2-3. We shall now apply this concept to a set functions of a real variable. A set of complex-valued functions $f_1, f_2, \cdots, f_n$ is said to be linearly dependent on the interval[1] $[t_1, t_2]$ over the field of complex numbers if there exist complex numbers $\alpha_1, \alpha_2, \cdots, \alpha_n$, not all zero, such that

$$\alpha_1 f_1(t) + \alpha_2 f_2(t) + \cdots + \alpha_n f_n(t) = 0 \qquad \text{for all } t \text{ in } [t_1, t_2] \tag{5-1}$$

Otherwise the set of functions is said to be linearly independent on $[t_1, t_2]$ over the field of complex numbers. In this definition, the specification of time interval is crucial.

### Example 1

Consider the two continuous functions $f_1$ and $f_2$ defined by

$$f_1(t) = t \qquad \text{for } t \text{ in } [-1, 1]$$

$$f_2(t) = \begin{cases} t & \text{for } t \text{ in } [0, 1] \\ -t & \text{for } t \text{ in } [-1, 0] \end{cases}$$

[1] The functions we study are mostly continuous functions; hence there is no substantial difference between using the open interval $(t_1, t_2)$ and the closed interval $[t_1, t_2]$. Every $[t_1, t_2]$ is assumed to be a nonzero interval.

It is clear that the functions $f_1$ and $f_2$ are linearly dependent on [0, 1], since if we choose $\alpha_1 = 1$, $\alpha_2 = -1$, then $\alpha_1 f_1(t) + \alpha_2 f_2(t) = 0$ for all $t$ in [0, 1]. The functions $f_1$ and $f_2$ are also linearly dependent on $[-1, 0]$. However $f_1$ and $f_2$ are linearly independent on $[-1, 1]$. ▮

From this example, we see that although a set of functions is linearly independent on an interval, it is not necessary that they are linearly independent on *any* subinterval. However, it is true that there exists a subinterval on which they are linearly independent. For example, in Example 1 the functions $f_1$ and $f_2$ are linearly independent on the subinterval $[-\epsilon, \epsilon]$ for any positive $\epsilon$. On the other hand, if a set of functions is linearly independent on an interval $[t_1, t_2]$, then the set of functions is linearly independent on *any* interval that contains $[t_1, t_2]$.

The concept of linear independence can be extended to vector-valued functions. Let $\mathbf{f}_i$, for $i = 1, 2, \cdots, n$, be $1 \times p$ complex-valued functions of $t$; then the $1 \times p$ complex-valued functions $\mathbf{f}_1, \mathbf{f}_2, \cdots, \mathbf{f}_n$ are linearly dependent on $[t_1, t_2]$ if there exist complex numbers $\alpha_1, \alpha_2, \cdots, \alpha_n$, not all zero, such that

$$\alpha_1 \mathbf{f}_1(t) + \alpha_2 \mathbf{f}_2(t) + \cdots + \alpha_n \mathbf{f}_n(t) = \mathbf{0} \qquad \text{for all } t \text{ in } [t_1, t_2] \tag{5-2}$$

Otherwise the **f**'s are linearly independent on $[t_1, t_2]$. Note that the zero vector in Equation (5-2) is a $1 \times p$ row vector $[0 \;\; 0 \;\; \cdots \;\; 0]$. As in Definition 2-4′, we may as well state that $\mathbf{f}_1, \mathbf{f}_2, \cdots, \mathbf{f}_n$ are linearly independent on $[t_1, t_2]$ if and only if

$$[\alpha_1 \;\; \alpha_2 \;\; \cdots \;\; \alpha_n] \begin{bmatrix} \mathbf{f}_1(t) \\ \mathbf{f}_2(t) \\ \cdot \\ \cdot \\ \cdot \\ \mathbf{f}_n(t) \end{bmatrix} \triangleq \boldsymbol{\alpha}\mathbf{F}(t) = \mathbf{0} \qquad \text{for all } t \text{ in } [t_1, t_2] \tag{5-3}$$

implies $\boldsymbol{\alpha} = \mathbf{0}$, where

$$\boldsymbol{\alpha} \triangleq [\alpha_1 \;\; \alpha_2 \;\; \cdots \;\; \alpha_n] \qquad \mathbf{F} \triangleq \begin{bmatrix} \mathbf{f}_1 \\ \mathbf{f}_2 \\ \cdot \\ \cdot \\ \cdot \\ \mathbf{f}_n \end{bmatrix}$$

Clearly, $\boldsymbol{\alpha}$ is a constant $1 \times n$ row vector and $\mathbf{F}$ is an $n \times p$ matrix function.

The linear independence of a set of functions is a property associated with an interval, hence in testing for linear independence, we have to consider the entire interval. Let $\mathbf{F}^*(t)$ be the complex conjugate transpose of $\mathbf{F}(t)$.

## Theorem 5-1

Let $\mathbf{f}_i$, for $i = 1, 2, \cdots, n$, be $1 \times p$ complex-valued continuous functions defined on $[t_1, t_2]$. Let $\mathbf{F}$ be the $n \times p$ matrix with $\mathbf{f}_i$ as its $i$th row. Define

$$\mathbf{W}(t_1, t_2) \triangleq \int_{t_1}^{t_2} \mathbf{F}(t)\mathbf{F}^*(t)\, dt$$

Then $\mathbf{f}_1, \mathbf{f}_2, \cdots, \mathbf{f}_n$ are linearly independent on $[t_1, t_2]$ if and only if the $n \times n$ constant matrix $\mathbf{W}(t_1, t_2)$ is nonsingular.

## Proof

The proof of this theorem is similar to that of Theorem 2-8. Recall that a matrix is nonsingular if and only if all the columns (rows) are linearly independent. We prove first the necessity of the theorem; we prove it by contradiction. Assume that the $\mathbf{f}_i$'s are linearly independent on $[t_1, t_2]$, but $\mathbf{W}(t_1, t_2)$ is singular. Then there exists a nonzero $1 \times n$ row vector $\boldsymbol{\alpha}$ such that $\boldsymbol{\alpha}\mathbf{W}(t_1, t_2) = \mathbf{0}$. This implies $\boldsymbol{\alpha}\mathbf{W}(t_1, t_2)\boldsymbol{\alpha}^* = 0$, or

$$\boldsymbol{\alpha}\mathbf{W}(t_1, t_2)\boldsymbol{\alpha}^* = \int_{t_1}^{t_2} (\boldsymbol{\alpha}\mathbf{F}(t))(\boldsymbol{\alpha}\mathbf{F}(t))^*\, dt = 0 \tag{5-4}$$

Since the integrand $(\boldsymbol{\alpha}\mathbf{F}(t))(\boldsymbol{\alpha}\mathbf{F}(t))^*$ is a continuous function and is nonnegative for all $t$ in $[t_1, t_2]$, Equation (5-4) implies that

$$\boldsymbol{\alpha}\mathbf{F}(t) = \mathbf{0} \qquad \text{for all } t \text{ in } [t_1, t_2]$$

This contradicts the linear independence assumption of the set of $\mathbf{f}_i$, $i = 1, 2, \cdots, n$. Hence if the $\mathbf{f}_i$'s are linearly independent on $[t_1, t_2]$, then $\det \mathbf{W}(t_1, t_2) \neq 0$.

We next prove the sufficiency of the theorem. Suppose that $\mathbf{W}(t_1, t_2)$ is nonsingular, but the $\mathbf{f}$'s are linearly dependent on $[t_1, t_2]$. Then, by definition, there exists a nonzero constant $1 \times n$ row vector $\boldsymbol{\alpha}$ such that $\boldsymbol{\alpha}\mathbf{F}(t) = \mathbf{0}$ for all $t$ in $[t_1, t_2]$. Consequently, we have

$$\boldsymbol{\alpha}\mathbf{W}(t_1, t_2) = \int_{t_1}^{t_2} \boldsymbol{\alpha}\mathbf{F}(t)\mathbf{F}^*(t)\, dt = \mathbf{0} \tag{5-5}$$

which contradicts the assumption that $\mathbf{W}(t_1, t_2)$ is nonsingular. Hence, if $\mathbf{W}(t_1, t_2)$ is nonsingular, then the $\mathbf{f}_i$'s are linearly independent on $[t_1, t_2]$. Q.E.D.

The determinant of $\mathbf{W}(t_1, t_2)$ is called the *Gram determinant* of the $\mathbf{f}_i$'s. In applying Theorem 5-1, the functions $\mathbf{f}_i$, for $i = 1, 2, \cdots, n$, are required to be continuous. If the functions $\mathbf{f}_i$, for $i = 1, 2, \cdots, n$, have continuous derivatives up to order $(n - 1)$, then we may use the following theorem.

### Theorem 5-2

Assume that the $1 \times p$ complex-valued functions $\mathbf{f}_1, \mathbf{f}_2, \cdots, \mathbf{f}_n$ have continuous derivatives up to order $(n-1)$ on the interval $[t_1, t_2]$. Let $\mathbf{F}$ be the $n \times p$ matrix with $\mathbf{f}_i$ as its $i$th row and let $\mathbf{F}^{(k)}$ be the $k$th derivative of $\mathbf{F}$. If there exists some $t_0$ in $[t_1, t_2]$ such that the $n \times np$ matrix

$$[\mathbf{F}(t_0) \vdots \mathbf{F}^{(1)}(t_0) \vdots \mathbf{F}^{(2)}(t_0) \vdots \cdots \vdots \mathbf{F}^{(n-1)}(t_0)] \tag{5-6}$$

has rank $n$, then the $\mathbf{f}_i$'s are linearly independent on $[t_1, t_2]$ over the field of complex numbers.[1a]

### Proof

We prove the theorem by contradiction. Suppose that there exists some $t_0$ in $[t_1, t_2]$ such that

$$\rho[\mathbf{F}(t_0) \vdots \mathbf{F}^{(1)}(t_0) \vdots \cdots \vdots \mathbf{F}^{(n-1)}(t_0)] = n$$

but the $\mathbf{f}_i$'s are linearly dependent on $[t_1, t_2]$. Then by definition, there exists a nonzero $1 \times n$ row vector $\boldsymbol{\alpha}$ such that

$$\boldsymbol{\alpha}\mathbf{F}(t) = \mathbf{0} \qquad \text{for all } t \text{ in } [t_1, t_2]$$

This implies that

$$\boldsymbol{\alpha}\mathbf{F}^{(k)}(t) = \mathbf{0} \qquad \text{for all } t \text{ in } [t_1, t_2] \text{ and } k = 1, 2, \cdots, n-1$$

Hence we have

$$\boldsymbol{\alpha}[\mathbf{F}(t) \vdots \mathbf{F}^{(1)}(t) \vdots \cdots \vdots \mathbf{F}^{(n-1)}(t)] = \mathbf{0} \qquad \text{for all } t \text{ in } [t_1, t_2]$$

In particular,

$$\boldsymbol{\alpha}[\mathbf{F}(t_0) \vdots \mathbf{F}^{(1)}(t_0) \vdots \cdots \vdots \mathbf{F}^{(n-1)}(t_0)] = \mathbf{0},$$

which implies that all the $n$ rows of

$$[\mathbf{F}(t_0) \vdots \mathbf{F}^{(1)}(t_0) \vdots \cdots \vdots \mathbf{F}^{(n-1)}(t_0)]$$

are linearly dependent. This contradicts the hypothesis that

$$[\mathbf{F}(t_0) \vdots \mathbf{F}^{(1)}(t_0) \vdots \cdots \vdots \mathbf{F}^{(n-1)}(t_0)]$$

has rank $n$. Hence the $\mathbf{f}$'s are linearly independent on $[t_1, t_2]$. Q.E.D.

The condition of Theorem 5-2 is sufficient but not necessary for a set of functions to be linearly independent. This can be seen from the following example.

---

[1a] If $t_0$ is at either $t_1$ or $t_2$, the end points of an interval, then $\mathbf{F}^{(k)}(t_0)$ is defined as $\mathbf{F}^{(k)}(t)$ with $t$ approaches $t_0$ from inside the interval.

### Example 2

Consider the two functions

$$f_1(t) = t^3$$
$$f_2(t) = |t^3|$$

defined over $[-1, 1]$. They are linearly independent on $[-1, 1]$; however,

$$\rho\begin{bmatrix} f_1(t) & f_1^{(1)}(t) \\ f_2(t) & f_2^{(1)}(t) \end{bmatrix} = \rho\begin{bmatrix} t^3 & 3t^2 \\ t^3 & 3t^2 \end{bmatrix} = 1 \qquad \text{for all } t \text{ in } (0, 1]$$

$$\rho\begin{bmatrix} f_1(t) & f_1^{(1)}(t) \\ f_2(t) & f_2^{(1)}(t) \end{bmatrix} = \rho\begin{bmatrix} t^3 & 3t^2 \\ -t^3 & -3t^2 \end{bmatrix} = 1 \qquad \text{for all } t \text{ in } [-1, 0)$$

and

$$\rho\begin{bmatrix} f_1(t) & f_1^{(1)}(t) \\ f_2(t) & f_2^{(1)}(t) \end{bmatrix} = 0 \qquad \text{at } t = 0$$

∎

To check the linear independence of a set of functions, if the functions are continuous we can employ Theorem 5-1, which requires an integration over an interval. If the functions are continuously differentiable up to certain order, then Theorem 5-2 can be used. It is clear that Theorem 5-2 is easier to use than Theorem 5-1; however, it gives only sufficient conditions. If the functions are analytic, then we can use Theorem 5-3, which is based on the fact that if a function is analytic on $[t_1, t_2]$, then the function is completely determinable from a point in $[t_1, t_2]$, if all the derivatives of the function at that point are known. (See Appendix A.)

### Theorem 5-3

Assume that for each $i$, $\mathbf{f}_i$ is analytic on $[t_1, t_2]$. Let $\mathbf{F}$ be the $n \times p$ matrix with $\mathbf{f}_i$ as its $i$th row and let $\mathbf{F}^{(k)}$ be the $k$th derivative of $\mathbf{F}$. Let $t_0$ be any fixed point in $[t_1, t_2]$. Then the $\mathbf{f}_i$'s are linearly independent on $[t_1, t_2]$ if and only if

$$\rho[\mathbf{F}(t_0) \,\vdots\, \mathbf{F}^{(1)}(t_0) \,\vdots\, \cdots \,\vdots\, \mathbf{F}^{(n-1)}(t_0) \,\vdots\, \cdots] = n \tag{5-7}$$

### Proof

The sufficiency of the theorem can be proved as in Theorem 5-2. Now we prove by contradiction the necessity of the theorem. Suppose that

$$\rho[\mathbf{F}(t_0) \,\vdots\, \mathbf{F}^{(1)}(t_0) \,\vdots\, \cdots \,\vdots\, \mathbf{F}^{(n-1)}(t_0) \,\vdots\, \cdots] < n$$

Then the rows of the infinite matrix

$$[\mathbf{F}(t_0) \,\vdots\, \mathbf{F}^{(1)}(t_0) \,\vdots\, \cdots \,\vdots\, \mathbf{F}^{(n-1)}(t_0) \,\vdots\, \cdots]$$

are linearly dependent. Consequently, there exists a nonzero $1 \times n$ row vector $\boldsymbol{\alpha}$ such that

$$\boldsymbol{\alpha}[\mathbf{F}(t_0) \,\vdots\, \mathbf{F}^{(1)}(t_0) \,\vdots\, \cdots \,\vdots\, \mathbf{F}^{(n-1)}(t_0) \,\vdots\, \cdots] = \mathbf{0} \tag{5-8}$$

The $\mathbf{f}_i$'s are analytic on $[t_1, t_2]$ by assumption; hence there exists an $\epsilon > 0$ such that, for all $t$ in $[t_0 - \epsilon, t_0 + \epsilon]$, $\mathbf{F}(t)$ can be represented as a Taylor series about the point $t_0$:

$$\mathbf{F}(t) = \sum_{n=0}^{\infty} \frac{(t - t_0)^n}{n!} \mathbf{F}^{(n)}(t_0) \qquad \text{for all } t \text{ in } [t_0 - \epsilon, t_0 + \epsilon] \tag{5-9}$$

Premultiplying $\boldsymbol{\alpha}$ on both sides of (5-9) and using (5-8), we obtain

$$\boldsymbol{\alpha}\mathbf{F}(t) = \mathbf{0} \qquad \text{for all } t \text{ in } [t_0 - \epsilon, t_0 + \epsilon] \tag{5-10}$$

Since the sum of analytic functions is an analytic function, the analyticity assumption of the $\mathbf{f}_i$'s implies that $\boldsymbol{\alpha}\mathbf{F}(t)$ as a row vector function is analytic over $[t_1, t_2]$. Consequently, Equation (5-10) implies that

$$\boldsymbol{\alpha}\mathbf{F}(t) = \mathbf{0} \qquad \text{for all } t \text{ in } [t_1, t_2]$$

or, equivalently, the $\mathbf{f}_i$'s are linearly dependent on $[t_1, t_2]$. This is a contradiction. Q.E.D.

A direct consequence of this theorem is that if a set of analytic functions is linearly independent on $[t_1, t_2]$, then

$$\rho[\mathbf{F}(t) \vdots \mathbf{F}^{(1)}(t) \vdots \cdots \vdots \mathbf{F}^{(n)}(t) \vdots \cdots] = n$$

for *all* $t$ in $[t_1, t_2]$. It follows that *if a set of analytic functions is linearly independent on* $[t_1, t_2]$, *then the set of analytic functions is linearly independent on every subinterval of* $[t_1, t_2]$. In this statement, the analyticity assumption is essential. The statement does not hold without it, as we have seen in Example 1.

Note that Theorem 5-3 is not true if the infinite matrix in (5-7) is replaced by

$$\rho[\mathbf{F}(t_0) \vdots \mathbf{F}^{(1)}(t_0) \vdots \cdots \vdots \mathbf{F}^{(n-1)}(t_0)] = n$$

This can be seen from the following example.

### Example 3

Let

$$\mathbf{F}(t) = \begin{bmatrix} \sin 1000t \\ \sin 2000t \end{bmatrix}$$

Then

$$[\mathbf{F}(t) \vdots \mathbf{F}^{(1)}(t)] = \begin{bmatrix} \sin 1000t & 10^3 \cos 1000t \\ \sin 2000t & 2 \times 10^3 \cos 2000t \end{bmatrix}$$

It is easy to verify that $\rho[\mathbf{F}(t) \vdots \mathbf{F}^{(1)}(t)] < 2$ at $t = 0, \pm 10^{-3}\pi, \cdots$. However,

$$\rho[\mathbf{F}(t) \vdots \mathbf{F}^{(1)}(t) \vdots \mathbf{F}^{(2)}(t) \vdots \mathbf{F}^{(3)}(t)] = \begin{bmatrix} \sin 10^3 t & 10^3 \cos 10^3 t & -10^6 \sin 10^3 t & -10^9 \cos 10^3 t \\ \sin 2 \times 10^3 t & 2 \times 10^3 \cos 2 \times 10^3 t & -4 \times 10^6 \sin 2 \times 10^3 t & -8 \times 10^9 \cos 2 \times 10^3 t \end{bmatrix} = 2$$

for all $t$. ∎

The matrix in (5-7) has $n$ rows but infinitely many columns. However, in many cases, it is not necessary to check all the derivatives of $\mathbf{F}$. For instance, in Example 3 we have to check only up to $\mathbf{F}^{(3)}$. If we use the matrix

$$[\mathbf{F}(t) \vdots \mathbf{F}^{(1)}(t) \vdots \cdots \vdots \mathbf{F}^{(n-1)}(t)]$$

then we have the following corollary.

### Corollary 5-3

Assume that, for each $i$, $\mathbf{f}_i$ is analytic on $[t_1, t_2]$. Then $\mathbf{f}_1, \mathbf{f}_2, \cdots, \mathbf{f}_n$ are linearly independent on $[t_1, t_2]$ if and only if

$$\rho[\mathbf{F}(t) \vdots \mathbf{F}^{(1)}(t) \vdots \cdots \vdots \mathbf{F}^{(n-1)}(t)] = n$$

for *almost* all $t$ in $[t_1, t_2]$.

This corollary will not be used in this book and therefore its proof is omitted.

## 5-3 Controllability of Linear Dynamical Equations

**Time-varying case.** In this section, we shall introduce the concept of controllability of linear dynamical equations. To be more precise, we study the *state* controllability of linear *state* equations. As will be seen immediately that the state controllability is a property of state equations only, output equations do not play any role here.

Consider the $n$-dimensional, linear state equation

$$E: \qquad \dot{\mathbf{x}} = \mathbf{A}(t)\mathbf{x}(t) + \mathbf{B}(t)\mathbf{u}(t) \tag{5-11}$$

where $\mathbf{x}$ is the $n \times 1$ state vector, $\mathbf{u}$ is the $p \times 1$ input vector, and $\mathbf{A}$ and $\mathbf{B}$ are, respectively, $n \times n$ and $n \times p$ matrices whose entries are continuous functions of $t$ defined over $(-\infty, \infty)$. The state space of the equation is an $n$-dimensional complex vector space and is denoted by $\Sigma$.

### Definition 5-1

The state equation $E$ is said to be (state) *controllable*[2] at time $t_0$, if for *any* state $\mathbf{x}(t_0)$ in the state space $\Sigma$, and any state $\mathbf{x}^1$ in $\Sigma$, there exist a finite time $t_1 > t_0$ and an input $\mathbf{u}_{[t_0, t_1]}$ that will transfer the state $\mathbf{x}(t_0)$ to the state $\mathbf{x}^1$ at time $t_1$. Otherwise, the state equation is said to be *uncontrollable* at time $t_0$. ∎

[2] In the literature, it is called completely controllable. For conciseness, the adverb "completely" is dropped in this book.

This definition requires only that the input $\mathbf{u}$ be capable of moving any state in the state space to any other state in a *finite* time; what trajectory the state should take is not specified. Furthermore, there is no constraint imposed on the input. Its magnitude can be as large as desired. We give some examples to illustrate the concept of controllability.

### Example 1

Consider the network shown in Figure 5-3. The state variable $x$ of the system is the voltage across the capacitor. If $x(t_0) = 0$, then $x(t) = 0$

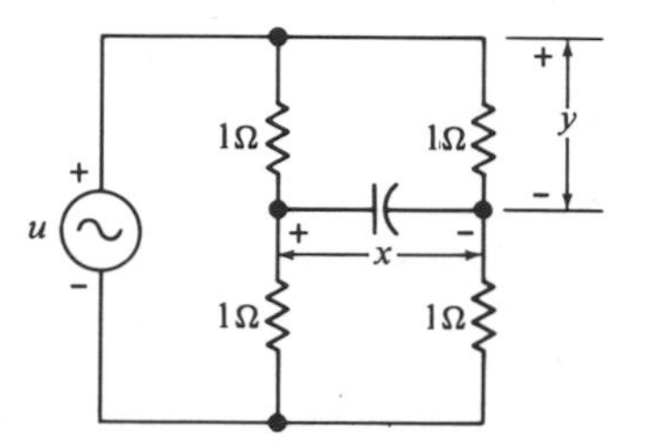


**Figure 5-3** An uncontrollable network.

for all $t \geq t_0$ no matter what input is applied. This is due to the symmetry of the network, and the input has no effect on the voltage across the capacitor. Hence the system—or, more precisely, the dynamical equation that describes the system—is not controllable at any $t_0$.

### Example 2

Consider the system shown in Figure 5-4. There are two state variables $x_1$ and $x_2$ in the system. The input can transfer $x_1$ *or* $x_2$ to any value;

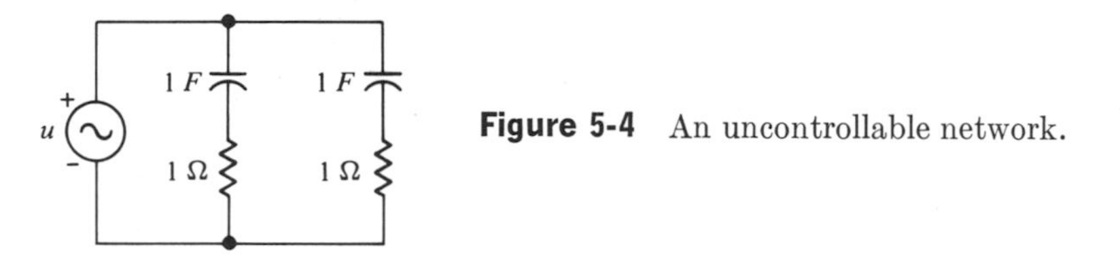


**Figure 5-4** An uncontrollable network.

however, it cannot transfer $x_1$ *and* $x_2$ to any values. For example, if $x_1(t_0) = 0$, $x_2(t_0) = 0$, then no matter what input is applied, $x_1(t)$ is always equal to $x_2(t)$, for all $t > t_0$. Hence the equation that describes the system is not controllable at any $t_0$. ▮

The solution of the state equation $E$ with $\mathbf{x}(t_0) = \mathbf{x}^0$ is given by

$$\mathbf{x}(t) = \boldsymbol{\phi}(t; t_0, \mathbf{x}^0, \mathbf{u}) = \boldsymbol{\Phi}(t, t_0)\mathbf{x}^0 + \int_{t_0}^{t} \boldsymbol{\Phi}(t, \tau)\mathbf{B}(\tau)\mathbf{u}(\tau)\, d\tau$$

$$= \boldsymbol{\Phi}(t, t_0)\left[\mathbf{x}^0 + \int_{t_0}^{t} \boldsymbol{\Phi}(t_0, \tau)\mathbf{B}(\tau)\mathbf{u}(\tau)\, d\tau\right] \qquad \textbf{(5-12)}$$

where $\mathbf{\Phi}(t, t_0) = \mathbf{\Psi}(t)\mathbf{\Psi}^{-1}(t_0)$; $\mathbf{\Psi}$ is a fundamental matrix of $\dot{\mathbf{x}} = \mathbf{A}(t)\mathbf{x}$ and is nonsingular for all $t$.

### Theorem 5-4

The state equation $E$ is controllable at time $t_0$ if and only if there exists a finite $t_1 > t_0$ such that the $n$ *rows* of the $n \times p$ matrix function $\mathbf{\Phi}(t_0, \cdot)\mathbf{B}(\cdot)$ are linearly independent on $[t_0, t_1]$.

### Proof

*Sufficiency:* If the rows of $\mathbf{\Phi}(t_0, \cdot)\mathbf{B}(\cdot)$ are linearly independent on $[t_0, t_1]$, from Theorem 5-1 the $n \times n$ constant matrix

$$\mathbf{W}(t_0, t_1) \triangleq \int_{t_0}^{t_1} \mathbf{\Phi}(t_0, \tau)\mathbf{B}(\tau)\mathbf{B}^*(\tau)\mathbf{\Phi}^*(t_0, \tau)\, d\tau \tag{5-13}$$

is nonsingular. Given any $\mathbf{x}(t_0) = \mathbf{x}^0$ and any $\mathbf{x}^1$, we claim that the input

$$\mathbf{u}(t) = -\mathbf{B}^*(t)\mathbf{\Phi}^*(t_0, t)\mathbf{W}^{-1}(t_0, t_1)[\mathbf{x}^0 - \mathbf{\Phi}(t_0, t_1)\mathbf{x}^1] \tag{5-14}$$

will transfer $\mathbf{x}^0$ to the state $\mathbf{x}^1$ at time $t_1$. Indeed, by substituting (5-14) into (5-12), we obtain

$$\begin{aligned}
\mathbf{x}(t_1) &= \mathbf{\Phi}(t_1, t_0)\left\{\mathbf{x}^0 - \int_{t_0}^{t_1} \mathbf{\Phi}(t_0, \tau)\mathbf{B}(\tau)\mathbf{B}^*(\tau)\mathbf{\Phi}^*(t_0, \tau)\, d\tau\, \mathbf{W}^{-1}(t_0, t_1)\right. \\
&\qquad\qquad \left. [\mathbf{x}^0 - \mathbf{\Phi}(t_0, t_1)\mathbf{x}^1]\right\} \\
&= \mathbf{\Phi}(t_1, t_0)\{\mathbf{x}^0 - \mathbf{W}(t_0, t_1)\mathbf{W}^{-1}(t_0, t_1)[\mathbf{x}^0 - \mathbf{\Phi}(t_0, t_1)\mathbf{x}^1]\} \\
&= \mathbf{\Phi}(t_1, t_0)\mathbf{\Phi}(t_0, t_1)\mathbf{x}^1 \\
&= \mathbf{x}^1
\end{aligned}$$

Thus we conclude that the equation $E$ is controllable. *Necessity:* By contradiction. Suppose $E$ is controllable at $t_0$, but the rows of $\mathbf{\Phi}(t_0, \cdot)\mathbf{B}(\cdot)$ are linearly dependent on $[t_0, t_1]$ for all $t_1 > t_0$. Then there exists a nonzero, constant $1 \times n$ row vector $\boldsymbol{\alpha}$ such that

$$\boldsymbol{\alpha}\mathbf{\Phi}(t_0, t)\mathbf{B}(t) = \mathbf{0} \qquad \text{for all } t \text{ in } [t_0, t_1] \tag{5-15}$$

Let us choose $\mathbf{x}(t_0) \triangleq \mathbf{x}^0 = \boldsymbol{\alpha}^*$. Then Equation (5-12) becomes

$$\mathbf{\Phi}(t_0, t_1)\mathbf{x}(t_1) = \boldsymbol{\alpha}^* + \int_{t_0}^{t_1} \mathbf{\Phi}(t_0, \tau)\mathbf{B}(\tau)\mathbf{u}(\tau)\, d\tau \tag{5-16}$$

Premultiplying both sides of (5-16) by $\boldsymbol{\alpha}$, we obtain

$$\boldsymbol{\alpha}\mathbf{\Phi}(t_0, t_1)\mathbf{x}(t_1) = \boldsymbol{\alpha}\boldsymbol{\alpha}^* + \int_{t_0}^{t_1} \boldsymbol{\alpha}\mathbf{\Phi}(t_0, \tau)\mathbf{B}(\tau)\mathbf{u}(\tau)\, d\tau \tag{5-17}$$

By hypothesis, $E$ is controllable at $t_0$; hence for any state—in particular, $\mathbf{x}^1 = \mathbf{0}$—there exists $\mathbf{u}_{[t_0, t_1]}$ such that $\mathbf{x}(t_1) = \mathbf{0}$. Since $\boldsymbol{\alpha}\boldsymbol{\Phi}(t_0, t)\mathbf{B}(t) = \mathbf{0}$, for all $t$ in $[t_0, t_1]$, Equation (5-17) reduces to

$$\boldsymbol{\alpha}\boldsymbol{\alpha}^* = 0$$

which, in turn, implies that $\boldsymbol{\alpha} = \mathbf{0}$. This is a contradiction. Q.E.D.

In the proof of this theorem, we also give in (5-14) an input $\mathbf{u}(t)$ that transfers $\mathbf{x}(t_0)$ to $\mathbf{x}^1$ at time $t_1$. Because of the continuity assumption of $\mathbf{A}$ and $\mathbf{B}$, the input $\mathbf{u}$ in (5-14) is a continuous function of $t$ in $[t_0, t_1]$.

If a linear dynamical equation is controllable, there are generally many different inputs $\mathbf{u}$ that can transfer $\mathbf{x}(t_0)$ to $\mathbf{x}^1$ at time $t_1$, because the trajectory between $\mathbf{x}(t_0)$ and $\mathbf{x}^1$ is not specified. Among these possible inputs that achieve the same mission, we may ask which input is the optimal one according to some criterion. If the total energy

$$\int_{t_0}^{t_1} \|\mathbf{u}(t)\|^2 \, dt$$

is used as a criterion, the input given in (5-14) will use the minimal energy in transferring $\mathbf{x}(t_0)$ to $\mathbf{x}^1$ at time $t_1$. This is proved in Appendix B.

In order to apply Theorem 5-4, a fundamental matrix $\boldsymbol{\Psi}$ or the state transition matrix $\boldsymbol{\Phi}(t, \tau)$ of $\dot{\mathbf{x}} = \mathbf{A}(t)\mathbf{x}$ has to be computed. As we mentioned earlier, this is generally a difficult task. Hence, Theorem 5-4 is not readily applicable. In the following we shall give a controllability criterion based on the matrices $\mathbf{A}$ and $\mathbf{B}$ without solving the state equation. However, in order to do so, we need some additional assumptions on $\mathbf{A}$ and $\mathbf{B}$. Assume that $\mathbf{A}$ and $\mathbf{B}$ are $(n - 1)$ times continuously differentiable. Define a sequence of $n \times p$ matrices $\mathbf{M}_0(\cdot)$, $\mathbf{M}_1(\cdot)$, $\cdots$ by the equation

$$\mathbf{M}_{k+1}(t) = -\mathbf{A}(t)\mathbf{M}_k(t) + \frac{d}{dt}\mathbf{M}_k(t) \qquad k = 0, 1, 2, \cdots, n-1 \tag{5-18a}$$

with

$$\mathbf{M}_0(t) = \mathbf{B}(t) \tag{5-18b}$$

Observe that

$$\boldsymbol{\Phi}(t_0, t)\mathbf{B}(t) = \boldsymbol{\Phi}(t_0, t)\mathbf{M}_0(t) \tag{5-19a}$$

$$\begin{aligned}\frac{\partial}{\partial t}\boldsymbol{\Phi}(t_0, t)\mathbf{B}(t) &= \boldsymbol{\Psi}(t_0)\left[\left(\frac{d}{dt}\boldsymbol{\Psi}^{-1}(t)\right)\mathbf{B}(t) + \boldsymbol{\Psi}^{-1}(t)\frac{d}{dt}\mathbf{B}(t)\right] \\ &= \boldsymbol{\Psi}(t_0)\boldsymbol{\Psi}^{-1}(t)\left[-\mathbf{A}(t)\mathbf{B}(t) + \frac{d}{dt}\mathbf{B}(t)\right] \\ &= \boldsymbol{\Phi}(t_0, t)\mathbf{M}_1(t)\end{aligned} \tag{5-19b}$$

and, in general,

$$\frac{\partial^k}{\partial t^k} \mathbf{\Phi}(t_0, t)\mathbf{B}(t) = \mathbf{\Phi}(t_0, t)\mathbf{M}_k(t) \qquad k = 0, 1, 2, \cdots, n-1 \tag{5-19c}$$

### Theorem 5-5

Assume that the matrices $\mathbf{A}(\cdot)$ and $\mathbf{B}(\cdot)$ in the $n$-dimensional state equation $E$ are $(n-1)$ times continuously differentiable. Then the state equation $E$ is controllable at $t_0$ if there exists a finite $t_1 > t_0$ such that

$$\rho[\mathbf{M}_0(t_1) \vdots \mathbf{M}_1(t_1) \vdots \cdots \vdots \mathbf{M}_{n-1}(t_1)] = n \tag{5-20}$$

### Proof

Define

$$\frac{\partial}{\partial t} \mathbf{\Phi}(t_0, t)\mathbf{B}(t)\bigg|_{t=t_1} \triangleq \frac{\partial}{\partial t_1} \mathbf{\Phi}(t_0, t_1)\mathbf{B}(t_1)$$

Then, from (5-19) and using (2-2), we have

$$\left[\mathbf{\Phi}(t_0, t_1)\mathbf{B}(t_1) \vdots \frac{\partial}{\partial t_1} \mathbf{\Phi}(t_0, t_1)\mathbf{B}(t_1) \vdots \cdots \vdots \frac{\partial^{n-1}}{\partial t_1^{n-1}} \mathbf{\Phi}(t_0, t_1)\mathbf{B}(t_1)\right]$$
$$= \mathbf{\Phi}(t_0, t_1)[\mathbf{M}_0(t_1) \vdots \mathbf{M}_1(t_1) \vdots \cdots \vdots \mathbf{M}_{n-1}(t_1)] \tag{5-21}$$

Since $\mathbf{\Phi}(t_0, t_1)$ is nonsingular, the assumption that

$$\rho[\mathbf{M}_0(t_1) \vdots \mathbf{M}_1(t_1) \vdots \cdots \vdots \mathbf{M}_{n-1}(t_1)] = n$$

implies that

$$\rho\left[\mathbf{\Phi}(t_0, t_1)\mathbf{B}(t_1) \vdots \frac{\partial}{\partial t_1} \mathbf{\Phi}(t_0, t_1)\mathbf{B}(t_1) \vdots \cdots \vdots \frac{\partial^{n-1}}{\partial t_1^{n-1}} \mathbf{\Phi}(t_0, t_1)\mathbf{B}(t_1)\right] = n$$

It follows from Theorem 5-2 that the rows of $\mathbf{\Phi}(t_0, \cdot)\mathbf{B}(\cdot)$ are linearly independent on $[t_2, t_0]$ for any $t_2 > t_1$. Thus, from Theorem 5-4, we conclude that the state equation $E$ is controllable. Q.E.D.

As in Theorem 5-2, the condition of Theorem 5-5 is sufficient but not necessary for the controllability of a state equation.

### Example 1

Consider

$$\begin{bmatrix} \dot{x}_1 \\ \dot{x}_2 \\ \dot{x}_3 \end{bmatrix} = \begin{bmatrix} t & 1 & 0 \\ 0 & t & 0 \\ 0 & 0 & t^2 \end{bmatrix} \begin{bmatrix} x_1 \\ x_2 \\ x_3 \end{bmatrix} + \begin{bmatrix} 0 \\ 1 \\ 1 \end{bmatrix} u \tag{5-22}$$

From (5-18), we have

$$\mathbf{M}_0(t) = \begin{bmatrix} 0 \\ 1 \\ 1 \end{bmatrix}$$

$$\mathbf{M}_1(t) = -\mathbf{A}(t)\mathbf{M}_0(t) + \frac{d}{dt}\mathbf{M}_0(t) = \begin{bmatrix} -1 \\ -t \\ -t^2 \end{bmatrix}$$

$$\mathbf{M}_2(t) = -\mathbf{A}(t)\mathbf{M}_1(t) + \frac{d}{dt}\mathbf{M}_1(t) = \begin{bmatrix} 2t \\ t^2 \\ t^4 \end{bmatrix} + \begin{bmatrix} 0 \\ -1 \\ -2t \end{bmatrix} = \begin{bmatrix} 2t \\ t^2 - 1 \\ t^4 - 2t \end{bmatrix}$$

Since the matrix $[\mathbf{M}_0(t) \vdots \mathbf{M}_1(t) \vdots \mathbf{M}_2(t)]$ has rank 3 for all $t \neq 0$, the dynamical equation is controllable at every $t$. ▮

### * Differential controllability, instantaneous controllability, and uniform controllability

The remainder of this subsection will be devoted to three distinct types of controllability. This material may be omitted with no loss of continuity.

#### Definition 5-2

A state equation $E$ is said to be *differentially* (completely) *controllable* at time $t_0$ if, for any state $\mathbf{x}(t_0)$ in the state space $\Sigma$ and any state $\mathbf{x}^1$ in $\Sigma$, there exists an input $\mathbf{u}$ that will transfer $\mathbf{x}(t_0)$ to the state $\mathbf{x}^1$ in an *arbitrarily* small interval of time. ▮

If every state in $\Sigma$ can be transferred to any other state in a finite time (no matter how long), the state equation is said to be controllable. If this can be achieved in an arbitrarily small interval of time, then the state equation is said to be differentially controllable. Clearly, differential controllability implies controllability. The condition for differential controllability can be easily obtained by slight modifications of Theorems 5-4 and 5-5. However, if the matrices $\mathbf{A}$ and $\mathbf{B}$ in the state equation $E$ are analytic on $(-\infty, \infty)$, then we have the following theorem.

#### Theorem 5-6

If the matrices $\mathbf{A}$ and $\mathbf{B}$ are analytic on $(-\infty, \infty)$, then the $n$-dimensional state equation $E$ is differentially controllable at every $t$ in $(-\infty, \infty)$ if and only if, for any fixed $t_0$ in $(-\infty, \infty)$,

$$\rho[\mathbf{M}_0(t_0) \vdots \mathbf{M}_1(t_0) \vdots \cdots \vdots \mathbf{M}_{n-1}(t_0) \vdots \cdots] = n \quad ▮$$

If the matrix $\mathbf{A}$ is analytic on $(-\infty, \infty)$, it can be shown that the state transition matrix $\mathbf{\Phi}(t_0, \cdot)$ of $\dot{\mathbf{x}} = \mathbf{A}(t)\mathbf{x}$ is also analytic on $(-\infty, \infty)$.

Since the product of two analytic functions is an analytic function, the assumption of Theorem 5-6 implies that $\mathbf{\Phi}(t_0, \cdot)\mathbf{B}(\cdot)$ is an analytic function. An implication of Theorem 5-3 is that a set of analytic functions is linearly independent on $(-\infty, \infty)$ if and only if the set of analytic functions is linearly independent on every subinterval (no matter how small) of $(-\infty, \infty)$. With this fact, Theorem 5-6 follows immediately from Theorems 5-3 and 5-4. Consequently, if a state equation with analytic $\mathbf{A}$ and $\mathbf{B}$ is controllable at any point at all, it is differentially controllable at every $t$ in $(-\infty, \infty)$.

### Definition 5-3

The linear dynamical equation $E$ is said to be *instantaneously controllable*[3] in $(-\infty, \infty)$ if and only if

$$\rho[\mathbf{M}_0(t) \vdots \mathbf{M}_1(t) \vdots \cdots \vdots \mathbf{M}_{n-1}(t)] = n \qquad \text{for all } t \text{ in } (-\infty, \infty)$$

where the $\mathbf{M}_i$'s are as defined in Equation (5-18). ∎

If a dynamical equation is instantaneously controllable, then the transfer of the states can be achieved instantaneously at any time by using an input that consists of $\delta$-functions and their derivatives up to an order of $(n-1)$. It is clear that instantaneous controllability implies differential controllability. The most important implication of instantaneous controllability is that in the case of a single input, the matrix $[\mathbf{M}_0(t) \quad \mathbf{M}_1(t) \quad \cdots \quad \mathbf{M}_{n-1}(t)]$ qualifies as an equivalence transformation (Definition 4-5). Consequently, many canonical-form equivalent dynamical equations can be obtained for instantaneously controllable dynamical equations. See References [10], [99], [103], and [110].

### Definition 5-4

The dynamical equation $E$ is said to be *uniformly controllable* if and only if there exist a positive $\sigma_c$ and positive $\alpha_i$ that depend on $\sigma_c$ such that

$$\mathbf{0} < \alpha_1(\sigma_c)\mathbf{I} \leq \mathbf{W}(t, t+\sigma_c) \leq \alpha_2(\sigma_c)\mathbf{I}$$

and

$$\mathbf{0} < \alpha_3(\sigma_c)\mathbf{I} \leq \mathbf{\Phi}(t+\sigma_c, t)\mathbf{W}(t, t+\sigma_c)\mathbf{\Phi}^*(t+\sigma_c, t) \leq \alpha_4(\sigma_c)\mathbf{I}$$

for all $t$ in $(-\infty, \infty)$, where $\mathbf{\Phi}$ is the state transition matrix and $\mathbf{W}$ is as defined in Equation (5-13). ∎

[3] In the engineering literature, it is called uniform controllability. However, this terminology was first used in Reference [56] to define a different kind of controllability (see Definition 5-4); hence we adopt the terminology "instantaneous controllability."

By $\mathbf{A} > \mathbf{B}$, we mean that the matrix $(\mathbf{A} - \mathbf{B})$ is a positive definite matrix (see Section 8-5). Uniform controllability insures that the transfer of the states can be achieved in the time interval $\sigma_c$. The concept of uniform controllability is needed in the stability study of optimal control systems. See References [56] and [102].

Instantaneous controllability and uniform controllability both imply controllability. However, instantaneous controllability neither implies nor is implied by uniform controllability.

### Example 2

Consider the one-dimensional linear dynamical equation

$$\dot{x} = e^{-|t|}u$$

Since $\rho(M_0(t)) = \rho(e^{-|t|}) = 1$ for all $t$, the dynamical equation is instantaneously controllable at every $t$. However, the dynamical equation is not uniformly controllable, because there exists no $\delta(\alpha_c)$ such that

$$W(t, t + \alpha_c) = \int_t^{t+\alpha_c} e^{-2t}\, dt = 0.5e^{-2t}(1 - e^{-2\alpha_c}) > \delta(\alpha_c)$$

for all $t > 0$.

### Example 3

Consider the one-dimensional dynamical equation

$$\dot{x} = b(t)u$$

with $b(t)$ defined as in Figure 5-5. The dynamical equation is not instantaneously controllable in the interval $(-1, 1)$. However, it is uniformly controllable in $(-\infty, \infty)$. This can be easily verified by choosing $\alpha_c = 5$.

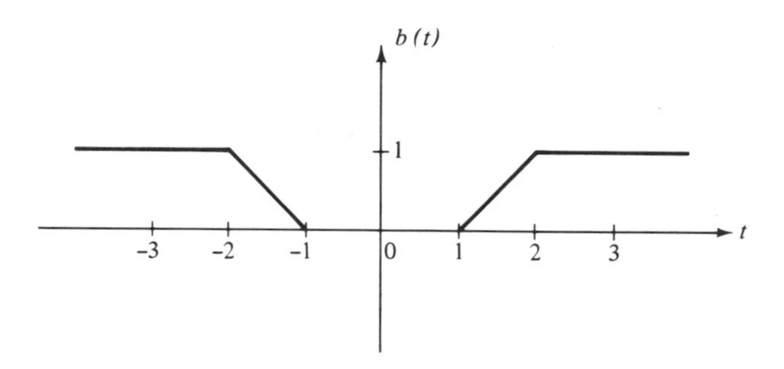


**Figure 5-5** The function $b(t)$.

A remark is in order concerning controllability, instantaneous (differential) controllability, and uniform controllability. If a dynamical equation is differentially controllable, a state can be transferred to any other state in an arbitrarily small interval of time. However, the magnitude of the input may become very large; in the extreme case (to achieve the transfer instantly), a delta-function input is required. If a dynamical equation is merely controllable, the transfer of the states may take a very long interval of time. However, if it is uniformly controllable, the transfer of the states can be achieved in the length of time $\alpha_c$; moreover, the magnitude of the control input will *not* be arbitrarily large [see (5-14) and note that the input is proportional to $\mathbf{W}^{-1}$]. In optimal control theory, the condition of uniform controllability is sometimes required to insure the stability of an optimal control system.

**Time-invariant case.** In this subsection, we study the controllability of the $n$-dimensional, linear, time-invariant state equation

$$FE: \qquad \dot{\mathbf{x}} = \mathbf{A}\mathbf{x} + \mathbf{B}\mathbf{u} \tag{5-23}$$

where $\mathbf{x}$ is the $n \times 1$ state vector, $\mathbf{u}$ is the $p \times 1$ input vector; $\mathbf{A}$ and $\mathbf{B}$ are $n \times n$ and $n \times p$ real constant matrices, respectively. For time-invariant dynamical equations, the time interval of interest is from the present time to infinity; that is, $[0, \infty)$.

Theorem 5-7

For the $n$-dimensional, linear, time-invariant dynamical equation $FE$, the following statements are equivalent:

1. $FE$ is controllable at every $t_0$ in $[0, \infty)$.
2. The rows of $e^{-\mathbf{A}t}\mathbf{B}$ (and consequently of $e^{\mathbf{A}t}\mathbf{B}$) are linearly independent on $[0, \infty)$ (over the field of complex numbers).[4]
3. $\mathbf{W}(t_0, t) = \int_{t_0}^{t} e^{\mathbf{A}(t-\tau)}\mathbf{B}\mathbf{B}^* e^{\mathbf{A}^*(t-\tau)}\, d\tau$ is nonsingular for any $t_0 > 0$, and any $t > t_0$.
4. The $n \times (np)$ matrix $[\mathbf{B} \vdots \mathbf{AB} \vdots \cdots \vdots \mathbf{A}^{n-1}\mathbf{B}]$ has rank $n$.
5. The rows of $(s\mathbf{I} - \mathbf{A})^{-1}\mathbf{B}$ are linearly independent over the field of complex numbers.[4]

Proof

The equivalence of statements 1, 2, and 3 follows directly from Theorems 5-1 and 5-4. Since the entries of $e^{-\mathbf{A}t}\mathbf{B}$ are analytic functions, Theorem 5-3 implies that the rows of $e^{-\mathbf{A}t}\mathbf{B}$ are linearly independent on $[0, \infty)$ if and only if

$$\rho[e^{-\mathbf{A}t}\mathbf{B} \vdots -e^{-\mathbf{A}t}\mathbf{AB} \vdots \cdots \vdots (-1)^{n-1}e^{-\mathbf{A}t}\mathbf{A}^{n-1}\mathbf{B} \vdots \cdots] = n \tag{5-24}$$

for *any* $t$ in $[0, \infty)$. Let $t = 0$; then Equation (5-24) reduces to

$$\rho[\mathbf{B} \vdots -\mathbf{AB} \vdots \cdots \vdots (-1)^{n-1}\mathbf{A}^{n-1}\mathbf{B} \vdots (-1)^{n}\mathbf{A}^{n}\mathbf{E} \vdots \cdots] = n$$

[4] Although all the entries of $\mathbf{A}$ and $\mathbf{B}$ are real numbers, we have agreed to consider them as elements of the field of complex numbers.

From the Cayley–Hamilton theorem, we know that $\mathbf{A}^m$ with $m \geq n$ can be written as a linear combination of $\mathbf{I}, \mathbf{A}, \cdots, \mathbf{A}^{n-1}$; hence the columns of $\mathbf{A}^m\mathbf{B}$ with $m \geq n$ are linearly dependent on the columns of $\mathbf{B}, \mathbf{AB}, \cdots, \mathbf{A}^{n-1}\mathbf{B}$. Consequently,

$$\rho[\mathbf{B} \vdots -\mathbf{AB} \vdots \cdots \vdots (-1)^{n-1}\mathbf{A}^{n-1}\mathbf{B} \vdots \cdots] = \rho[\mathbf{B} \vdots -\mathbf{AB} \vdots \cdots \vdots (-1)^{n-1}\mathbf{A}^{n-1}\mathbf{B}]$$

Since changing the sign will not change the linear independence, we conclude that the rows of $e^{-\mathbf{A}t}\mathbf{B}$ are linearly independent if and only if $\rho[\mathbf{B} \vdots \mathbf{AB} \vdots \cdots \vdots \mathbf{A}^{n-1}\mathbf{B}] = n$. This proves the equivalence of statements 2 and 4. In the foregoing argument we also proved that the rows of $e^{-\mathbf{A}t}\mathbf{B}$ are linearly independent if and only if the rows of $e^{\mathbf{A}t}\mathbf{B}$ are linearly independent on $[0, \infty)$ over the field of complex numbers. Next we show the equivalence of statements 2 and 5. Taking the Laplace transform of $e^{\mathbf{A}t}\mathbf{B}$, we have

$$\mathcal{L}[e^{\mathbf{A}t}\mathbf{B}] = (s\mathbf{I} - \mathbf{A})^{-1}\mathbf{B}$$

Since the Laplace transform is a one-to-one linear operator, if the rows of $e^{\mathbf{A}t}\mathbf{B}$ are linearly independent on $[0, \infty)$ over the field of complex unmbers, so are the rows of $(s\mathbf{I} - \mathbf{A})^{-1}\mathbf{B}$, and vice versa. Q.E.D.

For linear, time-invariant dynamical equations, because of the analyticity property of $e^{\mathbf{A}t}\mathbf{B}$, if it is controllable at some $t$, then it is controllable at every $t$ in $[0, \infty)$. Hence the specification of time will be dropped for the controllability of linear, time-invariant dynamical equations.

The matrix $\mathbf{U} \triangleq [\mathbf{B} \vdots \mathbf{AB} \vdots \cdots \vdots \mathbf{A}^{n-1}\mathbf{B}]$ is called the *controllability matrix* of the dynamical equation *FE*.

## Example 4

Consider the platform system shown in Figure 5-2. The mass of the platform is assumed to be zero. Consequently the two spring systems operate independently. Now if the initial displacements $x_1(0)$ and $x_2(0)$ are different from zero, the platform will vibrate and it will take an infinite time before the platform comes back to rest. Now the question is that if the initial displacements are $x_1(0) = x_2(0) = 10$, is it possible to apply a force to bring the platform to rest in 2 seconds? If yes, how?

The state equation of the platform system with the state-variable chosen as shown is

$$\begin{bmatrix} \dot{x}_1 \\ \dot{x}_2 \end{bmatrix} = \begin{bmatrix} -0.5 & 0 \\ 0 & -1 \end{bmatrix} \begin{bmatrix} x_1 \\ x_2 \end{bmatrix} + \begin{bmatrix} 0.5 \\ 1 \end{bmatrix} u$$

It is easy to verify that

$$\rho[\mathbf{B} \vdots \mathbf{AB}] = \rho \begin{bmatrix} 0.5 & -0.25 \\ 1 & -1 \end{bmatrix} = 2$$

hence the state equation is controllable. Consequently the displacements can be brought to zero in 2 seconds by applying a proper input $u$. Using Equations (5-13) and (5-14), we have

$$\mathbf{W}(0, 2) = \int_0^2 \begin{bmatrix} e^{0.5\tau} & 0 \\ 0 & e^{\tau} \end{bmatrix} \begin{bmatrix} 0.5 \\ 1 \end{bmatrix} [0.5 \quad 1] \begin{bmatrix} e^{0.5\tau} & 0 \\ 0 & e^{\tau} \end{bmatrix} d\tau = \begin{bmatrix} 1.6 & 6.33 \\ 6.33 & 27 \end{bmatrix}$$

and

$$u^1(t) = -[0.5 \quad 1] \begin{bmatrix} e^{0.5t} & 0 \\ 0 & e^{t} \end{bmatrix} \mathbf{W}^{-1}(0, 2) \begin{bmatrix} 10 \\ 10 \end{bmatrix} = -33.3e^{0.5t} + 15.25e^{t}$$

for $t$ in $[0, 2]$. If a force of the form $u^1$ is applied, the platform will come to rest at $t = 2$. The behavior of $x_1$, $x_2$, and of the input $u^1$ are plotted by using solid lines in Figure 5-6. ∎

In Figure 5-6 we also plot by using dotted lines the input $u^2(t)$ that transfers $x_1(0) = x_2(0) = 10$ to zero in 4 seconds. We see from Figure 5-6 that, in transferring $\mathbf{x}(0)$ to zero, the smaller the time interval the larger the magnitude of the input. If no restriction is imposed on the input $u$, then we can transfer $\mathbf{x}(0)$ to zero in an arbitrarily small interval of time; however, the magnitude of the input may become very large. If some restriction on the magnitude of $u$ is imposed, we might not be able to transfer $\mathbf{x}(0)$ to zero in an arbitrarily small interval of time. For example, if we require $|u(t)| \leq 5$ in Example 4, then we might not be able to transfer $\mathbf{x}(0)$ to zero in less than 4 seconds.

***Simplified controllability condition.** Define the $n \times (kp)$ matrix

$$\mathbf{U}_{k-1} \triangleq [\mathbf{B} \vdots \mathbf{AB} \vdots \cdots \vdots \mathbf{A}^{k-1}\mathbf{B}] \tag{5-25}$$

and

$$\mathbf{U} \triangleq \mathbf{U}_{n-1} = [\mathbf{B} \vdots \mathbf{AB} \vdots \cdots \vdots \mathbf{A}^{n-1}\mathbf{B}]$$

From Theorem 5-7, we know that if the controllability matrix $\mathbf{U} = \mathbf{U}_{n-1}$ has rank $n$, then the state equation $FE$ is controllable. To check controllability, in many instances we need not calculate $\mathbf{U}_{n-1}$ but only a matrix with a smaller number of columns. This is based on the following theorem.

### Theorem 5-8

If $j$ is the least integer such that

$$\rho(\mathbf{U}_j) = \rho(\mathbf{U}_{j+1})$$

then

$$\rho(\mathbf{U}_k) = \rho(\mathbf{U}_j) \qquad \text{for all integers } k > j$$

and

$$j \leq \min(n - r, \bar{n} - 1)$$

where $r$ is the rank of $\mathbf{B}$ and $\bar{n}$ is the degree of the minimal polynomial of $\mathbf{A}$.

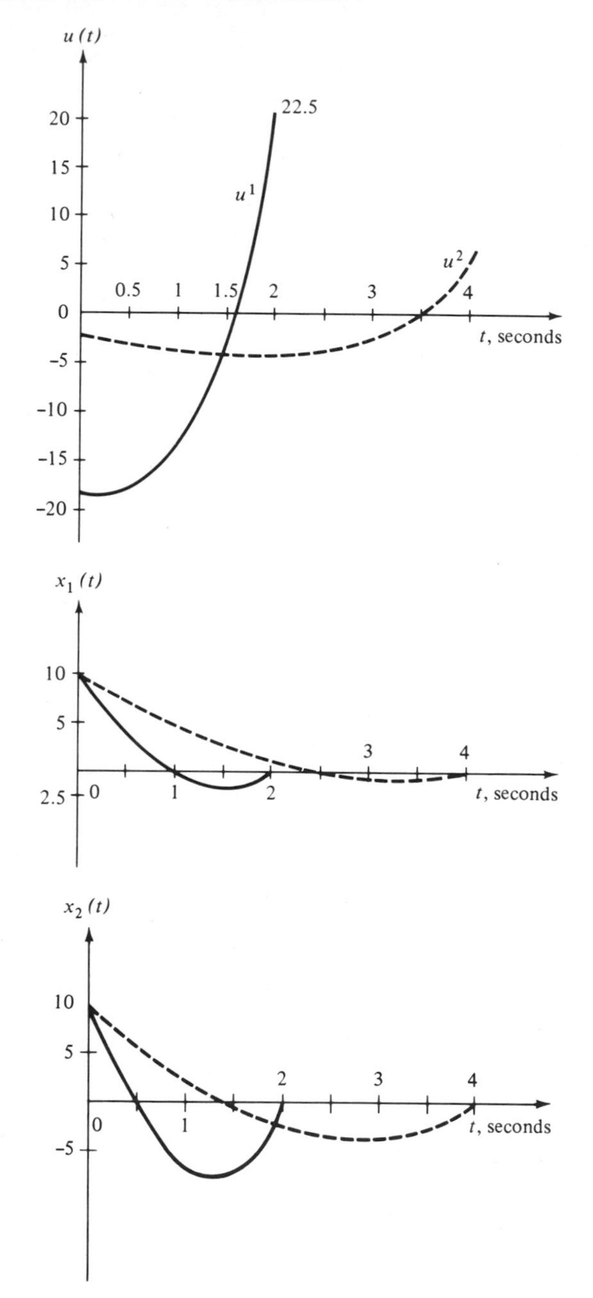


**Figure 5-6** Behavior of $x_1(t)$ and $x_2(t)$ and the waveform of $u$.

### Proof

Recall that $\rho(\mathbf{U}_j)$ is equal to the number of linearly independent columns in $\mathbf{U}_j$. Since all the columns of $\mathbf{U}_j$ are in $\mathbf{U}_{j+1}$, the condition $\rho(\mathbf{U}_j) = \rho(\mathbf{U}_{j+1})$ implies that every column of the matrix $\mathbf{A}^{j+1}\mathbf{B}$ is linearly dependent on the columns of the matrices $\mathbf{B}, \mathbf{AB}, \cdots, \mathbf{A}^j\mathbf{B}$. Consequently, every column of $\mathbf{A}^{j+2}\mathbf{B}$ is linearly dependent on the columns of the matrices $\mathbf{AB}, \mathbf{A}^2\mathbf{B}, \cdots, \mathbf{A}^{j+1}\mathbf{B}$. However, the columns of $\mathbf{A}^{j+1}\mathbf{B}$ are linearly dependent on the columns of $\mathbf{B}, \mathbf{AB}, \cdots, \mathbf{A}^j\mathbf{B}$; hence we conclude that every column of $\mathbf{A}^{j+2}\mathbf{B}$ is linearly dependent on the columns of $\mathbf{B}, \mathbf{AB}, \cdots, \mathbf{A}^j\mathbf{B}$. Proceeding forward, we can show that every column of $\mathbf{A}^k\mathbf{B}$ with $k \geq j$ is linearly dependent on the columns of $\mathbf{B}, \mathbf{AB}, \cdots, \mathbf{A}^j\mathbf{B}$. This proves that $\rho(\mathbf{U}_k) = \rho(\mathbf{U}_j)$ for all integers $k > j$.

The rank of the matrix $[\mathbf{B} \vdots \mathbf{AB} \vdots \cdots]$ must increase by at least 1 as each submatrix of the form $\mathbf{A}^k\mathbf{B}$ is added; otherwise its rank will cease to increase. The maximum rank of $[\mathbf{B} \vdots \mathbf{AB} \vdots \cdots]$ is $n$, and hence if the rank of $\mathbf{B}$ is $r$, in checking the maximum rank of $[\mathbf{B} \vdots \mathbf{AB} \vdots \cdots]$, it is sufficient to add at most $(n - r)$ submatrices of the form $\mathbf{A}^k\mathbf{B}$. Thus we have $j \leq n - r$. That $j \leq \bar{n} - 1$ follows from the definition of the minimal polynomial. Hence we conclude that $j \leq \min(n - r, \bar{n} - 1)$.

Q.E.D.

### Corollary 5-8

If the rank of $\mathbf{B}$ is $r$, then the state equation $FE$ is controllable if and only if

$$\rho(\mathbf{U}_{n-r}) = \rho[\mathbf{B} \vdots \mathbf{AB} \vdots \cdots \vdots \mathbf{A}^{n-r}\mathbf{B}] = n$$

or the $n \times n$ matrix

$$\mathbf{U}_{n-r}\mathbf{U}^*_{n-r}$$

is nonsingular. ∎

This corollary follows directly from Theorems 5-8 and 2-8.

### Example 5

Consider the three-dimensional state equation

$$\dot{\mathbf{x}} = \begin{bmatrix} 1 & 1 & 0 \\ 0 & 1 & 0 \\ 0 & 1 & 1 \end{bmatrix} \mathbf{x} + \begin{bmatrix} 0 & 1 \\ 1 & 0 \\ 0 & 1 \end{bmatrix} u$$

The rank of $\mathbf{B}$ is 2; therefore, we need to check $\mathbf{U}_1 = [\mathbf{B} \vdots \mathbf{AB}]$ in determining the controllability of the equation. Since

$$\rho[\mathbf{B} \vdots \mathbf{AB}] = \rho \begin{bmatrix} 0 & 1 & 1 & 1 \\ 1 & 0 & 1 & 0 \\ 0 & 1 & 1 & 1 \end{bmatrix} = 2 < 3$$

the state equation is not controllable. ∎

If a linear, time-invariant dynamical equation is controllable, then the input can excite all the modes of the equation. On the other hand, the input can also be used to suppress any undesirable mode.

Example 6

Consider the dynamical equation

$$\begin{bmatrix} \dot{x}_1 \\ \dot{x}_2 \end{bmatrix} = \begin{bmatrix} 0 & 1 \\ 2 & 1 \end{bmatrix} \mathbf{x} + \begin{bmatrix} 1 \\ 0 \end{bmatrix} u$$
$$y = [1 \quad 2]\mathbf{x}$$

The matrix $\mathbf{A}$ has eigenvalues $-1$ and 2; hence the equation has two modes $e^{-t}$ and $e^{2t}$. Since the mode $e^{2t}$ increases with time, we like to suppress it in the output. The controllability matrix of the equation

$$[\mathbf{b} \vdots \mathbf{Ab}] = \begin{bmatrix} 1 & 0 \\ 0 & 2 \end{bmatrix}$$

has rank 2, hence the equation is controllable. Consequently, the mode $e^{2t}$ can be suppressed. First we compute $e^{\mathbf{A}t}$:

$$e^{\mathbf{A}t} = \begin{bmatrix} \frac{1}{3}e^{2t} + \frac{2}{3}e^{-t} & \frac{1}{3}e^{2t} - \frac{1}{3}e^{-t} \\ \frac{2}{3}e^{2t} - \frac{2}{3}e^{-t} & \frac{1}{3}e^{-t} + \frac{2}{3}e^{2t} \end{bmatrix}$$

Now for any initial state $\mathbf{x}(0)$, we shall find an input $u$ such that the output $y$ does not contain the mode $e^{2t}$ after a certain instant of time, say $t_0$. Let the input $u$ be identically zero after $t_0$; then the output due to the state $\mathbf{x}$ at time $t_0$ is

$$\begin{aligned} y &= [1 \quad 2] \begin{bmatrix} \frac{1}{3}e^{2t} + \frac{2}{3}e^{-t} & \frac{1}{3}e^{2t} - \frac{1}{3}e^{-t} \\ \frac{2}{3}e^{2t} - \frac{2}{3}e^{-t} & \frac{2}{3}e^{2t} + \frac{1}{3}e^{-t} \end{bmatrix} \begin{bmatrix} x_1(t_0) \\ x_2(t_0) \end{bmatrix} \\ &= (x_1(t_0) + x_2(t_0))\tfrac{5}{3}e^{2t} + (x_2(t_0) - 2x_1(t_0))\tfrac{1}{3}e^{-t} \end{aligned}$$

We see that if $x_1(t_0) = -x_2(t_0)$, then the output will not contain the mode $e^{2t}$ after $t > t_0$. Now the dynamical equation is controllable, hence it is possible to choose an input $u_{[0, t_0]}$ that transfers $\mathbf{x}(0)$ to $\mathbf{x}(t_0)$. The required input may be computed by using (5-14). ▌

## 5-4 Observability of Linear Dynamical Equations

**Time-varying case.** The concept of observability is dual to that of controllability. Roughly speaking, controllability studies the possibility of steering the state from the input; observability studies the possibility of estimating the state from the output. If a dynamical equation is controllable, all the modes of the equation can be excited from the input; if a dynamical equation is observable, all the modes of the equation can be observed at the output. These two concepts are defined under the assumption that we have the complete knowledge of a dynamical equation; that is, the matrices **A**, **B**, **C**, and **D** are known beforehand. Hence, the problem of observability is different from the problem of realization or identification. The problem of identification is a problem of estimating the matrices **A**, **B**, **C**, and **D** from the information collected at the input and output terminals.

Consider the $n$-dimensional linear dynamical equation

$$E: \quad \dot{\mathbf{x}} = \mathbf{A}(t)\mathbf{x}(t) + \mathbf{B}(t)\mathbf{u}(t) \tag{5-26a}$$
$$\mathbf{y} = \mathbf{C}(t)\mathbf{x}(t) + \mathbf{D}(t)\mathbf{u}(t) \tag{5-26b}$$

where **A**, **B**, **C**, and **D** are $n \times n$, $n \times p$, $q \times n$, and $q \times p$ matrices whose entries are continuous functions of $t$ defined over $(-\infty, \infty)$.

### Definition 5-5

The dynamical equation $E$ is said to be (completely state) *observable* at $t_0$, if for any state $\mathbf{x}^0$ at time $t_0$ in the state space, there exists a finite $t_1 > t_0$ such that the knowledge of the input $\mathbf{u}_{[t_0,t_1]}$ and the output $\mathbf{y}_{[t_0,t_1]}$ over the time interval $[t_0, t_1]$ suffices to determine the state $\mathbf{x}^0$. Otherwise, the dynamical equation $E$ is said to be *unobservable* at $t_0$. ∎

### Example 1

Consider the network shown in Figure 5-7. If the input is zero, no matter what the initial voltage across the capacitor is, in view of the symmetry of the network, the output is identically zero. We know the input and output (both are identically zero), but we are not able to determine the initial condition of the capacitor; hence the system, or more precisely, the dynamical equation that describes the system, is not observable at any $t_0$.

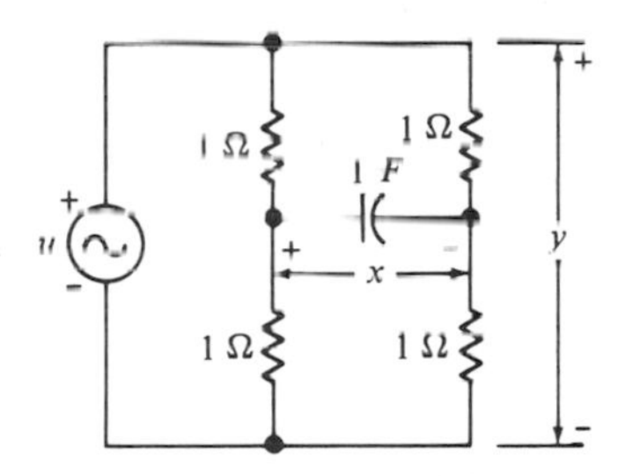


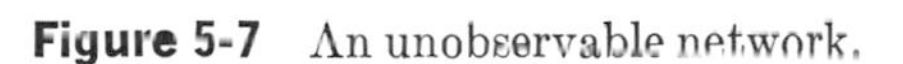

**Figure 5-7** An unobservable network.

### Example 2

Consider the network shown in Figure 5-8(a). If no input is applied, the network reduces to the one shown in Figure 5-8(b). Clearly the response

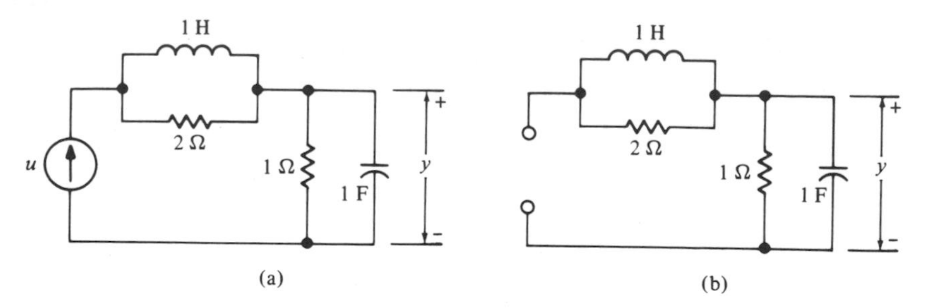


**Figure 5-8** An unobservable network.

to the initial current in the inductor can never appear at the output terminal. Therefore, there is no way of determining the initial current in the inductor from the input and the output terminals. Hence the system or its dynamical equation is not observable at any $t_0$. ∎

The response of the dynamical equation (5-26) is given by

$$\mathbf{y}(t) = \mathbf{C}(t)\mathbf{\Phi}(t, t_0)\mathbf{x}(t_0) + \mathbf{C}(t)\int_{t_0}^{t} \mathbf{\Phi}(t, \tau)\mathbf{B}(\tau)\mathbf{u}(\tau)\, d\tau + \mathbf{D}(t)\mathbf{u}(t) \tag{5-27}$$

where $\mathbf{\Phi}(t, \tau)$ is the state transition matrix of $\dot{\mathbf{x}} = \mathbf{A}(t)\mathbf{x}$. In the study of observability, the output $\mathbf{y}$ and the input $\mathbf{u}$ are assumed to be known, the initial $\mathbf{x}(t_0)$ is the only unknown, hence (5-27) can be written as

$$\bar{\mathbf{y}}(t) = \mathbf{C}(t)\mathbf{\Phi}(t, t_0)\mathbf{x}(t_0) \tag{5-28}$$

where

$$\bar{\mathbf{y}}(t) \triangleq \mathbf{y}(t) - \mathbf{C}(t)\int_{t_0}^{t} \mathbf{\Phi}(t, \tau)\mathbf{B}(\tau)\mathbf{u}(\tau)\, d\tau - \mathbf{D}(t)\mathbf{u}(t) \tag{5-29}$$

is a known function. Consequently, the observability problem is a problem of determining $\mathbf{x}(t_0)$ in (5-28) with the knowledge of $\bar{\mathbf{y}}$, $\mathbf{C}$, and $\mathbf{\Phi}(t, t_0)$. Note that the estimated state $\mathbf{x}(t_0)$ is the state not at time $t$, but at time $t_0$. However if $\mathbf{x}(t_0)$ is known, the state after $t_0$ can be computed from

$$\mathbf{x}(t) = \mathbf{\Phi}(t, t_0)\mathbf{x}(t_0) + \int_{t_0}^{t} \mathbf{\Phi}(t, \tau)\mathbf{B}(\tau)\mathbf{u}(\tau)\, d\tau + \mathbf{D}(t)\mathbf{u}(t) \tag{5-30}$$

### Theorem 5-9

The dynamical equation $E$ is observable at $t_0$ if and only if there exists a finite $t_1 > t_0$ such that the $n$ *columns* of the $q \times n$ matrix function $\mathbf{C}(\cdot)\mathbf{\Phi}(\cdot, t_0)$ are linearly independent on $[t_0, t_1]$.

### Proof

*Sufficiency:* Multiplying $\mathbf{\Phi}^*(t, t_0)\mathbf{C}^*(t)$ on both sides of (5-28) and integrating from $t_0$ to $t_1$, we obtain

$$\int_{t_0}^{t_1} \mathbf{\Phi}^*(t, t_0)\mathbf{C}^*(t)\bar{\mathbf{y}}(t)\, dt = \left(\int_{t_0}^{t_1} \mathbf{\Phi}^*(t, t_0)\mathbf{C}^*(t)\mathbf{C}(t)\mathbf{\Phi}(t, t_0)\, dt\right)\mathbf{x}^0 \triangleq \mathbf{V}(t_0, t_1)\mathbf{x}^0 \tag{5-31}$$

where

$$\mathbf{V}(t_0, t_1) \triangleq \int_{t_0}^{t_1} \mathbf{\Phi}^*(t, t_0)\mathbf{C}^*(t)\mathbf{C}(t)\mathbf{\Phi}(t, t_0)\, dt \tag{5-32}$$

From Theorem 5-1 and the assumption that all the columns of $\mathbf{C}(\cdot)\mathbf{\Phi}(\cdot, t_0)$ are linearly independent on $[t_0, t_1]$, we conclude that $\mathbf{V}(t_0, t_1)$ is nonsingular. Hence, from (5-31) we have

$$\mathbf{x}^0 = \mathbf{V}^{-1}(t_0, t_1) \int_{t_0}^{t_1} \mathbf{\Phi}^*(t, t_0)\mathbf{C}^*(t)\bar{\mathbf{y}}(t)\, dt \tag{5-33}$$

Thus, if the function $\bar{\mathbf{y}}_{[t_0, t_1]}$ is known, $\mathbf{x}^0$ can be computed from (5-33). *Necessity:* Prove by contradiction. Suppose $E$ is observable at $t_0$, but there exists *no* $t_1 > t_0$ such that the columns of $\mathbf{C}(\cdot)\mathbf{\Phi}(\cdot, t_0)$ are linearly independent on $[t_0, t_1]$. Then there exists an $n \times 1$ nonzero constant vector $\boldsymbol{\alpha}$ such that

$$\mathbf{C}(t)\mathbf{\Phi}(t, t_0)\boldsymbol{\alpha} = \mathbf{0} \qquad \text{for all } t > t_0$$

Let us choose $\mathbf{x}(t_0) = \boldsymbol{\alpha}$; then

$$\bar{\mathbf{y}}(t) = \mathbf{C}(t)\mathbf{\Phi}(t, t_0)\boldsymbol{\alpha} = \mathbf{0} \qquad \text{for all } t > t_0$$

Hence the initial state $\mathbf{x}(t_0) = \boldsymbol{\alpha}$ cannot be detected. This contradicts the assumption that $E$ is observable. Therefore, if $E$ is observable there exists a finite $t_1 > t_0$ such that the columns of $\mathbf{C}(\cdot)\mathbf{\Phi}(\cdot, t_0)$ are linearly independent on $[t_0, t_1]$. Q.E.D.

We see from this theorem that the observability of a linear dynamical equation depends only on $\mathbf{C}(t)$ and $\mathbf{\Phi}(t, t_0)$ or, equivalently, only on $\mathbf{C}$ and $\mathbf{A}$. This can also be deduced from Definition 5-5 by choosing $\mathbf{u} \equiv \mathbf{0}$. Hence in the observability study, it is sometimes convenient to assume $\mathbf{u} \equiv \mathbf{0}$ and study only $\dot{\mathbf{x}} = \mathbf{A}(t)\mathbf{x}$, $\mathbf{y} = \mathbf{C}(t)\mathbf{x}$.

The controllability of a dynamical equation is determined by the linear independence of the *rows* of $\mathbf{\Phi}(t_0, \cdot)\mathbf{B}(\cdot)$, whereas the observability is determined by the linear independence of the *columns* of $\mathbf{C}(\cdot)\mathbf{\Phi}(\cdot, t_0)$. The relationship between these two concepts is established in the following theorem.

#### Theorem 5-10 (theorem of duality)

Consider the dynamical equation $E$ in (5-26) and the dynamical equation $E^*$ defined by

$$E^*: \quad \dot{\mathbf{z}} = -\mathbf{A}^*(t)\mathbf{z} + \mathbf{C}^*(t)\mathbf{v} \tag{5-34a}$$

$$\boldsymbol{\gamma} = \mathbf{B}^*(t)\mathbf{z} + \mathbf{D}^*(t)\mathbf{v} \tag{5-34b}$$

where $\mathbf{A}^*$, $\mathbf{B}^*$, $\mathbf{C}^*$, and $\mathbf{D}^*$ are the complex conjugate transpose of $\mathbf{A}$, $\mathbf{B}$, $\mathbf{C}$, and $\mathbf{D}$ in $E$. The equation $E$ is controllable (observable) at $t_0$ if and only if the equation $E^*$ is observable (controllable) at $t_0$.

### Proof

From Theorem 5-4, the dynamical equation $E$ is controllable if and only if the *rows* of $\mathbf{\Phi}(t_0, t)\mathbf{B}(t)$ are linearly independent, in $t$, on $[t_0, t_1]$. From Theorem 5-9, the dynamical equation $E^*$ is observable if and only if the *columns* of $\mathbf{B}^*(t)\mathbf{\Phi}_1(t, t_0)$ are linearly independent, in $t$, on $[t_0, t_1]$; or equivalently, the *rows* of $[\mathbf{B}^*(t)\mathbf{\Phi}_1(t, t_0)]^* = \mathbf{\Phi}_1^*(t, t_0)\mathbf{B}(t)$ are linearly independent, in $t$, on $[t_0, t_1]$, where $\mathbf{\Phi}_1$ is the state transition matrix of $\dot{\mathbf{z}} = -\mathbf{A}^*(t)\mathbf{z}$. It is easy to show that $\mathbf{\Phi}_1^*(t, t_0) = \mathbf{\Phi}(t_0, t)$ (see Problem 4-9); hence $E$ is controllable if and only if $E^*$ is observable. Q.E.D.

We list in the following, for observability, Theorems 5-11–5-14 and Definitions 5-6–5-8, which are dual to Theorems 5-5–5-8 and Definitions 5-2–5-4 for controllability. Theorems 5-11–5-14 can be proved either directly or by applying Theorem 5-10 to Theorems 5-5–5-8. The interpretations in the controllability part also apply to the observability part.

### Theorem 5-11

Assume that the matrices $\mathbf{A}(\cdot)$ and $\mathbf{C}(\cdot)$ in the $n$-dimensional dynamical equation $E$ are $(n-1)$ times continuously differentiable. Then the dynamical equation $E$ is observable at $t_0$ if there exists a finite $t_1 > t_0$ such that

$$\rho \begin{bmatrix} \mathbf{N}_0(t_1) \\ \mathbf{N}_1(t_1) \\ \cdot \\ \cdot \\ \cdot \\ \mathbf{N}_{n-1}(t_1) \end{bmatrix} = n \tag{5-35}$$

where

$$\mathbf{N}_{k+1}(t) = \mathbf{N}_k(t)\mathbf{A}(t) + \frac{d}{dt}\mathbf{N}_k(t) \qquad k = 0, 1, 2, \cdots, n-1 \tag{5-36a}$$

with

$$\mathbf{N}_0(t) = \mathbf{C}(t) \tag{5-36b}$$

∎

### * Differential observability, instantaneous observability, and uniform observability

Differential, instantaneous, and uniform observabilities can be defined by using the theorem of duality; for example, we may define $\{\mathbf{A}, \mathbf{C}\}$ to be differentially observable if and only if $\{-\mathbf{A}^*, \mathbf{C}^*\}$ is differentially controllable. However, for ease of reference we shall define them explicitly in the following.

### Definition 5-6

The dynamical equation $E$ is said to be *differentially observable* at time $t_0$ if, for any state $\mathbf{x}(t_0)$ in the state space $\Sigma$, the knowledge of the input and the output over an arbitrarily small interval of time suffices to determine $\mathbf{x}(t_0)$.

**Theorem 5-12**

If the matrices **A** and **C** are analytic on $(-\infty, \infty)$, then the $n$-dimensional dynamical equation $E$ is differentially observable at every $t$ in $(-\infty, \infty)$ if and only if, for any fixed $t_0$ in $(-\infty, \infty)$,

$$\rho \begin{bmatrix} \mathbf{N}_0(t_0) \\ \mathbf{N}_1(t_0) \\ \cdot \\ \cdot \\ \cdot \\ \mathbf{N}_{n-1}(t_0) \\ \cdot \\ \cdot \\ \cdot \end{bmatrix} = n$$

∎

**Definition 5-7**

The linear dynamical equation $E$ is said to be *instantaneously observable* in $(-\infty, \infty)$ if and only if

$$\rho \begin{bmatrix} \mathbf{N}_0(t) \\ \mathbf{N}_1(t) \\ \cdot \\ \cdot \\ \cdot \\ \mathbf{N}_{n-1}(t) \end{bmatrix} = n \qquad \text{for all } t \text{ in } (-\infty, \infty)$$

where the $\mathbf{N}_i$'s are as defined in (5-36).

**Definition 5-8**

The linear dynamical equation $E$ is said to be *uniformly observable* in $(-\infty, \infty)$ if and only if there exist a positive $\sigma_o$ and positive $\beta_i$ that depend on $\sigma_o$ such that

$$\begin{aligned} 0 &< \beta_1(\sigma_o)\mathbf{I} \le \mathbf{V}(t, t+\sigma_o) \le \beta_2(\sigma_o)\mathbf{I} \\ 0 &< \beta_3(\sigma_o)\mathbf{I} \le \mathbf{\Phi}^*(t, t+\sigma_o)\mathbf{V}(t, t+\sigma_o)\mathbf{\Phi}(t, t+\sigma_o) \le \beta_4(\sigma_o)\mathbf{I} \end{aligned}$$

for all $t$; where $\mathbf{\Phi}$ is the state transition matrix and $\mathbf{V}$ is as defined in (5-32). ∎

**Linear, time-invariant dynamical equations.** Consider the linear, time-invariant dynamical equation

$$FE: \qquad \dot{\mathbf{x}} = \mathbf{A}\mathbf{x} + \mathbf{B}\mathbf{u} \tag{5-37a}$$

$$\mathbf{y} = \mathbf{C}\mathbf{x} + \mathbf{D}\mathbf{u} \tag{5-37b}$$

where **A**, **B**, **C** and **D** are $n \times n$, $n \times p$, $q \times n$ and $q \times p$ constant matrices. The time interval of interest is $[0, \infty)$.

### Theorem 5-13

For the $n$-dimensional, linear, time-invariant dynamical equation *FE*, the following statements are equivalent:

1. *FE* is observable at every $t_0$ in $[0, \infty)$.
2. The columns of $\mathbf{C}e^{\mathbf{A}t}$ are linearly independent on $[0, \infty)$ over the field of complex numbers.
3. $\mathbf{V}(t_0, t) = \int_{t_0}^{t} e^{\mathbf{A}^*(\tau - t_0)}C^*Ce^{\mathbf{A}(\tau - t_0)}\, d\tau$ is nonsingular for any $t_0 > 0$ and any $t > t_0$.
4. The $(nq \times n)$ matrix

$$\mathbf{V} \triangleq \begin{bmatrix} \mathbf{C} \\ \mathbf{CA} \\ \cdot \\ \cdot \\ \cdot \\ \mathbf{CA}^{n-1} \end{bmatrix} \tag{5-38}$$

has rank $n$.
5. The columns of $\mathbf{C}(s\mathbf{I} - \mathbf{A})^{-1}$ are linearly independent over the field of complex numbers.

The matrix **V** defined in (5-38) is called the *observability matrix* of *FE*. ∎

### * Simplified observability condition

### Theorem 5-14

Define

$$\mathbf{V}_{k-1} \triangleq \begin{bmatrix} \mathbf{C} \\ \mathbf{CA} \\ \cdot \\ \cdot \\ \cdot \\ \mathbf{CA}^{k-1} \end{bmatrix} \tag{5-39}$$

Let $j$ be the least integer such that $\rho(\mathbf{V}_j) = \rho(\mathbf{V}_{j+1})$; then $\rho(\mathbf{V}_k) = \rho(\mathbf{V}_j)$ for all integers $k > j$, and $j \leq \min(n - r, \bar{n} - 1)$, where $r$ is the rank of **C** and $\bar{n}$ is the degree of the minimal polynomial of **A**. The integer $j$ is called the *observability index.*

Corollary 5-14

If the rank of $\mathbf{C}$ is $r$, then the dynamical equation $FE$ is observable if and only if $\rho\mathbf{V}_{n-r} = n$ or the $n \times n$ matrix $\mathbf{V}^*_{n-r}\mathbf{V}_{n-r}$ is nonsingular. ∎

In the study of the observability of a linear dynamical equation, it is always assumed that the state variables are not available for direct measurements. If they are available for direct measurements, the estimation of the initial state is not needed; we just measure them directly.

Consider the linear, time-invariant dynamical equation

$$FE: \quad \begin{aligned} \dot{\mathbf{x}} &= \mathbf{A}\mathbf{x} + \mathbf{B}\mathbf{u} \\ \mathbf{y} &= \mathbf{C}\mathbf{x} + \mathbf{D}\mathbf{u} \end{aligned}$$

Suppose the state is not accessible; for any initial state $\mathbf{x}(t_0)$, is it possible to transfer the output to zero at time $t_2$ and keep it at zero thereafter? The answer is yes if the dynamical equation is controllable and observable. This can be achieved in two steps. First, we estimate the initial state and then apply a proper input to bring the state vector to zero. More specifically, we measure the output $\mathbf{y}$ from $t_0$ to $t_1$. Since the dynamical equation is observable, from $y_{[t_0,t_1]}$, we can compute $\mathbf{x}(t_0)$ and, consequently, $\mathbf{x}(t_1)$. As the second step, we apply an input $\mathbf{u}_{[t_1,t_2]}$ to transfer $\mathbf{x}(t_1)$ to the zero state at time $t_2$. This is always possible because of the controllability assumption. If the state at time $t_2$ is the zero state, and if no input is applied for $t > t_2$, then the output will be identically zero for all $t > t_2$. Note that this cannot be achieved if the dynamical equation is uncontrollable or unobservable.

## *5-5 Controllability and Observability of Jordan-Form Dynamical Equations[5]

We study in this section linear, time-invariant dynamical equations exclusively. Consider the $n$-dimensional, linear, time-invariant dynamical equation

$$FE: \quad \dot{\mathbf{x}} = \mathbf{A}\mathbf{x} + \mathbf{B}\mathbf{u} \tag{5-40a}$$
$$\mathbf{y} = \mathbf{C}\mathbf{x} + \mathbf{D}\mathbf{u} \tag{5-40b}$$

where $\mathbf{A}$, $\mathbf{B}$, $\mathbf{C}$, and $\mathbf{D}$ are $n \times n$, $n \times p$, $q \times n$ and $q \times p$ constant matrices, respectively. Let $\mathbf{P}$ be an $n \times n$ constant nonsingular matrix and let $\bar{\mathbf{x}} = \mathbf{P}\mathbf{x}$; then (5-40) becomes

$$F\bar{E}: \quad \dot{\bar{\mathbf{x}}} = \bar{\mathbf{A}}\bar{\mathbf{x}} + \bar{\mathbf{B}}\mathbf{u} \tag{5-41a}$$
$$\mathbf{y} = \bar{\mathbf{C}}\bar{\mathbf{x}} + \bar{\mathbf{D}}\mathbf{u} \tag{5-41b}$$

[5] Theorem 5-15 in this section will be used later. However, the rest of the section may be omitted with no loss of continuity.

where $\bar{\mathbf{A}} = \mathbf{PAP}^{-1}$, $\bar{\mathbf{B}} = \mathbf{PB}$, $\bar{\mathbf{C}} = \mathbf{CP}^{-1}$ and $\bar{\mathbf{D}} = \mathbf{D}$. The dynamical equations $FE$ and $F\bar{E}$ are said to be equivalent and the matrix $\mathbf{P}$ is called an equivalence transformation (Definition 4-3). Suppose the dynamical equation $FE$ is controllable, it is natural to ask whether the dynamical equation remains controllable after the equivalence transformation. The answer is yes. This is intuitively clear because an equivalence transformation changes only the basis of the state space. Therefore, the controllability of the state should not be affected.

### Theorem 5-15

The controllability and observability of a linear, time-invariant dynamical equation $FE$ are invariant under any equivalence transformation.

### Proof

We show that the equation $FE$ is controllable if and only if the equation $F\bar{E}$ is controllable. From Theorem 5-7, $FE$ is controllable if and only if $\rho[\mathbf{B} \vdots \mathbf{AB} \vdots \cdots \vdots \mathbf{A}^{n-1}\mathbf{B}] = n$ and $F\bar{E}$ is controllable if and only if $\rho[\bar{\mathbf{B}} \vdots \bar{\mathbf{A}}\bar{\mathbf{B}} \vdots \cdots \vdots \bar{\mathbf{A}}^{n-1}\bar{\mathbf{B}}] = n$. Now $\bar{\mathbf{A}} = \mathbf{PAP}^{-1}$ and $\bar{\mathbf{B}} = \mathbf{PB}$; hence

$$\begin{aligned}[\bar{\mathbf{B}} \vdots \bar{\mathbf{A}}\bar{\mathbf{B}} \vdots \cdots \vdots \bar{\mathbf{A}}^{n-1}\bar{\mathbf{B}}] &= [\mathbf{PB} \vdots \mathbf{PAB} \vdots \cdots \vdots \mathbf{PA}^{n-1}\mathbf{B}] \\ &= \mathbf{P}[\mathbf{B} \vdots \mathbf{AB} \vdots \cdots \vdots \mathbf{A}^{n-1}\mathbf{B}]\end{aligned}$$

Since the rank of a matrix does not change by the multiplication of a nonsingular matrix (Theorem 2-7), hence we have

$$\rho[\mathbf{B} \vdots \mathbf{AB} \vdots \cdots \vdots \mathbf{A}^{n-1}\mathbf{B}] = \rho[\bar{\mathbf{B}} \vdots \bar{\mathbf{A}}\bar{\mathbf{B}} \vdots \cdots \vdots \bar{\mathbf{A}}^{n-1}\bar{\mathbf{B}}]$$

Consequently $FE$ is controllable if and only if $F\bar{E}$ is controllable. The observability part can be similarly proved. Q.E.D.

This theorem can be extended to the linear, time-varying dynamical equation (see Problem 5-16). Since the controllability of a dynamical equation is preserved under any equivalence transformation, it is conceivable that we might obtain a simpler controllability criterion by transforming the dynamical equation into a special form. If we transform linear time-invariant equations into Jordan forms, their controllabilities and observabilities can be determined almost by inspection. Techniques for transforming a linear, time-invariant dynamical equation into a Jordan form are discussed in Sections 2-6 and 4-3; hence in this section we shall assume that all the dynamical equations are in the Jordan form.

Consider the $n$-dimensional, linear, time-invariant, Jordan-form dynamical equation

$$JFE: \quad \dot{\mathbf{x}} = \mathbf{Ax} + \mathbf{Bu} \qquad \text{(5-42a)}$$

$$\mathbf{y} = \mathbf{Cx} + \mathbf{Du} \qquad \text{(5-42b)}$$

where the matrices **A**, **B**, and **C** are assumed of the forms shown in Table 5-1. The $n \times n$ matrix **A** is in the Jordan form, with $m$ distinct eigen-

**Table 5-1** JORDAN-FORM DYNAMICAL EQUATION

$$\underset{(n \times n)}{\mathbf{A}} = \begin{bmatrix} \mathbf{A}_1 & & & & \\ & \mathbf{A}_2 & & & \\ & & \cdot & & \\ & & & \cdot & \\ & & & & \cdot \\ & & & & & \mathbf{A}_m \end{bmatrix} \qquad \underset{(n \times p)}{\mathbf{B}} = \begin{bmatrix} \mathbf{B}_1 \\ \mathbf{B}_2 \\ \cdot \\ \cdot \\ \cdot \\ \mathbf{B}_m \end{bmatrix}$$

$$\mathbf{C} = [\mathbf{C}_1 \quad \mathbf{C}_2 \quad \cdots \quad \mathbf{C}_m]$$

$$\underset{(n_i \times n_i)}{\mathbf{A}_i} = \begin{bmatrix} \mathbf{A}_{i1} & & & & \\ & \mathbf{A}_{i2} & & & \\ & & \cdot & & \\ & & & \cdot & \\ & & & & \cdot \\ & & & & & \mathbf{A}_{ir(i)} \end{bmatrix} \qquad \underset{(n_i \times p)}{\mathbf{B}_i} = \begin{bmatrix} \mathbf{B}_{i1} \\ \mathbf{B}_{i2} \\ \cdot \\ \cdot \\ \cdot \\ \mathbf{B}_{ir(i)} \end{bmatrix}$$

$$\underset{(q \times n_i)}{\mathbf{C}_i} = [\mathbf{C}_{i1} \quad \mathbf{C}_{i2} \quad \cdots \quad \mathbf{C}_{ir(i)}]$$

$$\underset{(n_{ij} \times n_{ij})}{\mathbf{A}_{ij}} = \begin{bmatrix} \lambda_i & 1 & & & \\ & \lambda_i & 1 & & \\ & & \cdot & \cdot & \\ & & & \cdot & \cdot \\ & & & & \cdot & \cdot \\ & & & & & \lambda_i & 1 \\ & & & & & & \lambda_i \end{bmatrix} \qquad \underset{(n_{ij} \times p)}{\mathbf{B}_{ij}} = \begin{bmatrix} \mathbf{b}_{1ij} \\ \mathbf{b}_{2ij} \\ \cdot \\ \cdot \\ \cdot \\ \mathbf{b}_{lij} \end{bmatrix}$$

$$\mathbf{C}_{ij} = [\mathbf{c}_{1ij} \quad \mathbf{c}_{2ij} \quad \cdots \quad \mathbf{c}_{lij}]$$

values $\lambda_1, \lambda_2, \cdots, \lambda_m$. $\mathbf{A}_i$ denotes all the Jordan blocks associated with the eigenvalue $\lambda_i$; $r(i)$ is the number of Jordan blocks in $\mathbf{A}_i$; and $\mathbf{A}_{ij}$ is the $j$th Jordan block in $\mathbf{A}_i$. Clearly,

$$\mathbf{A}_i = \text{diag}\,(\mathbf{A}_{i1}, \mathbf{A}_{i2}, \cdots, \mathbf{A}_{ir(i)}) \qquad \text{and} \qquad \mathbf{A} = \text{diag}\,(\mathbf{A}_1, \mathbf{A}_2, \cdots, \mathbf{A}_m)$$

Let $n_i$ and $n_{ij}$ be the order of $\mathbf{A}_i$ and $\mathbf{A}_{ij}$, respectively; then

$$n = \sum_{i=1}^{m} n_i = \sum_{i=1}^{m} \sum_{j=1}^{r(i)} n_{ij}$$

Corresponding to $\mathbf{A}_i$ and $\mathbf{A}_{ij}$, the matrices **B** and **C** are partitioned as shown. The first *row* and the last row of $\mathbf{B}_{ij}$ are denoted by $\mathbf{b}_{1ij}$ and $\mathbf{b}_{lij}$, respectively. The first *column* and the last column of $\mathbf{C}_{ij}$ are denoted by $\mathbf{c}_{1ij}$ and $\mathbf{c}_{lij}$.

### Theorem 5-16

The $n$-dimensional, linear, time-invariant, Jordan-form dynamical equation *JFE* is controllable if and only if for each $i = 1, 2, \cdots, m$, the rows of the $r(i) \times p$ matrix

$$\mathbf{B}_i^l \triangleq \begin{bmatrix} \mathbf{b}_{li1} \\ \mathbf{b}_{li2} \\ \cdot \\ \cdot \\ \cdot \\ \mathbf{b}_{lir(i)} \end{bmatrix} \tag{5-43a}$$

are linearly independent (over the field of complex numbers). $JFE$ is observable if and only if for each $i = 1, 2, \cdots, m$, the columns of the $q \times r(i)$ matrix

$$\mathbf{C}_i^1 = [\mathbf{c}_{1i1} \quad \mathbf{c}_{1i2} \quad \cdots \quad \mathbf{c}_{1ir(i)}] \tag{5-43b}$$

are linearly independent (over the field of complex numbers). ∎

### Example 1

Consider the Jordan-form dynamical equation

$$FE: \quad \dot{\mathbf{x}} = \begin{bmatrix} \lambda_1 & 1 & 0 & 0 & 0 & 0 & 0 \\ 0 & \lambda_1 & 0 & 0 & 0 & 0 & 0 \\ 0 & 0 & \lambda_1 & 0 & 0 & 0 & 0 \\ 0 & 0 & 0 & \lambda_1 & 0 & 0 & 0 \\ 0 & 0 & 0 & 0 & \lambda_2 & 1 & 0 \\ 0 & 0 & 0 & 0 & 0 & \lambda_2 & 1 \\ 0 & 0 & 0 & 0 & 0 & 0 & \lambda_2 \end{bmatrix} \mathbf{x} + \begin{bmatrix} 0 & 0 & 0 \\ 1 & 0 & 0 \\ 0 & 1 & 0 \\ 0 & 0 & 1 \\ 1 & 1 & 2 \\ 0 & 1 & 0 \\ 0 & 0 & 1 \end{bmatrix} \mathbf{u} \begin{matrix} \\ \leftarrow \mathbf{b}_{l11} \\ \leftarrow \mathbf{b}_{l12} \\ \leftarrow \mathbf{b}_{l13} \\ \\ \\ \leftarrow \mathbf{b}_{l21} \end{matrix} \tag{5-44a}$$

$$\mathbf{y} = \begin{bmatrix} 1 & 1 & 2 & 0 & 0 & 2 & 0 \\ 1 & 0 & 1 & 2 & 0 & 1 & 1 \\ 1 & 0 & 2 & 3 & 0 & 2 & 2 \end{bmatrix} \mathbf{x} \tag{5-44b}$$

$$\begin{matrix} \uparrow & & \uparrow & \uparrow & \uparrow \\ \mathbf{c}_{111} & & \mathbf{c}_{112} & \mathbf{c}_{113} & \mathbf{c}_{121} \end{matrix}$$

The matrix $\mathbf{A}$ has two distinct eigenvalues $\lambda_1$ and $\lambda_2$. There are three Jordan blocks associated with $\lambda_1$; hence, $r(1) = 3$. There is only one Jordan block associated with $\lambda_2$; hence $r(2) = 1$. The conditions for $JFE$ to be controllable are that the set $\{\mathbf{b}_{l11}, \mathbf{b}_{l12}, \mathbf{b}_{l13}\}$ and the set $\{\mathbf{b}_{l21}\}$ be, individually, linearly independent. This is the case, hence $JFE$ is controllable. The conditions for $JFE$ to be observable are that the set $\{\mathbf{c}_{111}, \mathbf{c}_{112}, \mathbf{c}_{113}\}$ and the set $\{\mathbf{c}_{121}\}$ be, individually, linearly independent. Although the set $\{\mathbf{c}_{111}, \mathbf{c}_{112}, \mathbf{c}_{113}\}$ is linearly independent, the set $\{\mathbf{c}_{121}\}$, which consists of a zero vector, is linearly dependent. Hence *JFE* is not observable. ▮

The conditions for controllability and observability in Theorem 5-16 require that each of the $m$ set of vectors be *individually* tested for linear independence. The linear dependence of one set on the other set is immaterial. Furthermore, the row vectors of $\mathbf{B}$ excluding the $\mathbf{b}_{lij}$'s do not play any role in determining the controllability of the equation.

The physical meaning of the conditions of Theorem 5-16 can be seen from the block diagram of the Jordan-form dynamical equation *JFE*. Instead of studying the general case, we draw in Figure 5-9 a block dia-

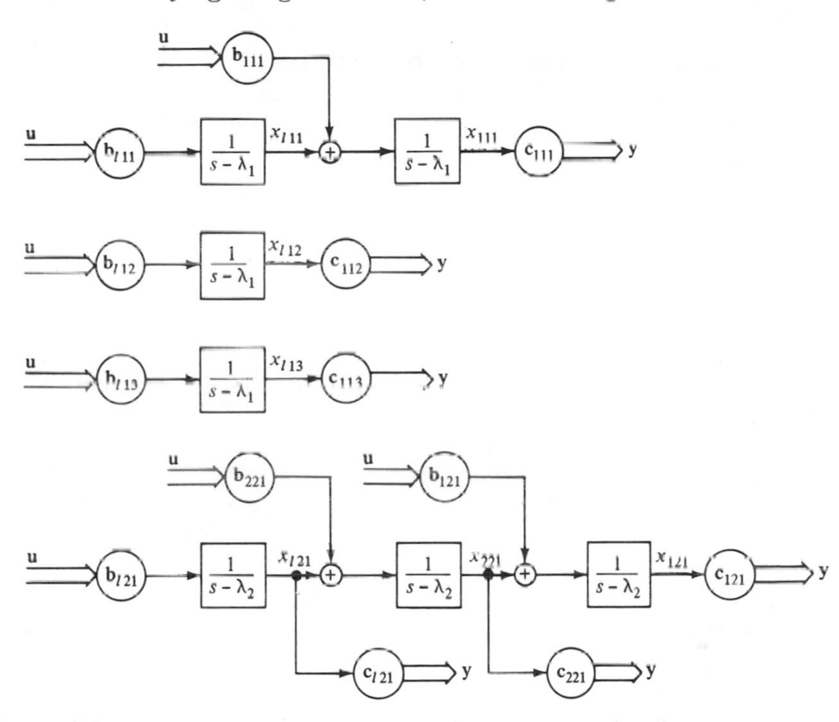


**Figure 5-9** Block diagram of the Jordan-form equation (5-44).

gram for the Jordan-form dynamical equation (5-44). Observe that each block consists of an integrator and a feedback path, as shown in Figure 5-10. The output of each block, or more precisely the output of each integrator, is assigned as a state variable. Each chain of blocks corresponds to a Jordan block in the equation. Consider the last chain of Figure 5-9. We see that if $\mathbf{b}_{l21} \neq \mathbf{0}$, then all state variables in that chain can be controlled; if $\mathbf{c}_{121} \neq \mathbf{0}$, then all state variables in that chain can be observed. If there are two or more chains associated with the same eigenvalue, then we require the linear independence of the first gain

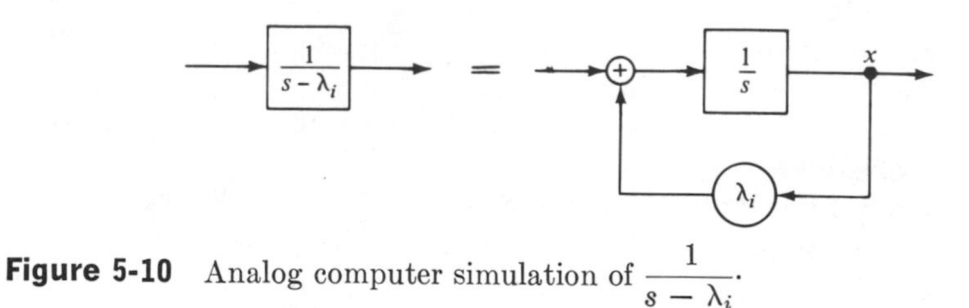


**Figure 5-10** Analog computer simulation of $\dfrac{1}{s - \lambda_i}$.

vectors of these chains. The chains associated with different eigenvalues can be studied independently.

### Proof of Theorem 5-16

We shall use the following fact to prove the theorem. A linear, time-invariant dynamical equation is controllable if and only if the rows of $(s\mathbf{I} - \mathbf{A})^{-1}\mathbf{B}$ are linearly independent over the field of complex numbers. Since the matrix $\mathbf{A}$ is of diagonal form, we have

$$(s\mathbf{I} - \mathbf{A})^{-1}\mathbf{B} = \begin{bmatrix} (s\mathbf{I} - \mathbf{A}_1)^{-1}\mathbf{B}_1 \\ (s\mathbf{I} - \mathbf{A}_2)^{-1}\mathbf{B}_2 \\ \cdot \\ \cdot \\ \cdot \\ (s\mathbf{I} - \mathbf{A}_m)^{-1}\mathbf{B}_m \end{bmatrix} = \begin{bmatrix} (s\mathbf{I} - \mathbf{A}_{11})^{-1}\mathbf{B}_{11} \\ (s\mathbf{I} - \mathbf{A}_{12})^{-1}\mathbf{B}_{12} \\ \cdot \\ \cdot \\ \cdot \\ (s\mathbf{I} - \mathbf{A}_{1r(1)})^{-1}\mathbf{B}_{1r(1)} \\ (s\mathbf{I} - \mathbf{A}_{21})^{-1}\mathbf{B}_{21} \\ \cdot \\ \cdot \\ \cdot \\ (s\mathbf{I} - \mathbf{A}_{mr(m)})^{-1}\mathbf{B}_{mr(m)} \end{bmatrix} \tag{5-45}$$

Since $\mathbf{A}_{ij}$ is a Jordan block, from (2-74) we have

$$(s\mathbf{I} - \mathbf{A}_{ij})^{-1}\mathbf{B}_{ij}$$

$$= \begin{bmatrix} \dfrac{1}{(s-\lambda_i)} & \dfrac{1}{(s-\lambda_i)^2} & \dfrac{1}{(s-\lambda_i)^3} & \cdots & \dfrac{1}{(s-\lambda_i)^{n_{ij}}} \\ 0 & \dfrac{1}{(s-\lambda_i)} & \dfrac{1}{(s-\lambda_i)^2} & \cdots & \dfrac{1}{(s-\lambda_i)^{n_{ij}-1}} \\ 0 & 0 & \dfrac{1}{(s-\lambda_i)} & \cdots & \dfrac{1}{(s-\lambda_i)^{n_{ij}-2}} \\ \cdot & \cdot & \cdot & & \cdot \\ \cdot & \cdot & \cdot & & \cdot \\ \cdot & \cdot & \cdot & & \cdot \\ 0 & 0 & 0 & \cdots & \dfrac{1}{(s-\lambda_i)} \end{bmatrix} \begin{bmatrix} \mathbf{b}_{1ij} \\ \mathbf{b}_{2ij} \\ \mathbf{b}_{3ij} \\ \cdot \\ \cdot \\ \cdot \\ \mathbf{b}_{lij} \end{bmatrix} \tag{5-46}$$

where $n_{ij}$ is the order of $\mathbf{A}_{ij}$. We see that all the rows of $(s\mathbf{I} - \mathbf{A}_i)^{-1}\mathbf{B}_i$ are linear combinations of the terms $1/(s - \lambda_i)^k$. Hence from the assumption that $\lambda_1, \lambda_2, \cdots, \lambda_m$ are distinct, if we show that for each $i$ all the rows of $(s\mathbf{I} - \mathbf{A}_i)^{-1}\mathbf{B}_i$ are linearly independent over the field of complex numbers, then we can conclude that all the rows of $(s\mathbf{I} - \mathbf{A})^{-1}\mathbf{B}$ are linearly independent over the field of complex numbers.

*Necessity:* Observe that in the matrix $(s\mathbf{I} - \mathbf{A}_i)^{-1}\mathbf{B}_i$, there are $r(i)$ rows of the form

$$\frac{1}{(s - \lambda_i)}\mathbf{b}_{li1}, \frac{1}{(s - \lambda_i)}\mathbf{b}_{li2}, \cdots, \frac{1}{(s - \lambda_i)}\mathbf{b}_{lir(i)}$$

Hence, if $\mathbf{b}_{li1}, \mathbf{b}_{li2}, \cdots, \mathbf{b}_{lir(i)}$ are not linearly independent, the rows of $(s\mathbf{I} - \mathbf{A}_i)^{-1}\mathbf{B}_i$ are not linearly independent. Consequently, if the rows of $\mathbf{B}_i^l$ are not linearly independent for some $i$, then the dynamical equation *JFE* is not controllable.

*Sufficiency:* We first show that the rows of $(s\mathbf{I} - \mathbf{A}_{ij})^{-1}\mathbf{B}_{ij}$ in (5-46) are linearly independent if and only if $\mathbf{b}_{lij} \neq \mathbf{0}$. Indeed, if $\mathbf{b}_{lij} = \mathbf{0}$, then the last row of $(s\mathbf{I} - \mathbf{A}_{ij})^{-1}\mathbf{B}_{ij}$ is identically zero, and hence the rows of $(s\mathbf{I} - \mathbf{A}_{ij})^{-1}\mathbf{B}_{ij}$ are linearly dependent. If $\mathbf{b}_{lij} \neq \mathbf{0}$, then the first row of $(s\mathbf{I} - \mathbf{A}_{ij})^{-1}\mathbf{B}_{ij}$ consists of the factor

$$\frac{1}{(s - \lambda_i)^{n_{ij}}}\mathbf{b}_{lij}$$

the second row,

$$\frac{1}{(s - \lambda_i)^{n_{ij}-1}}\mathbf{b}_{lij}$$

and so forth. Hence all the rows of $(s\mathbf{I} - \mathbf{A}_{ij})^{-1}\mathbf{B}_{ij}$ are linearly independent. If $\mathbf{b}_{li1}, \mathbf{b}_{li2}, \cdots, \mathbf{b}_{lir(i)}$ are linearly independent, it can be similarly shown that all the rows of $(s\mathbf{I} - \mathbf{A}_i)^{-1}\mathbf{B}_i$ are linearly independent. Consequently if the rows of $\mathbf{B}_i^l$ are linearly independent for each $i$, then the dynamical equation *JFE* is controllable. Q.E.D.

Observe that in order for the rows of an $r(i) \times p$ matrix $\mathbf{B}_i^l$ to be linearly independent, it is necessary that $r(i) \leq p$. Hence in the case of single input—that is, $p = 1$—it is necessary to have $r(i) = 1$ in order for the rows of $\mathbf{B}_i^l$ to be linearly independent. In words, a necessary condition for a single-input, Jordan-form dynamical equation to be controllable is that there is at most one Jordan block associated with one eigenvalue. In this case, the matrix $\mathbf{B}_i^l$ reduces to a vector. Thus we have the following corollary.

### Corollary 5-16

A single-input, linear, time-invariant, Jordan-form dynamical equation is controllable if and only if all the eigenvalues corresponding to each

Jordan block are pairwise distinct, and all the components of the column vector **B** that correspond to the last row of each Jordan block are different from zero.

A single-output linear, time-invariant, Jordan-form dynamical equation is observable if and only if all the eigenvalues corresponding to each Jordan block are pairwise distinct, and all the components of the row vector **C** that correspond to the first column of each Jordan block are different from zero. ∎

## Example 2

Consider the single-variable Jordan-form dynamical equation

$$\dot{\mathbf{x}} = \left[\begin{array}{ccc|c} 0 & 1 & 0 & 0 \\ 0 & 0 & 1 & 0 \\ 0 & 0 & 0 & 0 \\ \hline 0 & 0 & 0 & 1 \end{array}\right] \mathbf{x} + \left[\begin{array}{c} 10 \\ 9 \\ 0 \\ \hline 1 \end{array}\right] u$$

$$y = \left[\begin{array}{ccc|c} 1 & 0 & 0 & 1 \end{array}\right] \mathbf{x}$$

There are two Jordan blocks, one has eigenvalues 0, the other has 1. From Corollary 5-16, we conclude immediately that the equation is observable but not controllable.

## Example 3

Consider the following two Jordan-form state equations

$$\begin{bmatrix} \dot{x}_1 \\ \dot{x}_2 \end{bmatrix} = \begin{bmatrix} -1 & 0 \\ 0 & -2 \end{bmatrix} \mathbf{x} + \begin{bmatrix} 1 \\ 1 \end{bmatrix} u \tag{5-47}$$

and

$$\begin{bmatrix} \dot{x}_1 \\ \dot{x}_2 \end{bmatrix} = \begin{bmatrix} -1 & 0 \\ 0 & -2 \end{bmatrix} \mathbf{x} + \begin{bmatrix} e^{-t} \\ e^{-2t} \end{bmatrix} u \tag{5-48}$$

That the state equation (5-47) is controllable follows from Corollary 5-16. Equation (5-48) is a time-varying dynamical equation; however, since its **A** matrix is in the Jordan form and since the components of **b** are different from zero for all $t$, one might be tempted to conclude that (5-48) is controllable. Let us check this by using Theorem 5-4. For any fixed $t_0$, we have

$$\mathbf{\Phi}(t_0 - t)\mathbf{B}(t) = \begin{bmatrix} e^{-(t_0-t)} & 0 \\ 0 & e^{-2(t_0-t)} \end{bmatrix} \begin{bmatrix} e^{-t} \\ e^{-2t} \end{bmatrix} = \begin{bmatrix} e^{-t_0} \\ e^{-2t_0} \end{bmatrix}$$

It is clear that the rows of $\mathbf{\Phi}(t_0 - t)\mathbf{B}(t)$ are linearly dependent in $t$. Hence the state equation (5-48) is *not* controllable at any $t_0$. ∎

From this example we see that, in applying a theorem, all the conditions should be carefully checked; otherwise we might obtain a false conclusion.

## 5-6 Canonical Decomposition of a Linear, Time-Invariant Dynamical Equation

In this section we study again linear, time-invariant dynamical equations exclusively. Consider the following dynamical equation

$$FE: \quad \dot{\mathbf{x}} = \mathbf{A}\mathbf{x} + \mathbf{B}\mathbf{u} \tag{5-49a}$$

$$\mathbf{y} = \mathbf{C}\mathbf{x} + \mathbf{D}\mathbf{u} \tag{5-49b}$$

where $\mathbf{A}$, $\mathbf{B}$, $\mathbf{C}$, and $\mathbf{D}$ are $n \times n$, $n \times p$, $q \times n$, and $q \times p$ real constant matrices. We have introduced in the previous sections the concepts of controllability and observability. The conditions for the equation to be controllable and observable are also derived. A question that may be raised at this point is what can be said if the equation is uncontrollable and/or unobservable. In this section we shall study this problem. Before proceeding, the reader is advised to review the concept of equivalence transformation introduced in Section 4-3.

In the following, $c$ will be used to stand for controllable, $\bar{c}$ for uncontrollable, $o$ for observable, and $\bar{o}$ for unobservable.

### Theorem 5-17

Consider the $n$-dimensional, linear, time-invariant dynamical equation $FE$. If the controllability matrix of $FE$ has rank $n_1$ (where $n_1 < n$), then there exists an equivalence transformation $\bar{\mathbf{x}} = \mathbf{P}\mathbf{x}$, where $\mathbf{P}$ is a constant nonsingular matrix, which transforms $FE$ into

$$F\bar{E}: \quad \begin{bmatrix} \dot{\bar{\mathbf{x}}}^c \\ \dot{\bar{\mathbf{x}}}^{\bar{c}} \end{bmatrix} = \begin{bmatrix} \bar{\mathbf{A}}_c & \bar{\mathbf{A}}_{12} \\ \mathbf{0} & \bar{\mathbf{A}}_{\bar{c}} \end{bmatrix} \begin{bmatrix} \bar{\mathbf{x}}^c \\ \bar{\mathbf{x}}^{\bar{c}} \end{bmatrix} + \begin{bmatrix} \bar{\mathbf{B}}_c \\ \mathbf{0} \end{bmatrix} \mathbf{u} \tag{5-50a}$$

$$\mathbf{y} = [\bar{\mathbf{C}}_c \quad \bar{\mathbf{C}}_{\bar{c}}] \begin{bmatrix} \bar{\mathbf{x}}^c \\ \bar{\mathbf{x}}^{\bar{c}} \end{bmatrix} + \mathbf{D}\mathbf{u} \tag{5-50b}$$

and the $n_1$-dimensional subequation of $F\bar{E}$

$$F\bar{E}_c: \quad \dot{\bar{\mathbf{x}}}^c = \bar{\mathbf{A}}_c \bar{\mathbf{x}}^c + \bar{\mathbf{B}}_c \mathbf{u} \tag{5-51a}$$

$$\bar{\mathbf{y}} = \bar{\mathbf{C}}_c \bar{\mathbf{x}}^c + \mathbf{D}\mathbf{u} \tag{5-51b}$$

is controllable[5a] and has the same transfer function matrix as $F\bar{E}$. ▮

[5a] It is easy to show that if the equation $F\bar{E}$ is observable, then its subequation $F\bar{E}_c$ is also observable. (Try.)

### Proof

If the dynamical equation $FE$ is not controllable, then from Theorem 5-7 we have

$$\rho\mathbf{U} \triangleq \rho[\mathbf{B} \vdots \mathbf{AB} \vdots \cdots \vdots \mathbf{A}^{n-1}\mathbf{B}] = n_1 < n$$

Let $\mathbf{q}^1, \mathbf{q}^2, \cdots, \mathbf{q}^{n_1}$ be any $n_1$ linearly independent columns of $\mathbf{U}$. Note that for each $i = 1, 2, \cdots, n_1$, $\mathbf{A}\mathbf{q}^i$ can be written as a linear combination of $\{\mathbf{q}^1, \mathbf{q}^2, \cdots, \mathbf{q}^{n_1}\}$ (Why?). Define a nonsingular matrix

$$\mathbf{P}^{-1} \triangleq \mathbf{Q} \triangleq [\mathbf{q}^1 \quad \mathbf{q}^2 \quad \cdots \quad \mathbf{q}^{n_1} \quad \cdots \quad \mathbf{q}^n] \tag{5-51}$$

where the last $(n - n_1)$ columns of $\mathbf{Q}$ are entirely arbitrary so long as the matrix $\mathbf{Q}$ is nonsingular. We claim that the transformation $\bar{\mathbf{x}} = \mathbf{Px}$ will transform $FE$ into the form of (5-50). Recall from Figure 2-5 that in the transformation $\bar{\mathbf{x}} = \mathbf{Px}$ we are actually using the columns of $\mathbf{Q} \triangleq \mathbf{P}^{-1}$ as new basis vectors of the state space. The $i$th column of the new representation $\bar{\mathbf{A}}$ is the representation of $\mathbf{A}\mathbf{q}^i$ with respect to $\{\mathbf{q}^1, \mathbf{q}^2, \cdots, \mathbf{q}^n\}$. Now the vectors $\mathbf{A}\mathbf{q}^i$, for $i = 1, 2, \cdots, n_1$, are linearly dependent on the set $\{\mathbf{q}^1, \mathbf{q}^2, \cdots, \mathbf{q}^{n_1}\}$; hence the matrix $\bar{\mathbf{A}}$ has the form given in (5-50a). The columns of $\bar{\mathbf{B}}$ are the representations of the columns of $\mathbf{B}$ with respect to $\{\mathbf{q}^1, \mathbf{q}^2, \cdots, \mathbf{q}^n\}$. Now the columns of $\mathbf{B}$ depend only on $\{\mathbf{q}^1, \mathbf{q}^2, \cdots, \mathbf{q}^{n_1}\}$; hence $\bar{\mathbf{B}}$ is of the form shown in (5-50a).

Let $\mathbf{U}$ and $\bar{\mathbf{U}}$ be the controllability matrices of $FE$ and $F\bar{E}$, respectively. Then we have $\rho\mathbf{U} = \rho\bar{\mathbf{U}} = n_1$ (see Theorem 5-15). It is easy to verify that

$$\bar{\mathbf{U}} = \begin{bmatrix} \bar{\mathbf{B}}_c & \bar{\mathbf{A}}_c\bar{\mathbf{B}}_c & \cdots & \bar{\mathbf{A}}_c^{\,n-1}\bar{\mathbf{B}}_c \\ \mathbf{0} & \mathbf{0} & & \mathbf{0} \end{bmatrix}$$

$$= \begin{bmatrix} \bar{\mathbf{U}}_c & \bar{\mathbf{A}}_c^{\,n_1}\bar{\mathbf{B}}_c & \cdots & \bar{\mathbf{A}}_c^{\,n-1}\bar{\mathbf{B}}_c \\ \mathbf{0} & \mathbf{0} & & \mathbf{0} \end{bmatrix} \begin{matrix} \}\, n_1 \text{ rows} \\ \}\, (n - n_1) \text{ rows} \end{matrix}$$

where $\bar{\mathbf{U}}_c$ represents the controllability matrices of $F\bar{E}_c$. Since columns of $\bar{\mathbf{A}}_c^k\bar{\mathbf{B}}$ with $k \geq n_1$ are linearly dependent on the columns of $\bar{\mathbf{U}}_c$, the condition $\rho\bar{\mathbf{U}} = n_1$ implies $\rho\bar{\mathbf{U}}_c = n_1$. Hence the dynamical equation $F\bar{E}_c$ is controllable.

We show now that the dynamical equations $FE$ and $F\bar{E}_c$, or correspondingly $F\bar{E}$ and $F\bar{E}_c$, have the same transfer-function matrix. It is easy to verify that

$$\begin{bmatrix} s\mathbf{I} - \bar{\mathbf{A}}_c & -\bar{\mathbf{A}}_{12} \\ \mathbf{0} & s\mathbf{I} - \bar{\mathbf{A}}_{\bar{c}} \end{bmatrix} \begin{bmatrix} (s\mathbf{I} - \bar{\mathbf{A}}_c)^{-1} & (s\mathbf{I} - \bar{\mathbf{A}}_c)^{-1}\bar{\mathbf{A}}_{12}(s\mathbf{I} - \bar{\mathbf{A}}_{\bar{c}})^{-1} \\ \mathbf{0} & (s\mathbf{I} - \bar{\mathbf{A}}_{\bar{c}})^{-1} \end{bmatrix} = \mathbf{I} \tag{5-52}$$

Thus the transfer-function matrix of $F\bar{E}$ is

$$[\bar{\mathbf{C}}_c \quad \bar{\mathbf{C}}_{\bar{c}}] \begin{bmatrix} s\mathbf{I} - \bar{\mathbf{A}}_c & -\bar{\mathbf{A}}_{12} \\ \mathbf{0} & s\mathbf{I} - \bar{\mathbf{A}}_{\bar{c}} \end{bmatrix}^{-1} \begin{bmatrix} \bar{\mathbf{B}}_c \\ \mathbf{0} \end{bmatrix} + \mathbf{D}$$

$$= [\bar{\mathbf{C}}_c \quad \bar{\mathbf{C}}_{\bar{c}}] \begin{bmatrix} (s\mathbf{I} - \bar{\mathbf{A}}_c)^{-1} & (s\mathbf{I} - \bar{\mathbf{A}}_c)^{-1}\bar{\mathbf{A}}_{12}(s\mathbf{I} - \bar{\mathbf{A}}_{\bar{c}})^{-1} \\ \mathbf{0} & (s\mathbf{I} - \bar{\mathbf{A}}_{\bar{c}})^{-1} \end{bmatrix} \begin{bmatrix} \bar{\mathbf{B}}_c \\ \mathbf{0} \end{bmatrix} + \mathbf{D}$$

$$= \bar{\mathbf{C}}_c(s\mathbf{I} - \bar{\mathbf{A}}_c)^{-1}\bar{\mathbf{B}}_c + \mathbf{D}$$

which is the transfer-function matrix of $F\bar{E}_c$. Q.E.D.

In the equivalence transformation $\bar{\mathbf{x}} = \mathbf{P}\mathbf{x}$, the state space $\Sigma$ of $FE$ is divided into two subspaces. One is the $n_1$-dimensional subspace of $\Sigma$, denoted by $\Sigma_1$, which consists of all the vectors $\begin{bmatrix} \bar{\mathbf{x}}^c \\ \mathbf{0} \end{bmatrix}$; the other is the $(n - n_1)$-dimensional subspace, which consists of all the vectors $\begin{bmatrix} \mathbf{0} \\ \bar{\mathbf{x}}^{\bar{c}} \end{bmatrix}$. Since $F\bar{E}_c$ is controllable, all the vectors $\bar{\mathbf{x}}^c$ in $\Sigma_1$ are controllable. Equation (5-50a) shows that the state variables in $\bar{\mathbf{x}}^{\bar{c}}$ are not affected directly by the input $\mathbf{u}$ or indirectly through the state vector $\bar{\mathbf{x}}^c$; therefore, the state vector $\bar{\mathbf{x}}^{\bar{c}}$ is not controllable and is dropped in the reduced equation (5-51). Thus, if a linear, time-invariant dynamical equation is not controllable, by a proper choice of a basis, the state vector can be decomposed into two groups—one controllable, the other uncontrollable. By dropping the uncontrollable state vectors, we may obtain a controllable dynamical equation of lesser dimension that is zero-state equivalent to the original equation. See Problems 5-23 and 5-24.

### Example 1

Consider the network shown in Figure 5-11(a). With the state variables chosen as shown, its dynamical-equation description is given by

$$\begin{bmatrix} \dot{x}_1 \\ \dot{x}_2 \end{bmatrix} = \begin{bmatrix} -1 & 0 \\ 0 & -1 \end{bmatrix} \begin{bmatrix} x_1 \\ x_2 \end{bmatrix} + \begin{bmatrix} 1 \\ 1 \end{bmatrix} u$$

Its controllability matrix is

$$\mathbf{U} = \begin{bmatrix} 1 & -1 \\ 1 & -1 \end{bmatrix}$$

which has rank 1. The equivalence transformation $\mathbf{P}$ can be chosen as

$$\mathbf{P}^{-1} = \mathbf{Q} = \begin{bmatrix} 1 & 1 \\ 1 & 0 \end{bmatrix}$$

The first column of $\mathbf{Q}$ is the linearly independent column of $\mathbf{U}$; the second column of $\mathbf{Q}$ is chosen arbitrarily to make $\mathbf{Q}$ nonsingular. If we

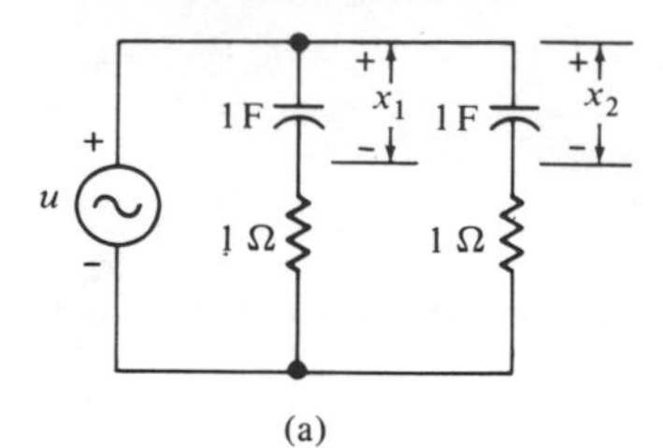


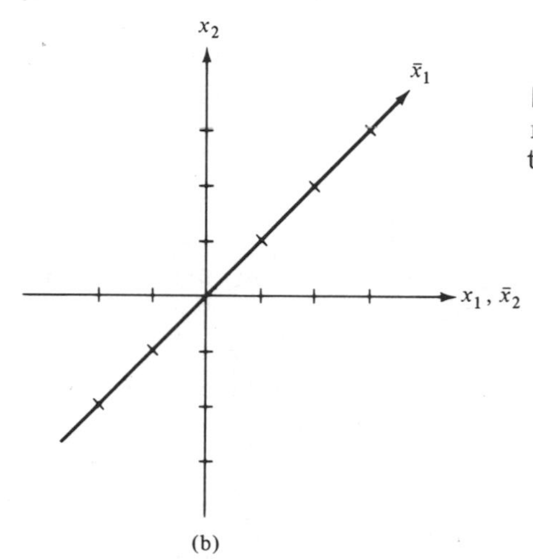


**Figure 5-11** (a) Uncontrollable network. (b) Canonical decomposition of the state space.

let $\bar{\mathbf{x}} = \mathbf{P}\mathbf{x}$, then its equivalent dynamical equation is

$$\begin{bmatrix} \dot{\bar{x}}_1 \\ \dot{\bar{x}}_2 \end{bmatrix} = \begin{bmatrix} -1 & 0 \\ 0 & -1 \end{bmatrix} \begin{bmatrix} \bar{x}_1 \\ \bar{x}_2 \end{bmatrix} + \begin{bmatrix} 1 \\ 0 \end{bmatrix} u$$

In the transformation $\bar{\mathbf{x}} = \mathbf{P}\mathbf{x}$, the basis of the state space changes from the set of the orthonormal vectors $\begin{bmatrix} 1 \\ 0 \end{bmatrix}, \begin{bmatrix} 0 \\ 1 \end{bmatrix}$ to the set of the vectors $\begin{bmatrix} 1 \\ 1 \end{bmatrix}, \begin{bmatrix} 1 \\ 0 \end{bmatrix}$ as shown in Figure 5-11(b). The state variable $\bar{x}_1$ is controllable. This fact can also be checked directly from the network; we see from Figure 5-11(b) that the state $\mathbf{x}$ can be controlled only if $x_1 = x_2$. ▮

### Theorem 5-18

Consider the $n$-dimensional, linear, time-invariant dynamical equation $FE$. If the observability matrix of $FE$ has rank $n_2 (n_2 < n)$, then there exists an equivalence transformation $\bar{\mathbf{x}} = \mathbf{P}\mathbf{x}$ that transforms $FE$ into

$$F\bar{E}: \quad \begin{bmatrix} \dot{\bar{\mathbf{x}}}^{o} \\ \dot{\bar{\mathbf{x}}}^{\bar{o}} \end{bmatrix} = \begin{bmatrix} \bar{\mathbf{A}}_o & \mathbf{0} \\ \bar{\mathbf{A}}_{21} & \bar{\mathbf{A}}_{\bar{o}} \end{bmatrix} \begin{bmatrix} \bar{\mathbf{x}}^{o} \\ \bar{\mathbf{x}}^{\bar{o}} \end{bmatrix} + \begin{bmatrix} \bar{\mathbf{B}}_o \\ \bar{\mathbf{B}}_{\bar{o}} \end{bmatrix} \mathbf{u} \tag{5-53a}$$

$$\mathbf{y} = [\bar{\mathbf{C}}_o \quad \mathbf{0}] \begin{bmatrix} \bar{\mathbf{x}}^{o} \\ \bar{\mathbf{x}}^{\bar{o}} \end{bmatrix} + \mathbf{D}\mathbf{u} \tag{5-53b}$$

and the $n_2$-dimensional subequation of $F\bar{E}$

$$F\bar{E}_o: \quad \dot{\bar{\mathbf{x}}}^o = \bar{\mathbf{A}}_o\bar{\mathbf{x}}^o + \bar{\mathbf{B}}_o\mathbf{u} \tag{5-54a}$$
$$\bar{\mathbf{y}} = \bar{\mathbf{C}}_o\bar{\mathbf{x}}^o + \mathbf{D}\mathbf{u} \tag{5-54b}$$

is observable and has the same transfer-function matrix as $FE$. ▮

This theorem can be deduced from Theorem 5-17 and 5-10. Equation (5-53) shows that the vector $\bar{\mathbf{x}}^{\bar{o}}$ does not appear directly in the output $\mathbf{y}$ nor indirectly through $\bar{\mathbf{x}}^o$. Hence the vector $\bar{\mathbf{x}}^{\bar{o}}$ is not observable and is dropped in the reduced equation.

Combining Theorems 5-17 and 5-18, we have the following very important theorem.

### Theorem 5-19 (canonical decomposition theorem)[6]

Consider the linear, time-invariant dynamical equation

$$FE: \quad \dot{\mathbf{x}} = \mathbf{A}\mathbf{x} + \mathbf{B}\mathbf{u}$$
$$\mathbf{y} = \mathbf{C}\mathbf{x} + \mathbf{D}\mathbf{u}$$

By equivalence transformations, $FE$ can be transformed into the following canonical form

$$FE: \quad \begin{bmatrix} \dot{\bar{\mathbf{x}}}^{c\bar{o}} \\ \dot{\bar{\mathbf{x}}}^{co} \\ \dot{\bar{\mathbf{x}}}^{\bar{c}} \end{bmatrix} = \begin{bmatrix} \bar{\mathbf{A}}_{c\bar{o}} & \bar{\mathbf{A}}_{12} & \bar{\mathbf{A}}_{13} \\ \mathbf{0} & \bar{\mathbf{A}}_{co} & \bar{\mathbf{A}}_{21} \\ \mathbf{0} & \mathbf{0} & \bar{\mathbf{A}}_{\bar{c}} \end{bmatrix} \begin{bmatrix} \bar{\mathbf{x}}^{c\bar{o}} \\ \bar{\mathbf{x}}^{co} \\ \bar{\mathbf{x}}^{\bar{c}} \end{bmatrix} + \begin{bmatrix} \bar{\mathbf{B}}_{c\bar{o}} \\ \bar{\mathbf{B}}_{co} \\ \mathbf{0} \end{bmatrix} \mathbf{u} \tag{5-55a}$$
$$\mathbf{y} = [\mathbf{0} \quad \bar{\mathbf{C}}_{co} \quad \bar{\mathbf{C}}_{\bar{c}}]\bar{\mathbf{x}} + \mathbf{D}\mathbf{u} \tag{5-55b}$$

where the vector $\bar{\mathbf{x}}^{c\bar{o}}$ is controllable but not observable, $\bar{\mathbf{x}}^{co}$ is controllable and observable, and $\bar{\mathbf{x}}^{\bar{c}}$ is not controllable. Furthermore, the transfer function of $FE$ is

$$\bar{\mathbf{C}}_{co}(s\mathbf{I} - \bar{\mathbf{A}}_{co})^{-1}\bar{\mathbf{B}}_{co} + \mathbf{D}$$

which depends solely on the controllable and observable part of the equation $FE$.

### Proof

If the dynamical equation $FE$ is not controllable, it can be transformed into the form of (5-50). Consider now the dynamical equation $F\bar{E}_c$ which is the controllable part of $FE$. If $F\bar{E}_c$ is not observable, then $F\bar{E}_c$ can be transformed into the form of (5-53), which can also be written as

$$\begin{bmatrix} \dot{\bar{\mathbf{x}}}^{c\bar{o}} \\ \dot{\bar{\mathbf{x}}}^{co} \end{bmatrix} = \begin{bmatrix} \bar{\mathbf{A}}_{c\bar{o}} & \bar{\mathbf{A}}_{21} \\ \mathbf{0} & \bar{\mathbf{A}}_{co} \end{bmatrix} \begin{bmatrix} \bar{\mathbf{x}}^{c\bar{o}} \\ \bar{\mathbf{x}}^{co} \end{bmatrix} + \begin{bmatrix} \bar{\mathbf{B}}_{c\bar{o}} \\ \bar{\mathbf{B}}_{co} \end{bmatrix} \mathbf{u}$$
$$\mathbf{y} = [\mathbf{0} \quad \bar{\mathbf{C}}_{co}]\bar{\mathbf{x}} + \mathbf{D}\mathbf{u}$$

[6] This is a simplified version of the canonical decomposition theorem. For the general form, see References [57], [60], and [116].

Combining these two transformations, we immediately obtain Equation (5-55). Following directly from Theorems 5-17 and 5-18, we conclude that the transfer function of $FE$ is given by $\bar{\mathbf{C}}_{co}(s\mathbf{I} - \bar{\mathbf{A}}_{co})^{-1}\bar{\mathbf{B}}_{co} + \mathbf{D}$.

Q.E.D.

This theorem can be illustrated symbolically as shown in Figure 5-12, in which the uncontrollable part is further decomposed into observable

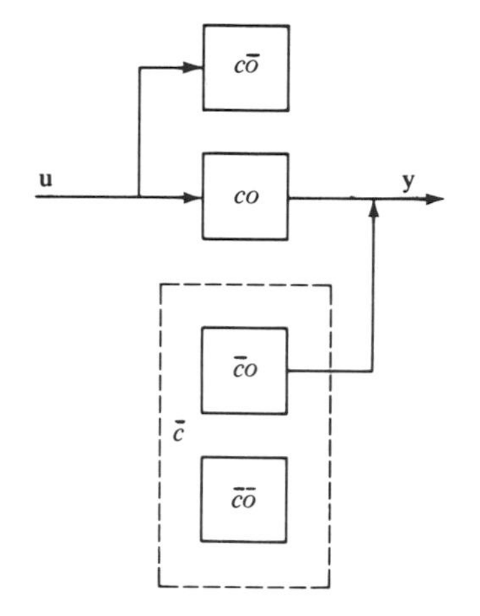


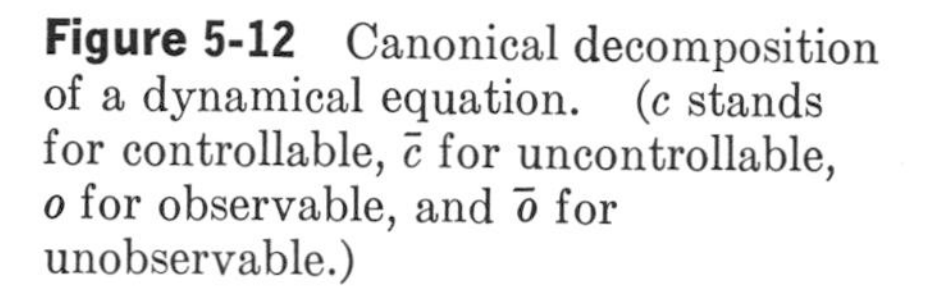

**Figure 5-12** Canonical decomposition of a dynamical equation. ($c$ stands for controllable, $\bar{c}$ for uncontrollable, $o$ for observable, and $\bar{o}$ for unobservable.)

and unobservable parts. We see that the transfer-function matrix or the impulse-response matrix of a dynamical equation depends solely on the controllable and observable part of the equation. In other words, *the impulse-response matrix (the input–output description) describes only the part of a system that is controllable and observable.* This is the most important relation between the input–output description and the state-variable description. This theorem tells us why the input–output description is sometimes insufficient to describe a system, for the uncontrollable and/or unobservable parts of the system do not appear in the transfer-function matrix description.

### Example 1

Consider the network shown in Figure 5-13(a). Because the input is a current source, the behavior due to the initial conditions in $C_1$ and $L_1$ can never be detected from the output. Hence the state variables associated with $C_1$ and $L_1$ are not observable (they may be controllable, but we don't care). Similarly the state variable associated with $L_2$ is not controllable. Because of the symmetry, the state variable associated with $C_2$ is uncontrollable and unobservable. By dropping the state variables that are either uncontrollable or unobservable, the network in Figure 5-13(a) is reduced to the form in Figure 5-13(b). Hence the transfer function of the network in Figure 5-12(a) is $\hat{g}(s) = 1$. ∎

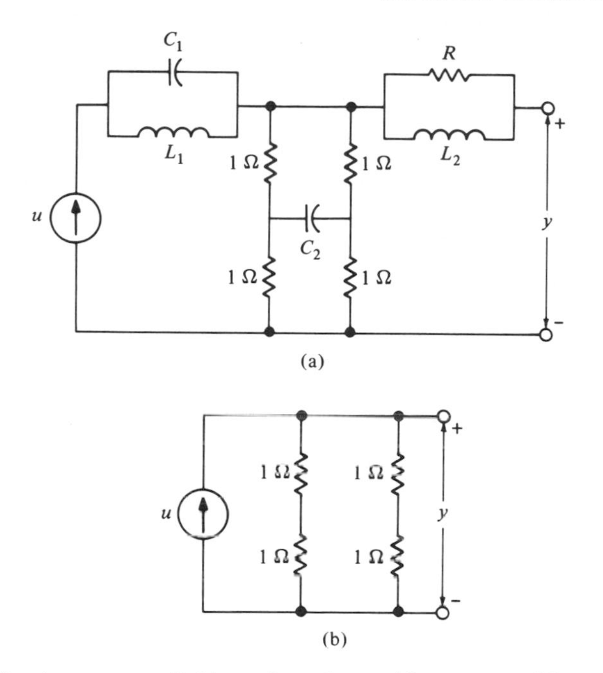


**Figure 5-13** An uncontrollable and unobservable system with transfer function 1.

**Irreducible dynamical equation.** We have seen from Theorems 5-17 and 5-18 that if a linear, time-invariant dynamical equation is either uncontrollable or unobservable, then there exists a dynamical equation of lesser dimension that has the same transfer-function matrix as the original dynamical equation. In other words, if a linear, time-invariant dynamical equation is either uncontrollable or unobservable, its dimension can be reduced such that the reduced equation still has the same zero-state response. This fact motivates the following definition.

Definition 5-9

A linear, time-invariant dynamical equation $FE$ is said to be *reducible* if and only if there exists a linear, time-invariant dynamical equation of lesser dimension that is zero-state equivalent to $FE$. Otherwise, the equation is said to be *irreducible*.

Theorem 5-20

A linear time-invariant dynamical equation $FE$ is irreducible if and only if $FE$ is controllable and observable.

Proof

If the dynamical equation $FE$ is either uncontrollable or unobservable, then $FE$ is reducible (Theorems 5-17 and 5-18). Hence what we have to prove is that if $FE$ is controllable and observable, then $FE$ is irreducible. We prove this by contradiction. Suppose that the $n$-dimensional equation $FE$ is controllable and observable and that there exists a linear time-invariant dynamical equation $F\bar{E}$,

$$F\bar{E}: \quad \dot{\bar{\mathbf{x}}} = \bar{\mathbf{A}}\bar{\mathbf{x}} + \bar{\mathbf{B}}\mathbf{u} \tag{5-56a}$$

$$\mathbf{y} = \bar{\mathbf{C}}\bar{\mathbf{x}} + \bar{\mathbf{D}}\mathbf{u} \tag{5-56b}$$

of lesser dimension, say $n_1 < n$, that is zero-state equivalent to $FE$. Then, by the definition of zero-state equivalent, we have

$$\mathbf{C}e^{\mathbf{A}t}\mathbf{B} + \mathbf{D}\delta(t) = \bar{\mathbf{C}}e^{\bar{\mathbf{A}}t}\bar{\mathbf{B}} + \bar{\mathbf{D}}\delta(t) \qquad \text{for all } t \text{ in } [0, \infty)$$

This implies that

$$\mathbf{CA}^k\mathbf{B} = \bar{\mathbf{C}}\bar{\mathbf{A}}^k\bar{\mathbf{B}} \qquad k = 0, 1, 2, \cdots \tag{5-57}$$

Consider now the product

$$\mathbf{VU} \triangleq \begin{bmatrix} \mathbf{C} \\ \mathbf{CA} \\ \cdot \\ \cdot \\ \cdot \\ \mathbf{CA}^{n-1} \end{bmatrix} [\mathbf{B} \quad \mathbf{AB} \quad \cdots \quad \mathbf{A}^{n-1}\mathbf{B}]$$

$$= \begin{bmatrix} \mathbf{CB} & \mathbf{CAB} & \cdots & \mathbf{CA}^{n-1}\mathbf{B} \\ \mathbf{CAB} & \mathbf{CA}^2\mathbf{B} & \cdots & \mathbf{CA}^n\mathbf{B} \\ \cdot & \cdot & & \cdot \\ \cdot & \cdot & & \cdot \\ \cdot & \cdot & & \cdot \\ \mathbf{CA}^{n-1}\mathbf{B} & \mathbf{CA}^n\mathbf{B} & \cdots & \mathbf{CA}^{2(n-1)}\mathbf{B} \end{bmatrix} \tag{5-58}$$

By (5-57), we may replace $\mathbf{CA}^k\mathbf{B}$ in (5-58) by $\bar{\mathbf{C}}\bar{\mathbf{A}}^k\bar{\mathbf{B}}$; consequently, we have

$$\mathbf{VU} = \bar{\mathbf{V}}_{n-1}\bar{\mathbf{U}}_{n-1} \tag{5-59}$$

where $\bar{\mathbf{V}}_{n-1}$ and $\bar{\mathbf{U}}_{n-1}$ are defined as in (5-39) and (5-25). Since $FE$ is controllable and observable, we have $\rho\mathbf{U} = n$ and $\rho\mathbf{V} = n$. It follows from Theorem 2-6 that $\rho(\mathbf{VU}) = n$. Now $\bar{\mathbf{V}}_{n-1}$ and $\bar{\mathbf{U}}_{n-1}$ are, respectively, $qn \times n_1$ and $n_1 \times np$ matrices; hence the matrix $\bar{\mathbf{V}}_{n-1}\bar{\mathbf{U}}_{n-1}$ can be at most of rank $n_1$. However, (5-59) implies that $\rho(\bar{\mathbf{V}}_{n-1}\bar{\mathbf{U}}_{n-1}) = n > n_1$. This is a contradiction. Hence if $FE$ is controllable and observable, then $FE$ is irreducible. Q.E.D.

Recall from Section 4-4 that if a dynamical equation $\{\mathbf{A}, \mathbf{B}, \mathbf{C}, \mathbf{D}\}$ has a prescribed transfer-function matrix $\hat{\mathbf{G}}(s)$, then the dynamical equation $\{\mathbf{A}, \mathbf{B}, \mathbf{C}, \mathbf{D}\}$ is called a realization of $\hat{\mathbf{G}}(s)$. Now if $\{\mathbf{A}, \mathbf{B}, \mathbf{C}, \mathbf{D}\}$ is controllable and observable, then $\{\mathbf{A}, \mathbf{B}, \mathbf{C}, \mathbf{D}\}$ is called an *irreducible realization* of $\hat{\mathbf{G}}(s)$. In the following we shall show that all the irreducible realizations of $\hat{\mathbf{G}}(s)$ are equivalent.

### Theorem 5-21

Let the dynamical equation $\{\mathbf{A}, \mathbf{B}, \mathbf{C}, \mathbf{D}\}$ be an irreducible realization of a $q \times p$ proper rational matrix $\hat{\mathbf{G}}(s)$. Then $\{\bar{\mathbf{A}}, \bar{\mathbf{B}}, \bar{\mathbf{C}}, \bar{\mathbf{D}}\}$ is also an irreducible realization of $\hat{\mathbf{G}}(s)$ if and only if $\{\mathbf{A}, \mathbf{B}, \mathbf{C}, \mathbf{D}\}$ and $\{\bar{\mathbf{A}}, \bar{\mathbf{B}}, \bar{\mathbf{C}}, \bar{\mathbf{D}}\}$ are equivalent; that is, there exists a nonsingular constant matrix $\mathbf{P}$ such that $\bar{\mathbf{A}} = \mathbf{PAP}^{-1}$, $\bar{\mathbf{B}} = \mathbf{PB}$, $\bar{\mathbf{C}} = \mathbf{CP}^{-1}$, and $\bar{\mathbf{D}} = \mathbf{D}$.

### Proof

The sufficiency follows directly from Theorems 4-6 and 5-15. We show now the necessity of the theorem. Let $\mathbf{U}$, $\mathbf{V}$ be the controllability and the observability matrices of $\{\mathbf{A}, \mathbf{B}, \mathbf{C}, \mathbf{D}\}$ and let $\bar{\mathbf{U}}$, $\bar{\mathbf{V}}$ be similarly defined for $\{\bar{\mathbf{A}}, \bar{\mathbf{B}}, \bar{\mathbf{C}}, \bar{\mathbf{D}}\}$. If $\{\mathbf{A}, \mathbf{B}, \mathbf{C}, \mathbf{D}\}$ and $\{\bar{\mathbf{A}}, \bar{\mathbf{B}}, \bar{\mathbf{C}}, \bar{\mathbf{D}}\}$ are realizations of the same $\hat{\mathbf{G}}(s)$, then from (5-57) and (5-58), we have $\mathbf{D} = \bar{\mathbf{D}}$,

$$\mathbf{VU} = \bar{\mathbf{V}}\bar{\mathbf{U}} \tag{5-60}$$

and

$$\mathbf{VAU} = \bar{\mathbf{V}}\bar{\mathbf{A}}\bar{\mathbf{U}} \tag{5-61}$$

The irreducibility assumption implies that $\rho\bar{\mathbf{V}} = n$; hence the matrix $(\bar{\mathbf{V}}^*\bar{\mathbf{V}})$ is nonsingular (Theorem 2-8). Consequently, from (5-60), we have

$$\bar{\mathbf{U}} = (\bar{\mathbf{V}}^*\bar{\mathbf{V}})^{-1}\bar{\mathbf{V}}^*\mathbf{VU} \triangleq \mathbf{PU} \tag{5-62}$$

where $\mathbf{P} \triangleq (\bar{\mathbf{V}}^*\bar{\mathbf{V}})^{-1}\bar{\mathbf{V}}^*\mathbf{V}$. From (5-62), we have $\rho\bar{\mathbf{U}} \leq \min(\rho\mathbf{P}, \rho\mathbf{U})$, which, together with $\rho\bar{\mathbf{U}} = n$, implies that $\rho\mathbf{P} = n$. Hence $\mathbf{P}$ qualifies as an equivalence transformation. The first $p$ columns of (5-62) give $\bar{\mathbf{B}} = \mathbf{PB}$. Since $\rho\mathbf{U} = n$, Equation (5-62) implies that

$$\mathbf{P} = (\bar{\mathbf{U}}\mathbf{U}^*)(\mathbf{UU}^*)^{-1}$$

With $\mathbf{P} = (\bar{\mathbf{V}}^*\bar{\mathbf{V}})^{-1}\bar{\mathbf{V}}^*\mathbf{V} = (\bar{\mathbf{U}}\mathbf{U}^*)(\mathbf{UU}^*)^{-1}$, it is easy to derive from (5-60) and (5-61) that $\mathbf{V} = \bar{\mathbf{V}}\mathbf{P}$ and $\mathbf{PA} = \bar{\mathbf{A}}\mathbf{P}$, which imply that $\mathbf{C} = \bar{\mathbf{C}}\mathbf{P}$ and $\bar{\mathbf{A}} = \mathbf{PAP}^{-1}$. Q.E.D.

This theorem implies that all the irreducible realizations of $\hat{\mathbf{G}}(s)$ have the same dimension. Physically, the dimension of an irreducible dynamical equation is the minimal number of integrators (if we simulate the equation in an analog computer) or the minimal number of energy-storage elements (if the system is an *RLC* network) required to generate the given transfer-function matrix.

We studied in this section only the canonical decomposition of linear, time-invariant dynamical equations. For the time-varying case, the interested reader is referred to References [106] and [108].

## *5-7 Output Controllability and Output Function Controllability

Similar to the (state) controllability of a dynamical equation, we may define controllability for the output vector of a system. Although these two concepts are the same except that one is defined for the *state* and the other for the *output*, the state controllability is a property of a dynamical equation, whereas the output controllability is a property of the impulse-response matrix of a system.

Consider a system with the input–output description

$$\mathbf{y}(t) = \int_{-\infty}^{t} \mathbf{G}(t, \tau)\mathbf{u}(\tau)\, d\tau$$

where $\mathbf{u}$ is the $p \times 1$ input vector, $\mathbf{y}$ is the $q \times 1$ output vector, $\mathbf{G}(t, \tau)$ is the $q \times p$ impulse-response matrix of the system. We assume for simplicity that $\mathbf{G}(t, \tau)$ does not contain $\delta$-functions and is continuous in $t$ and $\tau$ for $t > \tau$.

### Definition 5-10

A system with a continuous impulse-response matrix $\mathbf{G}(t, \tau)$ is said to be *output controllable* at time $t_0$ if, for any $\mathbf{y}^1$, there exist a finite $t_1 > t_0$ and an input $\mathbf{u}_{[t_0, t_1]}$ that transfers the output from $\mathbf{y}(t_0) = \mathbf{0}$ to $\mathbf{y}(t_1) = \mathbf{y}^1$.

### Theorem 5-22

A system with a continuous $\mathbf{G}(t, \tau)$ is output controllable at $t_0$ if and only if there exists a finite $t_1 > t_0$ such that the rows of $\mathbf{G}(t_1, \tau)$ are linearly independent in $\tau$ on $[t_0, t_1]$ over the field of complex numbers. ∎

The proof of this theorem is exactly the same as the one of Theorem 5-4 and is therefore omitted.

We study in the following the class of systems that also have linear, time-invariant, dynamical-equation descriptions. Consider the system that is describable by

$$FE: \quad \begin{aligned} \dot{\mathbf{x}} &= \mathbf{A}\mathbf{x} + \mathbf{B}\mathbf{u} \\ \mathbf{y} &= \mathbf{C}\mathbf{x} \end{aligned}$$

where $\mathbf{A}$, $\mathbf{B}$, and $\mathbf{C}$ are $n \times n$, $n \times p$, $q \times n$ real constant matrices. The impulse-response matrix of the system is

$$\mathbf{G}(t) = \mathbf{C}e^{\mathbf{A}t}\mathbf{B}$$

The transfer-function matrix of the system is

$$\hat{\mathbf{G}}(s) = \mathbf{C}(s\mathbf{I} - \mathbf{A})^{-1}\mathbf{B} \tag{5-63}$$

It is clear that $\hat{\mathbf{G}}(s)$ is a strictly proper rational function matrix.

### Corollary 5-22

A system whose transfer function is a strictly proper rational-function matrix is output controllable if and only if all the rows of $\hat{\mathbf{G}}(s)$ are linearly independent over the field of complex numbers or if and only if the $q \times np$ matrix

$$[\mathbf{CB} \vdots \mathbf{CAB} \vdots \cdots \vdots \mathbf{CA}^{n-1}\mathbf{B}] \tag{5-64}$$

has rank $q$. ▮

The proof of Theorem 5-7 can be applied here with slight modification. A trivial consequence of this corollary is that every single-output system is output controllable. We see that although the condition of output controllability can also be stated in terms of **A**, **B**, and **C**, compared with checking the linear independence of $\hat{\mathbf{G}}(s)$, the condition (5-64) seems more complicated.

The state controllability is defined for the dynamical equation, whereas the output controllability is defined for the input–output description; therefore, these two concepts are not necessarily related.

### Example 1

Consider the network shown in Figure 5-14. It is neither state controllable nor observable. But it is output controllable.

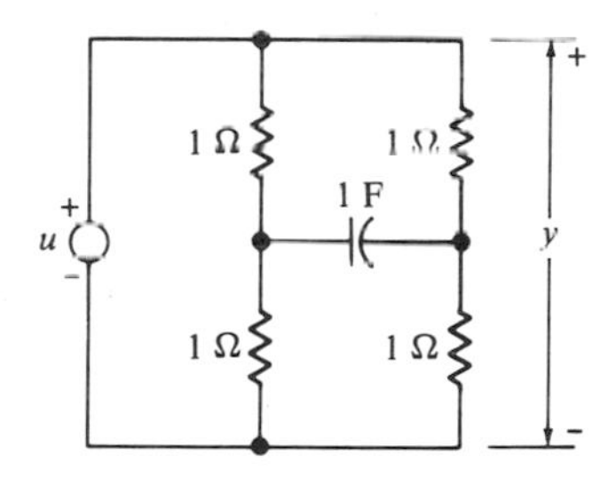


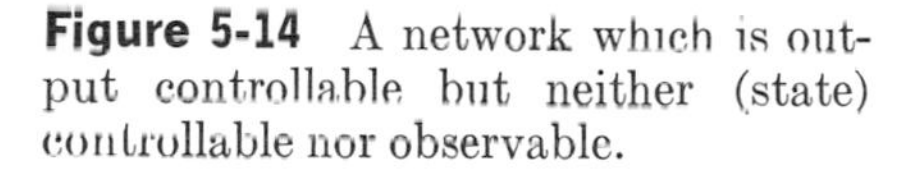

**Figure 5-14** A network which is output controllable but neither (state) controllable nor observable.

### Example 2

Consider the system shown in Figure 5-15. The transfer-function matrix of the system is

$$\begin{bmatrix} \dfrac{1}{s+1} \\[2mm] \dfrac{2}{s+1} \end{bmatrix}$$

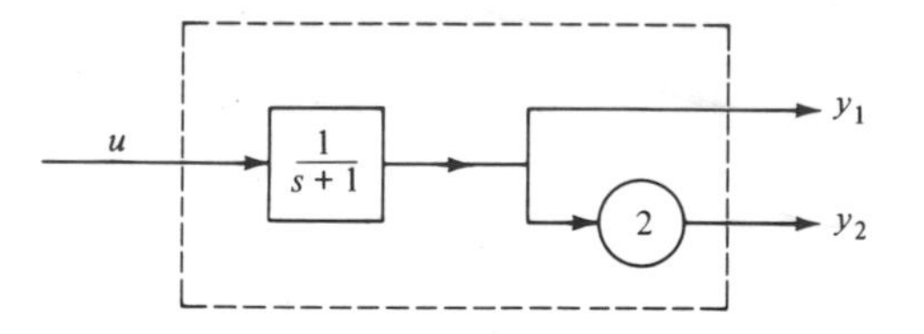


**Figure 5-15** A system which is controllable and observable but not output controllable.

the rows of which are linearly dependent. Hence the system is not output controllable. The dynamical equation of the system is

$$\dot{\mathbf{x}} = -x + u$$
$$\mathbf{y} = \begin{bmatrix} 1 \\ 2 \end{bmatrix} x$$

which is controllable and observable. ∎

If a system is output controllable, its output can be transferred to any desired value at certain *instant* of time. A related problem is whether it is possible to steer the output following a preassigned curve over any *interval* of time. A system whose output can be steered over any interval of time is said to be *output function controllable* or *functional reproducible.*

### Theorem 5-23

A system with a $q \times p$ proper rational-function matrix description is output function controllable if and only if $\rho\hat{\mathbf{G}}(s) = q$.

### Proof

If the system is initially relaxed, then we have

$$\hat{\mathbf{y}}(s) = \hat{\mathbf{G}}(s)\hat{\mathbf{u}}(s) \tag{5-65}$$

If $\rho\hat{\mathbf{G}}(s) = q$—that is, all the rows of $\hat{\mathbf{G}}(s)$ are linearly independent over the field of rational functions—then the $q \times q$ matrix $\hat{\mathbf{G}}(s)\hat{\mathbf{G}}^*(s)$ is nonsingular (Theorem 2-8). Consequently, for any $\hat{\mathbf{y}}(s)$, if we choose

$$\hat{\mathbf{u}}(s) = \hat{\mathbf{G}}^*(s)(\hat{\mathbf{G}}(s)\hat{\mathbf{G}}^*(s))^{-1}\hat{\mathbf{y}}(s) \tag{5-66}$$

then Equation (5-65) is satisfied. Consequently, if $\rho\hat{\mathbf{G}}(s) = q$, then the system is output function controllable. If $\rho\hat{\mathbf{G}}(s) < q$, we can always find a $\hat{\mathbf{y}}(s)$, not in the range of $\hat{\mathbf{G}}(s)$, for which there exists no solution $\hat{\mathbf{u}}(s)$ in (5-65) (Theorem 2-4). Q.E.D.

If the input is restricted to the class of piecewise continuous functions of $t$, then the given output function should be very smooth; otherwise the input computed from (5-66) will not be piecewise continuous. For example, if the given output has some discontinuity, an input containing $\delta$-functions may be needed to generate the discontinuity.

The condition for output function controllability can also be stated in terms of the matrices **A**, **B**, and **C**. However it is much more complicated. The interested reader is referred to Reference [8].

## 5-8 Concluding Remarks

In this chapter we have introduced the concepts of controllability and observability. Various theorems for linear dynamical equations to be controllable and observable were derived. We shall discuss briefly the relations among some of these theorems. We list first those theorems which are dual to each other:

| | | | | | | | | |
|---|---|---|---|---|---|---|---|---|
| *Controllability:* | Theorems | 5-4 | 5-5 | 5-6 | 5-7 | 5-8 | 5-17 | |
| | | ↕ | ↕ | ↕ | ↕ | ↕ | ↕ | Theorem of duality |
| *Observability:* | Theorems | 5-9 | 5-11 | 5-12 | 5-13 | 5-14 | 5-18 | |

The theorems in the observability part can be easily derived from the controllability part by applying Theorem 5-10 (theorem of duality), and vice versa.

Theorems 5-1 and 5-4 (or 5-9) are two fundamental results of this chapter. They are derived with the least assumption (continuity), and hence they are most widely applicable. If additional assumptions (continuous differentiability) are introduced, then we have Theorems 5-2 and 5-5 (or 5-11), which give only sufficient conditions but are easier to apply. If we have the analyticity assumption (the strongest possible assumption) on time-varying dynamical equations, then we have Theorems 5-3 and 5-6 (or 5-12). Theorem 5-7 (or 5-13), which follows directly from Theorems 5-1, 5-3, and 5-4, gives the necessary and sufficient conditions for a linear, time-invariant dynamical equation to be controllable.

The relationship between the transfer-function matrix and the linear, time-invariant dynamical equation was established in this chapter. This was achieved by decomposing a dynamical equation into four parts: (1) controllable and observable, (2) controllable but unobservable, (3) uncontrollable and unobservable, and (4) uncontrollable but observable. The transfer-function matrix depends only on the controllable and observable part of the dynamical equation. If a linear, time-invariant dynamical equation is not controllable and not observable, it can be reduced to a controllable and observable one by using the algorithms developed in Theorems 5-17 and 5-18.

The concepts of controllability and observability are essential in the study of the remainder of this book. They will be used in the realization of a rational matrix (Chapter 6) and the stability study of linear systems

(Chapter 8). Some practical implications of these concepts will be given in Chapter 7.

## PROBLEMS

**5-1** Which of the following sets are linearly independent over $(-\infty, \infty)$?

**a.** $\{t, t^2, e^t, e^{2t}, te^t\}$
**b.** $\{e^t, te^t, t^2e^t, te^{2t}, te^{3t}\}$
**c.** $\{\sin t, \cos t, \sin 2t\}$

**5-2** Check the controllability of the following dynamical equations:

**a.**
$$\begin{bmatrix} \dot{x}_1 \\ \dot{x}_2 \end{bmatrix} = \begin{bmatrix} 0 & 1 \\ 0 & t \end{bmatrix} \begin{bmatrix} x_1 \\ x_2 \end{bmatrix} + \begin{bmatrix} 0 \\ 1 \end{bmatrix} u$$
$$y = [0 \quad 1] \begin{bmatrix} x_1 \\ x_2 \end{bmatrix}$$

**b.**
$$\dot{\mathbf{x}} = \begin{bmatrix} 0 & 1 & 0 \\ 0 & 0 & 1 \\ -2 & -4 & -3 \end{bmatrix} \mathbf{x} + \begin{bmatrix} 1 & 0 \\ 0 & 1 \\ -1 & 1 \end{bmatrix} \boldsymbol{u}$$
$$\mathbf{y} = \begin{bmatrix} 0 & 1 & -1 \\ 1 & 2 & 1 \end{bmatrix} \mathbf{x}$$

**c.**
$$\dot{\mathbf{x}} = \begin{bmatrix} 0 & 4 & 3 \\ 0 & 20 & 16 \\ 0 & -25 & -20 \end{bmatrix} \mathbf{x} + \begin{bmatrix} -1 \\ 3 \\ 0 \end{bmatrix} u$$
$$y = [-1 \quad 3 \quad 0]\mathbf{x}$$

**5-3** Show that a linear dynamical equation is controllable at $t_0$ if and only if, for any $\mathbf{x}^0$, there exist a finite $t_1 > t_0$ and a $\mathbf{u}$ that transfers $\mathbf{x}^0$ to the zero state at time $t_1$.

**5-4** Show that if a linear dynamical equation is controllable at $t_0$, then it is controllable at any $t < t_0$. Is it true that if a linear dynamical equation is controllable at $t_0$, then it is controllable at any $t > t_0$? Why?

**5-5** Is it true that $\rho[\mathbf{B} \vdots \mathbf{AB} \vdots \cdots \vdots \mathbf{A}^{n-1}\mathbf{B}] = \rho[\mathbf{AB} \vdots \mathbf{A}^2\mathbf{B} \vdots \cdots \vdots \mathbf{A}^n\mathbf{B}]$? If not, under what condition will it be true?

**5-6** Show that if a linear, time-invariant dynamical equation is controllable, then it is uniformly controllable.

**5-7** Check the observability of the dynamical equations given in Problem 5-2.

**5-8** State (without proof) the necessary and sufficient condition for a linear dynamical equation $E$ to be differentially controllable and differentially observable at $t_0$.

**5-9** Is the following Jordan-form dynamical equation controllable and observable?

$$\dot{\mathbf{x}} = \begin{bmatrix} 2 & 1 & 0 & 0 & 0 & 0 & 0 \\ 0 & 2 & 0 & 0 & 0 & 0 & 0 \\ 0 & 0 & 2 & 0 & 0 & 0 & 0 \\ 0 & 0 & 0 & 2 & 0 & 0 & 0 \\ 0 & 0 & 0 & 0 & 1 & 1 & 0 \\ 0 & 0 & 0 & 0 & 0 & 1 & 0 \\ 0 & 0 & 0 & 0 & 0 & 0 & 1 \end{bmatrix} \mathbf{x} + \begin{bmatrix} 2 & 1 & 1 \\ 2 & 1 & 1 \\ 1 & 1 & 1 \\ 3 & 2 & 1 \\ -1 & 0 & 0 \\ 1 & 0 & 1 \\ 1 & 0 & 0 \end{bmatrix} \mathbf{u}$$

$$\mathbf{y} = \begin{bmatrix} 2 & 2 & 1 & 3 & -1 & 1 & 1 \\ 1 & 1 & 1 & 2 & 0 & 0 & 0 \\ 1 & 1 & 1 & 1 & 0 & 1 & 0 \end{bmatrix} \mathbf{x}$$

**5-10** Check the controllability of the following state equations:

**a.** $\dot{\mathbf{x}} = \begin{bmatrix} 0 & 0 \\ 0 & 1 \end{bmatrix} \mathbf{x} + \begin{bmatrix} 1 \\ 1 \end{bmatrix} u$

**b.** $\dot{\mathbf{x}} = \begin{bmatrix} 0 & 0 \\ 0 & 1 \end{bmatrix} \mathbf{x} + \begin{bmatrix} 1 \\ e^{-t} \end{bmatrix} u$

**c.** $\dot{\mathbf{x}} = \begin{bmatrix} 0 & 0 \\ 0 & 1 \end{bmatrix} \mathbf{x} + \begin{bmatrix} 0 \\ e^{-2t} \end{bmatrix} u$

**5-11** Is it possible to find a set of $b_{ij}$ and a set of $c_{ij}$ such that the following Jordan-form equation

$$\dot{\mathbf{x}} = \begin{bmatrix} 1 & 1 & 0 & 0 & 0 \\ 0 & 1 & 0 & 0 & 0 \\ 0 & 0 & 1 & 1 & 0 \\ 0 & 0 & 0 & 1 & 0 \\ 0 & 0 & 0 & 0 & 1 \end{bmatrix} \mathbf{x} + \begin{bmatrix} b_{11} & b_{12} \\ b_{21} & b_{22} \\ b_{31} & b_{32} \\ b_{41} & b_{42} \\ b_{51} & b_{52} \end{bmatrix} \mathbf{u}$$

$$\mathbf{y} = \begin{bmatrix} c_{11} & c_{12} & c_{13} & c_{14} & c_{15} \\ c_{21} & c_{22} & c_{23} & c_{24} & c_{25} \\ c_{31} & c_{32} & c_{33} & c_{34} & c_{35} \end{bmatrix} \mathbf{x}$$

is controllable? Observable?

**5-12** Given a *controllable*, linear, time-invariant, single-input dynamical equation

$$\dot{\mathbf{x}} = \mathbf{A}\mathbf{x} + \mathbf{b}u$$

where $\mathbf{A}$ is an $n \times n$ matrix, $\mathbf{b}$ is an $n \times 1$ column vector. What is the equivalent dynamical equation if the basis $\{\mathbf{b}, \mathbf{A}\mathbf{b}, \cdots, \mathbf{A}^{n-1}\mathbf{b}\}$ is chosen for the state space? Or, equivalently, if $\bar{\mathbf{x}} = \mathbf{P}\mathbf{x}$ and if $\mathbf{P}^{-1} = [\mathbf{b} \quad \mathbf{A}\mathbf{b} \quad \cdots \quad \mathbf{A}^{n-1}\mathbf{b}]$, what is the new state equation?

**5-13** Find the dynamical equations for the systems shown in Figures 5-3, 5-4, and 5-13 and check the controllability and observability of these equations.

**5-14** If a dynamical equation is controllable at $t_0$, then for any initial state we can transfer it to zero and keep it there for all $t$ thereafter. Is it possible to transfer it to any $\mathbf{x}^1 \neq \mathbf{0}$ and keep it at $\mathbf{x}^1$ thereafter?

**5-15** Consider the dynamical equation

$$\dot{\mathbf{x}} = \begin{bmatrix} -1 & 1 & 0 \\ 0 & -1 & 0 \\ 0 & 0 & -2 \end{bmatrix} \mathbf{x} + \begin{bmatrix} 0 \\ 1 \\ 1 \end{bmatrix} u$$
$$y = [1 \quad 1 \quad 1]\mathbf{x}$$

Is it possible to choose an initial state at $t = 0$ such that the output of the dynamical equation is of the form $y(t) = te^{-t}$ for $t > 0$?

**5-16** Show that controllability and observability of linear time-varying dynamical equations are invariant under any equivalence transformation $\bar{\mathbf{x}} = \mathbf{P}(t)\mathbf{x}$, where $\mathbf{P}$ is nonsingular for all $t$ and continuously differentiable in $t$.

**5-17** Consider the dynamical equation in Problem 5-15. It is assumed that the initial state of the equation is not known. Is it possible to find an input $u_{[0,\infty)}$ such that the output $y$ is of the form $y(t) = \sin(t + \theta)$ for all $t > 1$, where $\theta$ is a constant?

**5-18** Find the equivalence transformation required in transforming a dynamical equation into the form given in Equation (5-53).

**5-19** Show that the state of an observable, $n$-dimensional, linear, time-invariant dynamical equation can be determined instantaneously from the output and its derivatives up to $(n - 1)$ order. *Hint:* Compute $y(t), \dot{y}(t), \ldots, y^{(n-1)}(t)$.

**5-20** Find the equivalence transformation used in Theorem 5-19.

**5-21** Consider the linear, time-invariant dynamical equation

$$\dot{\mathbf{x}} = \mathbf{A}\mathbf{x} + \mathbf{B}\mathbf{u}$$
$$\mathbf{y} = \mathbf{C}\mathbf{x}$$

Every element of $e^{At}$ is a linear combination of the form $t^i e^{\lambda_j t}$. Each $t^i e^{\lambda_j t}$ is called a mode of the equation. For example, if

$$\mathbf{A}_1 = \begin{bmatrix} -1 & 1 \\ 0 & -1 \end{bmatrix} \qquad \mathbf{A}_2 = \begin{bmatrix} -1 & 0 \\ 0 & -1 \end{bmatrix}$$

then the modes of $\mathbf{A}_1$ are $e^{-t}$ and $te^{-t}$; the mode of $\mathbf{A}_2$ is $e^{-t}$. Show that if the dynamical equation is observable, then all the modes of the equation will appear at the output $\mathbf{y}$. Show also that even if all the modes appear at the output, the equation is not necessarily observable.

**5-22** Transform the following dynamical equation

$$\dot{\mathbf{x}} = \begin{bmatrix} \lambda_1 & 1 & 0 & 0 & 0 \\ 0 & \lambda_1 & 1 & 0 & 0 \\ 0 & 0 & \lambda_1 & 0 & 0 \\ 0 & 0 & 0 & \lambda_2 & 1 \\ 0 & 0 & 0 & 0 & \lambda_2 \end{bmatrix} \mathbf{x} + \begin{bmatrix} 0 \\ 1 \\ 0 \\ 0 \\ 1 \end{bmatrix} u$$

$$y = [0 \quad 1 \quad 1 \quad 0 \quad 1]\mathbf{x}$$

into the form of (5-55).

**5-23** Consider the $n$-dimensional, linear, time-invariant dynamical equation

$$FE: \quad \begin{aligned} \dot{\mathbf{x}} &= \mathbf{Ax} + \mathbf{Bu} \\ \mathbf{y} &= \mathbf{Cx} + \mathbf{Du} \end{aligned}$$

The rank of its controllability matrix,

$$\mathbf{U} = [\mathbf{B} \vdots \mathbf{AB} \vdots \cdots \vdots \mathbf{A}^{n-1}\mathbf{B}]$$

is assumed to be $n_1$ ($< n$). Let $\mathbf{Q}_1$ be an $n \times n_1$ matrix whose columns are any $n_1$ linearly independent columns of $\mathbf{U}$. Let $\mathbf{P}_1$ be an $n_1 \times n$ matrix such that $\mathbf{P}_1\mathbf{Q}_1 = \mathbf{I}_{n_1}$, where $\mathbf{I}_{n_1}$ is an $n_1 \times n_1$ unit matrix. Show that the following $n_1$ dimensional dynamical equation

$$F\bar{E}: \quad \begin{aligned} \dot{\bar{\mathbf{x}}}^1 &= \mathbf{P}_1\mathbf{A}\mathbf{Q}_1\bar{\mathbf{x}}^1 + \mathbf{P}_1\mathbf{Bu} \\ \bar{\mathbf{y}} &= \mathbf{C}\mathbf{Q}_1\bar{\mathbf{x}}^1 + \mathbf{Du} \end{aligned}$$

is controllable and is zero-state equivalent to $FE$. In other words, $FE$ is reducible to $F\bar{E}$. *Hint:* Use Theorem 5-17.

**5-24** In Problem 5-23, the reduction procedure reduces to solving for $\mathbf{P}_1$ in $\mathbf{P}_1\mathbf{Q}_1 = \mathbf{I}_{n_1}$. Find some methods for solving $\mathbf{P}_1$ in $\mathbf{P}_1\mathbf{Q}_1 = \mathbf{I}_{n_1}$.

**5-25** Develop a similar statement as Problem 5-23 for an unobservable, linear, time-invariant dynamical equation.

**5-26** Prove Corollary 5-22.

**5-27** In Corollary 5-22 we have shown that a dynamical equation with $\mathbf{D} = 0$ is output controllable if and only if $\rho[\mathbf{CB} \vdots \mathbf{CAB} \vdots \cdots \vdots \mathbf{CA}^{n-1}\mathbf{B}] = q$. Show that if $\mathbf{D} \neq 0$, then the dynamical equation is output controllable if and only if

$$\rho[\mathbf{CB} \vdots \mathbf{CAB} \vdots \cdots \vdots \mathbf{CA}^{n-1}\mathbf{B} \vdots \mathbf{D}] = q$$

**5-28** Consider a linear, time-invariant dynamical equation with $\mathbf{D} = \mathbf{0}$. Under what condition on $\mathbf{C}$ will the (state) controllability imply output controllability?

**5-29** Consider two systems with the transfer-function matrices

$$\hat{\mathbf{G}}_1 = \begin{bmatrix} \dfrac{1}{s+1} & \dfrac{(s+3)}{(s+2)(s+1)} \\ \dfrac{1}{(s+1)} & \dfrac{(s+2)}{(s+3)(s+1)} \end{bmatrix} \qquad \hat{\mathbf{G}}_2 = \begin{bmatrix} \dfrac{1}{(s+2)} \\ \dfrac{(s+1)}{(s+2)} \end{bmatrix}$$

Are they output controllable? Output function controllable?

**5-30** Consider the linear time-invariant discrete-time dynamical equation

$$\begin{aligned} \mathbf{x}(n+1) &= \mathbf{A}\mathbf{x}(n) + \mathbf{B}\mathbf{u}(n) \\ \mathbf{y}(n) &= \mathbf{C}\mathbf{x}(n) + \mathbf{D}\mathbf{u}(n) \end{aligned}$$

where $\mathbf{x}$ is the $n \times 1$ state vector, $\mathbf{u}(n)$ is the $p \times 1$ input vector and $\mathbf{y}$ is the $q \times 1$ output vector. The controllability and observability of discrete-time equation are similarly defined as in the continuous case. Show that the discrete-time equation is controllable if and only if

$$\rho[\mathbf{B} \vdots \mathbf{AB} \vdots \cdots \vdots \mathbf{A}^{n-1}\mathbf{B}] = n$$

The discrete-time equation is observable if and only if

$$\rho \begin{bmatrix} \mathbf{C} \\ \mathbf{CA} \\ \cdot \\ \cdot \\ \cdot \\ \mathbf{CA}^{n-1} \end{bmatrix} = n$$

*Hint:* See Problem 2-18.

**5-31** Consider the $n$-dimensional, single-variable, linear, time-invariant dynamical equation

$$FE_1: \qquad \begin{aligned} \dot{\mathbf{x}} &= \mathbf{A}\mathbf{x} + \mathbf{b}u \\ y &= \mathbf{c}\mathbf{x} \end{aligned}$$

If the input is piecewise constant on interval of constant length $T > 0$, then we have the following discrete-time state equation (Problem 4-28)

$$DFE_1: \quad \mathbf{x}(n+1) = \bar{\mathbf{A}}\mathbf{x}(n) + \bar{\mathbf{b}}u(n)$$
$$y(n) = \bar{\mathbf{c}}\mathbf{x}(n)$$

Show that the controllability of $FE_1$ implies the controllability of $DFE_1$ if and only if

$$\operatorname{Im}[\lambda_i(\mathbf{A}) - \lambda_j(\mathbf{A})] \neq \frac{2\pi q}{T} \qquad q = \pm 1, \pm 2, \cdots$$

whenever $\operatorname{Re}[\lambda_i(\mathbf{A}) - \lambda_j(\mathbf{A})] = 0$, where $\lambda_i(\mathbf{A})$ stands for an eigenvalue of **A**. (See Appendix C.)

**5-32** Consider the network shown in Figure P5-32. Suppose $v_c(0) = 0$ and $i_L(0) = 0$ and the switch $S_0$ is closed for $t \geq 0$. Determine a

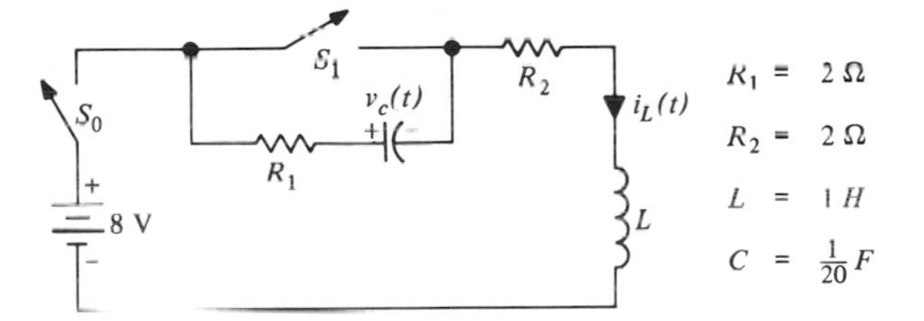


**Figure P5-32**

sequence of control (closed or open) on the switch $S_1$ such that it will bring $i_L(t)$ to zero in the shortest possible time and keep it identically zero thereafter.

**5-33** Verify that the answer to the question posed in Problem 3-36 is affirmative if the system has a controllable and observable dynamical-equation description.

CHAPTER

# 6

# Irreducible Realizations of Rational Transfer-Function Matrices

## 6-1 Introduction

Network synthesis is one of the important disciplines in electrical engineering. It is mainly concerned with determining a passive network that has a prescribed impedance or transfer function. The subject matter we shall introduce in this chapter is along the same line—that is, to determine a linear, time-invariant dynamical equation that has a prescribed rational matrix. Hence this chapter might be viewed as a modern version of network synthesis.

A linear, time-invariant dynamical equation that has a prescribed rational matrix $\hat{\mathbf{G}}(s)$ is called a realization of $\hat{\mathbf{G}}(s)$. The term "realization" is justified by the fact that, by using the dynamical equation, the system with the transfer-function matrix $\hat{\mathbf{G}}(s)$ can be built in the real world by using an operational amplifier circuit. Although we have proved in Theorem 4-10 that every proper rational matrix has a finite-dimensional, linear, time-invariant, dynamical-equation realization, we have not yet shown *how* to realize it. In this chapter we shall study this realization problem.

We study the realization problem for the following reasons: First, there are many design techniques and many computational algorithms developed exclusively for dynamical equations. In order to apply these techniques and algorithms, transfer-function matrices must be realized into dynamical equations. Second, in the design of complex system it is always desirable to simulate the system on an analog or a digital computer to check its performance before the system is built. A system cannot be simulated efficiently if the transfer function is the only available description. After a dynamical equation realization is obtained, by assigning the outputs of integrators as state variables, the system can be readily simulated. Finally, this technique provides a method to synthesize transfer functions by the use of operational amplifier circuits.

For every realizable transfer function matrix $\hat{\mathbf{G}}(s)$, there is an unlimited number of linear time-invariant dynamical equation realizations. Therefore a major problem in the realization is to find a "good" realization. It is clear that a dynamical-equation realization with the least possible dimension is a good realization. We claim that a realization of $\hat{\mathbf{G}}(s)$ with the least possible dimension must be a controllable and observable dynamical equation. Indeed, if a linear, time-invariant, dynamical-equation realization of $\hat{\mathbf{G}}(s)$ is found, and if the equation is not controllable and observable, then following from Theorem 5-20 it is possible to reduce the realization to a lesser dimensional equation that still has $\hat{\mathbf{G}}(s)$ as its transfer-function matrix. This reduction is impossible only if the equation is controllable and observable. Therefore, we conclude that a realization of $\hat{\mathbf{G}}(s)$ with the least possible dimension is a controllable and observable dynamical equation, or equivalently, an irreducible dynamical equation. Such a realization is called a *minimal-dimensional* or *irreducible realization*. In this chapter we study only irreducible realizations of a rational matrix for the following reasons: (1) A rational transfer-function matrix describes only the controllable and observable part of a dynamical equation, hence a faithful realization should be an irreducible one. (2) When an irreducible realization is used to synthesize a network, the number of integrators needed will be minimal. This is desirable for reasons of economy and sensitivity. Note that if an irreducible realization is found, any other irreducible realization can be obtained by applying an equivalence transformation (Theorem 5-21).

This chapter is organized as follows: In Section 6-2 we introduce the concepts of the degree and the characteristic polynomial of a proper rational matrix. We study in Section 6-3 various irreducible realizations of scalar transfer functions. Controllable canonical-form and observable canonical-form dynamical equations are introduced. In Section 6-4 Ho and Kalman's algorithm for realizing rational matrices is introduced. In Section 6-5 we introduce an irreducible Jordan-form realization; the realization is achieved in two steps.

We study in this chapter only realizations of proper rational-function matrices. The material is based in part on References [15], [42], [47], [60], [62], [67], [68], [83], [89], [98], and [115]. For the realizations of impulse-response matrices $\mathbf{G}(t, \tau)$, the interested reader is referred to References [32], [100], [101], and [114]. We note that all of the realization procedures introduced in this chapter can be applied without any modification to sampled transfer-function matrices (see Section 3-7). The only difference is that the resulting equations are discrete-time dynamical equations instead of continuous-time dynamical equations.

To conclude this introductory section, we emphasize that a transfer matrix is an input–output description of a system, whereas a dynamical equation describes not only the input–output relation but also the internal structure of a system. Thus, the realization problem may also be considered as an *identification problem:* a problem of identifying the internal structure of a system from the knowledge obtained through direct measurements at the input and output terminals.

## 6-2 The Characteristic Polynomial and the Degree of a Proper Rational Matrix

In this section, we shall introduce the concepts of the degree and the characteristic polynomial of a proper rational matrix. These two concepts are the extension to the matrix case of the denominator and its degree of a scalar transfer function. In order to see the significance of these two concepts, we introduce the following theorem.

### Theorem 6-1

Consider the single-variable, linear, time-invariant dynamical equation

$$FE_1: \quad \dot{\mathbf{x}} = \mathbf{A}\mathbf{x} + \mathbf{b}u \tag{6-1a}$$

$$y = \mathbf{c}\mathbf{x} + du, \tag{6-1b}$$

where $\mathbf{A}$, $\mathbf{b}$, $\mathbf{c}$, and $d$ are respectively $n \times n$, $n \times 1$, $1 \times n$, and $1 \times 1$ real constant matrices. Let $\Delta(s) \triangleq \det (s\mathbf{I} - \mathbf{A})$ and let

$$\mathbf{c}(s\mathbf{I} - \mathbf{A})^{-1}\mathbf{b} \triangleq \frac{N(s)}{\Delta(s)} \tag{6-2}$$

Then the equation $FE_1$ is irreducible (controllable and observable) if and only if the polynomials $\Delta(s)$ and $N(s)$ have no nontrivial common factor.

### Proof

*Sufficiency:* If $\Delta(s)$ and $N(s)$ have no common factor, then $FE_1$ is irreducible. We prove this by contradiction. Suppose $FE_1$ is either uncontrollable or unobservable, then $FE_1$ can be reduced to a dynamical equation $F\bar{E}_1$ of lesser dimension. Let $\hat{g}(s)$ be the transfer function of $F\bar{E}_1$. Then the degree of the denominator of $\hat{g}(s)$ is less than $n$. Since $FE_1$ and $F\bar{E}_1$ are zero-state equivalent, we have

$$\hat{g}(s) = \mathbf{c}(s\mathbf{I} - \mathbf{A})^{-1}\mathbf{b} + d = \frac{N(s)}{\Delta(s)} + d$$

Since the degree of the denominator of $\hat{g}(s)$ is less than $n$ and the degree of $\Delta(s)$ is $n$, there must be at least one common factor between $N(s)$ and $\Delta(s)$. This is a contradiction. *Necessity:* If $FE_1$ is irreducible, then $\Delta(s)$ and $N(s)$ have no common factor. We prove this by contradiction. Suppose $\Delta(s)$ and $N(s)$ have a common factor; then by using the reduced transfer function $(\bar{N}(s)/\bar{\Delta}(s)) + d$, one can obtain a dynamical equation $F\bar{E}_1$ that has a dimension less than $n$ and is zero-state equivalent to $FE_1$. This contradicts the assumption that $FE_1$ is irreducible. Q.E.D.

### Corollary 6-1

The dynamical equation $FE_1$ in Equation (6-1) is irreducible if and only if the denominator of the transfer function of $FE_1$ is equal to $\Delta(s)$, the characteristic polynomial of $\mathbf{A}$; or if and only if the degree of the denominator of the transfer function of $FE_1$ is equal to $n$, the dimension of $FE_1$. ∎

Recall that every rational function in this book is assumed to be irreducible; that is, there is no common factor between the denominator and the numerator. Hence when we mention a transfer function, it has been assumed to be irreducible; otherwise its denominator is not well-defined. It is easy to show that $N(s)$ and $\Delta(s)$ defined in (6-2) have no common factor if and only if $N(s) + d\Delta(s)$ and $\Delta(s)$ have no common factor (Problem 6-16). Using this fact and the fact that the transfer function of $FE_1$, $\hat{g}(s)$, is equal to $(N(s) + d\Delta(s))/\Delta(s)$ after canceling out the common factor, Corollary 6-1 follows directly from Theorem 6-1.

We see from Corollary 6-1 that the irreducibility of a single-variable, linear, time-invariant dynamical equation is completely determinable from the denominator of its transfer function, or simply from the degree of the denominator. It is of practical and theoretical interest to see whether this is also possible for the multivariable case. The answer is affirmative if a "denominator" can be defined for a rational matrix.

### Definition 6-1

The *characteristic polynomial* of a proper rational matrix $\hat{\mathbf{G}}(s)$ is defined to be the least common denominator of all minors of $\hat{\mathbf{G}}(s)$. The *degree* of $\hat{\mathbf{G}}(s)$, denoted by $\delta\hat{\mathbf{G}}(s)$, is defined to be the degree of the characteristic polynomial of $\hat{\mathbf{G}}(s)$.[1]

[1] If $\hat{\mathbf{G}}(s)$ is not proper, this definition cannot be used and its degree has to be defined in a different way. See References [35], [62], and [83].

### Example 1

Consider the rational-function matrices

$$\hat{\mathbf{G}}_1(s) = \begin{bmatrix} \dfrac{1}{s+1} & \dfrac{1}{s+1} \\ \dfrac{1}{s+1} & \dfrac{1}{s+1} \end{bmatrix} \qquad \hat{\mathbf{G}}_2(s) = \begin{bmatrix} \dfrac{2}{s+1} & \dfrac{1}{s+1} \\ \dfrac{1}{s+1} & \dfrac{1}{s+1} \end{bmatrix}$$

The minors of order 1 of $\hat{\mathbf{G}}_1(s)$ are $1/(s+1)$, $1/(s+1)$, $1/(s+1)$, and $1/(s+1)$. The minor of order 2 of $\hat{\mathbf{G}}_1(s)$ is 0. Hence the characteristic polynomial of $\hat{\mathbf{G}}_1(s)$ is $(s+1)$ and $\delta\hat{\mathbf{G}}_1(s) = 1$. The minors of order 1 of $\hat{\mathbf{G}}_2(s)$ are $2/(s+1)$, $1/(s+1)$, $1/(s+1)$, and $1/(s+1)$. The minor of order 2 of $\hat{\mathbf{G}}_2(s)$ is $1/(s+1)^2$. Hence the characteristic polynomial of $\hat{\mathbf{G}}_2(s)$ is $(s+1)^2$ and $\delta\hat{\mathbf{G}}_2(s) = 2$. ∎

From this example, we see that the characteristic polynomial of $\hat{\mathbf{G}}(s)$ is in general different from the denominator of the determinant of $\hat{\mathbf{G}}(s)$ [if $\hat{\mathbf{G}}(s)$ is a square matrix] and different from the least common denominator of all the entries of $\hat{\mathbf{G}}(s)$. If $\hat{\mathbf{G}}(s)$ is scalar (a $1 \times 1$ matrix), the characteristic polynomial of $\hat{\mathbf{G}}(s)$ reduces to the denominator of $\hat{\mathbf{G}}(s)$.

### Example 2

Consider the $2 \times 3$ rational-function matrix

$$\hat{\mathbf{G}}(s) = \begin{bmatrix} \dfrac{s}{s+1} & \dfrac{1}{(s+1)(s+2)} & \dfrac{1}{s+3} \\ \dfrac{-1}{s+1} & \dfrac{1}{(s+1)(s+2)} & \dfrac{1}{s} \end{bmatrix}$$

The minors of order 1 are the entries of $\hat{\mathbf{G}}(s)$. There are three minors of order 2. They are

$$\frac{s}{(s+1)^2(s+2)} + \frac{1}{(s+1)^2(s+2)} = \frac{s+1}{(s+1)^2(s+2)} = \frac{1}{(s+1)(s+2)} \tag{6-3}$$

$$\frac{s}{s+1}\cdot\frac{1}{s} + \frac{1}{(s+1)(s+3)} = \frac{s+4}{(s+1)(s+3)}$$

$$\frac{1}{(s+1)(s+2)s} - \frac{1}{(s+1)(s+2)(s+3)} = \frac{3}{s(s+1)(s+2)(s+3)}$$

Hence the characteristic polynomial of $\hat{\mathbf{G}}(s)$ is $s(s+1)(s+2)(s+3)$ and $\delta\hat{\mathbf{G}}(s) = 4$. ∎

Note that in computing the characteristic polynomial of a rational matrix, every minor of the matrix must be reduced to an irreducible one as we did in (6-3); otherwise we will obtain an erroneous answer. With the concepts of the characteristic polynomial and the degree of a rational matrix, we are ready to extend Corollary 6-1 to the multivariable dynamical equations.

### Theorem 6-2

An $n$-dimensional, multivariable, linear, time-invariant dynamical equation is irreducible (controllable and observable) if and only if the characteristic polynomial of its transfer-function matrix is equal to $\Delta(s) = \det(s\mathbf{I} - \mathbf{A})$, the characteristic polynomial of the matrix $\mathbf{A}$; or if and only if the degree of its transfer function matrix is equal to $n$. ∎

The proof of this theorem is very involved. The interested reader is referred to References [15] and [115]. The concepts of the degree and the characteristic polynomial of a rational-function matrix will be used constantly in the remainder of this chapter. They are also essential in the stability study of multivariable feedback systems (Chapter 9).

## 6-3 Irreducible Realizations of a Scalar Rational Transfer Function

**Irreducible realization of $\beta/D(s)$.** Before considering the general case, we study first the transfer function

$$\hat{g}(s) = \frac{\beta}{s^n + \alpha_1 s^{n-1} + \cdots + \alpha_{n-1}s + \alpha_n} \tag{6-4}$$

where $\beta$ and $\alpha_i$, for $i = 1, 2, \cdots, n$ are real constants. In the time domain, Equation (6-4) can be written as

$$(p^n + \alpha_1 p^{n-1} + \cdots + \alpha_n)y(t) = \beta u(t) \tag{6-5}$$

where $p^i$ stands for $d^i/dt^i$. By taking the Laplace transform of (6-5) and assuming the initial conditions to be zero, it is easy to verify that the transfer function from $\hat{u}(s)$ to $\hat{y}(s)$ in (6-5) is indeed the one given in (6-4). It is well known that in an $n$th-order differential equation, in order to have a unique solution for any $u$, we need $n$ number of initial conditions. Hence the state vector will consist of $n$ components. In this case the output $y$ and its derivatives up to the $(n-1)$th order qualify as state variables. Let

$$
\begin{aligned}
x_1(t) &\triangleq y(t) \\
x_2(t) &\triangleq \dot{y}(t) = py(t) = \dot{x}_1(t) \\
x_3(t) &\triangleq \ddot{y}(t) = p^2y(t) = \dot{x}_2(t) \\
&\cdot \\
&\cdot \\
&\cdot \\
x_n(t) &\triangleq y^{(n-1)}(t) = p^{n-1}y(t) = \dot{x}_{n-1}(t)
\end{aligned}
\tag{6-6}
$$

Differentiating $x_n(t)$ once and using (6-5) we obtain

$$
\dot{x}_n(t) = p^n y(t) = -\alpha_n x_1 - \alpha_{n-1}x_2 - \cdots - \alpha_1 x_n + \beta u \tag{6-7}
$$

Define

$$
\mathbf{x}(t) \triangleq \begin{bmatrix} x_1(t) \\ x_2(t) \\ \cdot \\ \cdot \\ \cdot \\ x_n(t) \end{bmatrix}
$$

Then the set of equations in (6-6) and (6-7) can be arranged in matrix form as

$$
\dot{\mathbf{x}} = \begin{bmatrix} 0 & 1 & 0 & \cdots & 0 \\ 0 & 0 & 1 & \cdots & 0 \\ 0 & 0 & 0 & \cdots & 0 \\ \cdot & \cdot & \cdot & & \cdot \\ \cdot & \cdot & \cdot & & \cdot \\ \cdot & \cdot & \cdot & & \cdot \\ 0 & 0 & 0 & \cdots & 1 \\ -\alpha_n & -\alpha_{n-1} & -\alpha_{n-2} & \cdots & -\alpha_1 \end{bmatrix} \mathbf{x} + \begin{bmatrix} 0 \\ 0 \\ 0 \\ \cdot \\ \cdot \\ \cdot \\ 0 \\ \beta \end{bmatrix} u \tag{6-8a}
$$

$$
y = [1 \quad 0 \quad 0 \quad \cdots \quad 0]\mathbf{x} \tag{6-8b}
$$

The block diagram of (6-8) is shown in Figure 6-1. The set of equations in (6-8) is derived from (6-5); hence the transfer function of (6-8) is equal to $\hat{g}(s)$ in (6-4). This can be verified either by applying to Figure 6-1

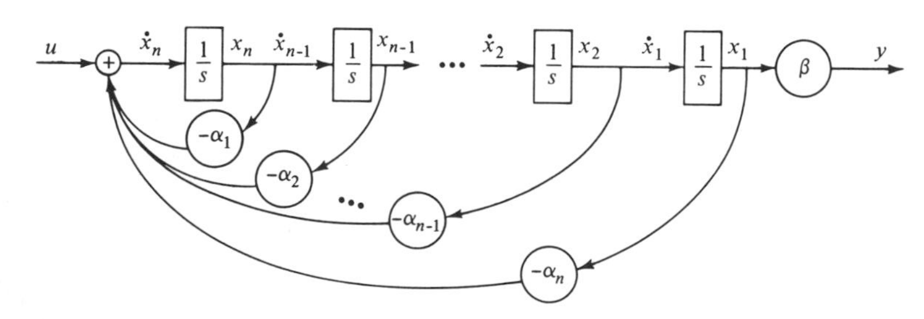


**Figure 6-1** The block diagram of Equation (6-8).

Mason's formula for signal-flow graphs (see pp. 151–152) or by using the formula

$$\hat{g}(s) = \mathbf{c}(s\mathbf{I} - \mathbf{A})^{-1}\mathbf{b}$$

where

$$(s\mathbf{I} - \mathbf{A})^{-1}\mathbf{b} = \begin{bmatrix} s & -1 & \cdots & 0 & 0 \\ 0 & s & \cdots & 0 & 0 \\ \cdot & \cdot & & \cdot & \cdot \\ \cdot & \cdot & & \cdot & \cdot \\ \cdot & \cdot & & \cdot & \cdot \\ 0 & 0 & \cdots & s & -1 \\ \alpha_n & \alpha_{n-1} & \cdots & \alpha_2 & s+\alpha_1 \end{bmatrix}^{-1} \begin{bmatrix} 0 \\ 0 \\ \cdot \\ \cdot \\ \cdot \\ 0 \\ \beta \end{bmatrix} \tag{6-9}$$

It is clear that $(s\mathbf{I} - \mathbf{A})^{-1}\mathbf{b}$ is equal to the last column of $(s\mathbf{I} - \mathbf{A})^{-1}\beta$; or equivalently, to the cofactors of the last row of $(\beta/\Delta(s))(s\mathbf{I} - \mathbf{A})$, where

$$\Delta(s) \triangleq \det (s\mathbf{I} - \mathbf{A}) = s^n + \alpha_1 s^{n-1} + \cdots + \alpha_n$$

(See Problem 2-20.) In view of the form of $(s\mathbf{I} - \mathbf{A})$, the cofactors of the last row of $(\beta/\Delta(s))(s\mathbf{I} - \mathbf{A})$ can be easily computed as

$$\frac{\beta}{\Delta(s)} [1 \quad s \quad s^2 \quad \cdots \quad s^{n-1}]'$$

Hence,

$$(s\mathbf{I} - \mathbf{A})^{-1}\mathbf{b} = \begin{bmatrix} s & -1 & \cdots & 0 & 0 \\ 0 & s & \cdots & 0 & 0 \\ \cdot & \cdot & & \cdot & \cdot \\ \cdot & \cdot & & \cdot & \cdot \\ \cdot & \cdot & & \cdot & \cdot \\ 0 & 0 & \cdots & s & -1 \\ \alpha_n & \alpha_{n-1} & \cdots & \alpha_2 & s+\alpha_1 \end{bmatrix}^{-1} \begin{bmatrix} 0 \\ 0 \\ \cdot \\ \cdot \\ \cdot \\ 0 \\ \beta \end{bmatrix} = \frac{\beta}{\Delta(s)} \begin{bmatrix} 1 \\ s \\ \cdot \\ \cdot \\ \cdot \\ s^{n-2} \\ s^{n-1} \end{bmatrix} \tag{6-10}$$

and

$$\hat{g}(s) = \mathbf{c}(s\mathbf{I} - \mathbf{A})^{-1}\mathbf{b} = [1 \quad 0 \quad \cdots \quad 0 \quad 0] \frac{\beta}{\Delta(s)} \begin{bmatrix} 1 \\ s \\ \cdot \\ \cdot \\ \cdot \\ s^{n-2} \\ s^{n-1} \end{bmatrix}$$

$$= \frac{\beta}{\Delta(s)} = \frac{\beta}{s^n + \alpha_1 s^{n-1} + \cdots + \alpha_n}$$

This verifies that (6-8) is indeed a realization of $\hat{g}(s)$ in (6-4). It follows from Corollary 6-1 that (6-8) is an irreducible realization. This can also be verified by showing that (6-8) is controllable (except for the trivial case $\beta = 0$) and observable. Observe that the realization (6-8) can be obtained directly from the coefficients of $\hat{g}(s)$ in (6-4).

There are many other methods of writing a dynamical equation for a transfer function $\hat{g}(s)$. One of the most common is to draw for $\hat{g}(s)$ a block diagram with transfer functions of the form $b/(s - a)$ as basic building blocks. A dynamical equation can then be written by assigning the output of each basic block as a state variable. As can be seen from Figure 6-2, the output of a basic block is in fact the output of an inte-

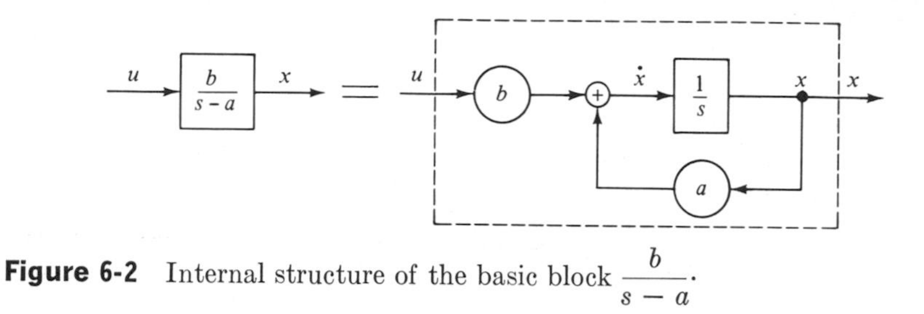


**Figure 6-2** Internal structure of the basic block $\dfrac{b}{s - a}$.

grator; therefore, it qualifies as a state variable. We shall apply this realization technique to the transfer function $\hat{g}(s)$ given in (6-4). Suppose the transfer function $\hat{g}(s)$ is factored as

$$\hat{g}(s) = \frac{\beta}{s^n + \alpha_1 s^{n-1} + \cdots + \alpha_n} = \frac{\beta}{(s - \lambda_1)(s - \lambda_2) \cdots (s - \lambda_n)} \tag{6-11}$$

Its block diagram is given in Figure 6-3. With the state variables chosen as shown, we obtain immediately

$$\begin{aligned} \dot{x}_1 &= \lambda_1 x_1 + \beta u \\ \dot{x}_i &= \lambda_i x_i + x_{i-1} \qquad i = 2, 3, \cdots, n \end{aligned}$$

**Figure 6-3** Block diagram of $\hat{g}(s)$ in Equation (6-11).

and

$$y = x_n$$

They can be arranged in matrix form as

$$\dot{\mathbf{x}} = \begin{bmatrix} \lambda_1 & 0 & 0 & \cdots & 0 & 0 \\ 1 & \lambda_2 & 0 & \cdots & 0 & 0 \\ 0 & 1 & \lambda_3 & \cdots & 0 & 0 \\ \cdot & \cdot & \cdot & & \cdot & \cdot \\ \cdot & \cdot & \cdot & & \cdot & \cdot \\ \cdot & \cdot & \cdot & & \cdot & \cdot \\ 0 & 0 & 0 & \cdots & 1 & \lambda_n \end{bmatrix} \mathbf{x} + \begin{bmatrix} \beta \\ 0 \\ 0 \\ \cdot \\ \cdot \\ \cdot \\ 0 \end{bmatrix} u \tag{6-12a}$$

$$y = [0 \;\; 0 \;\; 0 \;\; \cdots \;\; 0 \;\; 1]\mathbf{x} \tag{6-12b}$$

It is easy to verify that the dynamical equation (6-12) is controllable and observable; hence it is an irreducible realization of $\hat{g}(s)$. The characteristic polynomial of the matrix **A** in (6-12a) is again equal to the denominator of $\hat{g}(s)$.

If the transfer function in (6-11) is expanded by partial fraction expansion into a sum of terms of the form $1/(s - \lambda_i)$, then a Jordan-canonical-form dynamical-equation realization can be obtained. This will be discussed in the following subsection.

**Irreducible realizations of $\hat{g}(s) = N(s)/D(s)$.** Consider the following scalar proper transfer function

$$\hat{g}_1(s) = \frac{\bar{\beta}_0 s^n + \bar{\beta}_1 s^{n-1} + \cdots + \bar{\beta}_n}{\bar{\alpha}_0 s^n + \bar{\alpha}_1 s^{n-1} + \cdots + \bar{\alpha}_n}$$

where $\bar{\alpha}_i$ and $\bar{\beta}_i$, for $i = 0, 1, 2, \cdots, n$, are real constants. It is assumed that $\bar{\alpha}_0 \neq 0$. By long division, $\hat{g}_1(s)$ can be written as

$$\hat{g}_1(s) = \frac{\beta_1 s^{n-1} + \beta_2 s^{n-2} + \cdots + \beta_n}{s^n + \alpha_1 s^{n-1} + \cdots + \alpha_{n-1}s + \alpha_n} + d \triangleq \hat{g}(s) + d \qquad \textbf{(6-13)}$$

where $d = \hat{g}_1(\infty) = \bar{\beta}_0/\bar{\alpha}_0$. Since the constant $d$ immediately gives the direct transmission part of a realization, we need to consider in the following only the strictly proper rational function

$$\hat{g}(s) \triangleq \frac{N(s)}{D(s)} \triangleq \frac{\beta_1 s^{n-1} + \beta_2 s^{n-2} + \cdots + \beta_n}{s^n + \alpha_1 s^{n-1} + \cdots + \alpha_n} \qquad \textbf{(6-14)}$$

Since every rational function, $\hat{g}_1$ in particular, is assumed to be irreducible, it is easy to show from (6-13) that $N(s)$ and $D(s)$, defined in (6-14), have no common factor (see Problem 6-16). Equation (6-14) suggests that we consider the differential equation

$$D(p)y(t) = N(p)u(t) \qquad \textbf{(6-15)}$$

where $D(p) = p^n + \alpha_1 p^{n-1} + \cdots + \alpha_n$, $N(p) = \beta_1 p^{n-1} + \beta_2 p^{n-2} + \cdots + \beta_n$, and $p^i$ stands for $d^i/dt^i$. Clearly, the transfer function of (6-15) is $\hat{g}(s)$ in (6-14). In the following we shall give three different realizations of $N(s)/D(s)$: the observable canonical-form dynamical-equation realization, the controllable canonical-form dynamical-equation realization, and the Jordan canonical-form dynamical-equation realization.

### Observable canonical-form realization of $\hat{g}(s) = N(s)/D(s)$

Consider the $n$th-order differential equation $D(p)y(t) = N(p)u(t)$. It is well known that if we have $n$ initial conditions—for example, $y(t_0)$, $y^{(1)}(t_0)$, $\cdots$, $y^{(n-1)}(t_0)$—then for any input $u_{[t_0,t_1]}$, the output $y_{[t_0,t_1]}$ is completely determinable. In this case, however, if we choose $y(t)$, $y^{(1)}(t)$, $\cdots$, $y^{(n-1)}(t)$ as state variables as we did in (6-5), then we cannot obtain a dynamical equation of the form $\dot{\mathbf{x}} = \mathbf{A}\mathbf{x} + \mathbf{b}u$, $y = \mathbf{c}\mathbf{x}$. Instead, we will obtain an equation of the form $\dot{\mathbf{x}} = \mathbf{A}\mathbf{x} + \mathbf{b}u$, $y = \mathbf{c}\mathbf{x} + d_1u + d_2u^{(1)} + d_3u^{(2)} + \cdots$. Hence in order to realize $N(s)/D(s)$ in the form $\dot{\mathbf{x}} = \mathbf{A}\mathbf{x} + \mathbf{b}u$, $y = \mathbf{c}\mathbf{x}$, a different set of state variables has to be chosen.

Taking the Laplace transform of (6-15) and regrouping the terms associated with the same power of $s$, we finally obtain

$$\begin{aligned}\hat{y}(s) = \frac{N(s)}{D(s)}\hat{u}(s) + \frac{1}{D(s)}\{&y(0)s^{n-1} + [y^{(1)}(0) + \alpha_1 y(0) - \beta_1 u(0)]s^{n-2} \\ &+ \cdots + [y^{(n-1)}(0) + \alpha_1 y^{(n-2)}(0) - \beta_1 u^{(n-2)}(0) + \alpha_2 y^{(n-3)}(0) \\ &- \beta_2 u^{(n-3)}(0) + \cdots + \alpha_{n-1}y(0) - \beta_{n-1}u(0)]\}\end{aligned} \tag{6-16}$$

The term

$$\frac{N(s)}{D(s)}\hat{u}(s) = \hat{g}(s)\hat{u}(s)$$

in the right-hand side of (6-16) gives the response due to the input $\hat{u}(s)$; the remainder gives the response due to the initial conditions. Therefore, if all the coefficients associated with $s^{n-1}$, $s^{n-2}$, $\cdots$, $s^0$ in (6-16) are known, then for any $u$ a unique $y$ can be determined. Consequently, if we choose the state variables as

$$\begin{aligned}x_n(t) &\triangleq y(t) \\ x_{n-1}(t) &\triangleq y^{(1)}(t) + \alpha_1 y(t) - \beta_1 u(t) \\ x_{n-2}(t) &\triangleq y^{(2)}(t) + \alpha_1 y^{(1)}(t) - \beta_1 u^{(1)}(t) + \alpha_2 y(t) - \beta_2 u(t) \\ &\cdot \\ &\cdot \\ &\cdot \\ x_1(t) &\triangleq y^{(n-1)}(t) + \alpha_1 y^{(n-2)}(t) - \beta_1 u^{(n-2)}(t) + \cdots \\ &\qquad + \alpha_{n-1}y(t) - \beta_{n-1}u(t)\end{aligned} \tag{6-17}$$

then $\mathbf{x} = [x_1 \quad x_2 \quad \cdots \quad x_n]'$ qualifies as the state vector. The set of equations in (6-17) yields

$$\begin{aligned}y &= x_n \\ x_{n-1} &= \dot{x}_n + \alpha_1 x_n - \beta_1 u \\ x_{n-2} &= \dot{x}_{n-1} + \alpha_2 x_n - \beta_2 u \\ &\cdot \\ &\cdot \\ &\cdot \\ x_1 &= \dot{x}_2 + \alpha_{n-1}x_n - \beta_{n-1}u\end{aligned}$$

Differentiating $x_1$ in (6-17) once and using (6-15), we obtain

$$\dot{x}_1 = -\alpha_n x_n + \beta_n u$$

The foregoing equations can be arranged in matrix form as

$$\begin{bmatrix} \dot{x}_1 \\ \dot{x}_2 \\ \dot{x}_3 \\ \cdot \\ \cdot \\ \cdot \\ \\ \dot{x}_n \end{bmatrix} = \begin{bmatrix} 0 & 0 & 0 & \cdots & 0 & -\alpha_n \\ 1 & 0 & 0 & \cdots & 0 & -\alpha_{n-1} \\ 0 & 1 & 0 & \cdots & 0 & -\alpha_{n-2} \\ \cdot & \cdot & \cdot & & \cdot & \cdot \\ \cdot & \cdot & \cdot & & \cdot & \cdot \\ \cdot & \cdot & \cdot & & \cdot & \cdot \\ 0 & 0 & 0 & \cdots & 0 & -\alpha_2 \\ 0 & 0 & 0 & \cdots & 1 & -\alpha_1 \end{bmatrix} \begin{bmatrix} x_1 \\ x_2 \\ x_3 \\ \cdot \\ \cdot \\ \cdot \\ x_{n-1} \\ x_n \end{bmatrix} + \begin{bmatrix} \beta_n \\ \beta_{n-1} \\ \beta_{n-2} \\ \cdot \\ \cdot \\ \cdot \\ \beta_2 \\ \beta_1 \end{bmatrix} u \qquad \text{(6-18a)}$$

$$y = [0 \quad 0 \quad 0 \quad \cdots \quad 0 \quad 1]\mathbf{x} \qquad \text{(6-18b)}$$

A dynamical equation in the form of (6-18) is said to be in the *observable canonical form*. The block diagram of (6-18) is shown in Figure 6-4.

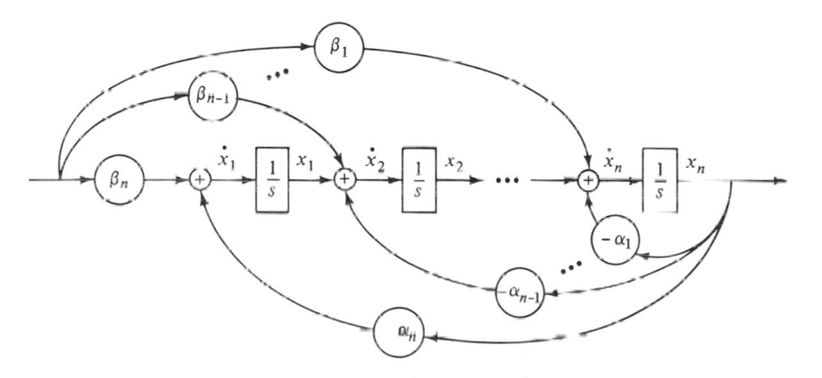


**Figure 6-4** Block diagram of the observable canonical-form dynamical equation (6-18).

Since (6-18) is derived from (6-15), the transfer function of (6-18) is $\hat{g}(s)$ in (6-14). This can be verified by either applying to Figure 6-4 Mason's formula for signal-flow graph or computing

$$\mathbf{c}(s\mathbf{I} - \mathbf{A})^{-1}\mathbf{b} = [\mathbf{c}(s\mathbf{I} - \mathbf{A})^{-1}\mathbf{b}]' = \mathbf{b}'(s\mathbf{I} - \mathbf{A}')^{-1}\mathbf{c}'$$

with the aid of (6-10), where the "prime" symbol denotes the transpose. The dynamical equation is an irreducible realization of $\hat{g}(s)$ in (6-14) following from Corollary 6-1. Note that this realization can be obtained directly from the coefficients of $\hat{g}(s)$.

### Controllable canonical-form realization of $\hat{g}(s) = N(s)/D(s)$

We shall now consider a different realization—called the controllable canonical-form dynamical-equation realization—of $\hat{g}(s) = N(s)/D(s)$ or, equivalently, of the differential equation

$$D(p)y(t) = N(p)u(t) \qquad \text{(6-19)}$$

Let us introduce a new variable $v(t)$ such that

$$D(p)v(t) = u(t) \tag{6-20}$$

Substituting (6-19) into (6-20) and employing the fact that $N(p)D(p) = D(p)N(p)$, we obtain

$$N(p)v(t) = y(t) \tag{6-21}$$

Observe that Equation (6-20) is in the form of (6-5). Hence, by defining $x_1 = v$, $x_2 = \dot{x}_1 = \dot{v}$, $\cdots$, $x_n = \dot{x}_{n-1} = v^{(n-1)}$, we immediately obtain the state equation

$$\begin{bmatrix} \dot{x}_1 \\ \dot{x}_2 \\ \dot{x}_3 \\ \cdot \\ \cdot \\ \cdot \\ \dot{x}_{n-1} \\ \dot{x}_n \end{bmatrix} = \begin{bmatrix} 0 & 1 & 0 & \cdots & 0 \\ 0 & 0 & 1 & \cdots & 0 \\ 0 & 0 & 0 & \cdots & 0 \\ \cdot & \cdot & \cdot & & \cdot \\ \cdot & \cdot & \cdot & & \cdot \\ \cdot & \cdot & \cdot & & \cdot \\ 0 & 0 & 0 & \cdots & 1 \\ -\alpha_n & -\alpha_{n-1} & -\alpha_{n-2} & \cdots & -\alpha_1 \end{bmatrix} \begin{bmatrix} x_1 \\ x_2 \\ x_3 \\ \cdot \\ \cdot \\ \cdot \\ x_{n-1} \\ x_n \end{bmatrix} + \begin{bmatrix} 0 \\ 0 \\ 0 \\ \cdot \\ \cdot \\ \cdot \\ 0 \\ 1 \end{bmatrix} u \tag{6-22a}$$

By substituting $x_i = v^{(i-1)}$, for $i = 1, 2, \cdots, n$, into (6-21), the output equation is obtained as

$$\begin{aligned} y(t) &= \beta_1 v^{(n-1)} + \beta_2 v^{(n-2)} + \cdots + \beta_{n-1}\dot{v} + \beta_n v \\ &= \beta_n x_1 + \beta_{n-1} x_2 + \cdots + \beta_1 x_n = [\beta_n \quad \beta_{n-1} \quad \cdots \quad \beta_1]\mathbf{x} \end{aligned} \tag{6-22b}$$

This is a novel way to write a dynamical equation. There are no simple relations between $x_i$, $u$, and $y$ as there were in (6-17). The dynamical equation (6-22) is an irreducible realization of $N(s)/D(s)$ following directly from Corollary 6-1. A dynamical equation in the form of (6-22) is said to be in the *controllable canonical form*. The block diagram of (6-22) is given in Figure 6-5.

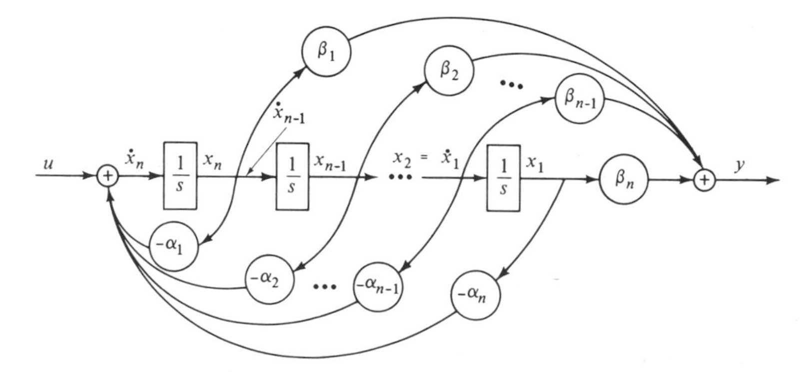


**Figure 6-5** Block diagram of the controllable canonical-form dynamical equation (6-22).

### Example 1

Consider the proper irreducible transfer function

$$\hat{g}(s) = \frac{4s^3 + 25s^2 + 45s + 34}{2s^3 + 12s^2 + 20s + 16}$$

By long division, $\hat{g}(s)$ can be written as

$$\hat{g}(s) = \frac{0.5s^2 + 2.5s + 1}{s^3 + 6s^2 + 10s + 8} + 2$$

Hence its controllable canonical-form realization is

$$\begin{bmatrix} \dot{x}_1 \\ \dot{x}_2 \\ \dot{x}_3 \end{bmatrix} = \begin{bmatrix} 0 & 1 & 0 \\ 0 & 0 & 1 \\ -8 & -10 & -6 \end{bmatrix} \begin{bmatrix} x_1 \\ x_2 \\ x_3 \end{bmatrix} + \begin{bmatrix} 0 \\ 0 \\ 1 \end{bmatrix} u$$

$$y = [1 \quad 2.5 \quad 0.5]\mathbf{x} + 2u$$

Its observable canonical-form realization is

$$\dot{\mathbf{x}} = \begin{bmatrix} 0 & 0 & -8 \\ 1 & 0 & -10 \\ 0 & 1 & -6 \end{bmatrix} \mathbf{x} + \begin{bmatrix} 1 \\ 2.5 \\ 0.5 \end{bmatrix} u$$

$$y = [0 \quad 0 \quad 1]\mathbf{x} + 2u$$

### Jordan-canonical-form realization of $\hat{g}(s) = N(s)/D(s)$

We use an example to illustrate the procedure to realize a transfer function into a Jordan-form dynamical equation. The idea can be easily extended to the general case. Assume that $D(s)$ consists of three distinct roots $\lambda_1$, $\lambda_2$, and $\lambda_3$, and assume that $D(s)$ can be factored as $D(s) = (s - \lambda_1)^3(s - \lambda_2)(s - \lambda_3)$. We also assume that $\hat{g}(s)$ can be expanded by partial fraction expansion into

$$\hat{g}(s) = \frac{e_{11}}{(s - \lambda_1)^3} + \frac{e_{12}}{(s - \lambda_1)^2} + \frac{e_{13}}{(s - \lambda_1)} + \frac{e_2}{(s - \lambda_2)} + \frac{e_3}{(s - \lambda_3)} \tag{6-23}$$

The block diagrams of $\hat{g}(s)$ are given in Figure 6-6. In Figure 6-6(a), the coefficients $e_{11}$, $e_{12}$, $e_{13}$, $e_2$, and $e_3$ are associated with the output. In Figure 6-6(b), they are associated with the input. With the state variables chosen as shown, from Figure 6-6(a) we can obtain the following dynamical equation:

$$\begin{bmatrix} \dot{x}_{11} \\ \dot{x}_{12} \\ \dot{x}_{13} \\ \dot{x}_2 \\ \dot{x}_3 \end{bmatrix} = \left[\begin{array}{ccc|cc} \lambda_1 & 1 & 0 & 0 & 0 \\ 0 & \lambda_1 & 1 & 0 & 0 \\ 0 & 0 & \lambda_1 & 0 & 0 \\ \hline 0 & 0 & 0 & \lambda_2 & 0 \\ 0 & 0 & 0 & 0 & \lambda_3 \end{array}\right] \mathbf{x} + \begin{bmatrix} 0 \\ 0 \\ 1 \\ 1 \\ 1 \end{bmatrix} u \tag{6-24a}$$

$$y = [e_{11} \quad e_{12} \quad e_{13} \quad e_2 \quad e_3]\mathbf{x} \tag{6-24b}$$

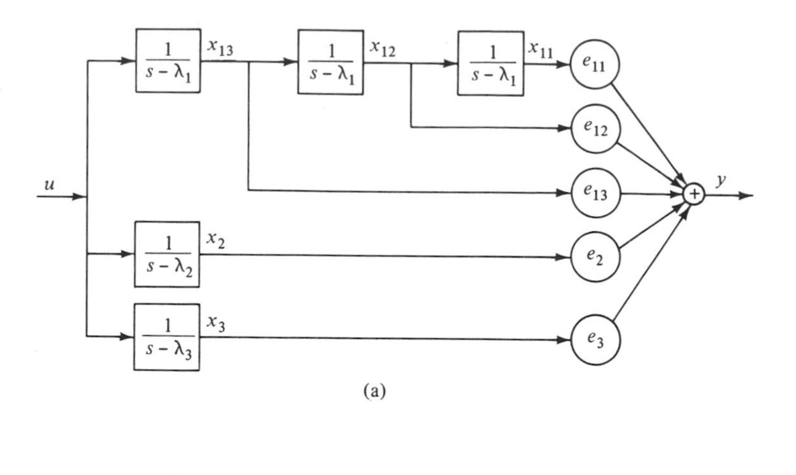


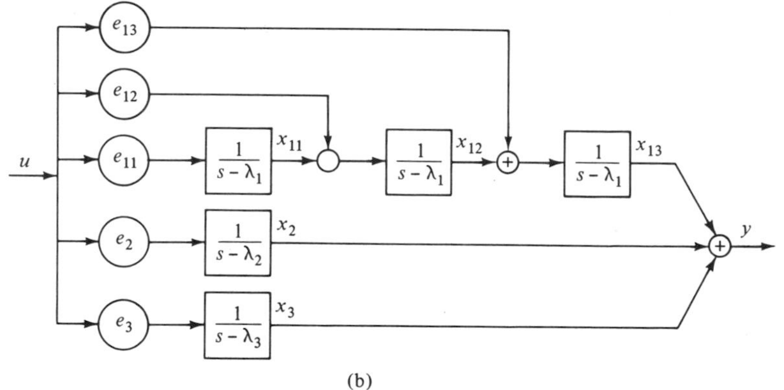


**Figure 6-6** Two different block diagrams of $\hat{g}(s)$ in Equation (6-23).

Equation (6-24) is in the Jordan canonical form. There is one Jordan block associated with each eigenvalue. The equation is clearly controllable (Corollary 5-16); it is also observable, except for the trivial cases $e_{11} = 0$, $e_2 = 0$, or $e_3 = 0$. Therefore the dynamical equation (6-24) is an irreducible realization of the $\hat{g}(s)$ in (6-23).

If the block diagram in Figure 6-6(b) is used and if the state variables are chosen as shown, then the dynamical equation is

$$\begin{bmatrix} \dot{x}_{11} \\ \dot{x}_{12} \\ \dot{x}_{13} \\ \dot{x}_2 \\ \dot{x}_3 \end{bmatrix} = \begin{bmatrix} \lambda_1 & 0 & 0 & 0 & 0 \\ 1 & \lambda_1 & 0 & 0 & 0 \\ 0 & 1 & \lambda_1 & 0 & 0 \\ 0 & 0 & 0 & \lambda_2 & 0 \\ 0 & 0 & 0 & 0 & \lambda_3 \end{bmatrix} \mathbf{x} + \begin{bmatrix} e_{11} \\ e_{12} \\ e_{13} \\ e_2 \\ e_3 \end{bmatrix} u \tag{6-25a}$$

$$y = [0 \quad 0 \quad 1 \quad 1 \quad 1] \tag{6-25b}$$

which is another irreducible Jordan-form realization of $\hat{g}(s)$. Note the differences in the assignment of the state variables in Figures 6-6(a) and 6-6(b).

There are two difficulties in realizing a transfer function into a Jordan canonical form. First, the denominator of the transfer function must be factored; or, correspondingly, the poles of the transfer function $\hat{g}(s)$ must be computed. This is a difficult task if the degree of $\hat{g}(s)$ is larger than 3. Second, if the transfer function has complex poles, the matrices $\mathbf{A}$, $\mathbf{b}$, and $\mathbf{c}$ will consist of complex numbers. In this case, the equation cannot be simulated in an analog computer, for complex numbers cannot be generated in the real world. However, this can be taken care of by introducing some equivalence transformation, as will be demonstrated in the following. Since all the coefficients of $\hat{g}(s)$ are assumed to be real, if the complex number $\lambda$ is a pole of $\hat{g}(s)$, then $\bar{\lambda}$, the complex conjugate of $\lambda$, is also a pole of $\hat{g}(s)$. Hence in the Jordan-form realization of $\hat{g}(s)$, we have the following subequation:

$$\begin{bmatrix} \dot{\mathbf{x}}^1 \\ \dot{\mathbf{x}}^2 \end{bmatrix} = \begin{bmatrix} \mathbf{A}_1 & \mathbf{0} \\ \mathbf{0} & \bar{\mathbf{A}}_1 \end{bmatrix} \begin{bmatrix} \mathbf{x}^1 \\ \mathbf{x}^2 \end{bmatrix} + \begin{bmatrix} \mathbf{b}_1 \\ \bar{\mathbf{b}}_1 \end{bmatrix} u \tag{6-26a}$$

$$y = [\mathbf{c}_1 \quad \bar{\mathbf{c}}_1] \begin{bmatrix} \mathbf{x}^1 \\ \mathbf{x}^2 \end{bmatrix} \tag{6-26b}$$

where $\mathbf{A}_1$ is the Jordan block associated with $\lambda$ and $\bar{\mathbf{A}}_1$ is the complex conjugate (no transpose) of $\mathbf{A}_1$. Clearly, $\bar{\mathbf{A}}_1$ is the Jordan block associated with $\bar{\lambda}$. Let us introduce the equivalence transformation $\bar{\mathbf{x}} = \mathbf{P}\mathbf{x}$, where

$$\mathbf{P} = \begin{bmatrix} \mathbf{I} & \mathbf{I} \\ i\mathbf{I} & -i\mathbf{I} \end{bmatrix} \qquad (i = \sqrt{-1})$$

and

$$\mathbf{P}^{-1} = \frac{1}{2} \begin{bmatrix} \mathbf{I} & -i\mathbf{I} \\ \mathbf{I} & i\mathbf{I} \end{bmatrix}$$

Then it can be easily verified that the dynamical equation (6-26) can be transformed into

$$\begin{bmatrix} \dot{\bar{\mathbf{x}}}^1 \\ \dot{\bar{\mathbf{x}}}^2 \end{bmatrix} = \begin{bmatrix} \operatorname{Re} \mathbf{A}_1 & \operatorname{Im} \mathbf{A}_1 \\ -\operatorname{Im} \mathbf{A}_1 & \operatorname{Re} \mathbf{A}_1 \end{bmatrix} \begin{bmatrix} \bar{\mathbf{x}}^1 \\ \bar{\mathbf{x}}^2 \end{bmatrix} + \begin{bmatrix} 2 \operatorname{Re} \mathbf{b}_1 \\ -2 \operatorname{Im} \mathbf{b}_1 \end{bmatrix} u \tag{6-27a}$$

$$y = [\operatorname{Re} \mathbf{c}_1 \quad \operatorname{Im} \mathbf{c}_1] \begin{bmatrix} \bar{\mathbf{x}}^1 \\ \bar{\mathbf{x}}^2 \end{bmatrix} \tag{6-27b}$$

where $\operatorname{Re} \mathbf{A}$ and $\operatorname{Im} \mathbf{A}$ denote the real part and the imaginary part of $\mathbf{A}$, respectively. Since all the coefficients in (6-27) are real, this equation can be used in computer simulations. Another convenient way to transform a Jordan-form dynamical equation into an equation with real coefficients is to use the transformation introduced in Problems 6-19 and 6-20.

Example

Consider the Jordan-form equation with complex eigenvalues.

$$\dot{\mathbf{x}} = \begin{bmatrix} 1+2i & 1 & 0 & 0 & 0 \\ 0 & 1+2i & 0 & 0 & 0 \\ 0 & 0 & 1-2i & 1 & 0 \\ 0 & 0 & 0 & 1-2i & 0 \\ 0 & 0 & 0 & 0 & 2 \end{bmatrix} x + \begin{bmatrix} 2-3i \\ 1 \\ 2+3i \\ 1 \\ 2 \end{bmatrix} u \qquad \text{(6-28)}$$

$$y = [1 \quad -i \;\vdots\; 1 \quad i \;\vdots\; 2]\mathbf{x}$$

Let $\bar{\mathbf{x}} = \mathbf{P}\mathbf{x}$, where

$$\mathbf{P} = \begin{bmatrix} 1 & 0 & 1 & 0 & 0 \\ 0 & 1 & 0 & 1 & 0 \\ i & 0 & -i & 0 & 0 \\ 0 & i & 0 & -i & 0 \\ 0 & 0 & 0 & 0 & 1 \end{bmatrix}$$

Then (6-28) can be transformed into

$$\dot{\bar{\mathbf{x}}} = \begin{bmatrix} 1 & 1 & 2 & 0 & 0 \\ 0 & 1 & 0 & 2 & 0 \\ -2 & 0 & 1 & 1 & 0 \\ 0 & -2 & 0 & 1 & 0 \\ 0 & 0 & 0 & 0 & 2 \end{bmatrix} \bar{\mathbf{x}} + \begin{bmatrix} 4 \\ 2 \\ 6 \\ 0 \\ 2 \end{bmatrix} u$$

$$y = [\;1 \quad 0 \;\vdots\; 0 \quad -1 \;\vdots\; 2]\bar{\mathbf{x}}$$

whose coefficients are all real. ∎

A remark is in order concerning various realizations of $\hat{g}(s)$. It is clear that the controllable or observable canonical-form realization is, in general, easier to obtain than the Jordan-form realization. However, the Jordan-form realization has the advantage that its eigenvalues are less sensitive to the variations of parameters. Hence if the sensitivity problem is a major concern, the Jordan-form realization should be used. See References [23], [81], and [87].

***Realization of $N(s)/D(s)$ where $N(s)$ and $D(s)$ have common factors.** We have shown that if $N(s)$ and $D(s)$ have no common factors, then irreducible realizations can be found for $N(s)/D(s)$. It is of interest to discuss the case where $N(s)$ and $D(s)$ have common factors. In this case, either controllable or observable dynamical-equation realization can be found for $N(s)/D(s)$, but the realization can never be both controllable and observable.

Let

$$D(s) = s^n + \alpha_1 s^{n-1} + \cdots + \alpha_n$$

and

$$N(s) = \beta_1 s^{n-1} + \beta_2 s^{n-2} + \cdots + \beta_n$$

It is assumed that $D(s)$ and $N(s)$ have at least one common factor. The observable canonical-form realization procedure and the controllable canonical-form realization procedure introduced in the previous subsections can be applied without any modification to $N(s)/D(s)$ even $N(s)$ and $D(s)$ have common factors. If we apply the controllable canonical-form realization procedure to $N(s)/D(s)$, we will obtain the following realization

$$\dot{\mathbf{x}} = \mathbf{A}\mathbf{x} + \mathbf{b}u \tag{6-29a}$$

$$y = \mathbf{c}\mathbf{x} \tag{6-29b}$$

where

$$\mathbf{A} = \begin{bmatrix} 0 & 1 & 0 & \cdots & 0 \\ 0 & 0 & 1 & \cdots & 0 \\ 0 & 0 & 0 & \cdots & 0 \\ \cdot & \cdot & \cdot & & \cdot \\ \cdot & \cdot & \cdot & & \cdot \\ \cdot & \cdot & \cdot & & \cdot \\ 0 & 0 & 0 & \cdots & 1 \\ -\alpha_n & -\alpha_{n-1} & -\alpha_{n-2} & \cdots & -\alpha_1 \end{bmatrix} \qquad \mathbf{b} = \begin{bmatrix} 0 \\ 0 \\ 0 \\ \cdot \\ \cdot \\ \cdot \\ 0 \\ 1 \end{bmatrix}$$

$$\mathbf{c} = [\beta_n \quad \beta_{n-1} \quad \beta_{n-2} \quad \cdots \quad \beta_1]$$

It is easy to show that the controllability matrix $[\mathbf{b} \quad \mathbf{Ab} \quad \cdots \quad \mathbf{A}^{n-1}\mathbf{b}]$ of (6-29) is nonsingular (in fact, its determinant is always equal to $-1$ no matter what the values of $\alpha_i$ are); hence the realization (6-29) is controllable. We show that if $D(s)$ and $N(s)$ have common factors, then (6-29) is not observable. Let $(s - \lambda)$ be a common factor of $N(s)$ and $D(s)$; then we have

$$D(\lambda) = \lambda^n + \alpha_1\lambda^{n-1} + \cdots + \alpha_{n-1}\lambda + \alpha_n = 0 \tag{6-30}$$

and

$$N(\lambda) - \beta_1\lambda^{n-1} + \beta_2\lambda^{n-2} + \cdots + \beta_n = 0 \tag{6-31}$$

Define the $n \times 1$ constant vector $\boldsymbol{\alpha}$ as $\boldsymbol{\alpha}' \triangleq [1 \quad \lambda \quad \lambda^2 \quad \cdots \quad \lambda^{n-1}]$. Then (6-31) can be written as $\mathbf{c}\boldsymbol{\alpha} = 0$, where $\mathbf{c}$ is defined in (6-29b). Using (6-30), it is easy to verify that $\mathbf{A}\boldsymbol{\alpha} = \lambda\boldsymbol{\alpha}$, $\mathbf{A}^2\boldsymbol{\alpha} = \lambda^2\boldsymbol{\alpha}$, $\cdots$, $\mathbf{A}^{n-1}\boldsymbol{\alpha} = \lambda^{n-1}\boldsymbol{\alpha}$. Hence the identity $\mathbf{c}\boldsymbol{\alpha} = 0$ implies that $\mathbf{c}\mathbf{A}^i\boldsymbol{\alpha} = 0$, for $i = 0, 1, \cdots, n-1$, which can be written as

$$\begin{bmatrix} \mathbf{c} \\ \mathbf{cA} \\ \cdot \\ \cdot \\ \cdot \\ \mathbf{cA}^{n-1} \end{bmatrix} \boldsymbol{\alpha} = \mathbf{0} \tag{6-32}$$

Since $\boldsymbol{\alpha}$ is a nonzero vector, (6-32) implies that the observability matrix of (6-29) has rank less than $n$. Hence the dynamical equation (6-29) is not observable.

Similarly, if we apply the observable canonical-form realization procedure to $N(s)/D(s)$, where $N(s)$ and $D(s)$ have common factors, then an observable but not controllable dynamical-equation realization can be obtained. Note that it is always possible to give $N(s)/D(s)$ an uncontrollable and unobservable realization (Problem 6-8).

***Realization of linear, time-varying differential equations.** Before concluding this section, we shall briefly discuss the setup of dynamical equations for linear, time-varying differential equations. If an $n$th-order, linear, time-varying differential equation is of the form

$$(p^n + \alpha_1(t)p^{n-1} + \cdots + \alpha_n(t))y(t) = \beta(t)u(t) \tag{6-33}$$

by choosing $y(t)$, $\dot{y}(t)$, $\cdots$, $y^{(n-1)}(t)$ as state variables, a dynamical equation of exactly the same form as (6-8) can be set up. However, if the right-hand side of (6-33) consists of the derivatives of $u$, although it can still be realized into a dynamical equation, the situation becomes very involved. Instead of giving a general formula, we give an example to illustrate the procedure.

Example 2

Consider the following second-order, time-varying differential equation

$$[p^2 + \alpha_1(t)p + \alpha_2(t)]y(t) = [\beta_0(t)p^2 + \beta_1(t)p + \beta_2(t)]u(t) \tag{6-34}$$

The procedure of formulating a dynamical equation for (6-34) is as follows: We first assume that (6-34) can be set into the form

$$\begin{bmatrix} \dot{x}_1 \\ \dot{x}_2 \end{bmatrix} = \begin{bmatrix} 0 & 1 \\ -\alpha_2(t) & -\alpha_1(t) \end{bmatrix} \begin{bmatrix} x_1 \\ x_2 \end{bmatrix} + \begin{bmatrix} b_1(t) \\ b_2(t) \end{bmatrix} u \tag{6-35a}$$

$$y = [1 \quad 0]\mathbf{x} + d(t)u \tag{6-35b}$$

and then verify this by computing the unknown time functions $b_1$, $b_2$, and $d$ in terms of the coefficients of (6-34). Differentiating (6-35b) and using (6-35a), we obtain

$$\begin{aligned} \dot{y} &= py = x_2 + b_1(t)u(t) + \dot{d}(t)u(t) + d(t)\dot{u}(t) \\ \ddot{y} &= p^2y = -\alpha_2 x_1 - \alpha_1 x_2 + b_2 u + \dot{b}_1 u + b_1 \dot{u} + \ddot{d}u + 2\dot{d}\dot{u} + d\ddot{u} \end{aligned}$$

Substituting these into (6-34) and equating the coefficients of $u$, $\dot{u}$, and $\ddot{u}$, we obtain

$$\begin{aligned} d(t) &= \beta_0(t) \\ b_1(t) &= \beta_1(t) - \alpha_1(t)\beta_0(t) - 2\dot{\beta}_0(t) \\ b_2(t) &= \beta_2(t) - \dot{b}_1(t) - \alpha_1(t)b_1(t) - \alpha_1(t)\dot{\beta}(t) - \alpha_2(t)\beta_0(t) - \ddot{\beta}_0 \end{aligned} \tag{6-36}$$

Since the time functions $b_1$, $b_2$ and $d$ can be solved from (6-36), we conclude that the differential equation (6-34) can be transformed into a dynamical equation of the form in (6-35). We see that even for a second-order differential equation, the relations between $\mathbf{b}$, $d$, and the $\alpha_i$'s and $\beta_i$'s are very complicated.

## *6-4 An Irreducible Realization of a Proper Rational Transfer-Function Matrix[2]

We showed in Section 4-4 that every rational-function matrix is realizable by a finite-dimensional, linear, time-invariant dynamical equation and also gave a realization procedure. Although the procedure is very simple, the resulting dynamical equation is generally uncontrollable and unobservable. Moreover, the dynamical equation is not internally coupled. Recall that in the realization procedure presented in Section 4-4, we realize each element of a transfer-function matrix at a time and then connect them together from the input and the output, as illustrated in Figure 6-7. We see that this kind of dynamical-equation realization

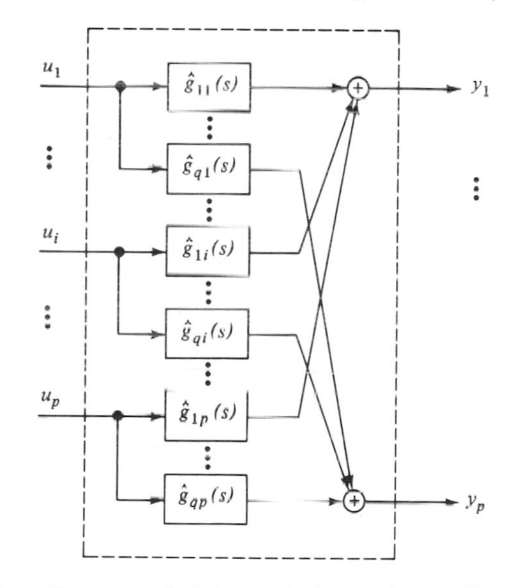


**Figure 6-7** Internally uncoupled dynamical equation realization.

is internally uncoupled. But the coupling or interacting of all variables is the very feature of every physical multivariable systems. Therefore the dynamical equation resulting from the realization procedure of Section 4-4 is not satisfactory. Although we may apply Theorems 5-17 and 5-18 to reduce the equation to an irreducible one, the computation required is very involved. Therefore it is desirable to introduce some

[2] This section closely follows Reference [47].

simpler realization procedures. In this and the following sections, we shall introduce two of them: Ho and Kalman's algorithm and an irreducible Jordan-form realization. Ho and Kalman's algorithm introduced in this section can also be used to reduce an uncontrollable and/or unobservable dynamical equation to an irreducible one.

Consider a $q \times p$ proper rational function matrix $\hat{\mathbf{G}}(s)$. We expand it into

$$\hat{\mathbf{G}}(s) = \hat{\mathbf{G}}(\infty) + \mathbf{H}_0 s^{-1} + \mathbf{H}_1 s^{-2} + \cdots \tag{6-37}$$

where $\mathbf{H}_i$, for $i = 0, 1, 2, \cdots$, are $q \times p$ constant matrices. Suppose the linear, time-invariant dynamical equation

$$FE: \quad \dot{\mathbf{x}} = \mathbf{A}\mathbf{x} + \mathbf{B}\mathbf{u} \tag{6-38a}$$
$$\mathbf{y} = \mathbf{C}\mathbf{x} + \mathbf{D}\mathbf{u} \tag{6-38b}$$

is a realization of $\hat{\mathbf{G}}(s)$; then, by definition,

$$\hat{\mathbf{G}}(s) = \mathbf{D} + \mathbf{C}(s\mathbf{I} - \mathbf{A})^{-1}\mathbf{B}$$

which, with the aid of (2-85), can be expanded into

$$\hat{\mathbf{G}}(s) = \mathbf{D} + \mathbf{C}\mathbf{B}s^{-1} + \mathbf{C}\mathbf{A}\mathbf{B}s^{-2} + \cdots \tag{6-39}$$

### Theorem 6-3

The dynamical equation $FE$ is a realization of $\hat{\mathbf{G}}(s)$ if and only if $\mathbf{D} = \hat{\mathbf{G}}(\infty)$ and $\mathbf{H}_i = \mathbf{C}\mathbf{A}^i\mathbf{B}$, for $i = 0, 1, 2, \cdots$. ∎

This theorem follows directly from (6-37) and (6-39). Since $\hat{\mathbf{G}}(\infty)$ immediately gives the matrix $\mathbf{D}$ of a dynamical-equation realization, in the following we study only strictly proper rational-function matrices. In view of Theorem 6-3, we may restate the irreducible realization problem as follows: Given $\mathbf{H}_i$, for $i = 0, 1, 2, \cdots$, find a triplet $\{\mathbf{A}, \mathbf{B}, \mathbf{C}\}$ such that $\mathbf{H}_i = \mathbf{C}\mathbf{A}^i\mathbf{B}$, for $i = 0, 1, 2, \cdots$, and $\{\mathbf{A}, \mathbf{B}, \mathbf{C}\}$ is controllable and observable. Before proceeding, we need some preliminary results.

Let the least common denominator of all the entries of $\hat{\mathbf{G}}(s)$ be

$$h(s) = s^r + a_1 s^{r-1} + \cdots + a_r \tag{6-40}$$

We expand $\hat{\mathbf{G}}(s)$ into

$$\hat{\mathbf{G}}(s) = \frac{\mathbf{R}_1 s^{r-1} + \mathbf{R}_2 s^{r-2} + \cdots + \mathbf{R}_r}{s^r + a_1 s^{r-1} + \cdots + a_r} \tag{6-41}$$

(see Problem 4-15). Since we have assumed that $\hat{\mathbf{G}}(s)$ is strictly proper, the highest power of the numerator of (6-41) is at most $r - 1$. Combining (6-37) and (6-41), and since $\hat{\mathbf{G}}(\infty) = \mathbf{0}$, we obtain

$$\mathbf{R}_1 s^{r-1} + \mathbf{R}_2 s^{r-2} + \cdots + \mathbf{R}_r$$
$$= (s^r + a_1 s^{r-1} + \cdots + a_r)(\mathbf{H}_0 s^{-1} + \mathbf{H}_1 s^{-2} + \cdots)$$

which implies, by identifying the coefficients of the same power of $s$, that

$$
\begin{aligned}
&\mathbf{H}_0 = \mathbf{R}_1 \\
&\mathbf{H}_1 + a_1\mathbf{H}_0 = \mathbf{R}_2 \\
&\vdots \\
&\mathbf{H}_{r-1} + a_1\mathbf{H}_{r-2} + \cdots + a_{r-1}\mathbf{H}_0 = \mathbf{R}_r \\
&\mathbf{H}_{r+i} + a_1\mathbf{H}_{r+i-1} + \cdots + a_r\mathbf{H}_i = \mathbf{0} \qquad i = 0, 1, 2, \cdots
\end{aligned}
\tag{6-42}
$$

This set of equations is essential in the following development.[3] We define the $qr \times qr$ and $pr \times pr$ matrices

$$
\mathbf{M} \triangleq \begin{bmatrix}
\mathbf{0} & \mathbf{I}_q & \mathbf{0} & \cdots & \mathbf{0} \\
\mathbf{0} & \mathbf{0} & \mathbf{I}_q & \cdots & \mathbf{0} \\
\vdots & \vdots & \vdots & & \vdots \\
\mathbf{0} & \mathbf{0} & \mathbf{0} & \cdots & \mathbf{I}_q \\
-a_r\mathbf{I}_q & -a_{r-1}\mathbf{I}_q & -a_{r-2}\mathbf{I}_q & \cdots & -a_1\mathbf{I}_q
\end{bmatrix}
\qquad
\mathbf{N} \triangleq \begin{bmatrix}
\mathbf{0} & \mathbf{0} & \cdots & \mathbf{0} & -a_r\mathbf{I}_p \\
\mathbf{I}_p & \mathbf{0} & \cdots & \mathbf{0} & -a_{r-1}\mathbf{I}_p \\
\mathbf{0} & \mathbf{I}_p & \cdots & \mathbf{0} & -a_{r-2}\mathbf{I}_p \\
\vdots & \vdots & & \vdots & \vdots \\
\mathbf{0} & \mathbf{0} & \cdots & \mathbf{I}_p & -a_1\mathbf{I}_p
\end{bmatrix}
\tag{6-43}
$$

where $\mathbf{I}_n$ is an $n \times n$ unit matrix. We also define the following two $qr \times pr$ matrices from the coefficient matrices $\mathbf{H}_i$ of $\hat{\mathbf{G}}(s)$

$$
\mathbf{T} \triangleq \begin{bmatrix}
\mathbf{H}_0 & \mathbf{H}_1 & \mathbf{H}_2 & \cdots & \mathbf{H}_{r-1} \\
\mathbf{H}_1 & \mathbf{H}_2 & \mathbf{H}_3 & \cdots & \mathbf{H}_r \\
\vdots & \vdots & \vdots & & \vdots \\
\mathbf{H}_{r-1} & \mathbf{H}_r & \mathbf{H}_{r+1} & \cdots & \mathbf{H}_{2r-2}
\end{bmatrix}
\tag{6-44a}
$$

$$
\tilde{\mathbf{T}} \triangleq \begin{bmatrix}
\mathbf{H}_1 & \mathbf{H}_2 & \mathbf{H}_3 & \cdots & \mathbf{H}_r \\
\mathbf{H}_2 & \mathbf{H}_3 & \mathbf{H}_4 & \cdots & \mathbf{H}_{r+1} \\
\vdots & \vdots & \vdots & & \vdots \\
\mathbf{H}_r & \mathbf{H}_{r+1} & \mathbf{H}_{r+2} & \cdots & \mathbf{H}_{2r-1}
\end{bmatrix}
\tag{6-44b}
$$

[3] The subsequent development applies to any integer $r$ for which a relation of form (6-42) holds; the integer $r$ need not be the degree of the least common denominator of all entries of $\hat{\mathbf{G}}(s)$ as defined in (6-40).

Then, by using (6-42), it is straightforward to verify that

$$\tilde{\mathbf{T}} = \mathbf{MT} = \mathbf{TN} \tag{6-45}$$

Thus, by multiplying $\mathbf{T}$ by $\mathbf{M}$ or $\mathbf{N}$, we have shifted $\mathbf{H}_1$ to the (1, 1) position. Similarly, we can show that by multiplying $\mathbf{T}$ by $\mathbf{M}^i$ or $\mathbf{N}^i$, the matrix $\mathbf{H}_i$ is shifted to the (1, 1) position. Let $\mathbf{I}_{n,m}$ be an $n \times m$ $(m \geq n)$ constant matrix of the form $\mathbf{I}_{n,m} = [\mathbf{I}_n \quad \mathbf{0}]$. Then we have[3a]

$$\mathbf{H}_i = \mathbf{I}_{q,qr}\mathbf{M}^i\mathbf{T}\mathbf{I}'_{p,pr} = \mathbf{I}_{q,qr}\mathbf{T}\mathbf{N}^i\mathbf{I}'_{p,pr} \qquad i = 0, 1, 2, \cdots \tag{6-46}$$

where the prime denotes the transpose of a matrix. Hence the triplet $\{\mathbf{M}, \mathbf{T}\mathbf{I}'_{p,pr}, \mathbf{I}_{q,qr}\}$ and the triplet $\{\mathbf{N}, \mathbf{I}'_{p,pr}, \mathbf{I}_{q,qr}\mathbf{T}\}$ are two realizations of $\hat{\mathbf{G}}(s)$ (see Theorem 6-3). These realizations are generally not irreducible. We give in the following an irreducible realization.

### Theorem 6-4

Given a $q \times p$ strictly proper rational-function matrix $\hat{\mathbf{G}}(s)$. Let $r$ be the degree of the least common denominator of all the entries of $\hat{\mathbf{G}}(s)$. We form the $qr \times pr$ matrices $\mathbf{T}$ and $\tilde{\mathbf{T}}$. Let the rank of $\mathbf{T}$ be $n$, and let $\mathbf{K}$ and $\mathbf{L}$ be $qr \times qr$ and $pr \times pr$ nonsingular constant matrices such that[4]

$$\mathbf{KTL} = \begin{bmatrix} \mathbf{I}_n & \mathbf{0} \\ \mathbf{0} & \mathbf{0} \end{bmatrix} = \mathbf{I}'_{n,qr}\mathbf{I}_{n,pr} \tag{6-47}$$

Then the triplet

$$\mathbf{A} = \mathbf{I}_{n,qr}\mathbf{K}\tilde{\mathbf{T}}\mathbf{L}\mathbf{I}'_{n,pr} \tag{6-48a}$$

$$\mathbf{B} = \mathbf{I}_{n,qr}\mathbf{K}\mathbf{T}\mathbf{I}'_{p,pr} \tag{6-48b}$$

$$\mathbf{C} = \mathbf{I}_{q,qr}\mathbf{T}\mathbf{L}\mathbf{I}'_{n,pr} \tag{6-48c}$$

is an irreducible realization of $\hat{\mathbf{G}}(s)$.

### Proof

We first introduce[5]

$$\mathbf{T}^\# \triangleq \mathbf{L}\mathbf{I}'_{n,pr}\mathbf{I}_{n,qr}\mathbf{K}$$

Using $\mathbf{T} = \mathbf{K}^{-1}\mathbf{I}'_{n,qr}\mathbf{I}_{n,pr}\mathbf{L}^{-1}$, which is obtained from (6-47), it is easy to show that

$$\mathbf{T}\mathbf{T}^\#\mathbf{T} = \mathbf{T} \tag{6-49}$$

---

[3a] The effect of premultiplying $\mathbf{I}_{q,qr}$ and postmultiplying $\mathbf{I}'_{p,pr}$ to a matrix is to retain the $q \times p$ submatrix at the upper left corner of the matrix.

[4] This can be achieved by a sequence of elementary operations, that is operations that (1) interchange two rows or columns, (2) multiply a row or column by a nonzero constant, or (3) add one row or column to another. See References [39] and [86].

[5] $\mathbf{T}^\#$ is called the pseudo inverse of $\mathbf{T}$. If $\mathbf{T}$ is a square nonsingular matrix, then $\mathbf{T}^\#$ reduces to the inverse of $\mathbf{T}$. See Reference [116].

Substituting (6-49) into (6-46), we obtain

$$\mathbf{H}_i = \mathbf{I}_{q,qr}\mathbf{M}^i\mathbf{T}\mathbf{I}'_{p,pr} = \mathbf{I}_{q,qr}\mathbf{M}^i\mathbf{T}\mathbf{T}^{\#}\mathbf{T}\mathbf{I}'_{p,pr}$$

By use of the fact that $\mathbf{M}^i\mathbf{T} = \mathbf{T}\mathbf{N}^i$, which can be deduced from (6-45), this equation can be rearranged as

$$\begin{aligned}\mathbf{H}_i &= \mathbf{I}_{q,qr}\mathbf{T}\mathbf{N}^i\mathbf{T}^{\#}\mathbf{T}\mathbf{I}'_{p,pr} = \mathbf{I}_{q,qr}\mathbf{T}\mathbf{T}^{\#}\mathbf{T}\mathbf{N}^i\mathbf{T}^{\#}\mathbf{T}\mathbf{I}'_{p,pr}\\ &= \mathbf{I}_{q,qr}\mathbf{T}\mathbf{T}^{\#}\mathbf{M}^i\mathbf{T}\mathbf{T}^{\#}\mathbf{T}\mathbf{I}'_{p,pr}\\ &= (\mathbf{I}_{q,qr}\mathbf{T}\mathbf{L}\mathbf{I}'_{n,pr})(\mathbf{I}_{n,qr}\mathbf{K}\mathbf{M}^i\mathbf{T}\mathbf{L}\mathbf{I}'_{n,pr})(\mathbf{I}_{n,qr}\mathbf{K}\mathbf{T}\mathbf{I}'_{p,pr})\\ &= (\mathbf{I}_{q,qr}\mathbf{T}\mathbf{L}\mathbf{I}'_{n,pr})(\mathbf{I}_{n,qr}\mathbf{K}\mathbf{M}\mathbf{T}\mathbf{L}\mathbf{I}'_{n,pr})^i(\mathbf{I}_{n,qr}\mathbf{K}\mathbf{T}\mathbf{I}'_{p,pr})\end{aligned} \tag{6-50}$$

The last step can be proved by noting that

$$\begin{aligned}\mathbf{I}_{n,qr}\mathbf{K}\mathbf{M}^2\mathbf{T}\mathbf{L}\mathbf{I}'_{n,pr} &= \mathbf{I}_{n,qr}\mathbf{K}\mathbf{M}\mathbf{T}\mathbf{N}\mathbf{L}\mathbf{I}'_{n,pr}\\ &= \mathbf{I}_{n,qr}\mathbf{K}\mathbf{M}\mathbf{T}\mathbf{T}^{\#}\mathbf{T}\mathbf{N}\mathbf{L}\mathbf{I}'_{n,pr}\\ &= (\mathbf{I}_{n,qr}\mathbf{K}\mathbf{M}\mathbf{T}\mathbf{L}\mathbf{I}'_{n,pr})(\mathbf{I}_{n,qr}\mathbf{K}\mathbf{M}\mathbf{T}\mathbf{L}\mathbf{I}'_{n,pr})\end{aligned}$$

Since (6-50) holds for $i = 0, 1, 2, \cdots$, the triplet defined in (6-48) is a realization of $\hat{\mathbf{G}}(s)$. We show now that the realization is irreducible (controllable and observable). Let $\mathbf{U}_{r-1}$ and $\mathbf{V}_{r-1}$ be defined as in (5-25) and (5-39). Note that $\mathbf{U}_{r-1}$ is an $n \times pr$ constant matrix and $\mathbf{V}_{r-1}$ is a $qr \times n$ constant matrix. It is easy to verify that $\mathbf{T} = \mathbf{V}_{r-1}\mathbf{U}_{r-1}$; see (5-58). Since $\mathbf{T}$ is assumed of rank $n$ and since $\rho\mathbf{T} \leq \min(\rho\mathbf{V}_{r-1}, \rho\mathbf{U}_{r-1})$, where $\rho(\cdot)$ denotes the rank of a matrix, we immediately obtain $\rho\mathbf{V}_{r-1} = n$ and $\rho\mathbf{U}_{r-1} = n$. Consequently, the dynamical equation defined in (6-48) is observable and controllable. Q.E.D.

The algorithm in Theorem 6-4 is generally too involved to be carried out by hand calculation. Fortunately it can be easily programmed in a digital computer; see Reference [67]. Theorem 6-4, in addition to Theorems 5-17 and 5-18, can also be used to reduce an uncontrollable and/or unobservable dynamical equation to an irreducible one. Let $\{\mathbf{A}, \mathbf{B}, \mathbf{C}\}$ be an $n$-dimensional reducible equation. Let $\mathbf{H}_i = \mathbf{C}\mathbf{A}^i\mathbf{B}$, for $i = 0, 1, 2, \cdots$. From the Cayley-Hamilton theorem, we have

$$\mathbf{H}_{n+i} + a_1\mathbf{H}_{n+i-1} + \cdots + a_n\mathbf{H}_i = \mathbf{0} \qquad i = 0, 1, \cdots$$

From the remark following Equation (6-42), we may replace $r$ by $n$ in Theorem 6-4 and obtain an irreducible dynamical equation that has the same transfer-function matrix as $\{\mathbf{A}, \mathbf{B}, \mathbf{C}\}$.

We shall now comment briefly on the rank of $\mathbf{T}$ in (6-44) and the degree of the rational matrix $\hat{\mathbf{G}}(s)$. From Theorems 6-2 and 6-4, we conclude that *the degree of* $\hat{\mathbf{G}}(s)$ *is equal to the rank of* $\mathbf{T}$. *Hence the matrix* $\mathbf{T}$ *in* (6-44) *offers an alternative way to compute the degree of* $\hat{\mathbf{G}}(s)$. If the degree of $\hat{\mathbf{G}}(s)$ can be computed (say $n$) beforehand, then the integer

$r$ in Theorem 6-4 can be chosen to be the least integer such that the rank of $\mathbf{T}$ is $n$.

## *6-5 Irreducible Jordan-Form Realization of a Proper Rational Transfer-Function Matrix[6]

The irreducible realization of a proper rational matrix introduced in the previous section is not in any canonical form. Furthermore, the orders of the matrices in the manipulation are very large; hence the realization method is beyond the reach of hand calculation and has to be carried out with a digital computer. In this section we shall introduce another irreducible realization which can be achieved by hand calculation. In addition, the realization is in the Jordan canonical form. However, the realization method has two disadvantages: first, the denominator of each entry of $\hat{\mathbf{G}}(s)$ must be factored; second, an irreducible realization cannot, in general, be achieved in one step.

Before considering the general case, we study first the $q \times p$ strictly proper rational-function matrix $\hat{\mathbf{G}}(s)$, which has only one pole $\lambda$. We assume, for simplicity, that the multiplicity of the pole $\lambda$ in $\hat{\mathbf{G}}(s)$ is equal to 3. We give in the following an irreducible Jordan-form realization of $\hat{\mathbf{G}}(s)$. This is accomplished in two steps: (1) form a controllable dynamical equation, and (2) reduce it to a controllable as well as observable dynamical equation by applying some simple rules.

**Controllable Jordan-form dynamical-equation realization of $\hat{\mathbf{G}}(s)$.** The $q \times p$ rational matrix $\hat{\mathbf{G}}(s)$ is expanded by partial fraction expansion into, say,

$$\hat{\mathbf{G}}(s) = \frac{1}{(s-\lambda)^3}\mathbf{M}_1 + \frac{1}{(s-\lambda)^2}\mathbf{M}_2 + \frac{1}{(s-\lambda)}\mathbf{M}_3 \tag{6-51}$$

where $\mathbf{M}_1$, $\mathbf{M}_2$, and $\mathbf{M}_3$ are $q \times p$ constant matrices. Let $\rho\mathbf{M}$ denote the rank of $\mathbf{M}$ and let $T(\mathbf{M})$ denote a set of linearly independent rows that generate all the rows of $\mathbf{M}$. We compute the following:

$$\begin{aligned}
\rho\mathbf{M}_1 &= r_1 & T_1(\mathbf{M}_1) &= \{\mathbf{b}_{l1}, \mathbf{b}_{l2}, \cdots, \mathbf{b}_{lr_1}\} \\
\rho\begin{bmatrix}\mathbf{M}_1\\ \mathbf{M}_2\end{bmatrix} &= r_2 & T_2\begin{pmatrix}\mathbf{M}_1\\ \mathbf{M}_2\end{pmatrix} &= \{\mathbf{b}_{l1}, \cdots, \mathbf{b}_{lr_1}, \mathbf{b}_{l(r_1+1)}, \cdots, \mathbf{b}_{lr_2}\} \\
\rho\begin{bmatrix}\mathbf{M}_1\\ \mathbf{M}_2\\ \mathbf{M}_3\end{bmatrix} &= r_3 & T_3\begin{pmatrix}\mathbf{M}_1\\ \mathbf{M}_2\\ \mathbf{M}_3\end{pmatrix} &= \{\mathbf{b}_{l1}, \cdots, \mathbf{b}_{lr_1}, \mathbf{b}_{l(r_1+1)}, \\
&&&\qquad \cdots, \mathbf{b}_{lr_2}, \mathbf{b}_{l(r_2+1)}, \cdots, \mathbf{b}_{lr_3}\}
\end{aligned} \tag{6-52}$$

[6] This section closely follows References [42] and [89].

Clearly $r_1 \leq r_2 \leq r_3 \leq p$. Let us express the matrix $\mathbf{M}_1$ in terms of the linearly independent rows $\mathbf{b}_{l1} \quad \mathbf{b}_{l2} \quad \cdots \quad \mathbf{b}_{lr_1}$ as

$$\mathbf{M}_1 = \mathbf{c}_{11}\mathbf{b}_{l1} + \mathbf{c}_{12}\mathbf{b}_{l2} + \cdots + \mathbf{c}_{1r_1}\mathbf{b}_{lr_1}$$
$$= [\mathbf{c}_{11} \quad \mathbf{c}_{12} \quad \cdots \quad \mathbf{c}_{1r_1}]\begin{bmatrix} \mathbf{b}_{l1} \\ \mathbf{b}_{l2} \\ \cdot \\ \cdot \\ \cdot \\ \mathbf{b}_{lr_1} \end{bmatrix} \triangleq \mathbf{C}_{r_1}{}^1\mathbf{B}_{r_1}{}^l \tag{6-53}$$

The $q \times 1$ column vectors $\mathbf{c}_{11}, \mathbf{c}_{12}, \cdots, \mathbf{c}_{1r_1}$, or correspondingly, the $q \times r_1$ matrix $\mathbf{C}_{r_1}{}^1$ can be computed as follows: Since all the rows of the $r_1 \times p$ matrix $\mathbf{B}_{r_1}{}^l$ are linearly independent by assumption, the $r_1 \times r_1$ matrix $\mathbf{B}_{r_1}{}^l(\mathbf{B}_{r_1}{}^l)^*$ is nonsingular. Hence, from (6-53), we have

$$\mathbf{C}_{r_1}{}^1 = \mathbf{M}_1(\mathbf{B}_{r_1}{}^l)^*[\mathbf{B}_{r_1}{}^l(\mathbf{B}_{r_1}{}^l)^*]^{-1} \tag{6-54}$$

It is easy to see from (6-53) that the matrix $\mathbf{C}_{r_1}{}^1$ has rank $r_1$; therefore all the column vectors $\mathbf{c}_{11}, \mathbf{c}_{12}, \cdots, \mathbf{c}_{1r_1}$ are linearly independent. Similarly, we can express $\mathbf{M}_2$, $\mathbf{M}_3$ in terms of $\mathbf{b}_{l1}, \mathbf{b}_{l2}, \cdots, \mathbf{b}_{lr_3}$ as

$$\mathbf{M}_2 = [\mathbf{c}_{21} \quad \mathbf{c}_{22} \quad \cdots \quad \mathbf{c}_{2r_1} \quad \mathbf{c}_{1(r_1+1)} \quad \cdots \quad \mathbf{c}_{1r_2}]\begin{bmatrix} \mathbf{b}_{l1} \\ \mathbf{b}_{l2} \\ \cdot \\ \cdot \\ \cdot \\ \mathbf{b}_{lr_1} \\ \mathbf{b}_{l(r_1+1)} \\ \cdot \\ \cdot \\ \cdot \\ \mathbf{b}_{lr_2} \end{bmatrix} \tag{6-55}$$

$$\mathbf{M}_3 = [\mathbf{c}_{31} \quad \cdots \quad \mathbf{c}_{3r_1} \quad \mathbf{c}_{2(r_1+1)} \quad \cdots \quad \mathbf{c}_{2r_2} \quad \mathbf{c}_{1(r_2+1)} \quad \cdots \quad \mathbf{c}_{1r_3}]\begin{bmatrix} \mathbf{b}_{l1} \\ \cdot \\ \cdot \\ \cdot \\ \mathbf{b}_{lr_2} \\ \mathbf{b}_{l(r_2+1)} \\ \cdot \\ \cdot \\ \cdot \\ \mathbf{b}_{lr_3} \end{bmatrix} \tag{6-56}$$

Note that although the set $\{\mathbf{c}_{11}, \mathbf{c}_{12}, \cdots, \mathbf{c}_{1r_1}\}$ is linearly independent, the set of column vectors $\mathbf{c}$ in (6-55) and the set of column vectors $\mathbf{c}$ in (6-56) are generally not linearly independent. With these preliminaries, we are ready to draw a block diagram and then to write a dynamical equation for $\hat{\mathbf{G}}(s)$. Consider first the part

$$\frac{1}{(s-\lambda)^3}\mathbf{M}_1 = \frac{1}{(s-\lambda)^3}[\mathbf{c}_{11}\mathbf{b}_{l1} + \mathbf{c}_{12}\mathbf{b}_{l2} + \cdots + \mathbf{c}_{1r_1}\mathbf{b}_{lr_1}] \tag{6-57}$$

It is obvious that the block diagram in Figure 6-8 has transfer-function matrix

$$\frac{1}{(s-\lambda)^3}\mathbf{M}_1$$

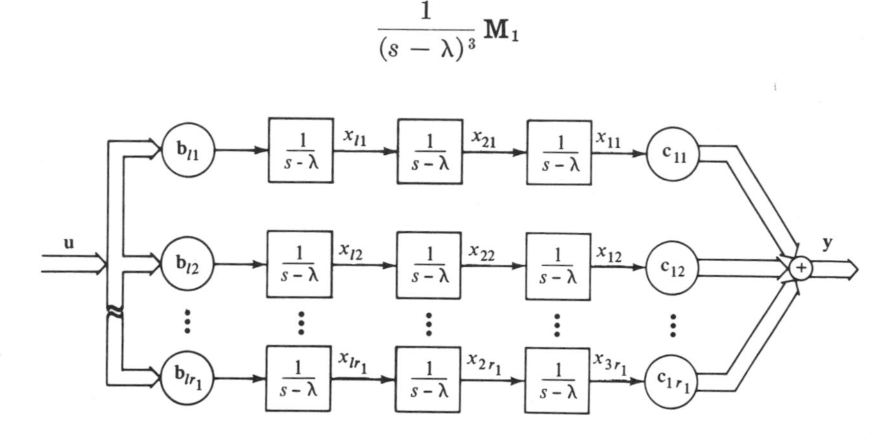


**Figure 6-8** The block diagram of $\frac{1}{(s-\lambda)^3}\mathbf{M}_1$.

In Figure 6-8, there are $r_1$ chains of blocks; each chain consists of three basic blocks with transfer function $1/(s-\lambda)$. The block diagram of

$$\frac{1}{(s-\lambda)^3}\mathbf{M}_1 + \frac{1}{(s-\lambda)^2}\mathbf{M}_2$$

is shown in Figure 6-9. We see that the first $r_1$ chains in Figure 6-9 are the same as those of Figure 6-8 except that additional outputs are added to generate

$$\frac{1}{(s-\lambda)^2}[\mathbf{c}_{21}\mathbf{b}_{l1} + \mathbf{c}_{22}\mathbf{b}_{l2} + \cdots + \mathbf{c}_{2r_1}\mathbf{b}_{lr_1}]$$

The remainder of

$$\frac{1}{(s-\lambda)^2}\mathbf{M}_2$$

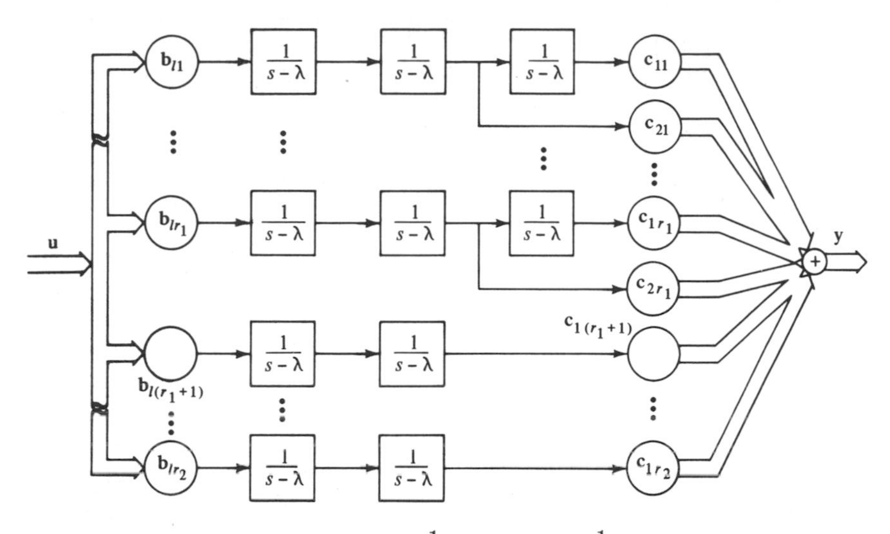


**Figure 6-9** The block diagram of $\dfrac{1}{(s-\lambda)^3}\mathbf{M}_1 + \dfrac{1}{(s-\lambda)^2}\mathbf{M}_2$.

is obtained by introducing additional $(r_2 - r_1)$ chains of blocks, each chain consists of two basic blocks. By adding

$$\frac{1}{(s-\lambda)}\mathbf{M}_3$$

to the block diagram in Figure 6-9, we finally obtain in Figure 6-10 a complete block diagram of $\hat{\mathbf{G}}(s)$. If a dynamical equation is written up by using the state variables chosen as shown, then each chain will constitute a Jordan block in the dynamical equation. For example, the dynamical equation of the first two chains is

$$\begin{bmatrix} \dot{x}_{11} \\ \dot{x}_{21} \\ \dot{x}_{l1} \\ \dot{x}_{12} \\ \dot{x}_{22} \\ \dot{x}_{l2} \end{bmatrix} = \left[\begin{array}{ccc|ccc} \lambda & 1 & 0 & & & \\ 0 & \lambda & 1 & & & \\ 0 & 0 & \lambda & & & \\ \hline & & & \lambda & 1 & 0 \\ & & & 0 & \lambda & 1 \\ & & & 0 & 0 & \lambda \end{array}\right]\mathbf{x} + \begin{bmatrix} \mathbf{0} \\ \mathbf{0} \\ \mathbf{b}_{l1} \\ \mathbf{0} \\ \mathbf{0} \\ \mathbf{b}_{l2} \end{bmatrix}\mathbf{u}$$

$$\mathbf{y} = [\mathbf{c}_{11} \quad \mathbf{c}_{21} \quad \mathbf{c}_{31} \qquad \mathbf{c}_{12} \quad \mathbf{c}_{22} \quad \mathbf{c}_{32}]\mathbf{x}$$

If a complete dynamical equation is written up for the block diagram in Figure 6-10 (Problem 6-14), then there are $r_3$ Jordan blocks in the equation. The first $r_1$ Jordan blocks are of dimension 3, the multiplicity of the pole associated with $\mathbf{M}_1$; the second $(r_2 - r_1)$ blocks are of dimension 2, and the

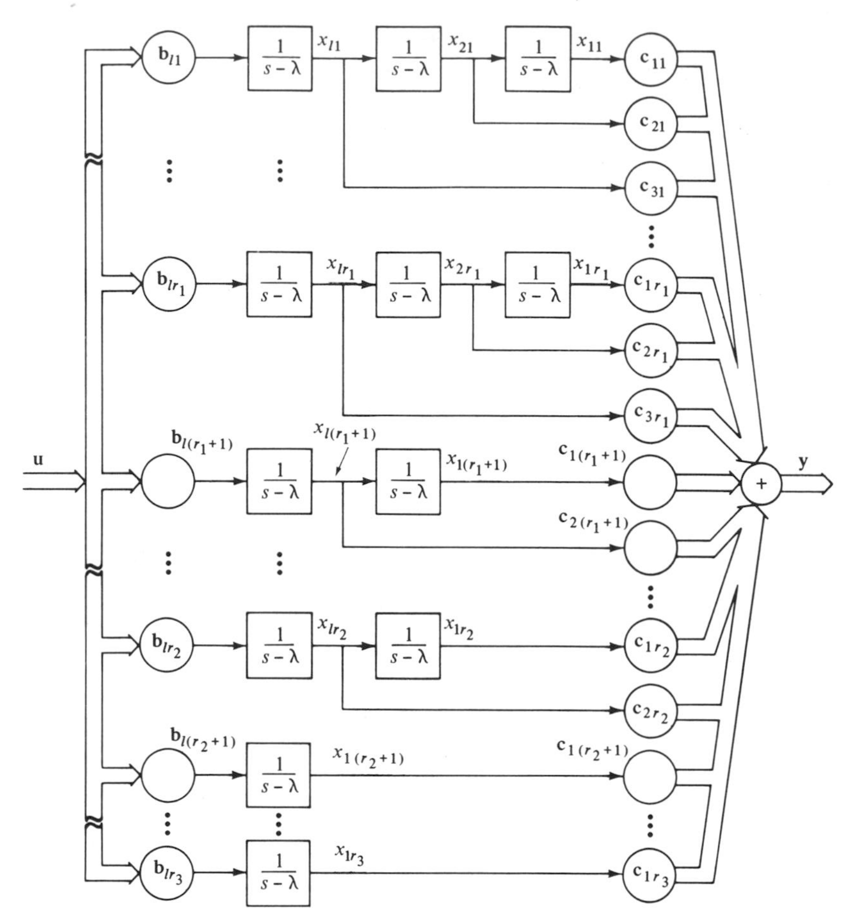


**Figure 6-10** The block diagram of $\hat{\mathbf{G}}(s)$.

last $(r_3 - r_2)$ blocks are of dimension 1. In this realization, the vectors $\mathbf{b}_{l1}, \mathbf{b}_{l2}, \cdots, \mathbf{b}_{lr_3}$ are chosen to be linearly independent; therefore, the dynamical equation is controllable (Theorem 5-16). The condition for the equation to be observable is that the set of vectors $\mathbf{c}_{11}, \mathbf{c}_{12}, \cdots, \mathbf{c}_{1r_1}, \mathbf{c}_{1(r_1+1)}, \cdots, c_{1r_3}$ is a linearly independent set. In this realization, this condition is not insured; therefore, the dynamical equation is not necessarily observable.

We give an example to illustrate the realization procedure.

## Example 1

Consider the rational-function matrix

$$\hat{\mathbf{G}}(s) = \frac{1}{s^4}\begin{bmatrix} s^3 - s^2 + 1 & 1 & -s^3 + s^2 - 2 \\ 1.5s + 1 & s + 1 & -1.5s - 2 \\ s^3 - 9s^2 - s + 1 & -s^2 + 1 & s^3 - s - 2 \end{bmatrix} \tag{6-58}$$

which has only one pole at $s = 0$. We first expand $\hat{\mathbf{G}}(s)$ into

$$\hat{\mathbf{G}}(s) = \frac{1}{s^4}\begin{bmatrix} 1 & 1 & -2 \\ 1 & 1 & -2 \\ 1 & 1 & -2 \end{bmatrix} + \frac{1}{s^3}\begin{bmatrix} 0 & 0 & 0 \\ 1.5 & 1 & -1.5 \\ -1 & 0 & -1 \end{bmatrix} + \frac{1}{s^2}\begin{bmatrix} -1 & 0 & 1 \\ 0 & 0 & 0 \\ -9 & -1 & 0 \end{bmatrix} + \frac{1}{s}\begin{bmatrix} 1 & 0 & -1 \\ 0 & 0 & 0 \\ 1 & 0 & 1 \end{bmatrix}$$

Next we write the coefficient matrices in terms of linearly independent row vectors as follows:

$$\mathbf{M}_1 \triangleq \begin{bmatrix} 1 & 1 & -2 \\ 1 & 1 & -2 \\ 1 & 1 & -2 \end{bmatrix} = \begin{bmatrix} 1 \\ 1 \\ 1 \end{bmatrix} [1 \quad 1 \quad -2]$$

$$\mathbf{M}_2 \triangleq \begin{bmatrix} 0 & 0 & 0 \\ 1.5 & 1 & -1.5 \\ -1 & 0 & -1 \end{bmatrix} = \left[\begin{array}{c:c} 0 & 0 \\ 1 & -0.5 \\ 0 & 1 \end{array}\right] \left[\begin{array}{ccc} 1 & 1 & -2 \\ \hdashline -1 & 0 & -1 \end{array}\right]$$

$$\mathbf{M}_3 \triangleq \begin{bmatrix} -1 & 0 & 1 \\ 0 & 0 & 0 \\ -9 & -1 & 0 \end{bmatrix} = \left[\begin{array}{c:c:c} 0 & 0 & 1 \\ 0 & 0 & 0 \\ -1 & 5 & 3 \end{array}\right] \left[\begin{array}{ccc} 1 & 1 & -2 \\ \hdashline -1 & 0 & -1 \\ \hdashline -1 & 0 & 1 \end{array}\right]$$

$$\mathbf{M}_4 \triangleq \begin{bmatrix} 1 & 0 & -1 \\ 0 & 0 & 0 \\ 1 & 0 & 1 \end{bmatrix} = \left[\begin{array}{c:c:c} 0 & 0 & -1 \\ 0 & 0 & 0 \\ 0 & -1 & 0 \end{array}\right] \left[\begin{array}{ccc} 1 & 1 & -2 \\ \hdashline -1 & 0 & 1 \\ \hdashline -1 & 0 & 1 \end{array}\right]$$

Since $r_1 = 1$, $r_2 = 2$, $r_3 = r_4 = 3$, there are three Jordan blocks. The number of Jordan blocks of order 4 is $r_1 = 1$; of order 3 is $r_2 - r_1 = 1$; of order 2 is $r_3 - r_2 = 1$; of order 1 is $r_4 - r_3 = 0$. The dynamical equation realization is

$$\dot{\mathbf{x}} = \left[\begin{array}{cccc:ccc:cc} 0 & 1 & 0 & 0 & & & & & \\ & 0 & 1 & 0 & & & & & \\ & & 0 & 1 & & & & & \\ & & & 0 & & & & & \\ \hdashline & & & & 0 & 1 & & & \\ & & & & & 0 & 1 & & \\ & & & & & & 0 & & \\ \hdashline & & & & & & & 0 & 1 \\ & & & & & & & & 0 \end{array}\right]\mathbf{x} + \left[\begin{array}{ccc} 0 & 0 & 0 \\ 0 & 0 & 0 \\ 0 & 0 & 0 \\ 1 & 1 & -2 \\ \hdashline 0 & 0 & 0 \\ 0 & 0 & 0 \\ -1 & 0 & -1 \\ \hdashline 0 & 0 & 0 \\ -1 & 0 & 1 \end{array}\right]\mathbf{u} \tag{6-59a}$$

$$\mathbf{y} = \left[\begin{array}{cccc:ccc:cc} 1 & 0 & 0 & 0 & 0 & 0 & 0 & 1 & -1 \\ 1 & 1 & 0 & 0 & -0.5 & 0 & 0 & 0 & 0 \\ 1 & 0 & -1 & 0 & 1 & 5 & -1 & 3 & 0 \end{array}\right]\mathbf{x} \tag{6-59b}$$

This dynamical equation is controllable; however, since the vectors

$$\mathbf{c}_{11} = \begin{bmatrix} 1 \\ 1 \\ 1 \end{bmatrix} \qquad \mathbf{c}_{12} = \begin{bmatrix} 0 \\ -0.5 \\ 1 \end{bmatrix} \qquad \mathbf{c}_{13} = \begin{bmatrix} 1 \\ 0 \\ 3 \end{bmatrix}$$

are linearly dependent, the dynamical equation (6-59) is not observable. Consequently, the dynamical is reducible. ∎

Although this realization gives a controllable but generally not observable dynamical equation, in the case where the pole is simple (no multiple pole), and in the case where $r_1 = r_2 = r_3$ in (6-52), the realization will be irreducible. In both cases, the dynamical equation consists of $r_1$ Jordan blocks, and since the vectors $\mathbf{c}_{11}, \mathbf{c}_{12}, \cdots, \mathbf{c}_{1r_1}$ are linearly independent, the dynamical equation is controllable and observable.

**Reduction of Jordan-form dynamical equations.** If the realization just introduced is not observable, the dynamical equation has to be reduced. The reduction procedure introduced in Sections 5-6 and 6-4 can be applied here. However, in the present case, we shall take full advantage of the Jordan form and introduce a new reduction procedure. The procedure is based on the following three rules. Let $\hat{\mathbf{A}}_r$ be a Jordan block of order $r$ associated with $\lambda$ and let $\mathbf{F}_r \triangleq (s\mathbf{I} - \hat{\mathbf{A}}_r)^{-1}$. Note that $\mathbf{F}_r$ is a triangular matrix of the following form [see (2-74)]:

$$\mathbf{F}_r = \begin{bmatrix} (s-\lambda)^{-1} & (s-\lambda)^{-2} & \cdots & (s-\lambda)^{-r} \\ 0 & (s-\lambda)^{-1} & \cdots & (s-\lambda)^{-r-1} \\ \cdot & \cdot & & \cdot \\ \cdot & \cdot & & \cdot \\ \cdot & \cdot & & \cdot \\ 0 & 0 & \cdots & (s-\lambda)^{-1} \end{bmatrix} \tag{6-60}$$

*Combination rule:* For any constants $\alpha_2$ and $\alpha_3$,

$$\begin{aligned} \mathbf{C}_1\mathbf{F}_r\mathbf{B}_1 + \mathbf{C}_2\mathbf{F}_r\mathbf{B}_2 + \mathbf{C}_3\mathbf{F}_r\mathbf{B}_3 &= (\mathbf{C}_1 + \alpha_2\mathbf{C}_2 + \alpha_3\mathbf{C}_3)\mathbf{F}_r\mathbf{B}_1 \\ &\quad + \mathbf{C}_2\mathbf{F}_r(\mathbf{B}_2 - \alpha_2\mathbf{B}_1) + \mathbf{C}_3\mathbf{F}_r(\mathbf{B}_3 - \alpha_3\mathbf{B}_1) \end{aligned} \tag{6-61}$$

This rule can be employed to make some column of $\mathbf{C}_1$ equal to zero.

*Shifting rule:*

$$[0 \quad \mathbf{c}_2 \quad \mathbf{c}_3 \quad \cdots \quad \mathbf{c}_r]\mathbf{F}_r \begin{bmatrix} \mathbf{b}_1 \\ \mathbf{b}_2 \\ \cdot \\ \cdot \\ \cdot \\ \mathbf{b}_{r-1} \\ \mathbf{b}_r \end{bmatrix} = [\mathbf{c}_2 \quad \mathbf{c}_3 \quad \cdots \quad \mathbf{c}_r \quad 0]\mathbf{F}_r \begin{bmatrix} \mathbf{b}_2 \\ \mathbf{b}_3 \\ \cdot \\ \cdot \\ \cdot \\ \mathbf{b}_r \\ 0 \end{bmatrix} \tag{6-62}$$

This rule interchanges unobservability and uncontrollability of a Jordan block.

*Reduction rule:*

$$[\mathbf{c}_1 \quad \mathbf{c}_2 \quad \cdots \quad \mathbf{c}_{r-1} \quad \mathbf{c}_r]\mathbf{F}_r \begin{bmatrix} \mathbf{b}_1 \\ \mathbf{b}_2 \\ \cdot \\ \cdot \\ \cdot \\ \mathbf{b}_{r-1} \\ \mathbf{0} \end{bmatrix} = [\mathbf{c}_1 \quad \mathbf{c}_2 \quad \cdots \quad \mathbf{c}_{r-1}]\mathbf{F}_{r-1} \begin{bmatrix} \mathbf{b}_1 \\ \mathbf{b}_2 \\ \cdot \\ \cdot \\ \cdot \\ \mathbf{b}_{r-1} \end{bmatrix} \tag{6-63}$$

If $\mathbf{b}_r = \mathbf{0}$, then the equation can be reduced. Whether or not $\mathbf{c}_r$ is zero is immaterial.

These three rules can be verified directly by equating both sides of (6-61)–(6-63). Note that (6-61) holds even if $\mathbf{F}_r$ is not in the form of (6-60). We shall now apply the rules to reduce a controllable equation to an irreducible one. We use an example to illustrate the procedure. Consider the dynamical equation

$$\dot{\mathbf{x}} = \begin{bmatrix} \lambda & 1 & 0 & & & & \\ 0 & \lambda & 1 & & & & \\ 0 & 0 & \lambda & & & & \\ & & & \lambda & 1 & & \\ & & & 0 & \lambda & & \\ & & & & & \lambda & 1 \\ & & & & & & \lambda \end{bmatrix} \mathbf{x} + \begin{bmatrix} \mathbf{b}_{11} \\ \mathbf{b}_{21} \\ \mathbf{b}_{31} \\ \mathbf{b}_{12} \\ \mathbf{b}_{22} \\ \mathbf{b}_{13} \\ \mathbf{b}_{23} \end{bmatrix} \mathbf{u} \tag{6-64a}$$

$$\mathbf{y} = [\mathbf{c}_{11} \quad \mathbf{c}_{21} \quad \mathbf{c}_{31} \vdots \mathbf{c}_{12} \quad \mathbf{c}_{22} \vdots \mathbf{c}_{13} \quad \mathbf{c}_{23}]\mathbf{x} \tag{6-64b}$$

It is assumed that the dynamical equation (6-64) is controllable but not observable. The transfer-function matrix of (6-64) is clearly equal to

$$\hat{\mathbf{G}}(s) = [\mathbf{c}_{11} \quad \mathbf{c}_{21} \quad \mathbf{c}_{31}]\mathbf{F}_3 \begin{bmatrix} \mathbf{b}_{11} \\ \mathbf{b}_{21} \\ \mathbf{b}_{31} \end{bmatrix} + [\mathbf{c}_{12} \quad \mathbf{c}_{22} \quad \mathbf{0}]\mathbf{F}_3 \begin{bmatrix} \mathbf{b}_{12} \\ \mathbf{b}_{22} \\ \mathbf{0} \end{bmatrix} + [\mathbf{c}_{13} \quad \mathbf{c}_{23} \quad \mathbf{0}]\mathbf{F}_3 \begin{bmatrix} \mathbf{b}_{13} \\ \mathbf{b}_{23} \\ \mathbf{0} \end{bmatrix} \tag{6-65}$$

Since (6-64) is assumed to be unobservable, the set of vectors $\{\mathbf{c}_{11}, \mathbf{c}_{12}, \mathbf{c}_{13}\}$ is linearly dependent. We may now apply the combination rule to make $\mathbf{c}_{12}$ or $\mathbf{c}_{13}$ (but not $\mathbf{c}_{11}$) equal to zero. Suppose $\mathbf{c}_{12} + \alpha_1\mathbf{c}_{11} + \alpha_2\mathbf{c}_{13} = 0$, then by applying the combination and shifting rules, we obtain

$$
\begin{aligned}
\hat{\mathbf{G}}(s) &= [\mathbf{c}_{11} \quad \mathbf{c}_{21} \quad \mathbf{c}_{31}]\mathbf{F}_3 \begin{bmatrix} \mathbf{b}_{11} - \alpha_1\mathbf{b}_{12} \\ \mathbf{b}_{22} - \alpha_1\mathbf{b}_{22} \\ \mathbf{b}_{31} \end{bmatrix} \\
&\quad + [\mathbf{0} \quad \mathbf{c}'_{22} \quad \mathbf{c}'_{23}]\mathbf{F}_3 \begin{bmatrix} \mathbf{b}_{12} \\ \mathbf{b}_{22} \\ \mathbf{0} \end{bmatrix} + [\mathbf{c}_{13} \quad \mathbf{c}_{23} \quad \mathbf{0}]\mathbf{F}_3 \begin{bmatrix} \mathbf{b}_{13} - \alpha_2\mathbf{b}_{12} \\ \mathbf{b}_{23} - \alpha_2\mathbf{b}_{22} \\ \mathbf{0} \end{bmatrix} \\
&= [\mathbf{c}_{11} \quad \mathbf{c}_{21} \quad \mathbf{c}_{31}]\mathbf{F}_3 \begin{bmatrix} \mathbf{b}_{11} - \alpha_1\mathbf{b}_{12} \\ \mathbf{b}_{22} - \alpha_1\mathbf{b}_{22} \\ \mathbf{b}_{31} \end{bmatrix} \\
&\quad + [\mathbf{c}'_{22} \quad \mathbf{c}'_{23} \quad \mathbf{0}]\mathbf{F}_3 \begin{bmatrix} \mathbf{b}_{22} \\ \mathbf{0} \\ \mathbf{0} \end{bmatrix} + [\mathbf{c}_{13} \quad \mathbf{c}_{23} \quad \mathbf{0}]\mathbf{F}_3 \begin{bmatrix} \mathbf{b}_{13} - \alpha_2\mathbf{b}_{12} \\ \mathbf{b}_{23} - \alpha_2\mathbf{b}_{22} \\ \mathbf{0} \end{bmatrix}
\end{aligned} \tag{6-66}
$$

where $\mathbf{c}'_{22} = \mathbf{c}_{22} + \alpha_1\mathbf{c}_{21} + \alpha_2\mathbf{c}_{23}$ and $\mathbf{c}'_{23} = \alpha_1\mathbf{c}_{31}$. The key idea in this process is to make as many zero row vectors as possible in the lower part of a matrix $\mathbf{B}$ (see the reduction rule). Note that if we make $\mathbf{c}_{11}$ equal to zero in (6-66), although the last row of $\mathbf{B}_1$ become zeros after shifting, all the rows of $\mathbf{B}_2$ and $\mathbf{B}_3$ may become nonzero; hence it is of no use to make $\mathbf{c}_{11}$ zero. Now if $\mathbf{c}_{11}$, $\mathbf{c}'_{22}$ and $\mathbf{c}_{13}$ are linearly independent, we apply the reduction rule to (6-66) and obtain the following observable dynamical equation.

$$
\dot{\bar{\mathbf{x}}} = \left[\begin{array}{ccc|c|cc} \lambda & 1 & 0 & & & \\ 0 & \lambda & 1 & & & \\ 0 & 0 & \lambda & & & \\ \hline & & & \lambda & & \\ \hline & & & & \lambda & 1 \\ & & & & 0 & \lambda \end{array}\right] \bar{\mathbf{x}} + \left[\begin{array}{c} \mathbf{b}_{11} - \alpha_1\mathbf{b}_{12} \\ \mathbf{b}_{22} - \alpha_1\mathbf{b}_{22} \\ \mathbf{b}_{31} \\ \hline \mathbf{b}_{22} \\ \hline \mathbf{b}_{13} - \alpha_2\mathbf{b}_{12} \\ \mathbf{b}_{23} - \alpha_2\mathbf{b}_{22} \end{array}\right] u \tag{6-67a}
$$

$$
\mathbf{y} = [\mathbf{c}_{11} \quad \mathbf{c}_{21} \quad \mathbf{c}_{31} \quad \mathbf{c}'_{22} \quad \mathbf{c}_{13} \quad \mathbf{c}_{23}]\bar{\mathbf{x}} \tag{6-67b}
$$

The controllability of (6-64) (linear independence of $\{\mathbf{b}_{31}, \mathbf{b}_{22}, \mathbf{b}_{23}\}$) implies the controllability of (6-67) (linear independence of $\{\mathbf{b}_{31}, \mathbf{b}_{22}, \mathbf{b}_{23} - \alpha_2\mathbf{b}_{22}\}$). Hence (6-67) is irreducible. If the set $\{\mathbf{c}_{11}, \mathbf{c}'_{22}, \mathbf{c}_{13}\}$ is linearly dependent, we may try to make $\mathbf{c}'_{22}$ equal to zero; if this is not possible, try to make $\mathbf{c}_{13}$ equal to zero. In conclusion, we apply repeatedly the combination and shifting rules until all the first columns of $\mathbf{C}_i$ are linearly independent and then apply the reduction rule to obtain an irreducible dynamical equation.

### Example 2

Reduce the dynamical equation in Example 1 to an irreducible one. The transfer function of (6-59) is

$$\hat{\mathbf{G}}(s) = \begin{bmatrix} 1 & 0 & 0 & 0 \\ 1 & 1 & 0 & 0 \\ 1 & 0 & -1 & 0 \end{bmatrix} \mathbf{F}_4 \begin{bmatrix} 0 & 0 & 0 \\ 0 & 0 & 0 \\ 0 & 0 & 0 \\ 1 & 1 & -2 \end{bmatrix} + \begin{bmatrix} 0 & 0 & 0 & 0 \\ -0.5 & 0 & 0 & 0 \\ 1 & 5 & -1 & 0 \end{bmatrix} \mathbf{F}_4 \begin{bmatrix} 0 & 0 & 0 \\ 0 & 0 & 0 \\ -1 & 0 & -1 \\ 0 & 0 & 0 \end{bmatrix} + \begin{bmatrix} 1 & -1 & 0 & 0 \\ 0 & 0 & 0 & 0 \\ 3 & 0 & 0 & 0 \end{bmatrix} \mathbf{F}_4 \begin{bmatrix} 0 & 0 & 0 \\ -1 & 0 & 1 \\ 0 & 0 & 0 \\ 0 & 0 & 0 \end{bmatrix}$$

Since the vectors $\mathbf{c}_{11} = [1 \;\; 1 \;\; 1]'$, $\mathbf{c}_{12} = [0 \;\; -0.5 \;\; 1]'$ and $\mathbf{c}_{13} = [1 \;\; 0 \;\; 3]'$ are linearly dependent and since $\mathbf{c}_{13} - \mathbf{c}_{11} - 2\mathbf{c}_{12} = \mathbf{0}$, by applying the combination rule and the shifting rule we obtain

$$\hat{\mathbf{G}}(s) = \begin{bmatrix} 1 & 0 & 0 & 0 \\ 1 & 1 & 0 & 0 \\ 1 & 0 & -1 & 0 \end{bmatrix} \mathbf{F}_4 \begin{bmatrix} 0 & 0 & 0 \\ -1 & 0 & 1 \\ 0 & 0 & 0 \\ 1 & 1 & -2 \end{bmatrix} + \begin{bmatrix} 0 & 0 & 0 & 0 \\ -0.5 & 0 & 0 & 0 \\ 1 & 5 & -1 & 0 \end{bmatrix} \mathbf{F}_4 \begin{bmatrix} 0 & 0 & 0 \\ -2 & 0 & 2 \\ -1 & 0 & -1 \\ 0 & 0 & 0 \end{bmatrix} + \begin{bmatrix} 0 & -1 & 0 & 0 \\ 0 & -1 & 0 & 0 \\ 0 & -10 & 1 & 0 \end{bmatrix} \mathbf{F}_4 \begin{bmatrix} 0 & 0 & 0 \\ -1 & 0 & 1 \\ 0 & 0 & 0 \\ 0 & 0 & 0 \end{bmatrix}$$

$$= \begin{bmatrix} 1 & 0 & 0 & 0 \\ 1 & 1 & 0 & 0 \\ 1 & 0 & -1 & 0 \end{bmatrix} \mathbf{F}_4 \begin{bmatrix} 0 & 0 & 0 \\ -1 & 0 & 1 \\ 0 & 0 & 0 \\ 1 & 1 & -2 \end{bmatrix} + \begin{bmatrix} 0 & 0 & 0 & 0 \\ -0.5 & 0 & 0 & 0 \\ 1 & 5 & -1 & 0 \end{bmatrix} \mathbf{F}_4 \begin{bmatrix} 0 & 0 & 0 \\ -2 & 0 & 2 \\ -1 & 0 & -1 \\ 0 & 0 & 0 \end{bmatrix} + \begin{bmatrix} -1 & 0 & 0 & 0 \\ -1 & 0 & 0 & 0 \\ -10 & 1 & 0 & 0 \end{bmatrix} \mathbf{F}_4 \begin{bmatrix} -1 & 0 & 1 \\ 0 & 0 & 0 \\ 0 & 0 & 0 \\ 0 & 0 & 0 \end{bmatrix}$$

Now the vector $[1 \quad 1 \quad 1]'$, $[0 \quad -0.5 \quad 1]'$ and $[-1 \quad -1 \quad -10]'$ are linearly independent; hence we may apply the reduction rule and obtain

$$\hat{\mathbf{G}}(s) = \begin{bmatrix} 1 & 0 & 0 & 0 \\ 1 & 1 & 0 & 0 \\ 1 & 0 & -1 & 0 \end{bmatrix} \mathbf{F}_4 \begin{bmatrix} 0 & 0 & 0 \\ -1 & 0 & 1 \\ 0 & 0 & 0 \\ 1 & 1 & -2 \end{bmatrix}$$
$$+ \begin{bmatrix} 0 & 0 & 0 \\ -0.5 & 0 & 0 \\ 1 & 5 & -1 \end{bmatrix} \mathbf{F}_3 \begin{bmatrix} 0 & 0 & 0 \\ -2 & 0 & 2 \\ -1 & 0 & -1 \end{bmatrix} + \begin{bmatrix} -1 \\ -1 \\ -10 \end{bmatrix} \mathbf{F}_1 [-1 \quad 0 \quad 1]$$

Hence an irreducible realization is

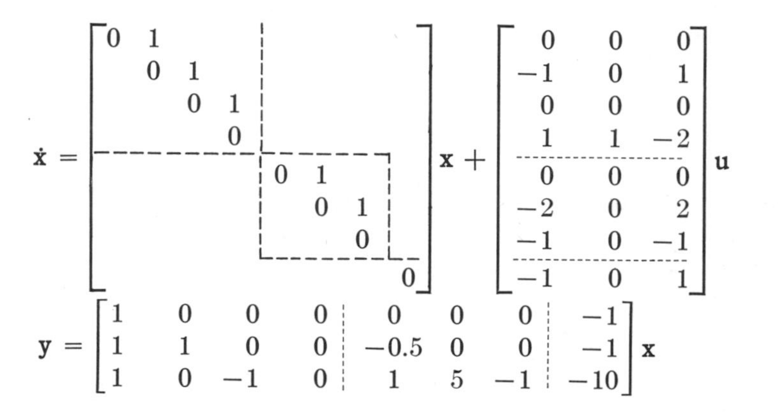

$$\dot{\mathbf{x}} = \begin{bmatrix} 0 & 1 & & & & & & \\ & 0 & 1 & & & & & \\ & & 0 & 1 & & & & \\ & & & 0 & & & & \\ & & & & 0 & 1 & & \\ & & & & & 0 & 1 & \\ & & & & & & 0 & \\ & & & & & & & 0 \end{bmatrix} \mathbf{x} + \begin{bmatrix} 0 & 0 & 0 \\ -1 & 0 & 1 \\ 0 & 0 & 0 \\ 1 & 1 & -2 \\ 0 & 0 & 0 \\ -2 & 0 & 2 \\ -1 & 0 & -1 \\ -1 & 0 & 1 \end{bmatrix} \mathbf{u}$$
$$\mathbf{y} = \begin{bmatrix} 1 & 0 & 0 & 0 & 0 & 0 & 0 & -1 \\ 1 & 1 & 0 & 0 & -0.5 & 0 & 0 & -1 \\ 1 & 0 & -1 & 0 & 1 & 5 & -1 & -10 \end{bmatrix} \mathbf{x}$$

∎

We are ready to give an irreducible realization of a general proper rational-function matrix. Consider the $q \times p$ proper rational-function matrix $\hat{\mathbf{G}}(s)$ with $m$ distinct poles $\lambda_1, \lambda_2, \cdots, \lambda_m$. By partial fraction expansion, $\hat{\mathbf{G}}(s)$ can be written as

$$\hat{\mathbf{G}}(s) = \hat{\mathbf{G}}_1(s) + \hat{\mathbf{G}}_2(s) + \cdots + \hat{\mathbf{G}}_m(s) + \hat{\mathbf{G}}(\infty) \tag{6-68}$$

where $\hat{\mathbf{G}}_i(s)$ is associated with only one pole $\lambda_i$, and $\hat{\mathbf{G}}(\infty)$ is a constant matrix. Define $\mathbf{D} \triangleq \hat{\mathbf{G}}(\infty)$ and let

$$\dot{\mathbf{x}}^i = \mathbf{A}_i \mathbf{x}^i + \mathbf{B}_i \mathbf{u}$$
$$\mathbf{y}^i = \mathbf{C}_i \mathbf{x}^i$$

be an irreducible Jordan-form realization of $\hat{\mathbf{G}}_i(s)$; then a Jordan-form realization of $\hat{\mathbf{G}}(s)$ is

$$\begin{bmatrix} \dot{\mathbf{x}}^1 \\ \dot{\mathbf{x}}^2 \\ \cdot \\ \cdot \\ \cdot \\ \dot{\mathbf{x}}^m \end{bmatrix} = \begin{bmatrix} \mathbf{A}_1 & \mathbf{0} & \cdots & \mathbf{0} \\ \mathbf{0} & \mathbf{A}_2 & \cdots & \mathbf{0} \\ \cdot & \cdot & & \cdot \\ \cdot & \cdot & & \cdot \\ \cdot & \cdot & & \cdot \\ \mathbf{0} & \mathbf{0} & \cdots & \mathbf{A}_m \end{bmatrix} \mathbf{x} + \begin{bmatrix} \mathbf{B}_1 \\ \mathbf{B}_2 \\ \cdot \\ \cdot \\ \cdot \\ \mathbf{B}_m \end{bmatrix} \mathbf{u} \tag{6-69a}$$

$$\mathbf{y} = [\mathbf{C}_1 \quad \mathbf{C}_2 \quad \cdots \quad \mathbf{C}_m]\mathbf{x} + \mathbf{D}\mathbf{u} \tag{6-69b}$$

The transfer-function matrix of (6-69) can be easily checked to be $\hat{\mathbf{G}}(s) = \hat{\mathbf{G}}_1(s) + \hat{\mathbf{G}}_2(s) + \cdots + \hat{\mathbf{G}}_m(s) + \hat{\mathbf{G}}(\infty)$. Since the $\mathbf{A}_i$'s are associated with different eigenvalues, from Theorem 5-16 we conclude that the equation is controllable and observable. Hence the realization is irreducible.

### Example 3

Consider the rational function

$$\hat{\mathbf{G}}(s) = \begin{bmatrix} \dfrac{1}{(s+1)^2} - \dfrac{1}{s+1} & \dfrac{1}{s+1} \\ \dfrac{1}{s} + \dfrac{1}{s+1} & \dfrac{1}{s+1} \end{bmatrix}$$

$$= \frac{1}{(s+1)^2}\begin{bmatrix} 1 & 0 \\ 0 & 0 \end{bmatrix} + \frac{1}{s+1}\begin{bmatrix} -1 & 1 \\ 1 & 1 \end{bmatrix} + \frac{1}{s}\begin{bmatrix} 0 & 0 \\ 1 & 0 \end{bmatrix}$$

The matrix $\hat{\mathbf{G}}(s)$ has two poles $-1$ and $0$. Let $\hat{\mathbf{G}}_1(s)$ denote the part associated with the pole $-1$, and $\hat{\mathbf{G}}_2(s)$ the part associated with the pole $0$. Consider first $\hat{\mathbf{G}}_1(s)$. Since

$$\mathbf{M}_1 = \begin{bmatrix} 1 & 0 \\ 0 & 0 \end{bmatrix} = \begin{bmatrix} 1 \\ 0 \end{bmatrix}[1 \quad 0]$$

$$\mathbf{M}_2 = \begin{bmatrix} -1 & 1 \\ 1 & 1 \end{bmatrix} = \left[\begin{array}{c|c} 0 & 1 \\ 2 & 1 \end{array}\right]\left[\begin{array}{cc} 1 & 0 \\ \hline -1 & 1 \end{array}\right]$$

it is easy to show that a realization of $\hat{\mathbf{G}}_1(s)$ is

$$\begin{bmatrix} \dot{x}_1 \\ \dot{x}_2 \\ \dot{x}_3 \end{bmatrix} = \left[\begin{array}{cc|c} -1 & 1 & 0 \\ 0 & -1 & 0 \\ \hline 0 & 0 & -1 \end{array}\right]\mathbf{x} + \left[\begin{array}{cc} 0 & 0 \\ 1 & 0 \\ \hline -1 & 1 \end{array}\right]\mathbf{u}$$

$$\mathbf{y} = \left[\begin{array}{cc|c} 1 & 0 & 1 \\ 0 & 2 & 1 \end{array}\right]\mathbf{x}$$

which is controllable and observable. An irreducible realization of $\hat{\mathbf{G}}_2(s)$ is

$$\dot{x}_4 = [1 \quad 0]\mathbf{u}$$

$$y = \begin{bmatrix} 0 \\ 1 \end{bmatrix} x_4$$

Hence an irreducible realization of $\hat{\mathbf{G}}(s)$ is

$$\begin{bmatrix} \dot{x}_1 \\ \dot{x}_2 \\ \dot{x}_3 \\ \dot{x}_4 \end{bmatrix} = \left[\begin{array}{ccc|c} -1 & 1 & 0 & 0 \\ 0 & -1 & 0 & 0 \\ 0 & 0 & -1 & 0 \\ \hline 0 & 0 & 0 & 0 \end{array}\right]\mathbf{x} + \left[\begin{array}{cc} 0 & 0 \\ 1 & 0 \\ -1 & 1 \\ \hline 1 & 0 \end{array}\right]u$$

$$\mathbf{y} = \left[\begin{array}{ccc|c} 1 & 0 & 1 & 0 \\ 0 & 2 & 1 & 1 \end{array}\right]\mathbf{x}$$

A block diagram of $\hat{\mathbf{G}}(s)$ is given in Figure 6-11.

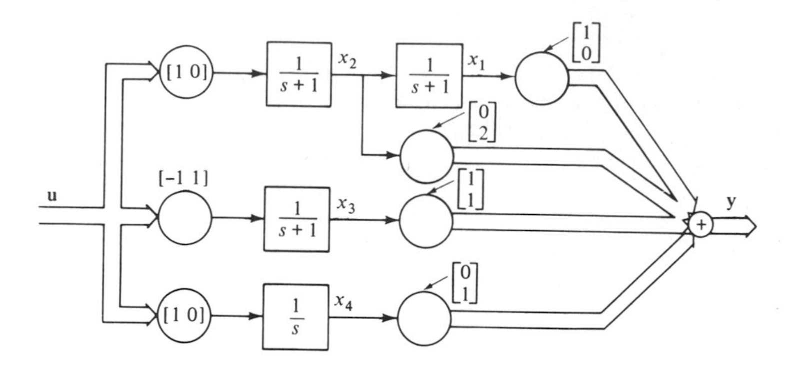


**Figure 6-11** The block diagram of Example 3.

## 6-6 Concluding Remarks

Given a transfer function, the controllable canonical-form and the observable canonical-form realizations can be read off directly from the coefficients of the transfer function. If the transfer function is irreducible, the realization will be controllable and observable; otherwise it can be either controllable or observable, but it cannot be both. The Jordan-form realization of a transfer function is comparatively more difficult to obtain. However, the realization has the advantage that its eigenvalues are less sensitive to the variations of parameters.

Two irreducible realization procedures for proper rational matrices were introduced. Theorem 6-4 can be easily programmed. A computer program called Automatic Synthesis Program (ASP) is given in Reference [67]. The algorithm is also written up by using Fortran IV, as described in Reference [74]. If a rational matrix is in a factor form, the realization procedure introduced in Section 6-5 can be applied.

In addition to the methods introduced in Sections 6-4 and 6-5, there are many other ways to realize a proper rational matrix. For example, in realizing the following matrix

$$\hat{G}(s) = \begin{bmatrix} \hat{G}_1(s) \\ \hat{G}_2(s) \end{bmatrix} = \begin{bmatrix} \hat{g}_{11}(s) & \hat{g}_{12}(s) & \hat{g}_{13}(s) \\ \hat{g}_{21}(s) & \hat{g}_{22}(s) & \hat{g}_{23}(s) \end{bmatrix} \tag{6-70}$$

we may first realize, $i = 1, 2$,

$$\hat{G}_i(s) = [\hat{g}_{i1}(s) \quad \hat{g}_{i2}(s) \quad \hat{g}_{i3}(s)] = \frac{1}{\Delta_i(s)} [N_{i1}(s) \quad N_{i2}(s) \quad N_{i3}(s)]$$

$$\overset{\text{say}}{=} \frac{1}{s^4 + 3s^3 + 2s^2 - s + 4} [s^3 + 2s + 1 \quad -s^2 + 3s + 1 \quad s - 2] \tag{6-71}$$

where $\Delta_i(s)$ is the least common denominator of $\hat{g}_{i1}$, $\hat{g}_{i2}$, and $\hat{g}_{i3}$ with leading coefficient 1, as

$$\dot{\mathbf{x}}^i = \mathbf{A}_i\mathbf{x}^i + \mathbf{B}_i\mathbf{u} = \begin{bmatrix} 0 & 0 & 0 & -4 \\ 1 & 0 & 0 & 1 \\ 0 & 1 & 0 & -2 \\ 0 & 0 & 1 & -3 \end{bmatrix}\mathbf{x}^i + \begin{bmatrix} 1 & 1 & -2 \\ 2 & 3 & 1 \\ 0 & -1 & 0 \\ 1 & 0 & 0 \end{bmatrix}\mathbf{u} \tag{6-72}$$

$$y_i = \mathbf{c}_i\mathbf{x}^i = [0 \quad 0 \quad 0 \quad 1]\mathbf{x}^i$$

This realization is an extension of the observable canonical-form realization of a rational function to the vector case. Note that the last column of $\mathbf{A}_i$ is taken from the coefficients of $\Delta_i$, the $j$th column of $\mathbf{B}_i$ is taken from the coefficients of $N_{ij}(s)$. Since $\Delta_i(s)$ in this case is the characteristic polynomial of $\hat{\mathbf{G}}_i(s)$, it follows from Theorem 6-2 that Equation (6-72) is an irreducible realization of $\hat{\mathbf{G}}_i(s)$. By using Equation (6-72), a realization (not necessarily irreducible) of $\hat{\mathbf{G}}(s)$ can be immediately written down as

$$\begin{bmatrix} \dot{\mathbf{x}}^1 \\ \dot{\mathbf{x}}^2 \end{bmatrix} = \begin{bmatrix} \mathbf{A}_1 & \mathbf{0} \\ \mathbf{0} & \mathbf{A}_2 \end{bmatrix}\begin{bmatrix} \mathbf{x}^1 \\ \mathbf{x}^2 \end{bmatrix} + \begin{bmatrix} \mathbf{B}_1 \\ \mathbf{B}_2 \end{bmatrix}\mathbf{u} \tag{6-73}$$

$$\mathbf{y} = \begin{bmatrix} y_1 \\ y_2 \end{bmatrix} = \begin{bmatrix} \mathbf{c}_1 & \mathbf{0} \\ \mathbf{0} & \mathbf{c}_2 \end{bmatrix}\begin{bmatrix} \mathbf{x}^1 \\ \mathbf{x}^2 \end{bmatrix}$$

Because of the form of $\mathbf{A}_i$ and $\mathbf{c}_i$, it is easy to check that the realization in Equation (6-73) is always observable. This realization is however generally not controllable. (A special case in which it will also be controllable is that $\Delta_1$ and $\Delta_2$ have no common roots.) If the realization is not controllable, by applying Theorem 5-17, an irreducible realization of $\hat{\mathbf{G}}(s)$ can be readily obtained. This realization procedure was discussed in W. A. Wolovich and P. L. Falb, "On the structure of multivariable systems," *SIAM J. Control*, vol. 7, pp. 437–451, 1969. For other irreducible realization procedures, the interested reader is referred to Y. L. Kuo, "On the irreducible Jordan form realization and the degree of a rational matrix," *IEEE Trans. on Circuit Theory*, vol. CT-17, August 1970; M. Heymann and J. A. Thorpe "Transfer equivalence of linear dynamical systems," *SIAM J. Control*, vol. 8, pp. 19–40, 1970 and References [62], [64], [98] and [115].

A remark is in order concerning the realization of a set of linear, time-invariant, high-order differential equations $\mathbf{D}(p)\mathbf{y}(t) = \mathbf{N}(p)\mathbf{u}(t)$, where $\mathbf{D}(p)$ and $\mathbf{N}(p)$ are $n \times n$ and $n \times p$ matrices whose entries are polynomials of $p \triangleq d/dt$. It is assumed that $\det \mathbf{D}(p) \not\equiv 0$. If the transfer-function matrix of the equation is first computed, an irreducible dynamical-equation realization can be found by using the procedure introduced in Sections 6-4 and 6-5. Note that this realization is only zero-state equivalent to the original differential equations. A dynamical-equation realization can also be obtained directly from $\mathbf{D}(p)\mathbf{y}(t) = \mathbf{N}(p)\mathbf{u}(t)$ without first computing the transfer-function matrix; the interested reader should study Reference [91], and C. T. Chen, "Irreducibility of dynamical equation realizations of sets of differential equations," *IEEE Trans. on Automatic Control*, vol. AC-15, 1970.

## PROBLEMS

**6-1** Find the degrees and the characteristic polynomials of the following proper rational matrices.

**a.** $$\begin{bmatrix} \dfrac{1}{(s+1)^2} & \dfrac{s+3}{s+2} & \dfrac{1}{s+5} \\ \dfrac{1}{(s+3)^2} & \dfrac{s+1}{s+4} & \dfrac{1}{s} \end{bmatrix}$$

**b.** $$\begin{bmatrix} \dfrac{1}{(s+1)^2} & \dfrac{1}{(s+1)(s+2)} \\ \dfrac{1}{(s+2)} & \dfrac{1}{(s+1)(s+2)} \end{bmatrix}$$

**c.** $$\begin{bmatrix} \dfrac{1}{s} & \dfrac{s+3}{s+1} \\ \dfrac{1}{s+3} & \dfrac{s}{s+1} \end{bmatrix}$$

**6-2** Find the dynamical-equation description of the block diagram with state variables chosen as shown in Figure P6-2.

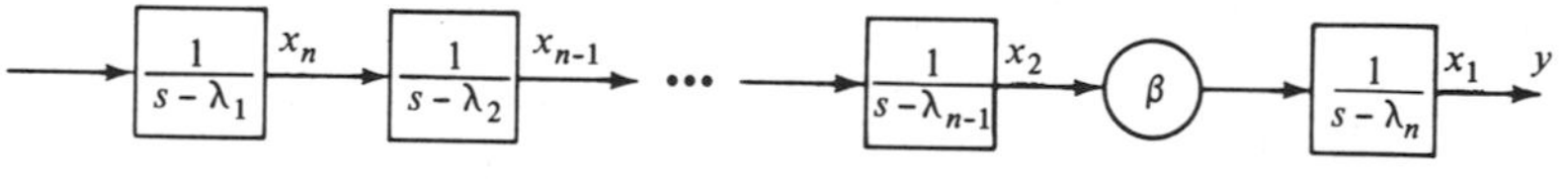


**Figure P6-2**

**6-3** Find the dynamical-equation realizations of the transfer functions

**a.** $$\frac{s^4+1}{4s^4+2s^3+2s+1}$$

**b.** $$\frac{s^2-s+1}{s^5-s^4+s^3-s^2+s-1}$$

Are these realizations irreducible? Find block diagrams for analog computer simulations of these transfer functions.

**6-4** Find Jordan-canonical-form dynamical-equation realizations of the transfer functions

**a.** $$\frac{s^2+1}{(s+1)(s+2)(s+3)}$$

**b.** $\dfrac{s^2 + 1}{(s + 2)^3}$

**c.** $\dfrac{s^2 + 1}{s^2 + 2s + 2}$

If the Jordan-form realizations consist of complex numbers, find their equivalent dynamical equations that do not contain any complex numbers.

**6-5** Write a dynamical equation for the feedback system shown in Figure P6-5. *Hint:* There are two ways to set up a dynamical equation

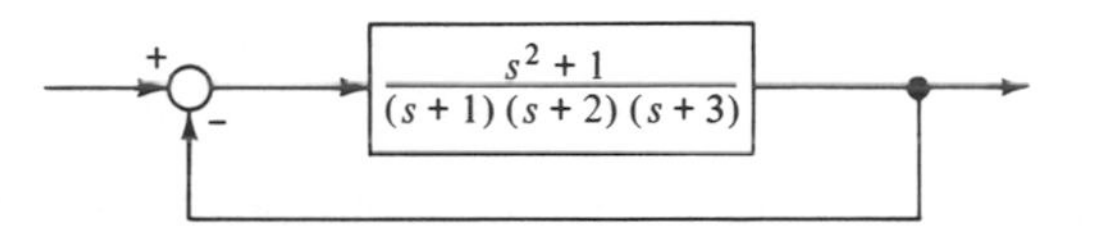


**Figure P6-5**

for the feedback system: (1) Find the closed loop transfer function and then proceed. (2) Write a dynamical equation for the open-loop transfer function and then make proper substitutions.

**6-6** Find the Jordan-form dynamical-equation description of the system shown in Figure P6-6.

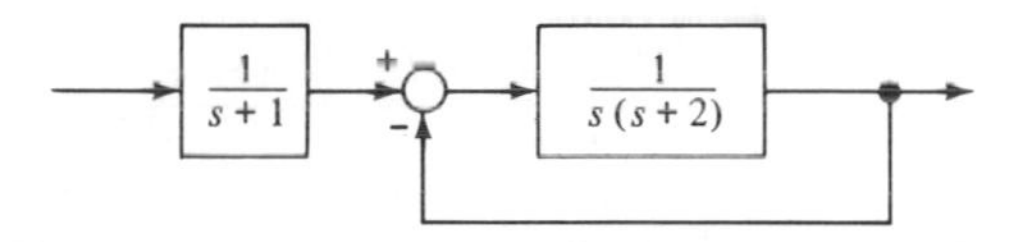


**Figure P6-6**

**6-7** Find the dynamical-equation descriptions of the nonlinear feedback systems shown in Figure P6-7.

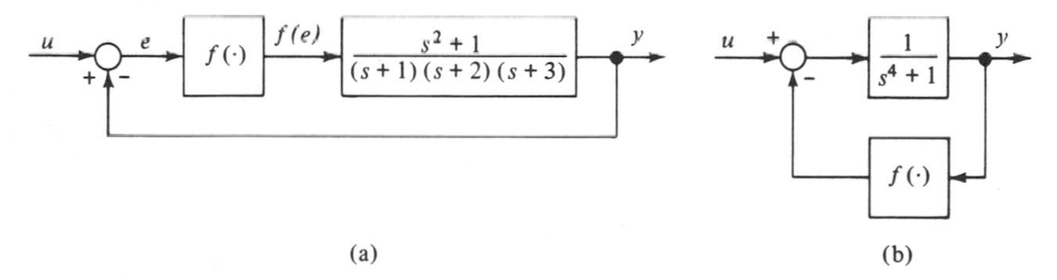


**Figure P6-7**

**6-8** Find an irreducible realization, an uncontrollable dynamical-equation realization, an unobservable dynamical-equation realization and an unobservable and uncontrollable dynamical-equation realization of $1/(s^3 + 1)$.

**6-9** Find the controllable canonical-form, the observable canonical-form, and the Jordan canonical-form dynamical-equation realizations of $1/s^4$.

**6-10** Set up a linear time-varying dynamical equation for the differential equation

$$(p^3 + \alpha_1(t)p^2 + \alpha_2(t)p + \alpha_3(t))y(t) = (\beta_0(t)p^2 + \beta_1(t)p + \beta_2(t))u(t)$$

where $p^i \triangleq d^i/dt^i$.

**6-11** Write an irreducible dynamical equation for the following simultaneous differential equation:

$$\begin{aligned} 2(p+1)y_1 + (p+1)y_2 &= pu_1 + u_2 \\ (p+1)y_1 + (p+1)y_2 &= (p-1)u_1 \end{aligned}$$

where $p \triangleq d/dt$. *Hint:* Find the transfer-function matrix from $\mathbf{u}$ to $\mathbf{y}$, and then proceed.

**6-12** Find an irreducible realization of the rational matrix

$$\begin{bmatrix} \dfrac{2+s}{s+1} & \dfrac{1}{s+3} \\ \dfrac{s}{s+1} & \dfrac{s+1}{s+2} \end{bmatrix}$$

**6-13** Find irreducible realizations of the rational matrix

$$\begin{bmatrix} \dfrac{s^2+1}{s^3} & \dfrac{2s+1}{s^2} \\ \dfrac{s+3}{s^2} & \dfrac{2}{s} \end{bmatrix}$$

by using the method introduced in Section 6-4 and the method introduced in Section 6-5. Which method do you think is simpler?

**6-14** Write a dynamical equation for the block diagram in Figure 6-10.

**6-15** Find the dynamical-equation description for the feedback system shown in Figure P6-15, where $\hat{\mathbf{G}}(s)$ is given in Problem 6-12 and

$$\mathbf{K} = \begin{bmatrix} k_{11} & k_{12} \\ k_{21} & k_{22} \end{bmatrix}$$

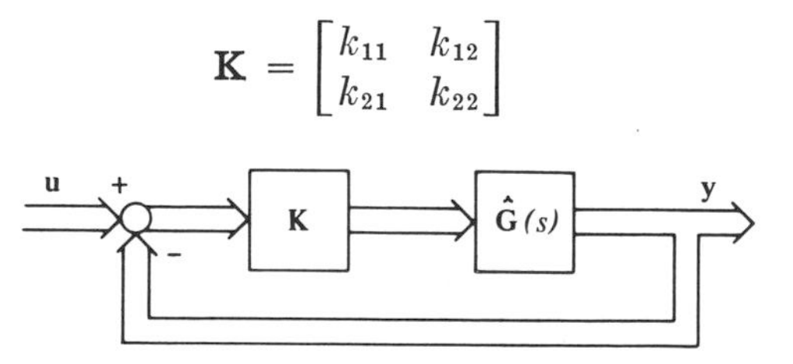


**Figure P6-15**

**6-16** Show that $N(s)$ and $D(s)$ have no common factors if and only if $N(s) + dD(s)$ and $D(s)$ have no common factors, where $N(s)$ and $D(s)$ are two polynomials and $d$ is a real constant.

**6-17** Find a linear, time-invariant, discrete-time dynamical equation whose sampled transfer function is

$$\frac{z^4 + 1}{4z^4 + 2z^3 + 2z + 1}$$

[See Problem 6-3(a).]

**6-18** Find an irreducible, discrete-time, dynamical-equation realization of the sampled transfer-function matrix

$$\begin{bmatrix} \dfrac{2+z}{z+1} & \dfrac{1}{z+3} \\ \dfrac{z}{z+1} & \dfrac{z+1}{z+2} \end{bmatrix}$$

**6-19** Consider

$$\dot{\mathbf{x}} = \begin{bmatrix} \lambda & 0 \\ 0 & \bar{\lambda} \end{bmatrix} \mathbf{x} + \begin{bmatrix} b_1 \\ \bar{b}_1 \end{bmatrix} u$$
$$y = [c_1 \quad \bar{c}_1]\mathbf{x}$$

where the overbar denotes the complex conjugate. Verify that by using the transformation $\mathbf{x} = \mathbf{Q}_1\tilde{\mathbf{x}}$, where

$$\mathbf{Q}_1 = \begin{bmatrix} -\bar{\lambda} b_1 & b_1 \\ -\lambda \bar{b}_1 & \bar{b}_1 \end{bmatrix}$$

the equation can be transformed into

$$\dot{\tilde{\mathbf{x}}} = \tilde{\mathbf{A}}\tilde{\mathbf{x}} + \tilde{\mathbf{b}}u$$
$$y = \tilde{\mathbf{c}}_1\tilde{\mathbf{x}}$$

where

$$\tilde{\mathbf{A}} = \begin{bmatrix} 0 & 1 \\ -\lambda\bar{\lambda} & \lambda + \bar{\lambda} \end{bmatrix} \qquad \tilde{\mathbf{b}} = \begin{bmatrix} 0 \\ 1 \end{bmatrix} \qquad \tilde{\mathbf{c}}_1 = [-2\,\mathrm{Re}\,(\bar{\lambda} b_1 c_1) \quad 2\,\mathrm{Re}\,(b_1 c_1)]$$

**6-20** Verify that the Jordan-form dynamical equation

$$\dot{\mathbf{x}} = \begin{bmatrix} \lambda & 1 & 0 & 0 & 0 & 0 \\ 0 & \lambda & 1 & 0 & 0 & 0 \\ 0 & 0 & \lambda & 0 & 0 & 0 \\ 0 & 0 & 0 & \bar{\lambda} & 1 & 0 \\ 0 & 0 & 0 & 0 & \bar{\lambda} & 1 \\ 0 & 0 & 0 & 0 & 0 & \bar{\lambda} \end{bmatrix} \mathbf{x} + \begin{bmatrix} b_1 \\ b_2 \\ b_3 \\ \bar{b}_1 \\ \bar{b}_2 \\ \bar{b}_3 \end{bmatrix} u$$
$$y = [c_1 \quad c_2 \quad c_3 \quad \bar{c}_1 \quad \bar{c}_2 \quad \bar{c}_3]\mathbf{x}$$

can be transformed into

$$\dot{\tilde{\mathbf{x}}} = \begin{bmatrix} \tilde{\mathbf{A}} & \mathbf{I}_2 & 0 \\ 0 & \tilde{\mathbf{A}} & \mathbf{I}_2 \\ 0 & 0 & \tilde{\mathbf{A}} \end{bmatrix} \tilde{\mathbf{x}} + \begin{bmatrix} \tilde{\mathbf{b}} \\ \tilde{\mathbf{b}} \\ \tilde{\mathbf{b}} \end{bmatrix} u$$

$$y = [\tilde{\mathbf{c}}_1 \quad \tilde{\mathbf{c}}_2 \quad \tilde{\mathbf{c}}_3]\tilde{\mathbf{x}}$$

where $\tilde{\mathbf{A}}$, $\tilde{\mathbf{b}}$, and $\tilde{\mathbf{c}}_i$ are defined in Problem 6-19 and $\mathbf{I}_2$ is a unit matrix of order 2. *Hint:* Change the order of state variables from $[x_1 \quad x_2 \quad x_3 \quad x_4 \quad x_5 \quad x_6]'$ to $[x_1 \quad x_4 \quad x_2 \quad x_5 \quad x_3 \quad x_6]'$ and then apply the equivalence transformation $\mathbf{x} = \mathbf{Q}\tilde{\mathbf{x}}$, where $\mathbf{Q} = \text{diag}\,(\mathbf{Q}_1 \quad \mathbf{Q}_2 \quad \mathbf{Q}_3)$.

**6-21** Find irreducible controllable or observable canonical-form realizations for the matrices

**a.** $$\begin{bmatrix} \dfrac{2s}{(s+1)(s+2)(s+3)} \\ \dfrac{s^2+2s+2}{s(s+1)^2(s+4)} \end{bmatrix}$$

**b.** $$\begin{bmatrix} \dfrac{2s+3}{(s+1)^2(s+2)} & \dfrac{s^2+2s+2}{s(s+1)^3} \end{bmatrix}$$

Is there any substantial difference between the process of realizing transfer functions and that of realizing vector transfer functions?

CHAPTER

# 7

# Canonical Forms, State Feedback, and State Estimators

## 7-1 Introduction

In engineering, design techniques are often developed from qualitative analyses of systems. For example, the design of feedback control systems by using Bode's plot was developed from the stability study of feedback systems. In Chapter 5 we introduced two qualitative properties of dynamical equations: controllability and observability. In this chapter we shall study their practical implications and develop some design techniques from them.

If an $n$-dimensional, linear, time-invariant dynamical equation is controllable, the controllability matrix $[\mathbf{B} \quad \mathbf{AB} \quad \cdots \quad \mathbf{A}^{n-1}\mathbf{B}]$ has $n$ linearly independent columns. By using these independent columns or their linear combinations as basis vectors of the state space, various canonical forms can be obtained. We introduce in Section 7-2 the most useful one: the controllable canonical form. We also introduce there the observable canonical-form dynamical equation. These two canonical forms are very useful in the designs of state feedback and state estimators.

The concept of feedback is basic in control theory. Roughly speaking, the control signal or input of a "feedback" system depends on the response of the system. If the response of the system does not differ very much from a desired one, the input signal need not be changed; otherwise, a proper change of the input signal is required to bring the response back to the desired one as fast as possible. Since the state of a system contains all the essential information concerning the system, it is conceivable that if the input of a feedback system is a function of the state, a reasonably good control can be achieved. In Section 7-3 we study the effects of introducing a linear state feedback of the form $\mathbf{u} = \mathbf{v} + \mathbf{Kx}$ on a dynamical equation. We show what can be achieved by the linear state feedback under the assumption of controllability. Stabilization of an uncontrollable dynamical equation is also discussed.

In state feedback, all the state variables are assumed to be available as outputs. This assumption generally does not hold in practice. Therefore if we want to introduce state feedback, the state vector has to be generated or estimated from the available information. In Section 7-4 we introduce asymptotic state estimators under the assumption of observability. The outputs of an asymptotic state estimator give an estimate of the state of the original equation. We also show that a state feedback and a state estimator can be independently designed.

In Section 7-5 we use the results of state feedback and state estimator to design compensators for single-variable feedback systems.

The inputs and outputs of a multivariable system are generally coupled; that is, every input controls more than one output, and every output is controlled by more than one input. If a compensator can be found such that every input controls one and only one output, then the multivariable system is said to be *decoupled*. We study in Section 7-6 the decoupling of a multivariable system by state feedback. The necessary and sufficient condition under which a system can be decoupled by linear state feedback is derived.

We study in this chapter only linear time-invariant dynamical equations. The material is based in part on References [1], [4], [7], [17], [18], [36], [41], [46], [51], [52], [68], [78]–[80], [84], [99], [111], and [112]. The results in this chapter can be extended to linear time-varying dynamical equations if they are instantaneously controllable (see Definition 5-3). The interested reader is referred to References [10], [98], [99], and [110].

## 7-2 Canonical-Form Dynamical Equations

**Single-variable case.** Consider the $n$-dimensional, linear, time-invariant, single-variable dynamical equation:

$$FE_1: \quad \dot{\mathbf{x}} = \mathbf{Ax} + \mathbf{b}u \qquad \textbf{(7-1a)}$$

$$y = \mathbf{cx} + du \qquad \textbf{(7-1b)}$$

where $\mathbf{A}$, $\mathbf{b}$, $\mathbf{c}$, and $d$ are respectively $n \times n$, $n \times 1$, $1 \times n$, and $1 \times 1$ real constant matrices. Note that the subscript 1 in $FE_1$ stands for single-variableness. In this section we shall derive various equivalent dynamical equations of $FE_1$ under the controllability or observability assumption.

Let the following dynamical equation $F\bar{E}_1$

$$F\bar{E}_1: \qquad \dot{\bar{\mathbf{x}}} = \bar{\mathbf{A}}\bar{\mathbf{x}} + \bar{\mathbf{b}}u \tag{7-2a}$$

$$y = \bar{\mathbf{c}}\bar{\mathbf{x}} + du \tag{7-2b}$$

be an equivalent dynamical equation of $FE_1$, which is obtained by introducing $\bar{\mathbf{x}} = \mathbf{P}\mathbf{x} = \mathbf{Q}^{-1}\mathbf{x}$, where $\mathbf{P} \triangleq \mathbf{Q}^{-1}$ is a nonsingular constant matrix. Then from (4-33) we have

$$\bar{\mathbf{A}} = \mathbf{PAP}^{-1} \qquad \bar{\mathbf{b}} = \mathbf{Pb} \qquad \bar{\mathbf{c}} = \mathbf{cP}^{-1}$$

The controllability matrices of $FE_1$ and $F\bar{E}_1$ are, respectively,

$$\mathbf{U} \triangleq [\mathbf{b} \quad \mathbf{Ab} \quad \cdots \quad \mathbf{A}^{n-1}\mathbf{b}] \tag{7-3}$$

$$\bar{\mathbf{U}} \triangleq [\bar{\mathbf{b}} \quad \bar{\mathbf{A}}\bar{\mathbf{b}} \quad \cdots \quad \bar{\mathbf{A}}^{n-1}\bar{\mathbf{b}}] = \mathbf{P}[\mathbf{b} \quad \mathbf{Ab} \quad \cdots \quad \mathbf{A}^{n-1}\mathbf{b}] = \mathbf{PU} = \mathbf{Q}^{-1}\mathbf{U} \tag{7-4}$$

Now if the dynamical equations $FE_1$ and, consequently, $F\bar{E}_1$ are controllable,[1] then the controllability matrices $\mathbf{U}$ and $\bar{\mathbf{U}}$ are nonsingular. Hence, from (7-4) we have

$$\mathbf{P} = \bar{\mathbf{U}}\mathbf{U}^{-1} \tag{7-5a}$$

or

$$\mathbf{Q} = \mathbf{U}\bar{\mathbf{U}}^{-1} \tag{7-5b}$$

Similarly, if $\mathbf{V}$ and $\bar{\mathbf{V}}$ are the observability matrices of $FE_1$ and $F\bar{E}_1$; that is,

$$\mathbf{V} = \begin{bmatrix} \mathbf{c} \\ \mathbf{cA} \\ \cdot \\ \cdot \\ \cdot \\ \mathbf{cA}^{n-1} \end{bmatrix} \qquad \bar{\mathbf{V}} = \begin{bmatrix} \bar{\mathbf{c}} \\ \bar{\mathbf{c}}\bar{\mathbf{A}} \\ \cdot \\ \cdot \\ \cdot \\ \bar{\mathbf{c}}\bar{\mathbf{A}}^{n-1} \end{bmatrix} = \mathbf{VP}^{-1} = \mathbf{VQ}$$

and if $FE_1$ and $F\bar{E}_1$ are observable, then

$$\mathbf{P} = \bar{\mathbf{V}}^{-1}\mathbf{V} \qquad \text{or} \qquad \mathbf{Q} = \mathbf{V}^{-1}\bar{\mathbf{V}} \tag{7-6}$$

Thus, if two same-dimensional, linear, time-invariant, single-variable dynamical equations are known to be equivalent and if they are either controllable or observable, then the equivalence transformation $\mathbf{P}$ between them can be computed by the use of either (7-5) or (7-6). This is also true for multivariable controllable or observable time-invariant dynamical

[1] See Theorem 5-15.

equations. For multivariable equivalent equations, the relation $\bar{\mathbf{U}} = \mathbf{PU}$ still holds. In this case $\mathbf{U}$ is not a square matrix; however, from Theorem 2-8, we have $\bar{\mathbf{U}}\mathbf{U}^* = \mathbf{PUU}^*$ and $\mathbf{P} = \bar{\mathbf{U}}\mathbf{U}^*(\mathbf{UU}^*)^{-1}$.

Let the characteristic polynomial of the matrix $\mathbf{A}$ in (7-1a) be

$$\Delta(\lambda) = \det(\lambda\mathbf{I} - \mathbf{A}) = \lambda^n + \alpha_1\lambda^{n-1} + \cdots + \alpha_{n-1}\lambda + \alpha_n$$

### Theorem 7-1

If the $n$-dimensional, linear, time-invariant, single-variable dynamical equation $FE_1$ is controllable, then it can be transformed, by an equivalence transformation, into the form

$CFE_1$:

$$\dot{\bar{\mathbf{x}}} = \begin{bmatrix} 0 & 1 & 0 & \cdots & 0 & 0 \\ 0 & 0 & 1 & \cdots & 0 & 0 \\ 0 & 0 & 0 & \cdots & 0 & 0 \\ \vdots & \vdots & \vdots & & \vdots & \vdots \\ 0 & 0 & 0 & \cdots & 0 & 1 \\ -\alpha_n & -\alpha_{n-1} & -\alpha_{n-2} & \cdots & -\alpha_2 & -\alpha_1 \end{bmatrix} \bar{\mathbf{x}} + \begin{bmatrix} 0 \\ 0 \\ 0 \\ \vdots \\ 0 \\ 1 \end{bmatrix} u \tag{7-7a}$$

$$y = [\beta_n \quad \beta_{n-1} \quad \beta_{n-2} \quad \cdots \quad \beta_2 \quad \beta_1]\bar{\mathbf{x}} + du \tag{7-7b}$$

where $\alpha_1, \alpha_2, \cdots, \alpha_n$ are the coefficients of the characteristic polynomial of $\mathbf{A}$, and the $\beta_i$'s are to be computed from $FE_1$. The dynamical equation (7-7) is said to be in the *controllable canonical form.* The transfer function of $FE_1$ is

$$\hat{g}(s) = \frac{\beta_1 s^{n-1} + \beta_2 s^{n-2} + \cdots + \beta_n}{s^n + \alpha_1 s^{n-1} + \cdots + \alpha_{n-1}s + \alpha_n} + d \tag{7-8}$$

### Proof

The dynamical equation $FE_1$ is controllable by assumption; hence the set of $n \times 1$ column vectors $\mathbf{b}, \mathbf{Ab}, \cdots, \mathbf{A}^{n-1}\mathbf{b}$ is linearly independent. Consequently the following set of $n \times 1$ vectors

$$\begin{aligned} \mathbf{q}^n &\triangleq \mathbf{b} \\ \mathbf{q}^{n-1} &\triangleq \mathbf{Aq}^n + \alpha_1\mathbf{q}^n = \mathbf{Ab} + \alpha_1\mathbf{b} \\ \mathbf{q}^{n-2} &\triangleq \mathbf{Aq}^{n-1} + \alpha_2\mathbf{q}^n = \mathbf{A}^2\mathbf{b} + \alpha_1\mathbf{Ab} + \alpha_2\mathbf{b} \\ &\vdots \\ \mathbf{q}^1 &\triangleq \mathbf{Aq}^2 + \alpha_{n-1}\mathbf{q}^n = \mathbf{A}^{n-1}\mathbf{b} + \alpha_1\mathbf{A}^{n-2}\mathbf{b} + \cdots + \alpha_{n-1}\mathbf{b} \end{aligned} \tag{7-9}$$

is linearly independent and qualifies as a basis of the state space of $FE_1$. Recall from Figure 2-5 that if the vectors $\{\mathbf{q}^1, \mathbf{q}^2, \cdots, \mathbf{q}^n\}$ are used as a basis, then the $i$th column of the new representation $\bar{\mathbf{A}}$ is the representation of $\mathbf{A}\mathbf{q}^i$ with respect to the basis $\{\mathbf{q}^1, \mathbf{q}^2, \cdots, \mathbf{q}^n\}$. Observe that

$$\mathbf{A}\mathbf{q}^1 = (\mathbf{A}^n + \alpha_1\mathbf{A}^{n-1} + \cdots + \alpha_{n-1}\mathbf{A} + \alpha_n\mathbf{I})\mathbf{b} - \alpha_n\mathbf{b}$$

$$= -\alpha_n\mathbf{b} = -\alpha_n\mathbf{q}^n = [\mathbf{q}^1 \quad \mathbf{q}^2 \quad \cdots \quad \mathbf{q}^n]\begin{bmatrix} 0 \\ 0 \\ \cdot \\ \cdot \\ \cdot \\ 0 \\ -\alpha_n \end{bmatrix}$$

$$\mathbf{A}\mathbf{q}^2 = \mathbf{q}^1 - \alpha_{n-1}\mathbf{q}^n = [\mathbf{q}^1 \quad \mathbf{q}^2 \quad \cdots \quad \mathbf{q}^n]\begin{bmatrix} 1 \\ 0 \\ \cdot \\ \cdot \\ \cdot \\ 0 \\ -\alpha_{n-1} \end{bmatrix}$$

$$\vdots$$

$$\mathbf{A}\mathbf{q}^n = \mathbf{q}^{n-1} - \alpha_1\mathbf{q}^n = [\mathbf{q}^1 \quad \mathbf{q}^2 \quad \cdots \quad \mathbf{q}^n]\begin{bmatrix} 0 \\ 0 \\ \cdot \\ \cdot \\ \cdot \\ 1 \\ -\alpha_1 \end{bmatrix}$$

Hence if we choose $\{\mathbf{q}^1, \mathbf{q}^2, \cdots, \mathbf{q}^n\}$ as a new basis of the state space, then $\mathbf{A}$ and $\mathbf{b}$ have new representations, of the form

$$\bar{\mathbf{A}} = \begin{bmatrix} 0 & 1 & 0 & \cdots & 0 & 0 \\ 0 & 0 & 1 & \cdots & 0 & 0 \\ 0 & 0 & 0 & \cdots & 0 & 0 \\ \cdot & \cdot & \cdot & & \cdot & \cdot \\ \cdot & \cdot & \cdot & & \cdot & \cdot \\ \cdot & \cdot & \cdot & & \cdot & \cdot \\ 0 & 0 & 0 & \cdots & 0 & 1 \\ -\alpha_n & -\alpha_{n-1} & -\alpha_{n-2} & \cdots & -\alpha_2 & -\alpha_1 \end{bmatrix} \qquad \bar{\mathbf{b}} = \begin{bmatrix} 0 \\ 0 \\ 0 \\ \cdot \\ \cdot \\ \cdot \\ 0 \\ 1 \end{bmatrix} \qquad \text{(7-10)}$$

The matrices $\bar{\mathbf{A}}$ and $\bar{\mathbf{b}}$ can also be obtained by using an equivalence transformation. Let $\mathbf{Q} \triangleq [\mathbf{q}^1 \quad \mathbf{q}^2 \quad \cdots \quad \mathbf{q}^n] \triangleq \mathbf{P}^{-1}$, and let $\bar{\mathbf{x}} = \mathbf{P}\mathbf{x}$ or $\mathbf{x} =$

$\mathbf{Q}\bar{\mathbf{x}}$, then the dynamical equation $FE_1$ can be transformed into

$$\begin{aligned}\dot{\bar{\mathbf{x}}} &= \mathbf{Q}^{-1}\mathbf{A}\mathbf{Q}\bar{\mathbf{x}} + \mathbf{Q}^{-1}\mathbf{b}u \\ y &= \mathbf{c}\mathbf{Q}\mathbf{x} + du\end{aligned}$$

The reader is advised to verify that $\mathbf{Q}^{-1}\mathbf{A}\mathbf{Q} = \bar{\mathbf{A}}$ or $\mathbf{A}\mathbf{Q} = \mathbf{Q}\bar{\mathbf{A}}$ and $\mathbf{Q}^{-1}\mathbf{b} = \bar{\mathbf{b}}$. The vector $\bar{\mathbf{c}}$ is to be computed from $\mathbf{c}\mathbf{Q}$. Let

$$\bar{\mathbf{c}} = \mathbf{c}\mathbf{Q} \triangleq [\beta_n \quad \beta_{n-1} \quad \beta_{n-2} \quad \cdots \quad \beta_1] \tag{7-11}$$

Hence the controllable dynamical equation $FE_1$ has the equivalent controllable canonical-form dynamical equation $CFE_1$. This proves the first part of the theorem.

The dynamical equations $FE_1$ and $CFE_1$ are equivalent; hence they have the same transfer function. It has been shown in Sections 4-4 and 6-3 that the transfer function of $CFE_1$ is equal to $\hat{g}(s)$ in Equation (7-8).

Q.E.D.

One may wonder how we obtain the set of basis vectors given in (7-9). This is derived in the following. Let $\bar{\mathbf{U}}$ be the controllability matrix of the controllable canonical-form dynamical equation, then

$$\bar{\mathbf{U}} \triangleq [\bar{\mathbf{b}} \quad \bar{\mathbf{A}}\bar{\mathbf{b}} \quad \cdots \quad \bar{\mathbf{A}}^{n-1}\bar{\mathbf{b}}] = \begin{bmatrix} 0 & 0 & 0 & \cdots & 1 \\ 0 & 0 & 0 & \cdots & e_1 \\ \cdot & \cdot & \cdot & & \cdot \\ \cdot & \cdot & \cdot & & \cdot \\ \cdot & \cdot & \cdot & & \cdot \\ 0 & 0 & 1 & \cdots & e_{n-3} \\ 0 & 1 & e_1 & \cdots & e_{n-2} \\ 1 & e_1 & e_2 & \cdots & e_{n-1} \end{bmatrix} \tag{7-12}$$

where

$$e_k = -\sum_{i=0}^{k-1} \alpha_{i+1} e_{k-i-1} \qquad k = 1, 2, \cdots, n-1; e_0 = 1$$

The controllability matrix $\bar{\mathbf{U}}$ is nonsingular for any $\alpha_1, \alpha_2, \cdots, \alpha_n$. Therefore the controllable canonical-form dynamical equation is always controllable, as one might expect. The inverse of $\bar{\mathbf{U}}$ has the following very simple form:

$$\bar{\mathbf{U}}^{-1} = \begin{bmatrix} \alpha_{n-1} & \alpha_{n-2} & \cdots & \alpha_1 & 1 \\ \alpha_{n-2} & \alpha_{n-3} & \cdots & 1 & 0 \\ \cdot & \cdot & & \cdot & \cdot \\ \cdot & \cdot & & \cdot & \cdot \\ \cdot & \cdot & & \cdot & \cdot \\ \alpha_1 & 1 & \cdots & 0 & 0 \\ 1 & 0 & \cdots & 0 & 0 \end{bmatrix} \tag{7-13}$$

This can be directly verified by showing that $\bar{\mathbf{U}}\bar{\mathbf{U}}^{-1} = \mathbf{I}$. The dynamical equations $FE_1$ and $CFE_1$ are related by the equivalence transformation $\mathbf{x} = \mathbf{Q}\bar{\mathbf{x}}$; hence $\mathbf{Q} = \mathbf{U}\bar{\mathbf{U}}^{-1}$. In the equivalence transformation $\mathbf{x} = \mathbf{Q}\bar{\mathbf{x}}$, we use the columns of $\mathbf{Q}$ as new basis vectors of the state space (see Figure 2-5). By computing $\mathbf{U}\bar{\mathbf{U}}^{-1}$, we see that the columns of $\mathbf{Q}$ are indeed those given in (7-9); that is,

$$\mathbf{Q} = [\mathbf{q}^1 \quad \mathbf{q}^2 \quad \cdots \quad \mathbf{q}^n] = [\mathbf{b} \quad \mathbf{Ab} \quad \cdots \quad \mathbf{A}^{n-1}\mathbf{b}]\bar{\mathbf{U}}^{-1}$$

We have the following theorem, which is similar to Theorem 7-1, for an observable dynamical equation.

### Theorem 7-2

If the $n$-dimensional, linear, time-invariant, single-variable dynamical equation $FE_1$ is observable, then it can be transformed, by an equivalence transformation, into the form

$$OFE_1: \quad \dot{\bar{\mathbf{x}}} = \begin{bmatrix} 0 & 0 & \cdots & 0 & -\alpha_n \\ 1 & 0 & \cdots & 0 & -\alpha_{n-1} \\ 0 & 1 & \cdots & 0 & -\alpha_{n-2} \\ \cdot & \cdot & & \cdot & \cdot \\ \cdot & \cdot & & \cdot & \cdot \\ \cdot & \cdot & & \cdot & \cdot \\ 0 & 0 & \cdots & 1 & -\alpha_1 \end{bmatrix} \bar{\mathbf{x}} + \begin{bmatrix} \beta_n \\ \beta_{n-1} \\ \beta_{n-2} \\ \cdot \\ \cdot \\ \cdot \\ \beta_1 \end{bmatrix} u \tag{7-14a}$$

$$y = [0 \quad 0 \quad \cdots \quad 0 \quad 1]\bar{\mathbf{x}} + du \tag{7-14b}$$

The dynamical equation (7-14) is said to be in the *observable canonical form;* moreover, the transfer function of $FE_1$ is

$$\hat{g}(s) = \frac{\beta_1 s^{n-1} + \beta_2 s^{n-2} + \cdots + \beta_{n-1}s + \beta_n}{s^n + \alpha_1 s^{n-1} + \cdots + \alpha_{n-1}s + \alpha_n} + d$$ ∎

This theorem can be proved either by a direct verification or by using the theorem of duality (Theorem 5-10). Its proof is left as an exercise. The equivalence transformation $\bar{\mathbf{x}} = \mathbf{Px}$ between (7-1) and (7-14) can be obtained by using Equation (7-6): $\mathbf{P} = \bar{\mathbf{V}}^{-1}\mathbf{V}$. It is easy to verify that the observability matrix $\bar{\mathbf{V}}$ of the observable canonical-form dynamical equation $OFE_1$ is the same as the matrix in (7-12). Hence the equivalence transformation $\bar{\mathbf{x}} = \mathbf{Px}$ between (7-1) and (7-14) is given by

$$\mathbf{P} = \begin{bmatrix} \alpha_{n-1} & \alpha_{n-2} & \cdots & \alpha_1 & 1 \\ \alpha_{n-2} & \alpha_{n-3} & \cdots & 1 & 0 \\ \cdot & \cdot & & \cdot & \cdot \\ \cdot & \cdot & & \cdot & \cdot \\ \cdot & \cdot & & \cdot & \cdot \\ \alpha_1 & 1 & \cdots & 0 & 0 \\ 1 & 0 & \cdots & 0 & 0 \end{bmatrix} \begin{bmatrix} \mathbf{c} \\ \mathbf{cA} \\ \cdot \\ \cdot \\ \cdot \\ \mathbf{cA}^{n-2} \\ \mathbf{cA}^{n-1} \end{bmatrix} \tag{7-15}$$

We see from (7-9) that the basis of the controllable canonical-form dynamical equation is obtained from a linear combination of the vectors $\{\mathbf{b}, \mathbf{Ab}, \cdots, \mathbf{A}^{n-1}\mathbf{b}\}$. One may wonder what form we shall obtain if we choose $\{\mathbf{b}, \mathbf{Ab}, \cdots, \mathbf{A}^{n-1}\mathbf{b}\}$ as the basis. Define $\mathbf{q}^1 \triangleq \mathbf{b}$, $\mathbf{q}^2 \triangleq \mathbf{Ab}$, $\cdots$, $\mathbf{q}^n \triangleq \mathbf{A}^{n-1}\mathbf{b}$ and let $\bar{\mathbf{x}} = \mathbf{Q}^{-1}\mathbf{x}$, where $\mathbf{Q} \triangleq [\mathbf{q}^1 \quad \mathbf{q}^2 \quad \cdots \quad \mathbf{q}^n]$; then we can obtain the following new representation (see Problem 7-2):

$$\dot{\bar{\mathbf{x}}} = \begin{bmatrix} 0 & 0 & \cdots & 0 & -\alpha_n \\ 1 & 0 & \cdots & 0 & -\alpha_{n-1} \\ 0 & 1 & \cdots & 0 & -\alpha_{n-2} \\ \cdot & \cdot & & \cdot & \cdot \\ \cdot & \cdot & & \cdot & \cdot \\ \cdot & \cdot & & \cdot & \cdot \\ 0 & 0 & \cdots & 1 & -\alpha_1 \end{bmatrix} \bar{\mathbf{x}} + \begin{bmatrix} 1 \\ 0 \\ 0 \\ \cdot \\ \cdot \\ \cdot \\ 0 \end{bmatrix} u \tag{7-16a}$$

$$y = \mathbf{cQ}\bar{\mathbf{x}} + du \tag{7-16b}$$

This equation looks very similar to the controllable canonical form and is easier to obtain because its equivalence transformation $\mathbf{Q}$ is simpler. However, the usefulness of an equation in the form of (7-16) is not known at present.

### Example 1

Transform the following controllable and observable single-variable dynamical equation

$$\dot{\mathbf{x}} = \begin{bmatrix} 1 & 2 & 0 \\ 3 & -1 & 1 \\ 0 & 2 & 0 \end{bmatrix} \mathbf{x} + \begin{bmatrix} 2 \\ 1 \\ 1 \end{bmatrix} u \tag{7-17a}$$

$$y = [0 \quad 0 \quad 1]\mathbf{x} \tag{7-17b}$$

into the controllable and observable canonical-form dynamical equations.

The characteristic polynomial of the matrix $\mathbf{A}$ in (7-17) is

$$\Delta(\lambda) = \det \begin{bmatrix} \lambda - 1 & -2 & 0 \\ -3 & \lambda + 1 & -1 \\ 0 & -2 & \lambda \end{bmatrix} = \lambda^3 - 9\lambda + 2$$

Hence $\alpha_3 = 2$, $\alpha_2 = -9$, and $\alpha_1 = 0$. The controllability matrix $\mathbf{U}$ is

$$\mathbf{U} = \begin{bmatrix} 2 & 4 & 16 \\ 1 & 6 & 8 \\ 1 & 2 & 12 \end{bmatrix}$$

From Equations (7-5b) and (7-13), we have

$$\mathbf{Q} = \mathbf{U}\bar{\mathbf{U}}^{-1} = \begin{bmatrix} 2 & 4 & 16 \\ 1 & 6 & 8 \\ 1 & 2 & 12 \end{bmatrix} \begin{bmatrix} -9 & 0 & 1 \\ 0 & 1 & 0 \\ 1 & 0 & 0 \end{bmatrix} = \begin{bmatrix} -2 & 4 & 2 \\ 1 & 6 & 1 \\ 3 & 2 & 1 \end{bmatrix}$$

The matrix $\mathbf{Q}$ can also be obtained from (7-9). From (7-11), we have

$$[\beta_3 \quad \beta_2 \quad \beta_1] = \mathbf{cQ} = [3 \quad 2 \quad 1]$$

Hence the equivalent controllable canonical-form and observable canonical-form dynamical equations are, respectively,

$$CFE_1: \qquad \dot{\bar{\mathbf{x}}} = \begin{bmatrix} 0 & 1 & 0 \\ 0 & 0 & 1 \\ -2 & 9 & 0 \end{bmatrix} \bar{\mathbf{x}} + \begin{bmatrix} 0 \\ 0 \\ 1 \end{bmatrix} u$$

$$y = [3 \quad 2 \quad 1]\bar{\mathbf{x}}$$

$$OFE_1: \qquad \dot{\bar{\mathbf{x}}} = \begin{bmatrix} 0 & 0 & -2 \\ 1 & 0 & 9 \\ 0 & 1 & 0 \end{bmatrix} \bar{\mathbf{x}} + \begin{bmatrix} 3 \\ 2 \\ 1 \end{bmatrix} u$$

$$y = [0 \quad 0 \quad 1]\bar{\mathbf{x}}$$

The transfer function of (7-17) is thus equal to

$$\hat{g}(s) = \frac{s^2 + 2s + 3}{s^3 - 9s + 2}$$

The controllable or observable canonical-form dynamical equation can also be obtained by first computing the transfer function of the dynamical equation. The coefficients of the transfer function give immediately the canonical-form equations. ∎

The controllable canonical-form and the observable canonical-form dynamical equations are essential in the study of state feedback and state estimator. They are also useful in the simulation of a dynamical equation on an analog computer. For example, in simulating the dynamical equation $FE_1$, we generally need $n^2$ amplifiers and attenuators (potentiometers) to simulate $\mathbf{A}$, and $2n$ amplifiers and attenuators to simulate $\mathbf{b}$ and $\mathbf{c}$. However, if $FE_1$ is transformed into the controllable or observable canonical form or into the form of (7-16), then the number of components required in the simulation is reduced from $n^2 + 2n$ to $2n$.

***Multivariable case.** Consider the $n$-dimensional, linear, time-invariant, multivariable dynamical equation

$$FE: \quad \dot{\mathbf{x}} = \mathbf{A}\mathbf{x} + \mathbf{B}\mathbf{u} \quad \text{(7-18a)}$$
$$\mathbf{y} = \mathbf{C}\mathbf{x} + \mathbf{D}\mathbf{u} \quad \text{(7-18b)}$$

where $\mathbf{A}$, $\mathbf{B}$, $\mathbf{C}$, and $\mathbf{D}$ are $n \times n$, $n \times p$, $q \times n$, and $q \times p$ real constant matrices, respectively. Let $\mathbf{b}_i$ be the $i$th column of $\mathbf{B}$; that is, $\mathbf{B} = [\mathbf{b}_1 \ \ \mathbf{b}_2 \ \ \cdots \ \ \mathbf{b}_p]$.

If the dynamical equation $FE$ is controllable, then the controllability matrix

$$\mathbf{U} = [\mathbf{b}_1 \ \ \mathbf{b}_2 \ \ \cdots \ \ \mathbf{b}_p \ \ \mathbf{A}\mathbf{b}_1 \ \ \cdots \ \ \mathbf{A}\mathbf{b}_p \ \ \cdots \ \ \mathbf{A}^{n-1}\mathbf{b}_1 \ \ \mathbf{A}^{n-1}\mathbf{b}_2 \ \ \cdots \ \ \mathbf{A}^{n-1}\mathbf{b}_p] \quad \text{(7-19)}$$

has rank $n$. Consequently, there are $n$ linearly independent column vectors in $\mathbf{U}$. There are many ways to choose $n$ linearly independent column vectors from the $n \times np$ composite matrix $\mathbf{U}$. In the following, we shall give two schemes for choosing $n$ linearly independent column vectors to form the bases of canonical-form dynamical equations.

### Scheme 1

We start with the vector $\mathbf{b}_1$ and then proceed to $\mathbf{A}\mathbf{b}_1$, $\mathbf{A}^2\mathbf{b}_1$, up to $\mathbf{A}^{\nu_1-1}\mathbf{b}_1$ until the vector $\mathbf{A}^{\nu_1}\mathbf{b}_1$ can be expressed as a linear combination of $\{\mathbf{b}_1, \cdots, \mathbf{A}^{\nu_1-1}\mathbf{b}_1\}$. If $\nu_1 = n$, the equation can be controlled by the first column of $\mathbf{B}$ alone. If $\nu_1 < n$, we select $\mathbf{b}_2$, $\mathbf{A}\mathbf{b}_2$, up to $\mathbf{A}^{\nu_2-1}\mathbf{b}_2$, until the vector $\mathbf{A}^{\nu_2}\mathbf{b}_2$ can be expressed as a linear combination of $\{\mathbf{b}_1, \cdots, \mathbf{A}^{\nu_1-1}\mathbf{b}_1, \mathbf{b}_2, \cdots, \mathbf{A}^{\nu_2-1}\mathbf{b}_2\}$ (see Problem 7-5). If $\nu_1 + \nu_2 < n$, we proceed to $\mathbf{b}_3$, $\mathbf{A}\mathbf{b}_3$, $\cdots$, $\mathbf{A}^{\nu_3-1}\mathbf{b}_3$ and so forth. Assume that $\nu_1 + \nu_2 + \nu_3 = n$, and the $n$ vectors

$$\{\mathbf{b}_1, \mathbf{A}\mathbf{b}_1, \cdots, \mathbf{A}^{\nu_1-1}\mathbf{b}_1; \mathbf{b}_2, \mathbf{A}\mathbf{b}_2, \cdots, \mathbf{A}^{\nu_2-1}\mathbf{b}_2; \mathbf{b}_3, \mathbf{A}\mathbf{b}_3, \cdots, \mathbf{A}^{\nu_3-1}\mathbf{b}_3\} \quad \text{(7-20)}$$

are linearly independent. An important property of this set is that the vector $\mathbf{A}^{\nu_i}\mathbf{b}_i$ can be expressed as a linear combination of the preceding vectors; for example, $\mathbf{A}^{\nu_2}\mathbf{b}_2$ can be expressed as a linear combination of $\{\mathbf{b}_1, \mathbf{A}\mathbf{b}_1, \cdots, \mathbf{A}^{\nu_1-1}\mathbf{b}_1, \mathbf{b}_2, \mathbf{A}\mathbf{b}_2, \cdots, \mathbf{A}^{\nu_2-1}\mathbf{b}_2\}$.

### Scheme 2

The linearly independent vectors are selected in the order of (7-19); that is, we start from $\mathbf{b}_1, \mathbf{b}_2, \cdots, \mathbf{b}_p$ and then $\mathbf{A}\mathbf{b}_1, \mathbf{A}\mathbf{b}_2, \cdots, \mathbf{A}\mathbf{b}_p$, and then $\mathbf{A}^2\mathbf{b}_1$, $\mathbf{A}^2\mathbf{b}_2$, and so forth, until we obtain $n$ linearly independent vectors. Note that if a vector, say $\mathbf{A}\mathbf{b}_2$ is skipped because of linear dependence on the vectors $\{\mathbf{b}_1, \mathbf{b}_2, \cdots, \mathbf{b}_p, \mathbf{A}\mathbf{b}_1\}$, then all vectors of the form $\mathbf{A}^k\mathbf{b}_2$, for $k \geq 1$, can also be skipped because they must also be dependent on the previous columns. After choosing $n$ linearly independent vectors in this order, we *rearrange* them as

$$\{\mathbf{b}_1, \cdots, \mathbf{A}^{\mu_1-1}\mathbf{b}_1; \mathbf{b}_2, \cdots, \mathbf{A}^{\mu_2-1}\mathbf{b}_2; \cdots; \mathbf{b}_p, \cdots, \mathbf{A}^{\mu_p-1}\mathbf{b}_p\} \tag{7-21}$$

where $\mu_1 + \mu_2 + \cdots + \mu_p = n$. Note that the main difference between this scheme and Scheme 1 is that in Equation (7-20) $\mathbf{A}^{\nu_1}\mathbf{b}_1$ can be expressed as a linear combination of $\{\mathbf{b}_1, \mathbf{A}\mathbf{b}_1, \cdots, \mathbf{A}^{\nu_1-1}\mathbf{b}_1\}$, whereas the vector $\mathbf{A}^{\mu_1}\mathbf{b}_1$ in (7-21) cannot be expressed as a linear combination of $\{\mathbf{b}_1, \cdots, \mathbf{A}^{\mu_1-1}\mathbf{b}_1\}$; $\mathbf{A}^{\mu_1}\mathbf{b}_1$ is generally linearly dependent on all vectors in (7-21). Similar remarks apply to $\mathbf{A}^{\mu_i}\mathbf{b}_i$, for $i = 1, 2, \cdots, p$.

Now if the set of vectors in (7-20) is chosen as a basis of the state space of $FE$ or, equivalently, let $\bar{\mathbf{x}} = \mathbf{Q}^{-1}\mathbf{x}$ where

$$\mathbf{Q} \triangleq [\mathbf{q}^1 \quad \mathbf{q}^2 \quad \cdots \quad \mathbf{q}^n] \triangleq [\mathbf{b}_1 \cdots \mathbf{A}^{\nu_1-1}\mathbf{b}_1 \quad \mathbf{b}_2 \cdots \mathbf{A}^{\nu_2-1}\mathbf{b}_2 \quad \mathbf{b}_3 \cdots \mathbf{A}^{\nu_3-1}\mathbf{b}_3]$$

then the matrices $\bar{\mathbf{A}}$ and $\bar{\mathbf{b}}$ will be of the forms

$$\bar{\mathbf{A}} = \left[\begin{array}{cccccc|cccccc|cccccc}
0 & 0 & 0 & \cdots & 0 & X & & & & & & X & & & & & & X \\
1 & 0 & 0 & \cdots & 0 & X & & & & & & X & & & & & & X \\
0 & 1 & 0 & \cdots & 0 & X & & & & & & X & & & & & & X \\
\cdot & \cdot & \cdot & & \cdot & \cdot & & & & & & \cdot & & & & & & \cdot \\
\cdot & \cdot & \cdot & & \cdot & \cdot & & & & & & \cdot & & & & & & \cdot \\
\cdot & \cdot & \cdot & & \cdot & \cdot & & & & & & \cdot & & & & & & \cdot \\
0 & 0 & 0 & \cdots & 1 & X & & & & & & X & & & & & & X \\
\hline
 & & (\nu_1 \times \nu_1) & & & & 0 & 0 & 0 & \cdots & 0 & X & & & & & & X \\
 & & & & & & 1 & 0 & 0 & \cdots & 0 & X & & & & & & X \\
 & & & & & & 0 & 1 & 0 & \cdots & 0 & X & & & & & & X \\
 & & & & & & \cdot & \cdot & \cdot & & \cdot & \cdot & & & & & & \cdot \\
 & & & & & & \cdot & \cdot & \cdot & & \cdot & \cdot & & & & & & \cdot \\
 & & & & & & \cdot & \cdot & \cdot & & \cdot & \cdot & & & & & & \cdot \\
 & & & & & & 0 & 0 & 0 & \cdots & 1 & X & & & & & & X \\
\hline
 & & & & & & & & (\nu_2 \times \nu_2) & & & & 0 & 0 & 0 & \cdots & 0 & X \\
 & & & & & & & & & & & & 1 & 0 & 0 & \cdots & 0 & X \\
 & & & & & & & & & & & & 0 & 1 & 0 & \cdots & 0 & X \\
 & & & & & & & & & & & & \cdot & \cdot & \cdot & & \cdot & \cdot \\
 & & & & & & & & & & & & \cdot & \cdot & \cdot & & \cdot & \cdot \\
 & & & & & & & & & & & & \cdot & \cdot & \cdot & & \cdot & \cdot \\
 & & & & & & & & & & & & 0 & 0 & 0 & \cdots & 1 & X \\
 & & & & & & & & & & & & & & (\nu_3 \times \nu_3) & & &
\end{array}\right]
\qquad
\bar{\mathbf{B}} = \left[\begin{array}{ccc}
1 & 0 & 0 \\
0 & 0 & 0 \\
0 & 0 & 0 \\
\cdot & \cdot & \cdot \\
\cdot & \cdot & \cdot \\
\cdot & \cdot & \cdot \\
0 & 0 & 0 \\
\hline
0 & 1 & 0 \\
0 & 0 & 0 \\
0 & 0 & 0 \\
\cdot & \cdot & \cdot \\
\cdot & \cdot & \cdot \\
\cdot & \cdot & \cdot \\
0 & 0 & 0 \\
\hline
0 & 0 & 1 \\
0 & 0 & 0 \\
0 & 0 & 0 \\
\cdot & \cdot & \cdot \\
\cdot & \cdot & \cdot \\
\cdot & \cdot & \cdot \\
0 & 0 & 0
\end{array}
\quad \bar{\mathbf{b}}_4 \cdots \bar{\mathbf{b}}_p\right] \tag{7-22}$$

where the X's denote possible nonzero elements. This can be easily verified by observing that the $i$th column of $\bar{\mathbf{A}}$ is the representation of $\mathbf{A}\mathbf{q}^i$ with respect to the basis vectors $[\mathbf{q}^1 \quad \mathbf{q}^2 \quad \cdots \quad \mathbf{q}^n]$.

If the set of vectors in (7-21) is used as a basis, then the new matrices $\tilde{\mathbf{A}}$ and $\tilde{\mathbf{b}}$ will be of the form

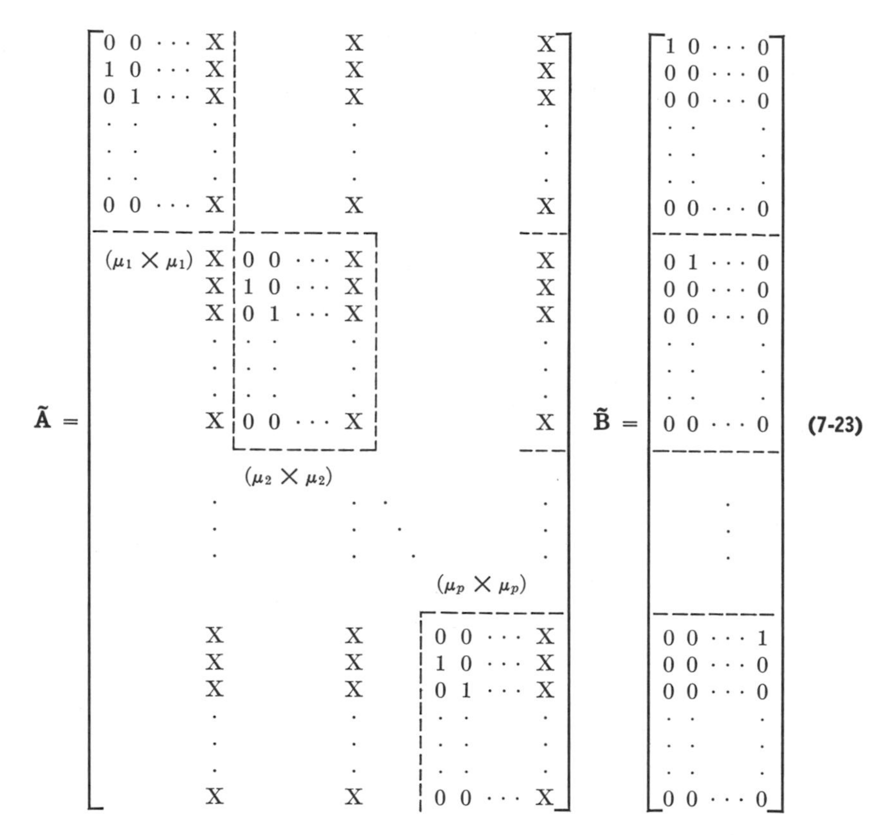

where the X's again denote possible nonzero elements. The matrices $\tilde{\mathbf{A}}$ and $\tilde{\mathbf{B}}$ can be verified by inspection. The matrix $\mathbf{C}$ in both cases are to be computed from $\mathbf{CQ}$.

By comparing (7-22) with (7-23), we see immediately the differences in the matrices $\mathbf{A}$ and $\mathbf{B}$ due to the different choices of basic vectors according to Schemes 1 and 2. The matrix $\bar{\mathbf{A}}$ has three blocks in the diagonal, whereas the matrix $\tilde{\mathbf{A}}$ has $p$ blocks. The first three columns of $\bar{\mathbf{B}}$ are very simple, whereas every column of $\tilde{\mathbf{B}}$ consists of only one nonzero element.

The usefulness of the canonical forms in (7-22) and (7-23) is not known at present. The purpose of introducing these two forms is to show that there is no additional conceptual difficulty in developing canonical forms for multivariable dynamical equations. By rearranging the vectors in (7-20) or (7-21), another canonical-form dynamical equations can be obtained. One of them will be discussed in Section 7-4.

## 7-3 State Feedback

In a control system, if the input or the commanding signal is predetermined and will not change no matter what the outcome of the control is, the system is said to be an *open-loop control system.* It is clear that an open-loop control system is not a good system. A desirable control signal should react to the behavior of the system. If the system behaves well, no change in the control signal is necessary; otherwise, a proper change in the control signal is required to bring the response of the system to a desired one. Such a system whose input signal depends on the outcome of the control is called a *feedback control system.*

If a dynamical-equation description of a system is available, it is reasonable to base the choice of the input on the value of the state, the reference input and possibly on $t$, because the state and the input determine completely the future behavior of the system. Hence a good control signal should be determined by an equation of the form $\mathbf{u}(t) = f(\mathbf{v}(t), \mathbf{x}(t), t)$. This relation is called a *control law.* Present-day optimal control theory is mainly concerned with how to find the best control law and implement it.

We study only linear, time-invariant dynamical equations. Therefore, it is reasonable to assume that the control law depends linearly on $\mathbf{v}$ and $\mathbf{x}$ and is of the form $\mathbf{u}(t) = \mathbf{v}(t) + \mathbf{Kx}(t)$, where $\mathbf{v}$ may stand for a desired reference input and $\mathbf{K}$ is some real constant matrix called a *feedback gain matrix.* Finding the best $\mathbf{K}$ is in the scope of optimal control theory and will not be studied here. We shall, instead, discuss the effect of introducing linear state feedback of the form $\mathbf{u} = \mathbf{v} + \mathbf{Kx}$, and study what can be achieved by introducing this state feedback.

The distinction between state feedback and output feedback should be made. In output feedback, the output $\mathbf{y}$ is fed back into the input; in state feedback, the state $\mathbf{x}$ is fed back into the input. Since the number of state variables is generally larger than the number of output variables, there is more room for manipulation in state feedback than in output feedback. In fact, what can be achieved by output feedback can always be achieved by state feedback, but the converse is not true (see Problem 7-8). In this section we shall study state feedback exclusively; the study of output feedback will be postponed until the last chapter.

In this section, all the state variables are assumed to be available as outputs.

**Single-variable case.** Consider the single-variable, linear, time-invariant dynamical equation

$$FE_1: \qquad \dot{\mathbf{x}} = \mathbf{Ax} + \mathbf{b}u \tag{7-24a}$$

$$y = \mathbf{cx} + du \tag{7-24b}$$

where $\mathbf{x}$ is the $n \times 1$ state vector, $u$ is the scalar input, $y$ is the scalar output; $\mathbf{A}$ is an $n \times n$ real constant matrix, $\mathbf{b}$ is an $n \times 1$ real constant column vector, and $\mathbf{c}$ is a $1 \times n$ real constant row vector. In state feedback, every state variable is multiplied by a gain and fed back into the input terminal. Let the gain between the $i$th state variable and the input be $k_i$. Define $\mathbf{k} \triangleq [k_1 \quad k_2 \quad \cdots \quad k_n]$. Then the dynamical equation of the state-feedback system shown in Figure 7-1 is

$$FE_1{}^f: \qquad \dot{\mathbf{x}} = (\mathbf{A} + \mathbf{bk})\mathbf{x} + \mathbf{b}v \tag{7-25a}$$

$$y = (\mathbf{c} + d\mathbf{k})\mathbf{x} + dv \tag{7-25b}$$

which is obtained by replacing $u$ in (7-24) by $v + \mathbf{kx}$, where $v$ is a desired or reference input. Note that the dynamical equations (7-24) and (7-25)

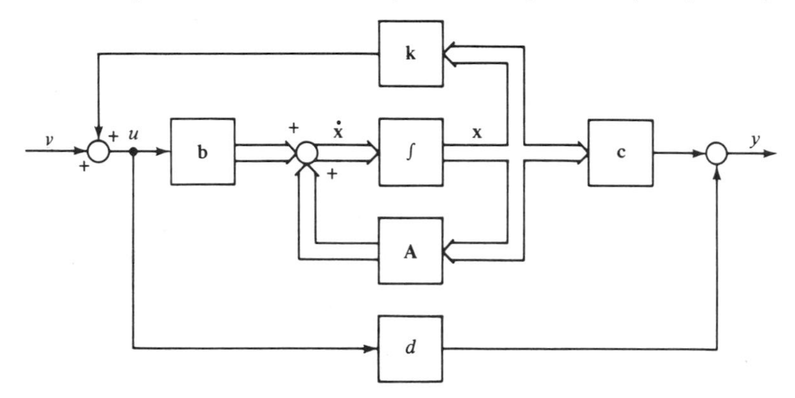


**Figure 7-1** A state feedback system.

have the same dimension and the same state space. Now we shall show that the controllability of a linear time-invariant dynamical equation is invariant under any linear state feedback.

### Theorem 7-3

The state feedback dynamical equation $FE_1{}^f$ in (7-25) is controllable for any $1 \times n$ real vector $\mathbf{k}$ if and only if the dynamical equation $FE_1$ in (7-24) is controllable.

### Proof

First we show that the controllability of $FE_1$ implies the controllability of $FE_1{}^f$. Let $\mathbf{x}^0$ and $\mathbf{x}^1$ be two arbitrary states in the state space $\Sigma$. By the controllability assumption of $FE_1$, there exists an input $u$ that will transfer $\mathbf{x}^0$ to $\mathbf{x}^1$ in a finite time. Now for the state-feedback dynamical equation, if we choose $v(t) = u(t) - \mathbf{kx}(t)$, then the input $v$ will transfer $\mathbf{x}^0$ to $\mathbf{x}^1$. Therefore we conclude that $FE_1{}^f$ is controllable.

We see from Figure 7-1 that the input $v$ does not control the state $\mathbf{x}$ directly, it generates $u$ to control $\mathbf{x}$. Therefore if $u$ cannot control $\mathbf{x}$, neither can $\mathbf{v}$. In other words, if $FE_1$ is not controllable neither is $FE_1{}^f$.

Q.E.D.

We see that in the proof, the assumptions of single-variableness and time-invariance are not used. Therefore we have the following corollary.

### Corollary 7-3

The controllability of a multivariable linear, time-varying dynamical equation is invariant under any state feedback of the form $\mathbf{u}(t) = \mathbf{v}(t) + \mathbf{K}(t)\mathbf{x}(t)$. ∎

Note that Theorem 7-3 can also be proved by showing that

$$\rho[\mathbf{b} \quad \mathbf{Ab} \quad \cdots \quad \mathbf{A}^{n-1}\mathbf{b}] = \rho[\mathbf{b} \quad (\mathbf{A} - \mathbf{bk})\mathbf{b} \quad \cdots \quad (\mathbf{A} - \mathbf{bk})^{n-1}\mathbf{b}] \tag{7-26}$$

for any $1 \times n$ real constant vector $\mathbf{k}$ (see Problem 7-10).

Although state feedback preserves the controllability of a dynamical equation, it is always possible to destroy the observability property of a dynamical equation by some choice of $\mathbf{k}$. For example, if $d \neq 0$ and if $\mathbf{k} = (-1/d)\mathbf{c}$, then the state-feedback equation (7-25) is not observable even if $FE_1$ is observable. If $d = 0$, it is still possible to choose some $\mathbf{k}$ such that the state-feedback dynamical equation will not preserve the observability property (see Problem 7-14).

### Example 1

Consider the controllable and observable dynamical equation

$$FE_1: \quad \dot{\mathbf{x}} = \begin{bmatrix} 1 & 2 \\ 3 & 1 \end{bmatrix} \mathbf{x} + \begin{bmatrix} 0 \\ 1 \end{bmatrix} u$$
$$y = [1 \quad 2]\mathbf{x}$$

If we introduce the state feedback

$$u = v + [-3 \quad -1]\mathbf{x}$$

then the state-feedback equation is

$$FE_1{}^f: \quad \dot{\mathbf{x}} = \begin{bmatrix} 1 & 2 \\ 0 & 0 \end{bmatrix} \mathbf{x} + \begin{bmatrix} 0 \\ 1 \end{bmatrix} v$$
$$y = [1 \quad 2]\mathbf{x}$$

which is controllable but not observable. ∎

An important property of state feedback is that it can be used to control the eigenvalues of a dynamical equation.

### Theorem 7-4

If the single-variable dynamical equation $FE_1$ given in (7-24) is controllable, then by the state feedback $u = v + \mathbf{kx}$, where $\mathbf{k}$ is a $1 \times n$ real vector, the eigenvalues of $(\mathbf{A} + \mathbf{bk})$ can be arbitrarily assigned.

Proof

If the dynamical equation $FE_1$ is controllable, by an equivalence transformation $\bar{\mathbf{x}} = \mathbf{Px}$, $FE_1$ can be transformed into the following controllable canonical form (Theorem 7-1):

$CFE_1$:

$$\dot{\bar{\mathbf{x}}} = \begin{bmatrix} 0 & 1 & 0 & \cdots & 0 & 0 \\ 0 & 0 & 1 & \cdots & 0 & 0 \\ 0 & 0 & 0 & \cdots & 0 & 0 \\ \cdot & \cdot & \cdot & & \cdot & \cdot \\ \cdot & \cdot & \cdot & & \cdot & \cdot \\ \cdot & \cdot & \cdot & & \cdot & \cdot \\ 0 & 0 & 0 & \cdots & 0 & 1 \\ -\alpha_n & -\alpha_{n-1} & -\alpha_{n-2} & \cdots & -\alpha_2 & -\alpha_1 \end{bmatrix} \bar{\mathbf{x}} + \begin{bmatrix} 0 \\ 0 \\ 0 \\ \cdot \\ \cdot \\ \cdot \\ 0 \\ 1 \end{bmatrix} u \tag{7-27a}$$

$$y = [\beta_n \quad \beta_{n-1} \quad \beta_{n-2} \quad \cdots \quad \beta_2 \quad \beta_1]\bar{\mathbf{x}} + du \tag{7-27b}$$

Let $\bar{\mathbf{A}}$ and $\bar{\mathbf{b}}$ denote the matrices in (7-27a), then $\bar{\mathbf{A}} = \mathbf{PAP}^{-1}$, $\bar{\mathbf{b}} = \mathbf{Pb}$. Because of the equivalence transformation, the state feedback becomes

$$u = v + \mathbf{kx} = v + \mathbf{kP}^{-1}\bar{\mathbf{x}} \triangleq v + \bar{\mathbf{k}}\bar{\mathbf{x}} \tag{7-28}$$

where $\bar{\mathbf{k}} \triangleq \mathbf{kP}^{-1}$. It is easy to see that the set of the eigenvalues of $(\mathbf{A} + \mathbf{bk})$ is equal to the set of the eigenvalues of $(\bar{\mathbf{A}} + \bar{\mathbf{b}}\bar{\mathbf{k}})$. Let the characteristic polynomial of the matrix $(\mathbf{A} + \mathbf{bk})$ or, correspondingly, of $(\bar{\mathbf{A}} + \bar{\mathbf{b}}\bar{\mathbf{k}})$ with desired eigenvalues be

$$s^n + \bar{\alpha}_1 s^{n-1} + \cdots + \bar{\alpha}_n$$

If $\bar{\mathbf{k}}$ is chosen as

$$\bar{\mathbf{k}} = (\alpha_n - \bar{\alpha}_n,\ \alpha_{n-1} - \bar{\alpha}_{n-1},\ \cdots,\ \alpha_1 - \bar{\alpha}_1) \tag{7-29}$$

then the state-feedback dynamical equation becomes

$CFE_1{}^f$:

$$\dot{\bar{\mathbf{x}}} = \begin{bmatrix} 0 & 1 & 0 & \cdots & 0 & 0 \\ 0 & 0 & 1 & \cdots & 0 & 0 \\ 0 & 0 & 0 & \cdots & 0 & 0 \\ \cdot & \cdot & \cdot & & \cdot & \cdot \\ \cdot & \cdot & \cdot & & \cdot & \cdot \\ \cdot & \cdot & \cdot & & \cdot & \cdot \\ 0 & 0 & 0 & \cdots & 0 & 1 \\ -\bar{\alpha}_n & -\bar{\alpha}_{n-1} & -\bar{\alpha}_{n-2} & \cdots & -\bar{\alpha}_2 & -\bar{\alpha}_1 \end{bmatrix} \bar{\mathbf{x}} + \begin{bmatrix} 0 \\ 0 \\ 0 \\ \cdot \\ \cdot \\ \cdot \\ 0 \\ 1 \end{bmatrix} v \tag{7-30a}$$

$$y = [\beta_n + d(\alpha_n - \bar{\alpha}_n) \quad \beta_{n-1} + d(\alpha_{n-1} - \bar{\alpha}_{n-1}) \quad \cdots \quad \beta_1 + d(\alpha_1 - \bar{\alpha}_1)]\bar{\mathbf{x}} + dv \tag{7-30b}$$

Since the characteristic polynomial of the $\mathbf{A}$ matrix in (7-30) is $s^n + \bar{\alpha}_1 s^{n-1} + \cdots + \bar{\alpha}_n$, we conclude that the state-feedback equation has the desired eigenvalues. Q.E.D.

The gain vector $\bar{\mathbf{k}}$ in (7-29) is chosen with respect to the state $\bar{\mathbf{x}}$; that is, $u = v + \bar{\mathbf{k}}\bar{\mathbf{x}}$. Therefore, with respect to the original state $\mathbf{x}$, we have to use $\mathbf{k} = \bar{\mathbf{k}}\mathbf{P}$, where the columns of $\mathbf{P}^{-1} = \mathbf{Q}$ are those vectors introduced in (7-9). The matrices $\mathbf{A}$, $\mathbf{b}$, and $\mathbf{k}$ are assumed to be real; hence, if a complex eigenvalue is assigned to the matrix $(\mathbf{A} + \mathbf{bk})$, its complex conjugate must also be assigned. The procedure of choosing $\mathbf{k}$ has nothing to do with the number of outputs; therefore it can be applied to any controllable single-input, multiple-output, linear, time-invariant dynamical equation. We summarize the procedure of choosing $\mathbf{k}$ in the following.

### Algorithm

Given a controllable $\{\mathbf{A}, \mathbf{b}\}$ and a set of eigenvalues $\bar{\lambda}_1, \bar{\lambda}_2, \cdots, \bar{\lambda}_n$. Find the $1 \times n$ real vector $\mathbf{k}$ such that the matrix $(\mathbf{A} + \mathbf{bk})$ has the set $\{\bar{\lambda}_1, \bar{\lambda}_2, \cdots, \bar{\lambda}_n\}$ as its eigenvalues.

1. Find the characteristic polynomial of $\mathbf{A}$: $\det(s\mathbf{I} - \mathbf{A}) = s^n + \alpha_1 s^{n-1} + \cdots + \alpha_n$.
2. Compute $(s - \bar{\lambda}_1)(s - \bar{\lambda}_2) \cdots (s - \bar{\lambda}_n) = s^n + \bar{\alpha}_1 s^{n-1} + \cdots + \bar{\alpha}_n$.
3. Compute $\bar{\mathbf{k}} = [\alpha_n - \bar{\alpha}_n \quad \alpha_{n-1} - \bar{\alpha}_{n-1} \quad \cdots \quad \alpha_1 - \bar{\alpha}_1]$.
4. Compute $\mathbf{q}^{n-i} = \mathbf{A}\mathbf{q}^{n-i+1} + \alpha_i \mathbf{q}^n$, for $i = 1, 2, \cdots, (n-1)$, with $\mathbf{q}^n = \mathbf{b}$.
5. Form $\mathbf{Q} = [\mathbf{q}^1 \quad \mathbf{q}^2 \quad \cdots \quad \mathbf{q}^n]$.
6. Find $\mathbf{P} \triangleq \mathbf{Q}^{-1}$.
7. $\mathbf{k} = \bar{\mathbf{k}}\mathbf{P}$. ∎

This has been written up in Reference [74] as a digital computer program by using Fortran VI. In this algorithm, the dynamical equation $\{\mathbf{A},\mathbf{b}\}$ is first transformed into the controllable canonical form. In the following we discuss a different method of computing $\mathbf{k}$—one that does not require any transformation. First we compute the characteristic polynomial of $(\mathbf{A} + \mathbf{bk})$ in terms of the $n$ unknown variables $k_1, k_2, \cdots, k_n$, where $k_i$ is the $i$th component of $\mathbf{k}$. If the matrix $(\mathbf{A} + \mathbf{bk})$ has the set $\{\bar{\lambda}_i, \text{ for } i = 1, 2, \cdots, n\}$ as its eigenvalues, its characteristic polynomial is equal to

$$\prod_{i=1}^{n} (s - \bar{\lambda}_i)$$

By equating the coefficients of $s^i$, for $i = 0, 1, \cdots, n-1$, a set of $n$ algebraic equations can be obtained and consequently the vector $\mathbf{k} = [k_1 \quad k_2 \quad \cdots \quad k_n]$ can be solved. The existence of the solution $\mathbf{k}$ is insured by the assumption that $\{\mathbf{A}, \mathbf{b}\}$ is controllable. Note that the set of $n$ equations is generally not linear; hence, in solving for $\mathbf{k}$, some searching by the use of a digital computer is generally unavoidable.

If a dynamical equation is controllable, then all the eigenvalues can be arbitrarily assigned by the introduction of state feedback. If a dynamical equation is not controllable, one may wonder how many eigenvalues can be controlled. It is shown in Theorem 5-17 that if a linear, time-invariant dynamical equation is not controllable, by a proper choice of basis vectors, the state equation can be transformed into

$$\dot{\bar{\mathbf{x}}} = \bar{\mathbf{A}}\bar{\mathbf{x}} + \bar{\mathbf{b}}u \tag{7-31}$$

where

$$\bar{\mathbf{A}} = \begin{bmatrix} \bar{\mathbf{A}}_{11} & \bar{\mathbf{A}}_{12} \\ \mathbf{0} & \bar{\mathbf{A}}_{22} \end{bmatrix} \qquad \bar{\mathbf{b}} = \begin{bmatrix} \bar{\mathbf{b}}_1 \\ \mathbf{0} \end{bmatrix} \tag{7-32}$$

and the reduced equation $\dot{\bar{\mathbf{x}}}^1 = \bar{\mathbf{A}}_{11}\bar{\mathbf{x}}^1 + \bar{\mathbf{b}}_1 u$ is controllable. Because of the form of $\bar{\mathbf{A}}$, the set of eigenvalues of $\bar{\mathbf{A}}$ is the union of the sets of the eigenvalues of $\bar{\mathbf{A}}_{11}$ and of $\bar{\mathbf{A}}_{22}$. In view of the form of $\bar{\mathbf{b}}$, it is easy to see that the matrix $\bar{\mathbf{A}}_{22}$ is not affected by the introduction of any state feedback of the form $u = v + \bar{\mathbf{k}}\bar{\mathbf{x}}$. Therefore all the eigenvalues of $\bar{\mathbf{A}}_{22}$ cannot be controlled. On the other hand, if $\{\bar{\mathbf{A}}_{11}, \bar{\mathbf{b}}_1\}$ is controllable, all the eigenvalues of $\bar{\mathbf{A}}_{11}$ can be arbitrarily assigned. This can be seen if we apply to (7-31) an equivalence transformation of the form

$$\mathbf{P} = \begin{bmatrix} \mathbf{P}_1 & \mathbf{0} \\ \mathbf{0} & \mathbf{I} \end{bmatrix} \tag{7-33}$$

in which $\mathbf{I}$ is a unit matrix and $\mathbf{P}_1$ is the equivalence transformation which transforms $\{\mathbf{A}_{11}, \mathbf{b}_1\}$ into the controllable canonical form. The rest of the argument follows directly from Theorem 7-4.

In the design of a system, sometimes it is required only to change unstable eigenvalues (the eigenvalues with nonnegative real parts) to stable eigenvalues (the eigenvalues with negative real parts). This is called *stabilization*. From the above discussion we see that if the matrix $\bar{\mathbf{A}}_{22}$ has unstable eigenvalues, then the equation cannot be stabilized. Whether a dynamical equation is stabilizable can also be seen from its equivalent Jordan-form dynamical equation. If all the Jordan blocks associated with unstable eigenvalues are controllable, then the equation is stabilizable.

Before concluding this subsection, we study the effect of state feedback on a transfer function. Consider the dynamical equation (7-24); its transfer function is $\hat{g}(s) = \mathbf{c}(s\mathbf{I} - \mathbf{A})^{-1}\mathbf{b} + \mathbf{d}$. Since every pole of $\hat{g}(s)$ is an eigenvalue of $\mathbf{A}$, from Theorem 7-4 we conclude that the poles of $\hat{g}(s)$ can be arbitrarily assigned by the introduction of state feedback. It is of interest to note that although the poles of $\hat{g}(s)$ are shifted by state feedback, the zeros of $\hat{g}(s)$ are not affected. That is, the zeros of $\hat{g}(s)$ remain unchanged after the introduction of state feedback. We prove this by showing that the numerator of the transfer function of $FE_1$ in

(7-24) is equal to the numerator of the transfer function of $CFE_1$ in (7-30). The transfer function of $FE_1$ is

$$\hat{g}(s) = \frac{\beta_1 s^{n-1} + \beta_2 s^{n-2} + \cdots + \beta_n}{s^n + \alpha_1 s^{n-1} + \cdots + \alpha_n} + d$$
$$= \frac{ds^n + (\beta_1 + d\alpha_1)s^{n-1} + \cdots + (\beta_n + d\alpha_n)}{s^n + \alpha_1 s^{n-1} + \cdots + \alpha_n} \qquad \textbf{(7-34)}$$

The transfer function of (7-30) is

$$\hat{g}_f(s) = \frac{[\beta_1 + d(\alpha_1 - \bar{\alpha}_1)]s^{n-1} + \cdots + [\beta_n + d(\alpha_n - \bar{\alpha}_n)]}{s^n + \bar{\alpha}_1 s^{n-1} + \cdots + \bar{\alpha}_n} + d$$
$$= \frac{ds^n + (\beta_1 + d\alpha_1)s^{n-1} + \cdots + (\beta_n + d\alpha_n)}{s^n + \bar{\alpha}_1 s^{n-1} + \cdots + \bar{\alpha}_n} \qquad \textbf{(7-35)}$$

which has the same numerator as $\hat{g}(s)$ in (7-34). This proves that the zeros of $\hat{g}(s)$ remain unchanged after the introduction of state feedback.

***Multivariable case.**[2] Consider the $n$-dimensional, linear, time-invariant, multivariable dynamical equation

$$FE: \quad \dot{\mathbf{x}} = \mathbf{A}\mathbf{x} + \mathbf{B}\mathbf{u} \qquad \textbf{(7-36a)}$$
$$\mathbf{y} = \mathbf{C}\mathbf{x} + \mathbf{D}\mathbf{u} \qquad \textbf{(7-36b)}$$

where $\mathbf{A}$, $\mathbf{B}$, $\mathbf{C}$, and $\mathbf{D}$ are, respectively, $n \times n$, $n \times p$, $q \times n$, and $q \times p$ constant real matrices. In state feedback, the input $\mathbf{u}$ in $FE$ is replaced by

$$\mathbf{u} = \mathbf{v} + \mathbf{K}\mathbf{x} \qquad \textbf{(7-37)}$$

where $\mathbf{v}$ stands for a reference input vector and $\mathbf{K}$ is a $p \times n$ real constant matrix, called the feedback gain matrix; and Equation (7-36) becomes

$$FE^f: \quad \dot{\mathbf{x}} = (\mathbf{A} + \mathbf{B}\mathbf{K})\mathbf{x} + \mathbf{B}\mathbf{v} \qquad \textbf{(7-38a)}$$
$$\mathbf{y} = (\mathbf{C} + \mathbf{D}\mathbf{K})\mathbf{x} + \mathbf{D}\mathbf{v} \qquad \textbf{(7-38b)}$$

In the following, we shall show that if the dynamical equation $FE$ is controllable, then the eigenvalues of $(\mathbf{A} + \mathbf{B}\mathbf{K})$ can be arbitrarily assigned by a proper choice of $\mathbf{K}$. This will be established in two steps: First we introduce a state feedback so that the resulting equation is controllable by a single component of $\mathbf{v}$, and then we apply the result established for the single-variable dynamical equation.

If the multivariable dynamical equation $FE$ is controllable, its controllability matrix

$$\mathbf{U} = [\mathbf{B} \quad \mathbf{A}\mathbf{B} \quad \cdots \quad \mathbf{A}^{n-1}\mathbf{B}]$$
$$= [\mathbf{b}_1 \quad \mathbf{b}_2 \quad \cdots \quad \mathbf{b}_p \quad \mathbf{A}\mathbf{b}_1 \quad \cdots \quad \mathbf{A}\mathbf{b}_p \quad \cdots \quad \mathbf{A}^{n-1}\mathbf{b}_1 \quad \cdots \quad \mathbf{A}^{n-1}\mathbf{b}_p]$$

[2] This section closely follows Reference [46].

where $\mathbf{b}_i$ is the $i$th column of $\mathbf{B}$, has rank $n$. If there exists $\mathbf{b}_i$ such that the matrix $[\mathbf{b}_i \quad \mathbf{A}\mathbf{b}_i \quad \cdots \quad \mathbf{A}^{n-1}\mathbf{b}_i]$ has rank $n$, then the control of the state of $FE$ can be achieved by using the $i$th component of $\mathbf{u}$ alone. If there exists no such $\mathbf{b}_i$, then the control of the state cannot be achieved by a single component of $\mathbf{u}$; it requires a joint effort of two or more components of $\mathbf{u}$. However, if we introduce a proper state feedback, the resulting multivariable dynamical equation can always be made to be controllable by a single component of input.

### Theorem 7-5

Let $\{\mathbf{A}, \mathbf{B}\}$ be controllable and let $\mathbf{b}_1, \mathbf{b}_2, \cdots, \mathbf{b}_p$ be the column vectors of $\mathbf{B}$. Then for any $i = 1, 2, \cdots, p$ (where $\mathbf{b}_i \neq \mathbf{0}$), there exists a $p \times n$ real constant matrix $\mathbf{K}_i$ such that $\{\mathbf{A} + \mathbf{B}\mathbf{K}_i, \mathbf{b}_i\}$ is controllable.

### Proof

Without loss of generality, we prove the theorem for $i = 1$. Since $\{\mathbf{A}, \mathbf{B}\}$ is controllable by assumption, the controllability matrix $\mathbf{U} \triangleq [\mathbf{B} \quad \mathbf{AB} \quad \cdots \quad \mathbf{A}^{n-1}\mathbf{B}]$ has rank $n$. Consequently, there are $n$ linearly independent columns in $\mathbf{U}$. Now we choose these $n$ columns according to Scheme 1 introduced in Section 7-2. Let the set of vectors in (7-20) be the vectors chosen. Define[3] the $n \times n$ nonsingular matrix $\mathbf{Q}$

$$\mathbf{Q} \triangleq [\mathbf{b}_1 \quad \mathbf{A}\mathbf{b}_1 \quad \cdots \quad \mathbf{A}^{\nu_1-1}\mathbf{b}_1 \quad \mathbf{b}_2 \quad \cdots \quad \mathbf{A}^{\nu_2-1}\mathbf{b}_2 \quad \mathbf{b}_3 \quad \cdots \quad \mathbf{A}^{\nu_3-1}\mathbf{b}_3] \tag{7-39}$$

and define the $p \times n$ matrix $\mathbf{S}$

$$\mathbf{S} \triangleq [\mathbf{0} \quad \mathbf{0} \quad \cdots \quad \mathbf{0} \quad \underset{\substack{\uparrow \\ \text{the } \nu_1\text{th} \\ \text{column}}}{\mathbf{e}_2} \quad \mathbf{0} \quad \cdots \quad \mathbf{0} \quad \underset{\substack{\uparrow \\ \text{the } (\nu_1+\nu_2)\text{th} \\ \text{column}}}{\mathbf{e}_3} \quad \mathbf{0} \quad \cdots \quad \mathbf{0} \quad \underset{\substack{\uparrow \\ \text{the } (\nu_1+\nu_2+\nu_3=n)\text{th} \\ \text{column}}}{\mathbf{0}}] \tag{7-40}$$

where $\mathbf{e}_i$ is the $i$th column of a $p \times p$ unit matrix. Now we claim that the $\mathbf{K}_1$ given by

$$\mathbf{K}_1 = \mathbf{S}\mathbf{Q}^{-1} \tag{7-41}$$

satisfies the theorem; that is, $\{\mathbf{A} + \mathbf{B}\mathbf{K}_1, \mathbf{b}_1\}$ is controllable. First we write $\mathbf{K}_1\mathbf{Q} = \mathbf{S}$ explicitly as

$$\begin{aligned}\mathbf{K}_1[\mathbf{b}_1 \quad \mathbf{A}\mathbf{b}_1 \quad \cdots \quad \mathbf{A}^{\nu_1-1}\mathbf{b}_1 \quad \mathbf{b}_2 \quad \cdots \quad \mathbf{A}^{\nu_2-1}\mathbf{b}_2 \quad \mathbf{b}_3 \quad \cdots \quad \mathbf{A}^{\nu_3-1}\mathbf{b}_3]& \\ = [\mathbf{0} \quad \mathbf{0} \quad \cdots \quad \mathbf{0} \quad \mathbf{e}_2 \quad \mathbf{0} \quad \cdots \quad \mathbf{0} \quad \mathbf{e}_3 \quad \mathbf{0} \quad \cdots \quad \mathbf{0} \quad \mathbf{0}]&\end{aligned} \tag{7-42}$$

To show that $\{\mathbf{A} + \mathbf{B}\mathbf{K}_1, \mathbf{b}_1\}$ is controllable is the same as showing that the vectors $\mathbf{b}_1, \bar{\mathbf{A}}\mathbf{b}_1, \cdots, \bar{\mathbf{A}}^{n-1}\mathbf{b}_1$, where $\bar{\mathbf{A}} \triangleq \mathbf{A} + \mathbf{B}\mathbf{K}_1$, are linearly

[3] If the column vectors of $\mathbf{Q}$ in (7-39) are chosen according to Scheme 2 in Section 7-2, the subsequent development may not hold.

independent. From (7-42), it is easy to verify that

$$
\begin{aligned}
\mathbf{b}_1 &= \mathbf{b}_1 \\
\bar{\mathbf{A}}\mathbf{b}_1 &= (\mathbf{A} + \mathbf{B}\mathbf{K}_1)\mathbf{b}_1 = \mathbf{A}\mathbf{b}_1 \\
\bar{\mathbf{A}}^2\mathbf{b}_1 &= (\mathbf{A} + \mathbf{B}\mathbf{K}_1)\mathbf{A}\mathbf{b}_1 = \mathbf{A}^2\mathbf{b}_1 \\
&\ \ \vdots \\
\bar{\mathbf{A}}^{\nu_1-1}\mathbf{b}_1 &= (\mathbf{A} + \mathbf{B}\mathbf{K}_1)\mathbf{A}^{\nu_1-2}\mathbf{b}_1 = \mathbf{A}^{\nu_1-1}\mathbf{b}_1 \\
\bar{\mathbf{A}}^{\nu_1}\mathbf{b}_1 &= (\mathbf{A} + \mathbf{B}\mathbf{K}_1)\mathbf{A}^{\nu_1-1}\mathbf{b}_1 = \mathbf{A}^{\nu_1}\mathbf{b}_1 + \mathbf{B}\mathbf{e}_2 = \mathbf{b}_2 + \cdots \\
\bar{\mathbf{A}}^{\nu_1+1}\mathbf{b}_1 &= (\mathbf{A} + \mathbf{B}\mathbf{K}_1)(\mathbf{b}_2 + \mathbf{A}^{\nu_1}\mathbf{b}_1) = \mathbf{A}\mathbf{b}_2 + \cdots \\
&\ \ \vdots \\
\bar{\mathbf{A}}^{n-1}\mathbf{b}_1 &= (\mathbf{A} + \mathbf{B}\mathbf{K}_1)(\mathbf{A}^{\nu_3-2}\mathbf{b}_3 + \cdots) = \mathbf{A}^{\nu_3-1}\mathbf{b}_3 + \cdots
\end{aligned}
$$

where, in the above expression, the ellipsis $(\cdot\ \cdot\ \cdot)$ is used to denote linear combinations of the preceding vectors. Clearly these vectors are linearly independent. Hence $\{\mathbf{A} + \mathbf{B}\mathbf{K}_1, \mathbf{b}_1\}$ is controllable. Q.E.D.

With this theorem, we are ready to extend Theorem 7-4 to the multivariable case. Since the matrices **A, B** and **K** in the following theorem are assumed to be real, if a complex number is assigned as an eigenvalue of $(\mathbf{A} + \mathbf{B}\mathbf{K})$, its complex conjugate must also be assigned.

### Theorem 7-6

If the dynamical equation *FE* in (7-36) is controllable, by a linear state feedback of the form $\mathbf{u} = \mathbf{v} + \mathbf{K}\mathbf{x}$, where **K** is a $p \times n$ real constant matrix, the eigenvalues of $(\mathbf{A} + \mathbf{B}\mathbf{K})$ can be arbitrarily assigned.

### Proof

By introducing the state feedback $\mathbf{u} = \mathbf{w} + \mathbf{K}_1\mathbf{x}$, the state equation of *FE* in (7-36) becomes

$$\dot{\mathbf{x}} = (\mathbf{A} + \mathbf{B}\mathbf{K}_1)\mathbf{x} + \mathbf{B}\mathbf{w} \tag{7-43}$$

Since $\{\mathbf{A}, \mathbf{B}\}$ is controllable by assumption, the gain matrix $\mathbf{K}_1$ in (7-43) can be chosen so that $\{\bar{\mathbf{A}}, \mathbf{b}_1\}$ is controllable, where $\bar{\mathbf{A}} = \mathbf{A} + \mathbf{B}\mathbf{K}_1$ and $\mathbf{b}_1$ is the first column of **B** (Theorem 7-5). We introduce, as shown in Figure 7-2, another state feedback $\mathbf{w} = \mathbf{v} + \mathbf{M}\mathbf{x}$, with **M** of the form

$$\mathbf{M} = \begin{bmatrix} \mathbf{m}_1 \\ \mathbf{0} \\ \vdots \\ \mathbf{0} \end{bmatrix} \tag{7-44}$$

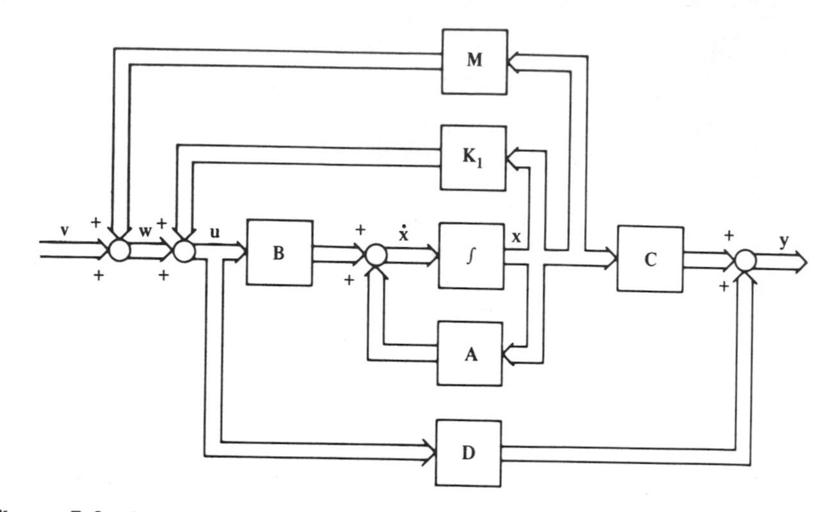


**Figure 7-2** State feedback of a multivariable dynamical equation.

where $\mathbf{m}_1$ and $\mathbf{0}$ are $1 \times n$ row vectors; then (7-43) becomes

$$\dot{\mathbf{x}} = (\bar{\mathbf{A}} + \mathbf{BM})\mathbf{x} + \mathbf{Bv} = (\bar{\mathbf{A}} + \mathbf{b}_1\mathbf{m}_1)\mathbf{x} + \mathbf{Bv}$$

Since $\{\bar{\mathbf{A}}, \mathbf{b}_1\}$ is controllable, the eigenvalues of $(\bar{\mathbf{A}} + \mathbf{b}_1\mathbf{m}_1)$ can be arbitrarily assigned by a proper choice of $\mathbf{m}_1$ (Theorem 7-4). By combining the state feedback $\mathbf{u} = \mathbf{w} + \mathbf{K}_1\mathbf{x}$ and the state feedback $\mathbf{w} = \mathbf{v} + \mathbf{Mx}$ as

$$\mathbf{u} = \mathbf{v} + (\mathbf{K}_1 + \mathbf{M})\mathbf{x} \triangleq \mathbf{v} + \mathbf{Kx} \tag{7-45}$$

the theorem is proved. Q.E.D.

The feedback gain matrices $\mathbf{K}_1$ and $\mathbf{M}$ can be computed using the algorithms developed in Theorems 7-5 and 7-4; hence the gain matrix $\mathbf{K} = \mathbf{K}_1 + \mathbf{M}$ can be computed by the use of a digital computer. A computer program has been written up in Reference [74]. The gain matrix $\mathbf{K}$ in Theorem 7-6 is not unique, there are many other possible ways to choose $\mathbf{K}$; for example, if a matrix $\mathbf{K}_2$ is chosen such that $\{\mathbf{A} + \mathbf{BK}_2, \mathbf{b}_2\}$ is controllable, then a different $\mathbf{K}$ will be obtained. In the engineering literature, there are some other methods to computer $\mathbf{K}$; see References [17] and [111]. However, the method introduced here seems to be the simplest.

We give a remark concerning the effect of state feedback on the zeros of the transfer function matrix $\hat{\mathbf{G}}(s) = \mathbf{C}(s\mathbf{I} - \mathbf{A})^{-1}\mathbf{B} + \mathbf{D}$ of a multivariable dynamical equation. We recall that in the single-variable case, the zeros (the roots of the numerator) of a transfer function are not affected by introducing state feedback. This is however not so in the multivariable case; the numerators of some elements of $\hat{\mathbf{G}}(s)$ will be changed after introducing state feedback. This can be seen by constructing a simple example. It is of interest to note that in the case of $p = q$, that is, $\hat{\mathbf{G}}(s)$ is a square matrix, although the numerators of some elements of $\hat{\mathbf{G}}(s)$ are changed after introducing state feedback, the polynomial $\Delta(s) \det \hat{\mathbf{G}}(s)$, where $\Delta(s)$ is the characteristic polynomial of $\hat{\mathbf{G}}(s)$, remains unchanged; that is, $\Delta(s) \det \hat{\mathbf{G}}(s) = \Delta_f(s) \det \hat{\mathbf{G}}_f(s)$, where $\hat{\mathbf{G}}_f(s)$ is the transfer function matrix of (7-38) and $\Delta_f(s)$ is the characteristic polynomial of $\hat{\mathbf{G}}_f(s)$. See Reference [41] and W. A. Wolovich and P. L. Falb, "On the structure of multivariable systems," *SIAM J. Control,* vol. 7, pp. 437–451, 1969.

## 7-4 State Estimators

In the previous section we introduced state feedback under the assumption that all the state variables are available as outputs. This assumption often does not hold in practice, however, either because the state variables are not accessible for direct measurement or because the number of measuring devices is limited. Thus, in order to apply state feedback to stabilize, to optimize, or to decouple (see Section 7-6) a system, a reasonable substitute for the state vector often has to be found. In this section we shall show how the available inputs and outputs of a dynamical equation can be used to drive a device so that the outputs of the device will approximate the state vector. The device that constructs an approximation of the state vector is called a *state estimator.* In this section, various state estimators will be introduced. We shall also show that in state feedback, the results will be the same even if an estimated state, instead of the real state, is used in the feedback.

The class of systems we study is assumed to have no noise. If there is noise in a system, what we shall discuss must be modified appropriately. However the basic equations of the estimators remain the same.

In this section we use the circumflex over a variable to denote an estimate of the variable. For example, $\hat{\mathbf{x}}$ is an estimate of $\mathbf{x}$; $\hat{\bar{\mathbf{x}}}$ is an estimate of $\bar{\mathbf{x}}$.

### Single-variable case

#### The $n$-dimensional estimator

Consider the linear, time-invariant single-variable dynamical equation

$$FE_1: \qquad \dot{\mathbf{x}} = \mathbf{A}\mathbf{x} + \mathbf{b}u \qquad \text{(7-46a)}$$

$$y = \mathbf{c}\mathbf{x} \qquad \text{(7-46b)}$$

where $\mathbf{x}$ is the $n \times 1$ state vector, $u$ is the scalar input, $y$ is the scalar output, $\mathbf{A}$ is an $n \times n$ real constant matrix, $\mathbf{b}$ is an $n \times 1$ real constant column vector, and $\mathbf{c}$ is a $1 \times n$ real constant row vector. Without loss of generality, the direct transmission part has been assumed to be zero. We assume now that the state variables are not accessible. Note that although the state variables are not accessible, the matrices $\mathbf{A}$, $\mathbf{b}$, and $\mathbf{c}$ are assumed to be completely known. Hence the problem is that of estimating or generating $\mathbf{x}(t)$ from the available input $u$ and the output $y$ with the knowledge of the matrices $\mathbf{A}$, $\mathbf{b}$, and $\mathbf{c}$. If we know the matrices $\mathbf{A}$ and $\mathbf{b}$, we can duplicate the original system as shown in Figure 7-3. We

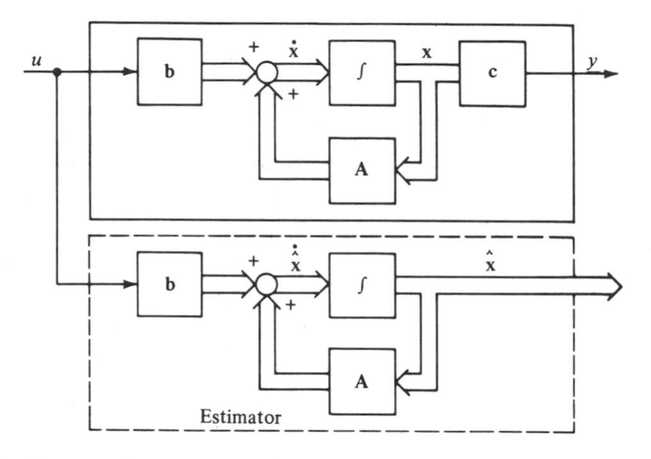


**Figure 7-3** An open-loop state estimator.

called the system an *open-loop estimator*. Now if the original equation $FE_1$ and the estimator have the same initial state and are driven by the same input, the output of the estimator $\hat{\mathbf{x}}(t)$ will be equal to the real state $\mathbf{x}(t)$ for all $t$. Therefore, the remaining question is how to find the initial state of $FE_1$ and set the initial state of the estimator to that state. This problem was solved in Section 5-4. It is shown there that if the dynamical equation $FE_1$ is observable, the initial state of $FE_1$ can be computed from its input and output. Consequently, if the equation $FE_1$ is observable, an open-loop estimator can be used to generate the state vector.

There are, however, two disadvantages in using an open-loop estimator. First, the initial state must be computed and set each time we use the estimator. This is very inconvenient. Second, and more seriously, if the matrix $\mathbf{A}$ has eigenvalues with positive real parts, then even for a very small difference between $\mathbf{x}(t_0)$ and $\hat{\mathbf{x}}(t_0)$ at some $t_0$, which may be caused by disturbance or incorrect estimation of the initial state, the difference between the real state $\mathbf{x}(t)$ and the estimated $\hat{\mathbf{x}}(t)$ will increase with time. Therefore an open-loop estimator is, in general, not very satisfactory.

Another possible way to generate the $n$-dimensional state vector is to differentiate the output and the input $(n - 1)$ times. If the dynamical equation is observable, then from $u(t)$, $y(t)$, and their derivatives, the state vector can be computed (see Problem 5-19). However, pure differentiators are not easy to build. Furthermore, the estimated state might be severely distorted by noise if pure differentiators are used.

We see from Figure 7-3 that although the input and the output of $FE_1$ are available, we use only the input in the open-loop estimator. It is conceivable that if both the output and input are utilized, the performance of an estimator can be improved.

Consider the state estimator shown in Figure 7-4. The estimator is driven by the input as well as the output of the original system. The

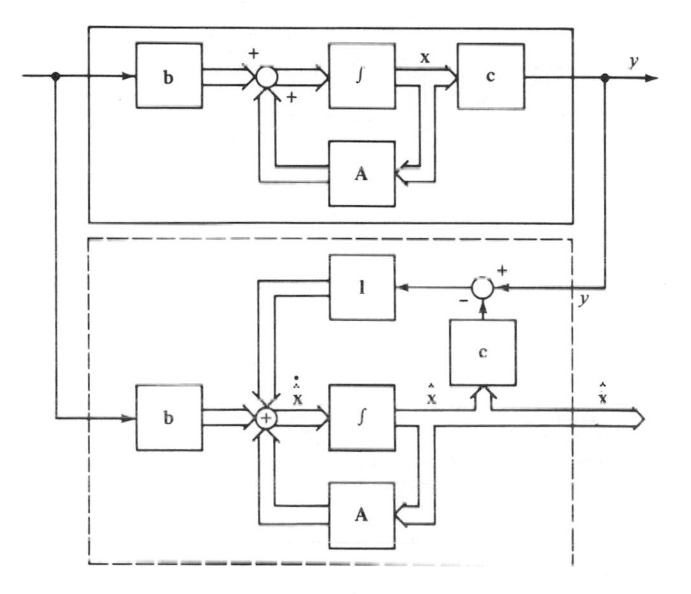


**Figure 7-4** An asymptotic state estimator.

output of $FE_1$, $y = \mathbf{cx}$, is compared with $\hat{y} \triangleq \mathbf{c}\hat{\mathbf{x}}$, and their difference is used to serve as a correct term. The difference of $y$ and $\mathbf{c}\hat{\mathbf{x}}$, $y - \mathbf{c}\hat{\mathbf{x}}$, is multiplied by an $n \times 1$ real constant column vector $\mathbf{l}$ and fed into the input of the integrators of the estimator. This estimator will be called an *asymptotic state estimator* for the reason to be seen later.

The dynamical equation of the asymptotic state estimator shown in Figure 7-4 is given by

$$\dot{\hat{\mathbf{x}}} = \mathbf{A}\hat{\mathbf{x}} + \mathbf{l}[y - \mathbf{c}\hat{\mathbf{x}}] + \mathbf{b}u \tag{7-47}$$

which can be written as

$$\dot{\hat{\mathbf{x}}} = (\mathbf{A} - \mathbf{lc})\hat{\mathbf{x}} + \mathbf{l}y + \mathbf{b}u \tag{7-48}$$

The asymptotic estimator in Figure 7-4 can be redrawn as in Figure 7-5 either by a block diagram manipulation or from (7-48). Define

$$\tilde{\mathbf{x}} \triangleq \mathbf{x} - \hat{\mathbf{x}} \tag{7-49}$$

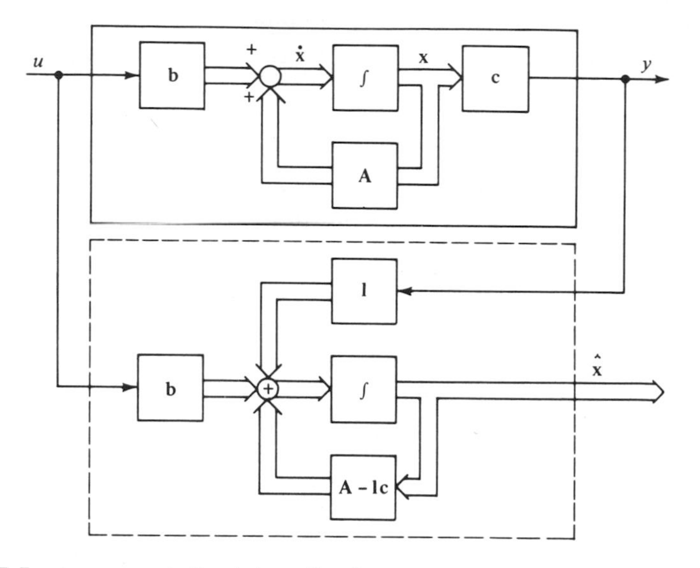


**Figure 7-5** An asymptotic state estimator.

Clearly $\tilde{\mathbf{x}}$ is the error between the real state and the estimated state. Subtracting (7-48) from (7-46), we obtain

$$\dot{\tilde{\mathbf{x}}} = (\mathbf{A} - \mathbf{lc})\tilde{\mathbf{x}} \tag{7-50}$$

If the eigenvalues of $(\mathbf{A} - \mathbf{lc})$ can be chosen arbitrarily, then the behavior of the error $\tilde{\mathbf{x}}$ can be controlled. For example, if all the eigenvalues of $(\mathbf{A} - \mathbf{lc})$ have negative real parts that are smaller than $-\sigma$, then all the components of $\tilde{\mathbf{x}}$ will approach zero in the rate of $e^{-\sigma t}$. Consequently, even if there is a large error between $\hat{\mathbf{x}}(t_0)$ and $\mathbf{x}(t_0)$ at initial time $t_0$, the vector $\hat{\mathbf{x}}$ will approach $\mathbf{x}$ rapidly. Thus, if the eigenvalues of $(\mathbf{A} - \mathbf{lc})$ can be chosen properly, an asymptotic state estimator is much more desirable than an open-loop estimator.

### Theorem 7-7

If the single-variable, linear, time-invariant dynamical equation $FE_1$ in (7-46) is observable, then an asymptotic state estimator with any eigenvalues can be constructed.

Proof

If the dynamical equation $FE_1$ is observable, by an equivalence transformation $\bar{\mathbf{x}} = \mathbf{P}\mathbf{x}$, it can be transformed into the following equivalent observable canonical-form dynamical equation (Theorem 7-2)

$$OFE_1: \quad \dot{\bar{\mathbf{x}}} = \begin{bmatrix} 0 & 0 & \cdots & 0 & -\alpha_n \\ 1 & 0 & \cdots & 0 & -\alpha_{n-1} \\ 0 & 1 & \cdots & 0 & -\alpha_{n-2} \\ \cdot & \cdot & & \cdot & \cdot \\ \cdot & \cdot & & \cdot & \cdot \\ \cdot & \cdot & & \cdot & \cdot \\ 0 & 0 & \cdots & 1 & -\alpha_1 \end{bmatrix} \bar{\mathbf{x}} + \begin{bmatrix} \beta_n \\ \beta_{n-1} \\ \beta_{n-2} \\ \cdot \\ \cdot \\ \cdot \\ \beta_1 \end{bmatrix} u \tag{7-51a}$$

$$y = [0 \quad 0 \quad \cdots \quad 0 \quad 1]\bar{\mathbf{x}} \tag{7-51b}$$

Let $\bar{\mathbf{A}}$, $\bar{\mathbf{b}}$, and $\bar{\mathbf{c}}$ denote the matrices in Equation (7-51). Clearly we have $\mathbf{A} = \mathbf{P}^{-1}\bar{\mathbf{A}}\mathbf{P}$, $\mathbf{b} = \mathbf{P}^{-1}\bar{\mathbf{b}}$ and $\mathbf{c} = \bar{\mathbf{c}}\mathbf{P}$. Define $\mathbf{l} \triangleq \mathbf{P}^{-1}\bar{\mathbf{l}}$. Then it is easy to show that $(\mathbf{A} - \mathbf{l}\mathbf{c}) = \mathbf{P}^{-1}(\bar{\mathbf{A}} - \bar{\mathbf{l}}\bar{\mathbf{c}})\mathbf{P}$, and that the characteristic polynomial of $(\mathbf{A} - \mathbf{l}\mathbf{c})$ is equal to the characteristic polynomial of $(\bar{\mathbf{A}} - \bar{\mathbf{l}}\bar{\mathbf{c}})$. Let the characteristic polynomial of $(\bar{\mathbf{A}} - \bar{\mathbf{l}}\bar{\mathbf{c}})$ with a set of desired eigenvalues be

$$s^n + \bar{\alpha}_1 s^{n-1} + \cdots + \bar{\alpha}_{n-1}s + \bar{\alpha}_n$$

If the vector $\bar{\mathbf{l}}'$, the transpose of $\bar{\mathbf{l}}$, is chosen as

$$\bar{\mathbf{l}}' = [\bar{\alpha}_n - \alpha_n \quad \bar{\alpha}_{n-1} - \alpha_{n-1} \quad \cdots \quad \bar{\alpha}_1 - \alpha_1] \tag{7-52}$$

then we have

$$\bar{\mathbf{A}} - \bar{\mathbf{l}}\bar{\mathbf{c}} = \begin{bmatrix} 0 & 0 & \cdots & 0 & -\bar{\alpha}_n \\ 1 & 0 & \cdots & 0 & -\bar{\alpha}_{n-1} \\ 0 & 1 & \cdots & 0 & -\bar{\alpha}_{n-2} \\ \cdot & \cdot & & \cdot & \cdot \\ \cdot & \cdot & & \cdot & \cdot \\ \cdot & \cdot & & \cdot & \cdot \\ 0 & 0 & \cdots & 1 & -\bar{\alpha}_1 \end{bmatrix}$$

which has the set of desired eigenvalues. Q.E.D.

If $\mathbf{l} = \mathbf{P}^{-1}\bar{\mathbf{l}}$ is computed as in the above proof, then a state estimator with a set of desired eigenvalues can be constructed by using Equation (7-48). In order to have an easier presentation of $(n-1)$-dimensional state estimators later, we shall discuss the construction of a state estimator by using the vector $\bar{\mathbf{l}}$ given in Equation (7-52). Recall that the vector $\bar{\mathbf{l}}$ is computed by using the observable canonical form dynamical equation

(7-51), hence if $\bar{\mathbf{l}}$ is used, the dynamical equation of the state estimator should be

$$\dot{\hat{\bar{\mathbf{x}}}} = (\bar{\mathbf{A}} - \bar{\mathbf{l}}\bar{\mathbf{c}})\hat{\bar{\mathbf{x}}} + \bar{\mathbf{l}}y + \bar{\mathbf{b}}u$$

This estimator gives an estimate of $\bar{\mathbf{x}}$ (not $\mathbf{x}$). Since $\mathbf{x}$ and $\bar{\mathbf{x}}$ are related by $\mathbf{x} = \mathbf{P}^{-1}\bar{\mathbf{x}}$, if $\hat{\bar{\mathbf{x}}}$ is multiplied by $\mathbf{P}^{-1}$, or equivalently, if $\hat{\bar{\mathbf{x}}}$ passes through a device with gain $\mathbf{P}^{-1}$, then the output $\hat{\mathbf{x}} = \mathbf{P}^{-1}\hat{\bar{\mathbf{x}}}$ gives an estimate of $\mathbf{x}$.

If the dynamical equation to be estimated is observable, the eigenvalues of its state estimator can be arbitrarily chosen. It is clear that if these eigenvalues are chosen to have negative real parts, then no matter what the initial state of the estimator is, the output of the estimator $\hat{\mathbf{x}}$ will approach the real state $\mathbf{x}$ asymptotically. This is the reason why we called this estimator an asymptotic state estimator. In using an asymptotic estimator, there is no need of setting its initial state, because no matter what the initial state is, its output will tend to the real state rapidly.

A remark is in order concerning the eigenvalues of an asymptotic estimator. It is clear that the larger the negative real parts of the eigenvalues of an estimator, the faster the estimated state approaches the real state. However, it is not clear whether or not these eigenvalues with large negative real parts will cause the system easier to saturate or make the system more susceptible to noises. In short, what are the best eigenvalues of an estimator is not known at present. For an attempt of solving this problem, see Reference [37].

#### Example 1

Consider the single-variable dynamical equation

$$\dot{\mathbf{x}} = \begin{bmatrix} 1 & 2 & 0 \\ 3 & -1 & 1 \\ 0 & 2 & 0 \end{bmatrix} \mathbf{x} + \begin{bmatrix} 2 \\ 1 \\ 1 \end{bmatrix} u$$
$$y = [0 \quad 0 \quad 1]\mathbf{x}$$

which has the following equivalent observable canonical-form dynamical equation (see Section 7-2, Example 1)

$$\dot{\bar{\mathbf{x}}} = \begin{bmatrix} 0 & 0 & -2 \\ 1 & 0 & 9 \\ 0 & 1 & 0 \end{bmatrix} \bar{\mathbf{x}} + \begin{bmatrix} 3 \\ 2 \\ 1 \end{bmatrix} u$$
$$y = [0 \quad 0 \quad 1]\bar{\mathbf{x}}$$

This equation is obtained by the equivalence transformation $\bar{\mathbf{x}} = \mathbf{P}\mathbf{x}$, where

$$\mathbf{Q} = \mathbf{P}^{-1} = \begin{bmatrix} 1/6 & 1/6 & 7/6 \\ 0 & 1/2 & 0 \\ 0 & 0 & 1 \end{bmatrix}$$

(See Problem 7-3.)

Let the desired eigenvalues of an asymptotic estimator be $-3$, $-4$, and $-5$. Then the characteristic polynomial of $(\bar{\mathbf{A}} - \bar{\mathbf{l}}\bar{\mathbf{c}})$ is

$$(s + 3)(s + 4)(s + 5) = s^3 + 12s^2 + 47s + 60$$

Consequently, the column vector $\bar{\mathbf{l}}$ must be chosen as

$$\bar{\mathbf{l}}' = [\bar{\alpha}_3 - \alpha_3 \quad \bar{\alpha}_2 - \alpha_2 \quad \bar{\alpha}_1 - \alpha_1] = [58 \quad 56 \quad 12]$$

and the dynamical equation of the estimator is

$$\begin{bmatrix} \dot{\hat{\bar{x}}}_1 \\ \dot{\hat{\bar{x}}}_2 \\ \dot{\hat{\bar{x}}}_3 \end{bmatrix} = \begin{bmatrix} 0 & 0 & -60 \\ 1 & 0 & -47 \\ 0 & 1 & -12 \end{bmatrix} \begin{bmatrix} \hat{\bar{x}}_1 \\ \hat{\bar{x}}_2 \\ \hat{\bar{x}}_3 \end{bmatrix} + \begin{bmatrix} 58 \\ 56 \\ 12 \end{bmatrix} y + \begin{bmatrix} 3 \\ 2 \\ 1 \end{bmatrix} u$$

This estimator gives an estimate of $\bar{\mathbf{x}}$. If $\hat{\bar{\mathbf{x}}}$ is premultiplied by $\mathbf{Q} = \mathbf{P}^{-1}$, we obtain immediately an estimate of $\mathbf{x}$. The computer simulation block diagram of the estimator is given in Figure 7-6.

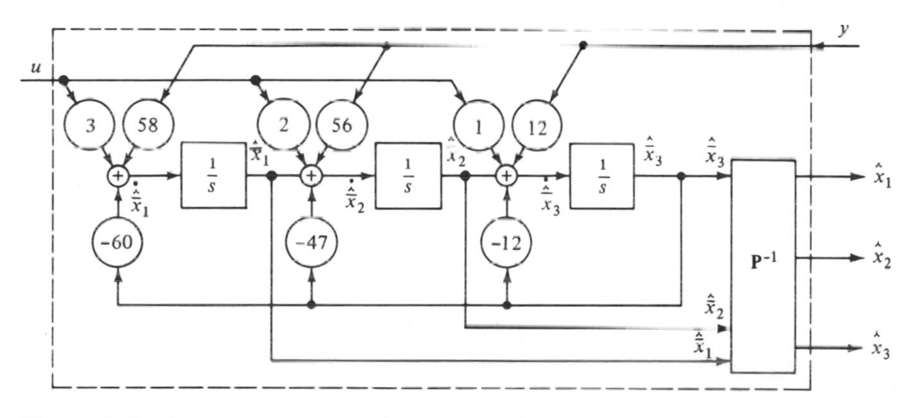


**Figure 7-6** An estimator.

## The separation property

We digress at this point to check whether there is any difference in state feedback between using the estimated state $\hat{\mathbf{x}}$ and the real state $\mathbf{x}$. Consider the controllable and observable $n$-dimensional, linear, time-invariant dynamical equation

$$FE_1: \qquad \dot{\mathbf{x}} = \mathbf{A}\mathbf{x} + \mathbf{b}u \tag{7-53a}$$
$$y = \mathbf{c}\mathbf{x} \tag{7-53b}$$

It is assumed that, by state feedback, a gain vector $\mathbf{k}$ is found such that the matrix $(\mathbf{A} + \mathbf{b}\mathbf{k})$ has a set of desired eigenvalues, or correspondingly, a desired characteristic polynomial $\Delta_1(s)$. Suppose now the state vector $\mathbf{x}$ is not available and we construct the asymptotic state estimator

$$\dot{\hat{\mathbf{x}}} = (\mathbf{A} - \mathbf{l}\mathbf{c})\hat{\mathbf{x}} + \mathbf{l}y + \mathbf{b}u \tag{7-54}$$

with the characteristic polynomial $\Delta_2(s) = \det(s\mathbf{I} - \mathbf{A} + \mathbf{lc})$, to generate an estimate of the state vector. Since the state $\mathbf{x}$ is not available, we use

$$u = v + \mathbf{k}\hat{\mathbf{x}} \tag{7-55}$$

instead of $u = v + \mathbf{kx}$ in the state feedback. Two questions may be raised in using this estimator: (1) The gain vector $\mathbf{k}$ is designed with respect to the real state, now we use the estimated state in the feedback; will this give the same result? In other words, will the state feedback system still have the desired characteristic polynomial $\Delta_1(s)$? (2) What is the effect of introducing the state estimator in the system? Will the eigenvalues of the state estimator appear in the entire system without any change? It turns out that there is no difference in state feedback between using $\hat{\mathbf{x}}$ or $\mathbf{x}$ and that the eigenvalues of the asymptotic state estimator will appear in the entire system without any change. This will be proved in the following.

By substituting (7-53b) and (7-55) into (7-53a) and (7-54), the dynamical equation of the entire system shown in Figure 7-7 can be

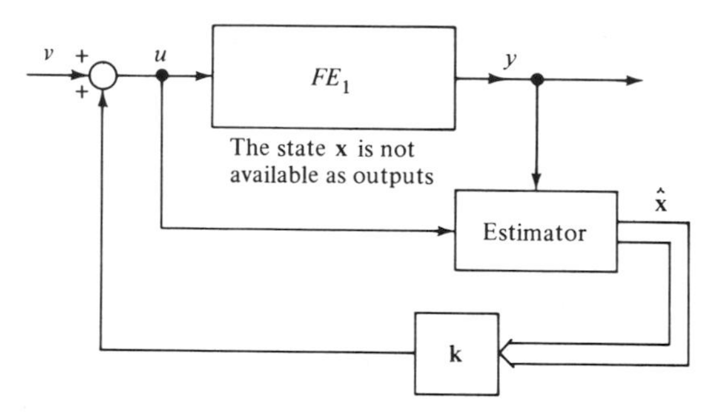


**Figure 7-7** State feedback introduced through an estimator.

obtained as

$$\begin{bmatrix} \dot{\mathbf{x}} \\ \dot{\hat{\mathbf{x}}} \end{bmatrix} = \begin{bmatrix} \mathbf{A} & \mathbf{bk} \\ \mathbf{lc} & \mathbf{A} - \mathbf{lc} + \mathbf{bk} \end{bmatrix} \begin{bmatrix} \mathbf{x} \\ \hat{\mathbf{x}} \end{bmatrix} + \begin{bmatrix} \mathbf{b} \\ \mathbf{b} \end{bmatrix} v \tag{7-56}$$

By the following equivalence transformation

$$\begin{bmatrix} \mathbf{x} \\ \tilde{\mathbf{x}} \end{bmatrix} \triangleq \begin{bmatrix} \mathbf{I} & \mathbf{0} \\ \mathbf{I} & -\mathbf{I} \end{bmatrix} \begin{bmatrix} \mathbf{x} \\ \hat{\mathbf{x}} \end{bmatrix} = \begin{bmatrix} \mathbf{x} \\ \mathbf{x} - \hat{\mathbf{x}} \end{bmatrix}$$

equation (6-56) becomes

$$\begin{bmatrix} \dot{\mathbf{x}} \\ \dot{\tilde{\mathbf{x}}} \end{bmatrix} = \begin{bmatrix} \mathbf{A} + \mathbf{bk} & -\mathbf{bk} \\ \mathbf{0} & \mathbf{A} - \mathbf{lc} \end{bmatrix} \begin{bmatrix} \mathbf{x} \\ \tilde{\mathbf{x}} \end{bmatrix} + \begin{bmatrix} \mathbf{b} \\ 0 \end{bmatrix} v \tag{7-57}$$

From (7-57), we see immediately that the characteristic polynomial of the entire system is the product of those of $(\mathbf{A} + \mathbf{bk})$ and $(\mathbf{A} - \mathbf{lc})$.

This proves our assertion. Hence we conclude that as far as the eigenvalues are concerned, there is no difference in state feedback between using the estimated state $\hat{\mathbf{x}}$ and the real state $\mathbf{x}$. Consequently, the design of state feedback and the design of a state estimator can be carried out independently, and the characteristic polynomial of the entire system will be the product of those of state feedback and state estimator. This property is often called the *separation property.*

### The $(n-1)$-dimensional estimator

Let us go back to the asymptotic state estimator. The estimator introduced in Theorem 7-7 has the same dimension as the dynamical equation to be estimated. If we examine carefully the dynamical equation (7-51), we see that the output $y$ already gives the state variable $\bar{x}_n$, the last component of the state vector $\bar{\mathbf{x}}$. Therefore we need to estimate only the first $(n-1)$ components of $\bar{\mathbf{x}}$—namely, $\bar{x}_1$, $\bar{x}_2$, . . . , $\bar{x}_{n-1}$. We shall show that if the original dynamical equation is observable, these $(n-1)$ state variables can be estimated by using an $(n-1)$-dimensional asymptotic state estimator with a set of arbitrarily chosen eigenvalues. This is based on the following theorem.

#### Theorem 7-8

The observable canonical-form dynamical equation (7-51) is transformable into the form

$$
\begin{bmatrix} \dot{\check{x}}_1 \\ \dot{\check{x}}_2 \\ \dot{\check{x}}_3 \\ \vdots \\ \dot{\check{x}}_{n-1} \\ \dot{\check{x}}_n \end{bmatrix}
=
\begin{bmatrix}
0 & 0 & \cdots & 0 & -\hat{\alpha}_{n-1} & -\hat{\alpha}_{n-1}\hat{\alpha}_1 - \alpha_n + \hat{\alpha}_{n-1}\alpha_1 \\
1 & 0 & \cdots & 0 & -\hat{\alpha}_{n-2} & \hat{\alpha}_{n-1} - \hat{\alpha}_{n-2}\hat{\alpha}_1 - \alpha_{n-1} + \hat{\alpha}_{n-2}\alpha_1 \\
0 & 1 & \cdots & 0 & -\hat{\alpha}_{n-3} & \hat{\alpha}_{n-2} - \hat{\alpha}_{n-3}\hat{\alpha}_1 - \alpha_{n-2} + \hat{\alpha}_{n-3}\alpha_1 \\
\vdots & \vdots & & \vdots & \vdots & \vdots \\
0 & 0 & \cdots & 1 & -\hat{\alpha}_1 & \hat{\alpha}_2 - \hat{\alpha}_1\hat{\alpha}_1 - \alpha_2 + \hat{\alpha}_1\alpha_1 \\
0 & 0 & \cdots & 0 & 1 & -\alpha_1 + \hat{\alpha}_1
\end{bmatrix}
\check{\mathbf{x}}
+
\begin{bmatrix}
\beta_n - \hat{\alpha}_{n-1}\beta_1 \\
\beta_{n-1} - \hat{\alpha}_{n-2}\beta_1 \\
\beta_{n-2} - \hat{\alpha}_{n-3}\beta_1 \\
\vdots \\
\beta_2 - \hat{\alpha}_1\beta_1 \\
\beta_1
\end{bmatrix} u \qquad \text{(7-58a)}
$$

$$
y = [0 \quad 0 \quad \cdots \quad 0 \quad 1]\check{\mathbf{x}} \qquad \text{(7-58b)}
$$

by the equivalence transformation $\check{\mathbf{x}} = \mathbf{P}_1\bar{\mathbf{x}}$, where

$$\mathbf{P}_1 = \begin{bmatrix} 1 & 0 & \cdots & 0 & -\hat{\alpha}_{n-1} \\ 0 & 1 & \cdots & 0 & -\hat{\alpha}_{n-2} \\ \cdot & \cdot & & \cdot & \cdot \\ \cdot & \cdot & & \cdot & \cdot \\ \cdot & \cdot & & \cdot & \cdot \\ 0 & 0 & \cdots & 1 & -\hat{\alpha}_1 \\ 0 & 0 & \cdots & 0 & 1 \end{bmatrix} \tag{7-59}$$

and $\hat{\alpha}_1, \hat{\alpha}_2, \cdots, \hat{\alpha}_{n-1}$ are arbitrary real numbers.

Proof

Because of the form of $\mathbf{P}_1$, the inverse of $\mathbf{P}_1$ is

$$\mathbf{P}_1^{-1} = \begin{bmatrix} 1 & 0 & \cdots & 0 & \hat{\alpha}_{n-1} \\ 0 & 1 & \cdots & 0 & \hat{\alpha}_{n-2} \\ \cdot & \cdot & & \cdot & \cdot \\ \cdot & \cdot & & \cdot & \cdot \\ \cdot & \cdot & & \cdot & \cdot \\ 0 & 0 & \cdots & 1 & \hat{\alpha}_1 \\ 0 & 0 & \cdots & 0 & 1 \end{bmatrix}$$

With this, the theorem can be directly verified. Q.E.D.

Every observable single-variable linear time-invariant dynamical equation can be transformed into the observable canonical form; therefore we conclude from Theorem 7-8 that every observable single-variable dynamical equation can be transformed into the form of (7-58).

If the state vector $\check{\mathbf{x}}$ is to be estimated, since $y = \check{x}_n$, we have to estimate only $\check{x}_1, \check{x}_2, \cdots, \check{x}_{n-1}$. This can be achieved by using an $(n-1)$-dimensional estimator. The dynamical-equation description of this $(n-1)$ dimensional asymptotic state estimator will be chosen as

$$\begin{bmatrix} \dot{\hat{\check{x}}}_1 \\ \dot{\hat{\check{x}}}_2 \\ \dot{\hat{\check{x}}}_3 \\ \cdot \\ \cdot \\ \cdot \\ \dot{\hat{\check{x}}}_{n-1} \end{bmatrix} = \begin{bmatrix} 0 & 0 & \cdots & 0 & -\hat{\alpha}_{n-1} \\ 1 & 0 & \cdots & 0 & -\hat{\alpha}_{n-2} \\ 0 & 1 & \cdots & 0 & -\hat{\alpha}_{n-3} \\ \cdot & \cdot & & \cdot & \cdot \\ \cdot & \cdot & & \cdot & \cdot \\ \cdot & \cdot & & \cdot & \cdot \\ 0 & 0 & \cdots & 1 & -\hat{\alpha}_1 \end{bmatrix} \begin{bmatrix} \hat{\check{x}}_1 \\ \hat{\check{x}}_2 \\ \hat{\check{x}}_3 \\ \cdot \\ \cdot \\ \cdot \\ \hat{\check{x}}_{n-1} \end{bmatrix}$$
$$+ \begin{bmatrix} -\hat{\alpha}_{n-1}\hat{\alpha}_1 - \alpha_n + \hat{\alpha}_{n-1}\alpha_1 \\ \hat{\alpha}_{n-1} - \hat{\alpha}_{n-2}\hat{\alpha}_1 - \alpha_{n-1} + \hat{\alpha}_{n-2}\alpha_1 \\ \hat{\alpha}_{n-2} - \hat{\alpha}_{n-3}\hat{\alpha}_1 - \alpha_{n-2} + \hat{\alpha}_{n-3}\alpha_1 \\ \cdot \\ \cdot \\ \cdot \\ \hat{\alpha}_2 - \hat{\alpha}_1^2 - \alpha_2 + \hat{\alpha}_1\alpha_1 \end{bmatrix} y + \begin{bmatrix} \beta_n - \hat{\alpha}_{n-1}\beta_1 \\ \beta_{n-1} - \hat{\alpha}_{n-2}\beta_1 \\ \beta_{n-2} - \hat{\alpha}_{n-3}\beta_1 \\ \cdot \\ \cdot \\ \cdot \\ \beta_2 - \hat{\alpha}_1\beta_1 \end{bmatrix} u \tag{7-60}$$

Then from the first $(n-1)$ equations of (7-58a) and (7-60) we have

$$\begin{bmatrix} \dot{\tilde{x}}_1 \\ \dot{\tilde{x}}_2 \\ \cdot \\ \cdot \\ \cdot \\ \dot{\tilde{x}}_{n-1} \end{bmatrix} = \begin{bmatrix} 0 & 0 & \cdots & 0 & -\hat{\alpha}_{n-1} \\ 1 & 0 & \cdots & 0 & -\hat{\alpha}_{n-2} \\ \cdot & \cdot & & \cdot & \cdot \\ \cdot & \cdot & & \cdot & \cdot \\ \cdot & \cdot & & \cdot & \cdot \\ 0 & 0 & \cdots & 1 & -\hat{\alpha}_1 \end{bmatrix} \begin{bmatrix} \tilde{x}_1 \\ \tilde{x}_2 \\ \cdot \\ \cdot \\ \cdot \\ \tilde{x}_{n-1} \end{bmatrix} \tag{7-61}$$

where $\tilde{x}_i = \breve{x}_i - \hat{\breve{x}}_i$, for $i = 1, 2, \cdots, n-1$. Since $\hat{\alpha}_1, \hat{\alpha}_2, \cdots, \hat{\alpha}_{n-1}$ can be arbitrarily chosen, the estimator can be designed to respond as fast as desired. Note that the output of the $(n-1)$-dimensional estimator and the output of the original system give an estimate of $\hat{\breve{\mathbf{x}}}$. In order to obtain an estimate of the original state $\mathbf{x}$, $\hat{\breve{\mathbf{x}}}$ must be transformed by $\mathbf{P}_1^{-1}$ to give an estimate of $\bar{\mathbf{x}}$, and then further transformed by $\mathbf{P}^{-1}$ to give an estimate of $\mathbf{x}$ as shown in Figure 7-8.

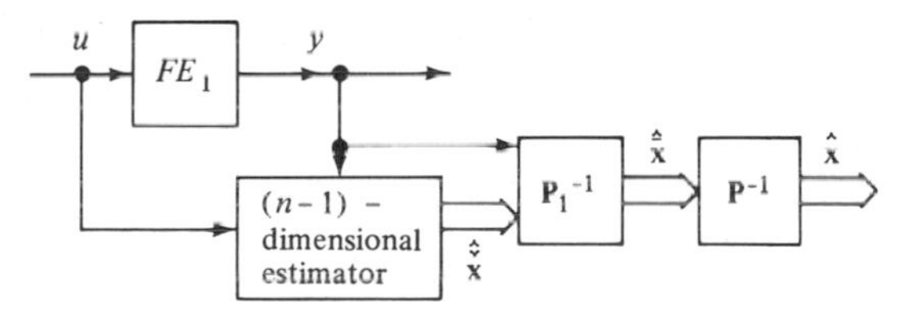


**Figure 7-8** An $(n-1)$-dimensional state estimator.

In the design of asymptotic estimators, the input does not play any role. Therefore what we have discussed is applicable to multiple-input single-output systems. Similar to the $n$-dimensional estimator, it can be shown that the design of a state feedback and the design of an $(n-1)$-dimensional estimator can be carried out independently; that is, the separation property still hold when an $(n-1)$-dimensional estimator is used in the state feedback.

### Example 2

Design a two-dimensional estimator for the dynamical equation in Example 1.

From Example 1, we have $\alpha_3 = 2$, $\alpha_2 = -9$, $\alpha_1 = 0$, $\beta_3 = 3$, $\beta_2 = 2$, and $\beta_1 = 1$. Let $-3$ and $-4$ be the eigenvalues of the two-dimensional estimator to be designed. Then $(s+3)(s+4) = s^2 + 7s + 12$ and $\hat{\alpha}_2 = 12$, $\hat{\alpha}_1 = 7$. From (7-60), we obtain immediately the dynamical equation of the estimator

$$\begin{bmatrix} \dot{\hat{\breve{x}}}_1 \\ \dot{\hat{\breve{x}}}_2 \end{bmatrix} = \begin{bmatrix} 0 & -12 \\ 1 & -7 \end{bmatrix} \begin{bmatrix} \hat{\breve{x}}_1 \\ \hat{\breve{x}}_2 \end{bmatrix} + \begin{bmatrix} -86 \\ -28 \end{bmatrix} y + \begin{bmatrix} -9 \\ -5 \end{bmatrix} u \tag{7-62}$$

Hence an estimate of the original state $\mathbf{x}$ is given by

$$\hat{\mathbf{x}} = \mathbf{P}^{-1}\mathbf{P}_1^{-1}\begin{bmatrix} \hat{\hat{x}}_1 \\ \hat{\hat{x}}_2 \\ y \end{bmatrix} = \begin{bmatrix} 1/6 & 1/6 & 7/6 \\ 0 & 1/2 & 0 \\ 0 & 0 & 1 \end{bmatrix} \begin{bmatrix} 1 & 0 & -12 \\ 0 & 1 & -7 \\ 0 & 0 & 1 \end{bmatrix}^{-1} \begin{bmatrix} \hat{\hat{x}}_1 \\ \hat{\hat{x}}_2 \\ y \end{bmatrix}$$

where the matrix $\mathbf{P}^{-1}$ has been computed in Example 1 and the matrix $\mathbf{P}_1$ is obtained from (7-59).

***Multivariable case.** In this subsection we shall study the design of asymptotic state estimators for observable, linear, time-invariant, multivariable dynamical equations. From the fact that observability and controllability are dual to each other, we know that the idea used in the design of state feedback for multivariable dynamical equations can be employed to design estimators. If we use that idea or procedure, the resulting estimator will be of dimension $n$ or $(n - 1)$. However, if a different approach is used, it is possible to design an estimator of dimension less than $n - 1$. Clearly, it is economically desirable to design an estimator with the least possible dimension. We shall show that if the rank of the matrix $\mathbf{C}$ is $q$, then an asymptotic estimator of dimension $(n - q)$ can be designed to generate all the state variables. This will be achieved in two steps. First we transform the dynamical equation into a canonical form such that the new equation can be considered as consisting of a set of single-output subequations. Then for each subequation we design an estimator by applying the procedure introduced in the previous subsection.

Consider the $n$-dimensional, linear, time-invariant dynamical equation

$$FE: \quad \dot{\mathbf{x}} = \mathbf{A}\mathbf{x} + \mathbf{B}\mathbf{u} \tag{7-63a}$$
$$\mathbf{y} = \mathbf{C}\mathbf{x} \tag{7-63b}$$

where $\mathbf{x}$ is the $n \times 1$ state vector, $\mathbf{u}$ is the $p \times 1$ input vector and $\mathbf{y}$ is the $q \times 1$ output vector; $\mathbf{A}$, $\mathbf{B}$, and $\mathbf{C}$ are $n \times n$, $n \times p$, and $q \times n$ real constant matrices, respectively. It is assumed that the dynamical equation $FE$ is observable. If the matrix $\mathbf{C}$ is written as

$$\mathbf{C} = \begin{bmatrix} \mathbf{c}_1 \\ \mathbf{c}_2 \\ \cdot \\ \cdot \\ \cdot \\ \mathbf{c}_q \end{bmatrix}$$

then a set of $n$ linearly independent row vectors, say

$$\mathbf{M} \triangleq \begin{bmatrix} \mathbf{c}_1 \\ \mathbf{c}_1\mathbf{A} \\ \cdot \\ \cdot \\ \cdot \\ \mathbf{c}_1\mathbf{A}^{\mu_1-1} \\ \mathbf{c}_2 \\ \mathbf{c}_2\mathbf{A} \\ \cdot \\ \cdot \\ \cdot \\ \mathbf{c}_2\mathbf{A}^{\mu_2-1} \\ \cdot \\ \cdot \\ \cdot \\ \mathbf{c}_q \\ \cdot \\ \cdot \\ \cdot \\ \mathbf{c}_q\mathbf{A}^{\mu_q-1} \end{bmatrix} \tag{7-64}$$

can be found. This set of vectors is obtained from the observability matrix of $FE$ by using Scheme 2 introduced in Section 7-2.[4] That is, we retain the linearly independent row vectors in the order of $\mathbf{c}_1, \mathbf{c}_2, \cdots, \mathbf{c}_q, \mathbf{c}_1\mathbf{A}, \cdots, \mathbf{c}_q\mathbf{A}, \mathbf{c}_1\mathbf{A}^2, \cdots,$ and then rearrange them to obtain the matrix in (7-64). If all the rows of $\mathbf{C}$ are linearly independent, then the matrix $\mathbf{M}$ in (7-64) contains all the $\mathbf{c}_i$'s. Otherwise, the linearly dependent vector will not appear in $\mathbf{M}$. We have assumed in (7-64) that $\mu_1 + \mu_2 + \cdots + \mu_q = n$. Define

$$\mathbf{M}^{-1} \triangleq [\mathbf{e}_{11} \quad \mathbf{e}_{12} \quad \cdots \quad \mathbf{e}_{1\mu_1} \quad \mathbf{e}_{21} \quad \mathbf{e}_{22} \quad \cdots \quad \mathbf{e}_{2\mu_2} \quad \cdots \quad \mathbf{e}_{q1} \quad \mathbf{e}_{q2} \quad \cdots \quad \mathbf{e}_{q\mu_q}] \tag{7-65}$$

From the fact that $\mathbf{MM}^{-1} = \mathbf{I}$, where $\mathbf{I}$ is a unit matrix, it can be easily verified that

$$\begin{aligned} \mathbf{c}_l\mathbf{A}^k\mathbf{e}_{ij} &= 1 \qquad \text{if } i = l \text{ and } j = k + 1 \\ &= 0 \qquad \text{otherwise} \end{aligned} \tag{7-66}$$

Now define

$$\mathbf{Q} \triangleq [\mathbf{e}_{1\mu_1} \quad \mathbf{A}\mathbf{e}_{1\mu_1} \quad \cdots \quad \mathbf{A}^{\mu_1-1}\mathbf{e}_{1\mu_1} \quad \mathbf{e}_{2\mu_2} \quad \mathbf{A}\mathbf{e}_{2\mu_2} \quad \cdots \quad \mathbf{A}^{\mu_2-1}\mathbf{e}_{2\mu_2} \quad \cdots \quad \mathbf{e}_{q\mu_q} \quad \cdots \quad \mathbf{A}^{\mu_q-1}\mathbf{e}_{q\mu_q}] \tag{7-67}$$

[4] Scheme 1 in Section 7-2 can also be used in the subsequent development. The only difference between using these two schemes is that the dimension of the estimator resulted by using Scheme 1 is generally larger than that by using Scheme 2.

Using (7-66), the columns of **Q** can be shown to be linearly independent (see Problem 7-17). If the columns of **Q** are used as basis vectors, then we will obtain the following equivalent dynamical equation:

$$
\begin{bmatrix} \dot{\bar{x}}_{11} \\ \dot{\bar{x}}_{12} \\ \vdots \\ \dot{\bar{x}}_{1\mu_1} \\ \dot{\bar{x}}_{21} \\ \dot{\bar{x}}_{22} \\ \vdots \\ \dot{\bar{x}}_{2\mu_2} \\ \vdots \\ \dot{\bar{x}}_{q1} \\ \dot{\bar{x}}_{q2} \\ \vdots \\ \dot{\bar{x}}_{q\mu_q} \end{bmatrix}
=
\left[\begin{array}{ccccc|ccccc|c|ccccc}
0 & 0 & \cdots & 0 & X & & & & & X & & & & & & X \\
1 & 0 & \cdots & 0 & X & & & & & X & & & & & & X \\
\vdots & \vdots & & \vdots & \vdots & & & & & \vdots & & & & & & \vdots \\
0 & 0 & \cdots & 1 & X & & & & & X & & & & & & X \\
\hline
 & & & & X & 0 & 0 & \cdots & 0 & X & & & & & & X \\
 & & & & X & 1 & 0 & \cdots & 0 & X & & & & & & X \\
 & & & & \vdots & \vdots & \vdots & & \vdots & \vdots & & & & & & \vdots \\
 & & & & X & 0 & 0 & \cdots & 1 & X & & & & & & X \\
\hline
 & & & & \vdots & & & & & \vdots & \ddots & & & & & \vdots \\
\hline
 & & & & X & & & & & X & & 0 & 0 & \cdots & 0 & X \\
 & & & & X & & & & & X & & 1 & 0 & \cdots & 0 & X \\
 & & & & \vdots & & & & & \vdots & & \vdots & \vdots & & \vdots & \vdots \\
 & & & & X & & & & & X & & 0 & 0 & \cdots & 1 & X
\end{array}\right] \bar{\mathbf{x}} + \bar{\mathbf{B}}\mathbf{u} \tag{7-68a}
$$

$$
\begin{bmatrix} y_1 \\ y_2 \\ \vdots \\ y_q \end{bmatrix}
=
\left[\begin{array}{ccccc|ccccc|c|ccccc}
0 & 0 & \cdots & 0 & 1 & 0 & 0 & \cdots & 0 & 0 & \cdots & 0 & 0 & \cdots & 0 & 0 \\
0 & 0 & \cdots & 0 & 0 & 0 & 0 & \cdots & 0 & 1 & \cdots & 0 & 0 & \cdots & 0 & 0 \\
\vdots & \vdots & & \vdots & \vdots & \vdots & \vdots & & \vdots & \vdots & & \vdots & \vdots & & \vdots & \vdots \\
0 & 0 & \cdots & 0 & 0 & 0 & 0 & \cdots & 0 & 0 & \cdots & 0 & 0 & \cdots & 0 & 1
\end{array}\right] \bar{\mathbf{x}} \tag{7-68b}
$$

where the X's denote possible nonzero elements. That the matrix $\bar{\mathbf{A}}$ in (7-68) is similar to the one in (7-23) becomes self-explanatory if we compare the matrix **Q** in (7-67) with the basis vectors in (7-21). The $\bar{\mathbf{C}}$ matrix in (7-68b) is equal to $\mathbf{CP}^{-1} = \mathbf{CQ}$ and is reduced to the form of (7-68b) by the use of (7-66). Equation (7-68) can be looked upon as consisting of $q$ number of multiple-input single-output equations of the form

$$
\begin{bmatrix} \dot{\bar{x}}_{i1} \\ \dot{\bar{x}}_{i2} \\ \vdots \\ \dot{\bar{x}}_{i\mu_i} \end{bmatrix}
=
\begin{bmatrix}
0 & 0 & \cdots & 0 & X \\
1 & 0 & \cdots & 0 & X \\
\vdots & \vdots & & \vdots & \vdots \\
0 & 0 & \cdots & 1 & X
\end{bmatrix} \bar{\mathbf{x}}_i + \bar{\mathbf{E}}_i\mathbf{y} + \bar{\mathbf{B}}_i\mathbf{u} \tag{7-69a}
$$

$$
y_i = [0 \quad 0 \quad \cdots \quad 0 \quad 1]\bar{\mathbf{x}}_i \tag{7-69b}
$$

Therefore, what we have discussed in the previous subsection can be applied here. For Equation (7-69), we can construct an asymptotic estimator of dimension $\mu_i - 1$. Consequently, for all of Equation (7-68), we need $q$ estimators with total dimension

$$\sum_{i=1}^{q} (\mu_i - 1) = n - q$$

The $q$ estimators generate $(n - q)$ state variables; the output vector gives another $q$ state variables. The eigenvalues of the estimators can be arbitrarily chosen. We summarize this as a theorem.

### Theorem 7-9

If the $n$-dimensional, linear, time-invariant dynamical equation $FE$ is observable, an $(n - q)$-dimensional asymptotic state estimator with any desired eigenvalues can be constructed, where $q$ is the rank of the matrix $\mathbf{C}$ of $FE$. ∎

If all the $n$ state variables are to be reconstructed, an estimator of dimension $(n - q)$ is required. In many cases, however, it is not necessary to reconstruct all the state variables; what is needed is to generate some function of the state—for example, to generate $\mathbf{Kx}$ in the state feedback. In these cases, the dimensions of the estimators can be smaller than $(n - q)$. For example, to generate $\mathbf{Kx}$, the dimension of the required estimator is only $\mu - 1$, where $\mu = \max_i (\mu_i)$. This can be argued intuitively as follows. Consider Equation (7-69) for $i = 1, 2, \cdots, q$. Let $\mu = \max (\mu_i, i = 1, 2, \cdots, q)$. It is clear that the dimension of the estimator of each equation will be at most $\mu - 1$. Now if the eigenvalues of the estimator of each equation are chosen from the same set, then an estimator of dimension $\mu - 1$ can be used to generate the function $\mathbf{Kx}$. Hence in this case, the dimension of the estimator will be $\mu - 1 = \max_i (\mu_i - 1)$, instead of $\sum_i (\mu_i - 1) = n - q$. We observe from the process of chosing $\mu_i$ in (7-64) that $\mu = \max_i (\mu_i)$ is the least integer such that the matrix

$$\begin{bmatrix} \mathbf{C} \\ \mathbf{CA} \\ \cdot \\ \cdot \\ \cdot \\ \mathbf{CA}^{\mu-1} \end{bmatrix}$$

has rank $n$. In the literature, $\mu$ is called the *observability index*. Note that we have $\mu \geq [n/q]$, where $[\cdot]$ denotes the least integer larger than or equal to $n/q$.

## 7-5 Design of Feedback Systems: An Example

In this section we shall apply the results obtained in the previous sections to the design of feedback systems. Consider a given system $S_1$ shown in Figure 7-9(a). The system might be a simplified model of an armature-controlled dc motor driving a mechanical load (Example 2 in Section 3-4 with $L_a = 0$). It is required to design a compensator such that the feedback system shown in Figure 7-9(b) or (c) has poles at $(-1 + i)$ and $(-1 - i)$. The reason for choosing the poles of the feedback system at $-1 + i$ and $-1 - i$ is to have a faster response. How to choose the best poles for feedback systems is studied in optimal control theory.

If the dynamical-equation description of the system $S_1$ in Figure 7-9(a) is found, the results in Sections 7-3 and 7-4 can be applied to

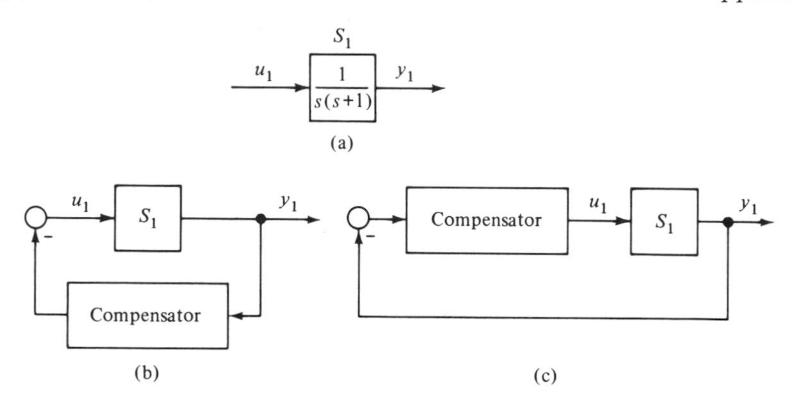


**Figure 7-9** The design of a feedback system.

design a compensator. The design procedure consists of two steps: first, to move the poles of $S_1$ to $-1 + i$ and $-1 - i$ by introducing state feedback (see Section 7-3) and then to design an asymptotic state estimator to generate the state vector (see Section 7-4). Let us first realize $1/s(s + 1)$ in the following controllable canonical-form dynamical equation (see Section 6-3).

$$\begin{bmatrix} \dot{x}_1 \\ \dot{x}_2 \end{bmatrix} = \begin{bmatrix} 0 & 1 \\ 0 & -1 \end{bmatrix} \begin{bmatrix} x_1 \\ x_2 \end{bmatrix} + \begin{bmatrix} 0 \\ 1 \end{bmatrix} u_1 \tag{7-70a}$$

$$y_1 = [1 \quad 0]\mathbf{x} \tag{7-70b}$$

We compute now

$$(\lambda + 1 + i)(\lambda + 1 - i) = \lambda^2 + 2\lambda + 2$$

Hence if we introduce the state feedback $u = v + \mathbf{kx}$ where $\mathbf{k} = [-2 \quad 1-2] = [-2 \quad -1]$, then the poles of the state-feedback system are $-1 - i$ and $-1 + i$. Since the reference input $v$ does not play any role in shaping the poles of the feedback system, we assume in the following that $v = 0$.

The state variables of the system $S_1$ are not available; hence, before state feedback is introduced, a state estimator has to be designed. In order to design a state estimator, we first transform the dynamical equation (7-70) into the observable canonical form by using the equivalence transformation (7-15), which in this case is

$$\mathbf{P} = \begin{bmatrix} \alpha_1 & 1 \\ 1 & 0 \end{bmatrix} \begin{bmatrix} \mathbf{c} \\ \mathbf{cA} \end{bmatrix} = \begin{bmatrix} 1 & 1 \\ 1 & 0 \end{bmatrix} \begin{bmatrix} 1 & 0 \\ 0 & 1 \end{bmatrix} = \begin{bmatrix} 1 & 1 \\ 1 & 0 \end{bmatrix} \tag{7-71}$$

Let $\bar{\mathbf{x}} = \mathbf{Px}$; then we have

$$\dot{\bar{\mathbf{x}}} = \begin{bmatrix} 0 & 0 \\ 1 & -1 \end{bmatrix} \bar{\mathbf{x}} + \begin{bmatrix} 1 \\ 0 \end{bmatrix} u_1 \tag{7-72a}$$

$$y_1 = [0 \quad 1]\bar{\mathbf{x}} \tag{7-72b}$$

By using this dynamical equation, a two dimensional asymptotic state estimator can be easily designed by employing Theorem 7-7. Instead of doing so, we shall use Theorem 7-8 and Equation (7-60) to design a one-dimensional state estimator. Let us choose $-2$ as the eigenvalue of the estimator. Then from (7-59), we have

$$\mathbf{P}_1 = \begin{bmatrix} 1 & -\hat{\alpha}_1 \\ 0 & 1 \end{bmatrix} = \begin{bmatrix} 1 & -2 \\ 0 & 1 \end{bmatrix}. \tag{7-73}$$

and the equivalence transformation $\check{\mathbf{x}} = \mathbf{P}_1\bar{\mathbf{x}}$ transforms (7-72) into

$$\dot{\check{\mathbf{x}}} = \begin{bmatrix} -2 & -2 \\ 1 & 1 \end{bmatrix} \check{\mathbf{x}} + \begin{bmatrix} 1 \\ 0 \end{bmatrix} u_1 \tag{7-74a}$$

$$y_1 = [0 \quad 1]\check{\mathbf{x}} \tag{7-74b}$$

Hence the one-dimensional state estimator is given by the first equation of (7-74a); namely,

$$\dot{\check{x}}_1 = -2\check{x}_1 - 2y_1 + u_1 \tag{7-75}$$

The solution of (7-75) and $y_1$ give an estimate of $\check{\mathbf{x}}$. In order to obtain an estimate of $\mathbf{x}$, we multiply $\check{\mathbf{x}}$ first by $\mathbf{P}_1^{-1}$ and then by $\mathbf{P}^{-1}$, as shown in Figure 7-10.

Now we shall combine state feedback and state estimator to give a compensator for the system $S_1$. Since the reference input $v$ is assumed to be zero, the input $u_1$ of $S_1$ is equal to $\mathbf{k}\hat{\mathbf{x}} = [-2 \quad -1]\hat{\mathbf{x}}$. Hence the entire compensator is the one shown in Figure 7-10. Its transfer function can be computed as follows:

$$\hat{u}_1 = [-2 \quad -1] \begin{bmatrix} 0 & 1 \\ 1 & -1 \end{bmatrix} \begin{bmatrix} 1 & 2 \\ 0 & 1 \end{bmatrix} \begin{bmatrix} \dfrac{1}{s+2}(\hat{u}_1 - 2\hat{y}_1) \\ \hat{y}_1 \end{bmatrix}$$

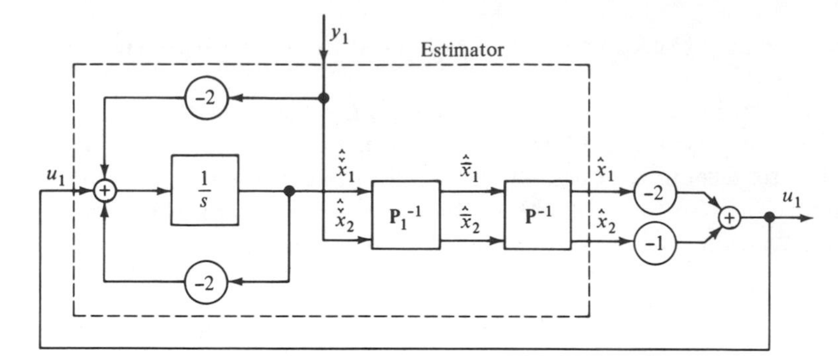


**Figure 7-10** A compensator.

which gives

$$\frac{\hat{u}_1(s)}{\hat{y}_1(s)} = -\frac{3s+4}{s+3} \tag{7-76}$$

Hence the transfer function of the compensator in Figure 7-9(b) or (c) is $(3s+4)/(s+3)$. It is easy to verify that the poles of the feedback system are $-2$, $-1+i$, $-1-i$ as expected. [See the discussion on the separation property in the preceding section.]

The discussion of design in this section is far from complete. For example, how to choose the poles of feedback systems and the problem of the realizability of compensators have not been covered. These are discussed in References [3], [12], [18], [37], and [56].

This design problem will again be studied in Section 9-5 by the use of the transfer-function approach. Comparisons between the approach used here and the transfer-function approach will also be discussed there. Section 9-5 is independent of the remainder of this book and may be studied at this point.

## *7-6 Decoupling by State Feedback

Consider a $p$-input $p$-output system with the linear, time-invariant, dynamical-equation description

$$FE: \qquad \dot{\mathbf{x}} = \mathbf{A}\mathbf{x} + \mathbf{B}\mathbf{u} \tag{7-77a}$$

$$\mathbf{y} = \mathbf{C}\mathbf{x} \tag{7-77b}$$

where $\mathbf{u}$ is the $p \times 1$ input vector, $\mathbf{y}$ is the $p \times 1$ output vector; $\mathbf{A}$, $\mathbf{B}$, and $\mathbf{C}$ are $n \times n$, $n \times p$, and $p \times n$ real constant matrices, respectively. It is assumed that $p \leq n$. The transfer function of the system is

$$\hat{\mathbf{G}}(s) = \mathbf{C}(s\mathbf{I} - \mathbf{A})^{-1}\mathbf{B} \tag{7-78}$$

Clearly $\hat{\mathbf{G}}(s)$ is a $p \times p$ rational-function matrix. If the system is initially in the zero state, its inputs and outputs are related by

$$
\begin{aligned}
\hat{y}_1(s) &= \hat{g}_{11}(s)\hat{u}_1(s) + \hat{g}_{12}(s)\hat{u}_2(s) + \cdots + \hat{g}_{1p}(s)\hat{u}_p(s) \\
\hat{y}_2(s) &= \hat{g}_{21}(s)\hat{u}_1(s) + \hat{g}_{22}(s)\hat{u}_2(s) + \cdots + \hat{g}_{2p}(s)\hat{u}_p(s) \\
&\quad \vdots \\
\hat{y}_p(s) &= \hat{g}_{p1}(s)\hat{u}_1(s) + \hat{g}_{p2}(s)\hat{u}_2(s) + \cdots + \hat{g}_{pp}(s)\hat{u}_p(s)
\end{aligned}
\tag{7-79}
$$

where $\hat{g}_{ij}$ is the $ij$th element of $\hat{\mathbf{G}}(s)$. We see from (7-79) that every input controls more than one output and that every output is controlled by more than one input. Because of this phenomenon, which is called *coupling* or *interacting*, it is generally very difficult to control a multivariable system. For example, suppose we like to control $\hat{y}_1(s)$ without affecting $\hat{y}_2(s)$, $\hat{y}_3(s)$, $\cdots$, $\hat{y}_p(s)$, the required inputs $\hat{u}_1(s)$, $\hat{u}_2(s)$, $\cdots$, $\hat{u}_p(s)$ cannot be readily found. Therefore, in some cases we like to introduce some compensator so that a coupled multivariable system may become decoupled in the sense that every input controls only one output and every output is controlled by only one input. Consequently, a decoupled system can be considered as consisting of a set of independent single-variable systems. It is clear that if the transfer-function matrix of a multivariable system is diagonal, then the system is decoupled.

**Definition 7-1**

A multivariable system is said to be *decoupled* if its transfer-function matrix is diagonal and nonsingular. ∎

In this section we shall study the problem of the decoupling of multivariable systems by linear state feedback of the form

$$
\mathbf{u}(t) = \mathbf{K}\mathbf{x} + \mathbf{H}\mathbf{v} \tag{7-80}
$$

where $\mathbf{K}$ is a $p \times n$ real constant matrix, $\mathbf{H}$ is a $p \times p$ real constant nonsingular matrix, and $\mathbf{v}$ represents a new $p \times 1$ input vector. The state feedback is shown in Figure 7-11. Substituting (7-80) into (7-77), we obtain

$$
FE^f: \quad \dot{\mathbf{x}} = (\mathbf{A} + \mathbf{B}\mathbf{K})\mathbf{x} + \mathbf{B}\mathbf{H}\mathbf{v} \tag{7-81a}
$$

$$
\mathbf{y} = \mathbf{C}\mathbf{x} \tag{7-81b}
$$

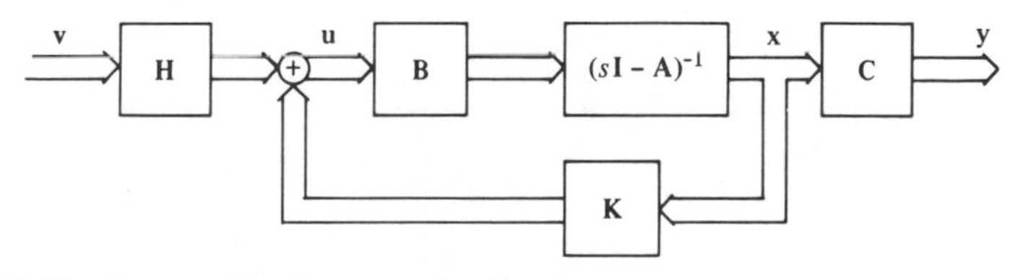

**Figure 7-11** Decoupling by state feedback.

The transfer function of the state-feedback system is

$$\hat{\mathbf{G}}_f(s, \mathbf{K}, \mathbf{H}) \triangleq \mathbf{C}(s\mathbf{I} - \mathbf{A} - \mathbf{BK})^{-1}\mathbf{BH} \tag{7-82}$$

We shall derive in the following the condition on $\hat{\mathbf{G}}(s)$ under which the system can be decoupled by state feedback. Define the nonnegative integer $d_i$ as

$d_i \triangleq$ min (the difference of the degree in $s$ of the denominator and the numerator of each entry of the $i$th row of $\hat{\mathbf{G}}(s)$) $- 1$

and the $1 \times p$ constant row vector $\mathbf{E}_i$ as

$$\mathbf{E}_i \triangleq \lim_{s \to \infty} s^{d_i+1}\hat{\mathbf{G}}_i(s)$$

where $\hat{\mathbf{G}}_i(s)$ is the $i$th row of $\hat{\mathbf{G}}(s)$.

### Example

Consider

$$\hat{\mathbf{G}}(s) = \begin{bmatrix} \dfrac{s+2}{s^2+s+1} & \dfrac{1}{s^2+s+2} \\ \dfrac{1}{s^2+2s+1} & \dfrac{3}{s^2+s+4} \end{bmatrix}$$

The differences in degree of the first row of $\hat{\mathbf{G}}(s)$ are 1 and 2; hence $d_1 = 0$ and

$$\mathbf{E}_1 = \lim_{s \to \infty} s \begin{bmatrix} \dfrac{s+2}{s^2+s+1} & \dfrac{1}{s^2+s+2} \end{bmatrix} = [1 \quad 0]$$

The differences in degree of the second row of $\hat{\mathbf{G}}(s)$ are 2 and 2, hence $d_2 = 1$ and

$$\mathbf{E}_2 = \lim_{s \to \infty} s^2 \begin{bmatrix} \dfrac{1}{s^2+2s+1} & \dfrac{3}{s^2+s+4} \end{bmatrix} = [1 \quad 3]$$

### Theorem 7-10

A system with the transfer-function matrix $\hat{\mathbf{G}}(s)$ can be decoupled by state feedback of the form $\mathbf{u} = \mathbf{Kx} + \mathbf{Hv}$ if and only if the constant matrix

$$\mathbf{E} = \begin{bmatrix} \mathbf{E}_1 \\ \mathbf{E}_2 \\ \cdot \\ \cdot \\ \cdot \\ \mathbf{E}_p \end{bmatrix}$$

is nonsingular. ∎

We see from this theorem that whether or not a system can be decoupled is a property of its transfer-function matrix. The dynamical-equation description comes into play a role only when the gain matrix **K** is to be found. Therefore the controllability and observability of the dynamical-equation description of the system are immaterial here. Let $\hat{\mathbf{G}}_i(s)$ and $\mathbf{C}_i$ be the $i$th row of $\hat{\mathbf{G}}(s)$ and $\mathbf{C}$, respectively. Then, from (7-78) and using (2-3), we have

$$\hat{\mathbf{G}}_i(s) = \mathbf{C}_i(s\mathbf{I} - \mathbf{A})^{-1}\mathbf{B} \tag{7-83}$$

In the following, we shall establish the relations between the integer $d_i$, the vector $\mathbf{E}_i$ and the matrices $\mathbf{C}_i$, $\mathbf{A}$, and $\mathbf{B}$. First, we expand (7-83) into (see Problem 4-15)

$$\begin{aligned}\hat{\mathbf{G}}_i(s) &= \frac{1}{\Delta(s)}\mathbf{C}_i[s^{n-1}\mathbf{I} + \mathbf{R}_1 s^{n-2} + \cdots + \mathbf{R}_{n-1}]\mathbf{B} \\ &= \frac{1}{\Delta(s)}[\mathbf{C}_i\mathbf{B}s^{n-1} + \mathbf{C}_i\mathbf{R}_1\mathbf{B}s^{n-2} \\ &\qquad + \cdots + \mathbf{C}_i\mathbf{R}_{d_i}\mathbf{B}s^{n-d_i-1} + \cdots + \mathbf{C}_i\mathbf{R}_{n-1}\mathbf{B}]\end{aligned} \tag{7-84}$$

where

$$\Delta(s) \triangleq \det(s\mathbf{I} - \mathbf{A}) \triangleq s^n + \alpha_1 s^{n-1} + \cdots + \alpha_n \tag{7-85}$$

and

$$\begin{aligned}\mathbf{R}_1 &= \mathbf{A} + \alpha_1\mathbf{I} \\ \mathbf{R}_2 &= \mathbf{A}\mathbf{R}_1 + \alpha_2\mathbf{I} = \mathbf{A}^2 + \alpha_1\mathbf{A} + \alpha_2\mathbf{I} \\ &\ \ \vdots \\ \mathbf{R}_{n-1} &= \mathbf{A}\mathbf{R}_{n-2} + \alpha_{n-1}\mathbf{I} = \mathbf{A}^{n-1} + \alpha_1\mathbf{A}^{n-2} + \cdots + \alpha_{n-1}\mathbf{I}\end{aligned} \tag{7-86}$$

Since $(d_i + 1)$ is the smallest difference in degree between the numerator and the denominator of each entry of $\hat{\mathbf{G}}_i(s)$, we conclude from (7-84) that the coefficients associated with $s^{n-d_i}$, $s^{n-d_i+1}$, . . . , and $s^{n-1}$ vanish, but the coefficient matrix associated with $s^{n-d_i-1}$ is different from zero. Hence we have

$$\mathbf{C}_i\mathbf{B} = \mathbf{0},\ \mathbf{C}_i\mathbf{R}_1\mathbf{B} = \mathbf{0},\ \cdots,\ \mathbf{C}_i\mathbf{R}_{d_i-1}\mathbf{B} = \mathbf{0} \tag{7-87}$$

and

$$\mathbf{E}_i = \mathbf{C}_i\mathbf{R}_{d_i}\mathbf{B} \neq \mathbf{0} \tag{7-88}$$

From (7-86), it is easy to see that conditions (7-87) imply that

$$\mathbf{C}_i\mathbf{B} = \mathbf{0},\ \mathbf{C}_i\mathbf{A}\mathbf{B} = \mathbf{0},\ \cdots,\ \mathbf{C}\mathbf{A}^{d_i-1}\mathbf{B} = \mathbf{0} \tag{7-89}$$

and Equation (7-88) becomes

$$\mathbf{E}_i = \mathbf{C}_i\mathbf{A}^{d_i}\mathbf{B} \neq \mathbf{0} \tag{7-90}$$

Therefore, $d_i$ can also be defined from a dynamical equation as the smallest integer such that $\mathbf{C}_i\mathbf{A}^{d_i}\mathbf{B} \neq \mathbf{0}$. If $\mathbf{C}_i\mathbf{A}^k\mathbf{B} = \mathbf{0}$, for all $k \leq n$, then $d_i \triangleq n - 1$. (See Problem 7-22.)

Similarly, we may define $\bar{d}_i$ and $\bar{\mathbf{E}}_i$ for the transfer-function matrix $\hat{\mathbf{G}}_f$ in (7-82). If we expand $\hat{\mathbf{G}}_{fi}$ into

$$\hat{\mathbf{G}}_{fi} = \frac{1}{\bar{\Delta}(s)} [\mathbf{C}_i\mathbf{BH}s^{n-1} + \mathbf{C}_i\bar{\mathbf{R}}_1\mathbf{BH}s^{n-2} + \cdots + \mathbf{C}_i\bar{\mathbf{R}}_{n-1}\mathbf{BH}] \tag{7-91}$$

where

$$\bar{\Delta}(s) \triangleq \det(s\mathbf{I} - \mathbf{A} - \mathbf{BK}) \triangleq s^n + \bar{\alpha}_1 s^{n-1} + \cdots + \bar{\alpha}_n \tag{7-92}$$

and

$$\begin{aligned} \bar{\mathbf{R}}_1 &= (\mathbf{A} + \mathbf{BK}) + \bar{\alpha}_1\mathbf{I} \\ \bar{\mathbf{R}}_2 &= (\mathbf{A} + \mathbf{BK})^2 + \bar{\alpha}_1(\mathbf{A} + \mathbf{BK}) + \bar{\alpha}_2\mathbf{I} \\ &\cdot \\ &\cdot \\ &\cdot \\ \bar{\mathbf{R}}_{n-1} &= (\mathbf{A} + \mathbf{BK})^{n-1} + \bar{\alpha}_1(\mathbf{A} + \mathbf{BK})^{n-2} + \cdots + \bar{\alpha}_{n-1}\mathbf{I} \end{aligned} \tag{7-93}$$

then

$$\mathbf{CBH} = \mathbf{0},\ \mathbf{C}(\mathbf{A} + \mathbf{BK})\mathbf{BH} = \mathbf{0},\ \cdots,\ \mathbf{C}(\mathbf{A} + \mathbf{BK})^{\bar{d}_i - 1}\mathbf{BH} = \mathbf{0} \tag{7-94}$$

and

$$\bar{\mathbf{E}} = \mathbf{C}(\mathbf{A} + \mathbf{BK})^{\bar{d}_i}\mathbf{BH} \neq \mathbf{0} \tag{7-95}$$

At this point it is natural to ask what the effects on $d_i$ and $\mathbf{E}_i$ are by introducing the state feedback $\mathbf{u} = \mathbf{Kx} + \mathbf{Hv}$. If $\hat{\mathbf{G}}(s)$ is a scalar rational function, from the fact that its numerator is not affected by state feedback [see the remark preceding Equation (7-34)] we know that $d_i$ and $\mathbf{E}_i$ are independent of $\mathbf{K}$. It turns out that this is also true for the matrix case.

### Theorem 7-11

For any $\mathbf{K}$ and any nonsingular $\mathbf{H}$, we have $\bar{d}_i = d_i$ and $\bar{\mathbf{E}}_i = \mathbf{E}_i\mathbf{H}$.

### Proof

It is easy to verify that, for each $i$, the conditions $\mathbf{C}_i\mathbf{B} = \mathbf{0}$, $\mathbf{C}_i\mathbf{AB} = \mathbf{0}$, $\cdots$, $\mathbf{C}_i\mathbf{A}^{d_i-1}\mathbf{B} = \mathbf{0}$ imply that

$$\mathbf{C}_i(\mathbf{A} + \mathbf{BK})^k = \mathbf{C}_i\mathbf{A}^k \qquad k = 0, 1, 2, \cdots, d_i \tag{7-96}$$

and

$$\mathbf{C}_i(\mathbf{A} + \mathbf{BK})^k = \mathbf{C}_i\mathbf{A}^{d_i}(\mathbf{A} + \mathbf{BK})^{k-d_i} \qquad k = d_i + 1, d_i + 2, \cdots \tag{7-97}$$

Consequently, we have

$$\mathbf{C}_i(\mathbf{A} + \mathbf{BK})^k\mathbf{BH} = \mathbf{0} \qquad k = 0, 1, \cdots, d_i - 1$$

and

$$\mathbf{C}_i(\mathbf{A} + \mathbf{BK})^{d_i}\mathbf{BH} = \mathbf{C}_i\mathbf{A}^{d_i}\mathbf{BH} = \mathbf{E}_i\mathbf{H}$$

Since $\mathbf{H}$ is nonsingular by assumption, if $\mathbf{E}_i$ is nonzero, so is $\mathbf{E}_i\mathbf{H}$. Therefore we conclude that $\bar{d}_i = d_i$ and $\bar{\mathbf{E}}_i = \mathbf{E}_i\mathbf{H}$. Q.E.D.

With these preliminaries, we are ready to prove Theorem 7-10.

### Proof of Theorem 7-10

*Necessity:* Suppose that there exists $\mathbf{K}$ and $\mathbf{H}$ such that $\hat{\mathbf{G}}_f(s, \mathbf{K}, \mathbf{H})$ is diagonal and nonsingular. Then

$$\bar{\mathbf{E}} = \begin{bmatrix} \bar{\mathbf{E}}_1 \\ \bar{\mathbf{E}}_2 \\ \cdot \\ \cdot \\ \cdot \\ \bar{\mathbf{E}}_p \end{bmatrix}$$

is a diagonal constant matrix and nonsingular. Since $\bar{\mathbf{E}} = \mathbf{E}\mathbf{H}$ and since $\mathbf{H}$ is nonsingular by assumption, we conclude that $\mathbf{E}$ is nonsingular. *Sufficiency:* If the matrix $\mathbf{E}$ is nonsingular, then the system can be decoupled. Define

$$\mathbf{F} \triangleq \begin{bmatrix} \mathbf{C}_1\mathbf{A}^{d_1+1} \\ \mathbf{C}_2\mathbf{A}^{d_2+1} \\ \cdot \\ \cdot \\ \cdot \\ \mathbf{C}_p\mathbf{A}^{d_p+1} \end{bmatrix} \tag{7-98}$$

We show that if $\mathbf{K} = -\mathbf{E}^{-1}\mathbf{F}$ and $\mathbf{H} = \mathbf{E}^{-1}$, then the system can be decoupled and the transfer function of the decoupled system is

$$\hat{\mathbf{G}}_f(s, -\mathbf{E}^{-1}\mathbf{F}, \mathbf{E}^{-1}) = \begin{bmatrix} \frac{1}{s^{d_1+1}} & 0 & \cdots & 0 \\ 0 & \frac{1}{s^{d_2+1}} & \cdots & 0 \\ \cdot & \cdot & & \cdot \\ \cdot & \cdot & & \cdot \\ \cdot & \cdot & & \cdot \\ 0 & 0 & \cdots & \frac{1}{s^{d_p+1}} \end{bmatrix}$$

or, equivalently

$$\mathbf{C}_i(s\mathbf{I} - \mathbf{A} - \mathbf{B}\mathbf{K})^{-1}\mathbf{B}\mathbf{H} = \frac{1}{s^{d_i+1}}\mathbf{e}_i \tag{7-99}$$

where $\mathbf{e}_i$ is a row vector with 1 in the $i$th place and zeros elsewhere. First we show that $\mathbf{C}_i(\mathbf{A} + \mathbf{B}\mathbf{K})^{d_i+1} = \mathbf{0}$. From (7-90), (7-97) and (7-98) and using $\mathbf{K} = -\mathbf{E}^{-1}\mathbf{F}$ and $\mathbf{E}_i\mathbf{E}^{-1} = \mathbf{e}_i$, we obtain

$$\begin{aligned} \mathbf{C}_i(\mathbf{A} + \mathbf{B}\mathbf{K})^{d_i+1} &= \mathbf{C}_i\mathbf{A}^{d_i}(\mathbf{A} + \mathbf{B}\mathbf{K}) = \mathbf{C}_i\mathbf{A}^{d_i+1} + \mathbf{C}_i\mathbf{A}^{d_i}\mathbf{B}\mathbf{K} \\ &= \mathbf{F}_i - \mathbf{E}_i\mathbf{E}^{-1}\mathbf{F} = \mathbf{F}_i - \mathbf{e}_i\mathbf{F} = \mathbf{0} \end{aligned}$$

where $\mathbf{F}_i$ and $\mathbf{E}_i$ are the $i$th row of $\mathbf{F}$ and $\mathbf{E}$, respectively. Hence we conclude that

$$\mathbf{C}_i(\mathbf{A} + \mathbf{BK})^{d_i+k} = \mathbf{0} \qquad \text{for any positive integer } k \tag{7-100}$$

Since $\bar{d}_i = d_i$, Equation (7-91) reduces to

$$\mathbf{C}_i(s\mathbf{I} - \mathbf{A} - \mathbf{BK})^{-1}\mathbf{BH} = \frac{1}{\bar{\Delta}(s)} [\mathbf{C}_i\bar{\mathbf{R}}_{d_i}\mathbf{B}s^{n-d_i-1} + \mathbf{C}_i\bar{\mathbf{R}}_{d_i+1}\mathbf{B}s^{n-d_i-2} + \cdots + \mathbf{C}_i\bar{\mathbf{R}}_{n-1}\mathbf{B}]\mathbf{H} \tag{7-101}$$

Now, from (7-90), (7-93), (7-97), (7-100), and the fact $\mathbf{C}_i(\mathbf{A} + \mathbf{BK})^k\mathbf{B} = \mathbf{0}$, for $k = 0, 1, \cdots, \bar{d}_i - 1$, which follows from (7-94) and the nonsingularity of $\mathbf{H}$, it is straightforward to verify that

$$\begin{aligned}
\mathbf{C}_i\bar{\mathbf{R}}_{d_i}\mathbf{B} &= \mathbf{C}_i(\mathbf{A} + \mathbf{BK})^{d_i}\mathbf{B} = \mathbf{C}_i\mathbf{A}^{d_i}\mathbf{B} = \mathbf{E}_i \\
\mathbf{C}_i\bar{\mathbf{R}}_{d_i+1}\mathbf{B} &= \mathbf{C}_i[(\mathbf{A} + \mathbf{BK})^{d_i+1} + \bar{\alpha}_1(\mathbf{A} + \mathbf{BK})^{d_i}]\mathbf{B} = \bar{\alpha}_1\mathbf{E}_i \\
&\ \ \vdots \\
\mathbf{C}_i\bar{\mathbf{R}}_{n-1}\mathbf{B} &= \bar{\alpha}_{n-1-d_i}\mathbf{E}_i
\end{aligned}$$

Consequently, Equation (7-101) becomes

$$\mathbf{C}_i(s\mathbf{I} - \mathbf{A} - \mathbf{BK})^{-1}\mathbf{BH} = \frac{1}{\bar{\Delta}(s)} [s^{n-d_i-1} + \bar{\alpha}_1 s^{n-d_i-2} + \cdots + \bar{\alpha}_{n-1-d_i}]\mathbf{E}_i\mathbf{H} \tag{7-102}$$

What is left to be shown is that

$$\begin{aligned}
\bar{\Delta}(s) &= s^n + \bar{\alpha}_1 s^{n-1} + \bar{\alpha}_2 s^{n-2} + \cdots + \bar{\alpha}_n \\
&= s^{d_i+1}(s^{n-d_i-1} + \bar{\alpha}_1 s^{n-d_i-2} + \cdots + \bar{\alpha}_{n-1-d_i})
\end{aligned} \tag{7-103}$$

From the Cayley–Hamilton theorem, we have

$$(\mathbf{A} + \mathbf{BK})^n + \bar{\alpha}_1(\mathbf{A} + \mathbf{BK})^{n-1} + \cdots + \bar{\alpha}_n\mathbf{I} = \mathbf{0} \tag{7-104}$$

If $\mathbf{C}_i(\mathbf{A} + \mathbf{BK})^{d_i}$ is premultiplied to (7-104), from (7-100) we obtain $\bar{\alpha}_n\mathbf{C}_i(\mathbf{A} + \mathbf{BK})^{d_i} = \mathbf{0}$ which implies that $\bar{\alpha}_n = 0$. Next by multiplying $\mathbf{C}_i(\mathbf{A} + \mathbf{BK})^{d_i-1}$, we can show that $\bar{\alpha}_{n-1} = 0$. Proceed forward, we can prove Equation (7-103). By substituting (7-103) into (7-102) and using $\mathbf{E}_i\mathbf{H} = \mathbf{E}_i\mathbf{E}^{-1} = \mathbf{e}_i$, we immediately obtain Equation (7-99). Consequently, the system can be decoupled by using $\mathbf{K} = -\mathbf{E}^{-1}\mathbf{F}$ and $\mathbf{H} = \mathbf{E}^{-1}$.
Q.E.D.

Although a system can be decoupled by using $\mathbf{K} = -\mathbf{E}^{-1}\mathbf{F}$ and $\mathbf{H} = \mathbf{E}^{-1}$, the resulting system is not satisfactory because all the poles of the decoupled system are at the origin. However, we may introduce additional state feedback to move these poles to the desired location.

If the dynamical equation is controllable, then we can control all the eigenvalues by state feedback. Now, if in addition we like to decouple the system, the number of eigenvalues that can be controlled by state feedback will be reduced. For a complete discussion of this problem and others, such as how to compute **K** and **H** to have a decoupled system with desired poles, see References [36], [41], and [112].

## 7-7 Concluding Remarks

In this chapter we studied the practical implications of controllability and observability. We showed that if a linear, time-invariant dynamical equation is controllable, we can, by introducing state feedback, arbitrarily assign the eigenvalues of the resulting dynamical equation. This was achieved by the use of controllable canonical-form dynamical equations. When we introduce state feedback, all the state variables must be available. If they are not, then an asymptotic state estimator has to be constructed. If a dynamical equation is observable, a state estimator with a set of arbitrary eigenvalues can be constructed. The construction of state estimators is achieved by the use of observable canonical-form dynamical equation.

We studied in this chapter only the time-invariant case. The results in Section 7-2, 7-3, and 7-4 can be extended to the linear, time varying case if the equation is instantaneously controllable since, under the instantaneous controllability assumption, the controllability matrix of the linear, time-varying dynamical equation has full rank for all $t$. Hence by using it in the equivalence transformation, a controllable canonical-form dynamical equation can be obtained. Consequently the state feedback can be introduced. Similar remarks apply to the observability part.

We also studied the decoupling of a transfer function matrix by introducing state feedback. The concepts of controllability and observability are not required in this part of study. It was discussed in this chapter because the decoupling also used the state feedback as in Section 7-3.

The combination of state feedback and state estimator will again be studied in Section 9-5 directly from the transfer-function matrix. Comparisons between the state-variable approach and the transfer-function approach will also be discussed there.

## PROBLEMS

**7-1** The equivalence transformation $\mathbf{Q}$ of $\bar{\mathbf{x}} = \mathbf{P}\mathbf{x} = \mathbf{Q}^{-1}\mathbf{x}$ in Theorem 7-1 can be obtained either by using $\mathbf{Q} = [\mathbf{q}^1 \quad \mathbf{q}^2 \quad \cdots \quad \mathbf{q}^n]$, where the $\mathbf{q}^i$'s are given in (7-9) or by using $\mathbf{Q} = \mathbf{U}\bar{\mathbf{U}}^{-1}$ where $\mathbf{U}$ is the controllability matrix of $FE_1$ and $\bar{\mathbf{U}}^{-1}$ is given in (7-13). From a computational point of view, which one is easier to use?

**7-2** Consider the controllable dynamical equation

$$FE_1: \qquad \begin{aligned} \dot{\mathbf{x}} &= \mathbf{A}\mathbf{x} + \mathbf{b}u \\ y &= \mathbf{c}\mathbf{x} \end{aligned}$$

If the vectors $\{\mathbf{b} \quad \mathbf{A}\mathbf{b} \quad \cdots \quad \mathbf{A}^{n-1}\mathbf{b}\}$ are used as basis vectors, what is its equivalent dynamical equation? Does the new $\bar{\mathbf{c}}$ bear a simple relation to the coefficients of the transfer function of $FE_1$ as it does in the controllable canonical form? *Hint:* See Problem 2-15.

**7-3** Find the matrix $\mathbf{Q}$ which transforms (7-17) into the observable canonical-form dynamical equation.

**7-4** Transform the equation

$$\dot{\mathbf{x}} = \begin{bmatrix} -1 & -2 & -2 \\ 0 & -1 & 1 \\ 1 & 0 & -1 \end{bmatrix} \mathbf{x} + \begin{bmatrix} 2 \\ 0 \\ 1 \end{bmatrix} u$$
$$y = [1 \quad 1 \quad 0]\mathbf{x}$$

into the controllable canonical-form dynamical equation. What is its transfer function?

**7-5** Let $\nu_1$ be the largest integer such that $\{\mathbf{b}_1, \mathbf{A}\mathbf{b}_1, \cdots, \mathbf{A}^{\nu_1-1}\mathbf{b}_1\}$ is a linearly independent set. Let $\nu_2$ be the largest integer such that $\{\mathbf{b}_1, \mathbf{A}\mathbf{b}_1, \cdots, \mathbf{A}^{\nu_1-1}\mathbf{b}_1, \mathbf{b}_2, \mathbf{A}\mathbf{b}_2, \cdots, \mathbf{A}^{\nu_2-1}\mathbf{b}_2\}$ is a linearly independent set. Show that $\mathbf{A}^n\mathbf{b}_2$ for all $n \geq \nu_2$ is linearly dependent on $\{\mathbf{b}_1, \cdots, \mathbf{A}^{\nu_1-1}\mathbf{b}_1, \mathbf{b}_2, \cdots, \mathbf{A}^{\nu_2-1}\mathbf{b}_2\}$.

**7-6** Transform the dynamical equation in Problem 7-4 into the observable canonical-form dynamical equation.

**7-7** Consider the equivalent dynamical equations

$$\begin{cases} \dot{\mathbf{x}} = \mathbf{A}\mathbf{x} + \mathbf{B}\mathbf{u} \\ \mathbf{y} = \mathbf{C}\mathbf{x} \end{cases} \qquad \begin{cases} \dot{\bar{\mathbf{x}}} = \bar{\mathbf{A}}\bar{\mathbf{x}} + \bar{\mathbf{B}}\mathbf{u} \\ \mathbf{y} = \bar{\mathbf{C}}\bar{\mathbf{x}} \end{cases}$$

where $\bar{\mathbf{x}} = \mathbf{P}\mathbf{x}$. Their adjoint equations are, respectively,

$$\begin{cases} \dot{\mathbf{z}} = -\mathbf{A}^*\mathbf{z} + \mathbf{C}^*\mathbf{u} & \text{(1a)} \\ \mathbf{y} = \mathbf{B}^*\mathbf{z} & \text{(1b)} \end{cases}$$
$$\begin{cases} \dot{\bar{\mathbf{z}}} = -\bar{\mathbf{A}}^*\bar{\mathbf{z}} + \bar{\mathbf{C}}^*\mathbf{u} & \text{(2a)} \\ \mathbf{y} = \bar{\mathbf{B}}^*\mathbf{z} & \text{(2b)} \end{cases}$$

where $\mathbf{A}^*$, $\bar{\mathbf{A}}^*$ are the complex conjugate transposes of $\mathbf{A}$ and $\bar{\mathbf{A}}$, respectively. Show that Equations (1) and (2) are equivalent and that they are related by $\bar{\mathbf{z}} = (\mathbf{P}^{-1})^*\mathbf{z}$.

**7-8** Consider the state feedback and the output feedback systems shown in Figure P7-8. Show that for any constant matrix **H**, there exists a

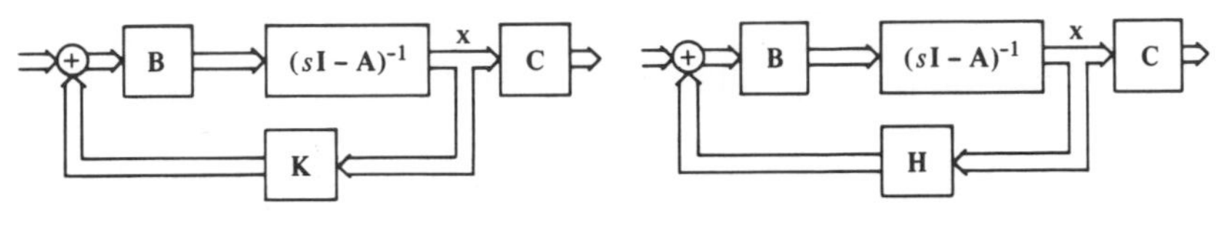


**Figure P7-8**

constant matrix **K** such that $\mathbf{Kx} = \mathbf{HCx}$. Under what condition on **C** will there exist a matrix **H** such that $\mathbf{K} = \mathbf{HC}$ for any **K**? (**C**, **K** and **H** are $q \times n$, $p \times n$, *and* $p \times q$ constant matrices, respectively. It is generally assumed that $n \geq q$.) *Answer:* A solution **H** exists in $\mathbf{K} = \mathbf{HC}$ for a given **K** if and only if $\rho \begin{bmatrix} \mathbf{C} \\ \mathbf{K} \end{bmatrix} = \rho \mathbf{C}$. Consequently, a solution **H** exists in $\mathbf{K} = \mathbf{HC}$ for *any* **K** if and only if **C** is square and nonsingular.

**7-9** Using state feedback to transfer the eigenvalues of the dynamical equation in Problem 7-4 to $-1$, $-2$, and $-2$. Draw a block diagram for the dynamical equation in Problem 7-4 and then add the required state feedback.

**7-10** Show that

$$\rho[\mathbf{b} \quad \mathbf{Ab} \quad \cdots \quad \mathbf{A}^{n-1}\mathbf{b}] = \rho[\mathbf{b} \quad (\mathbf{A} - \mathbf{bk})\mathbf{b} \quad \cdots \quad (\mathbf{A} - \mathbf{bk})^{n-1}\mathbf{b}]$$

for any $1 \times n$ constant vector **k**.

**7-11** Consider the Jordan-form dynamical equation

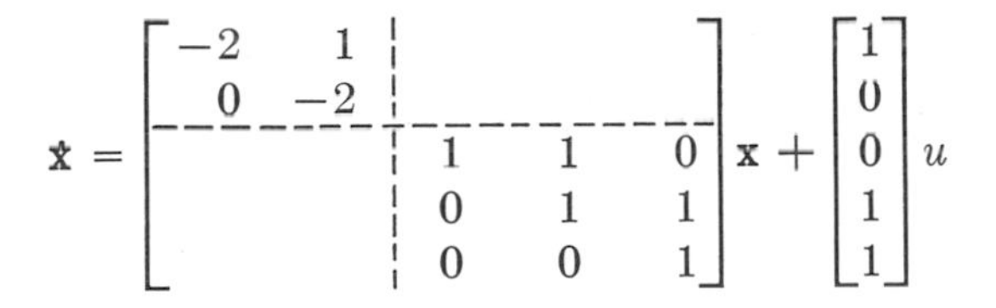

$$\dot{\mathbf{x}} = \left[\begin{array}{cc|ccc} -2 & 1 & & & \\ 0 & -2 & & & \\ \hline & & 1 & 1 & 0 \\ & & 0 & 1 & 1 \\ & & 0 & 0 & 1 \end{array}\right]\mathbf{x} + \begin{bmatrix} 1 \\ 0 \\ 0 \\ 1 \\ 1 \end{bmatrix} u$$

which has an unstable eigenvalue 1. The dynamical equation is not controllable but the subequation associated with eigenvalue 1 is controllable. Do you think it is possible to stabilize the equation by using state feedback? If yes, find the gain vector **k** such that the closed-loop equation has eigenvalues $-1$, $-1$, $-2$, $-2$, and $-2$.

**7-12** Given

$$\dot{\mathbf{x}} = \begin{bmatrix} 2 & 1 \\ -1 & 1 \end{bmatrix}\mathbf{x} + \begin{bmatrix} 1 \\ 2 \end{bmatrix} u$$

Find the gain vector $[k_1 \quad k_2]$ such that the state-feedback system has $-1$ and $-2$ as its eigenvalues. Compute $k_1$, $k_2$ directly without using any equivalence transformation.

**7-13** Consider the uncontrollable state equation

$$\dot{\mathbf{x}} = \begin{bmatrix} 2 & 1 & 0 & 0 \\ 0 & 2 & 0 & 0 \\ 0 & 0 & -1 & 0 \\ 0 & 0 & 0 & -1 \end{bmatrix} \mathbf{x} + \begin{bmatrix} 0 \\ 1 \\ 1 \\ 1 \end{bmatrix} u$$

Is it possible to find a gain vector $\mathbf{k}$ such that the equation with state feedback $u = \mathbf{kx} + v$ has eigenvalues $-2, -2, -1, -1$? Is it possible to have eigenvalues $-2, -2, -2, -1$? How about $-2, -2, -2, -2$? *Answers:* yes; yes; no.

**7-14** The observability of a dynamical equation is not invariant under any state feedback. Where does the argument fail if the argument used in the proof of Theorem 7-3 is used to prove the observability part?

**7-15** Find a three-dimensional asymptotic estimator with eigenvalues $-2, -2, -3$ for the dynamical equation in Problem 7-4.

**7-16** Find a two-dimensional asymptotic estimator with eigenvalues $-2$, $-3$ for the dynamical equation in Problem 7-4.

**7-17** Show that all the columns of $\mathbf{Q}$ in (7-67) are linearly independent. *Hint:* Let the constants $\alpha_{ij}$'s be such that

$$\sum_{i,j} \alpha_{ij} \mathbf{A}^j \mathbf{e}_{i\mu_i} = 0$$

Then use (7-66) to show that $\alpha_{ij} = 0$ for $i = 1, 2, \cdots, q$; $j = 1, 2, \cdots, \mu_i$.

**7-18** Consider a system with the transfer function

$$\frac{(s-1)(s+2)}{(s+2)(s-2)(s+3)}$$

Is it possible to change the transfer function to

$$\frac{s+1}{(s+2)(s+3)}$$

by state feedback? If yes, how?

**7-19** If the state variables of the system in Problem 7-18 are not available, is it still possible to do it? If an estimator with eigenvalues $-1$, $-1$ is used to generate the state variables, what is the complete transfer function?

**7-20** Design a compensator such that the feedback system in Figure P7-20 has the set $\{-2, -1 + i, -1 - i\}$ as its poles.

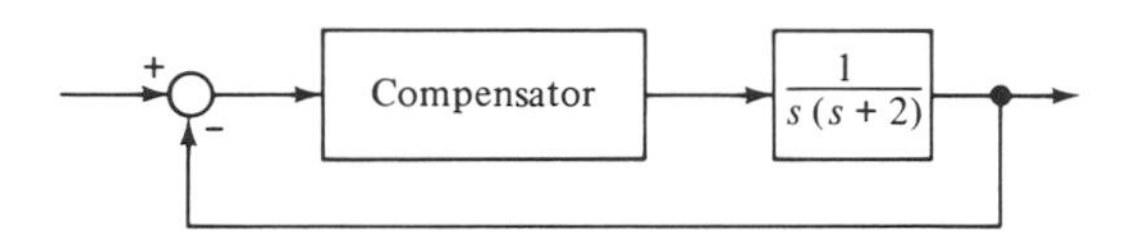


**Figure P7-20**

**7-21** Consider the example in Section 7-5. If the reference input $v$ is not assumed to be zero, show that the compensated system is of the form shown in Figure P7-21. What is the transfer function of the entire sys-

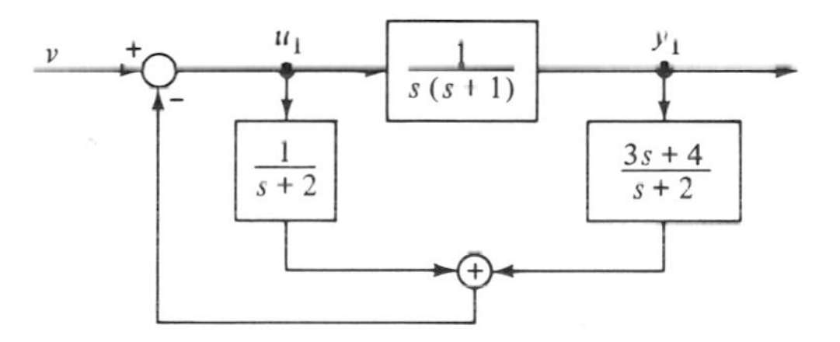


**Figure P7-21**

tem? Will the eigenvalue of the estimator appear in the transfer function of the entire system? If the dynamical-equation description is set up for the system, is the dynamical equation controllable and observable?

**7-22** Consider the irreducible single-variable dynamical equation

$$\dot{\mathbf{x}} = \mathbf{A}\mathbf{x} + \mathbf{b}u$$
$$y = \mathbf{c}\mathbf{x}$$

where $\mathbf{A}$ is an $n \times n$ matrix, $\mathbf{b}$ is an $n \times 1$ vector, and $\mathbf{c}$ is a $1 \times n$ vector. Let $\hat{g}$ be its transfer function. Show that $\hat{g}(s)$ has $m$ zeros—in other words, that the numerator of $\hat{g}(s)$ has degree $m$—if and only if

$$\mathbf{c}\mathbf{A}^i\mathbf{b} = 0 \qquad \text{for } i = 0, 1, 2, \cdots, n - m - 2$$

and $\mathbf{c}\mathbf{A}^{n-m-1}\mathbf{b} \neq 0$. Or equivalently, the difference between the degree of the denominator and the degree of the numerator of $\hat{g}(s)$ is $\bar{d} = n - m$ if and only if

$$\mathbf{c}\mathbf{A}^{\bar{d}-1}\mathbf{b} \neq 0 \qquad \mathbf{c}\mathbf{A}^i\mathbf{b} = 0 \qquad \text{for } i = 0, 1, 2, \cdots, \bar{d} - 2$$

*Hint:* Show that $\mathbf{c}\mathbf{A}^i\mathbf{b}$ is invariant under any equivalence transformation and then use the controllable canonical-form dynamical equation.

**7-23** **a.** Can a multivariable system with the transfer function

$$\begin{bmatrix} \dfrac{1}{s^3+1} & \dfrac{2}{s^2+1} \\ \dfrac{2s+1}{s^3+s+1} & \dfrac{1}{s} \end{bmatrix}$$

be decoupled by state feedback?

**b.** Can a system with the dynamical-equation description

$$\dot{\mathbf{x}} = \begin{bmatrix} 3 & 1 & 0 \\ 0 & 0 & -1 \\ 0 & 1 & -1 \end{bmatrix} \mathbf{x} + \begin{bmatrix} 0 & 0 \\ 1 & 0 \\ 0 & 1 \end{bmatrix} \mathbf{u}$$

$$\mathbf{y} = \begin{bmatrix} 2 & -1 & 1 \\ 0 & 2 & 1 \end{bmatrix} \mathbf{x}$$

be decoupled by state feedback? *Hint:* Use Equation (7-90).

CHAPTER

# 8

# Stability of Linear Systems

## 8-1 Introduction

Controllability and observability, introduced in Chapter 5, are two important qualitative properties of linear systems. In this chapter we shall introduce another qualitative property of systems—namely, stability. The concept of stability is extremely important, because almost every workable system is designed to be stable. If a system is not stable, it is usually of no use in practice.

We have introduced the input–output description and the dynamical-equation description of systems; hence it is natural to study stability in terms of these two descriptions separately. In Section 8-2 the bounded-input bounded-output (BIBO) stability of systems is introduced in terms of the input–output description. We show that if the impulse response of a system is absolutely integrable, then the system is BIBO stable. The stability condition in terms of rational transfer functions is also given in this section. In Section 8-3 we introduce the Routh–Hurwitz and the Liénard–Chipart criteria, which are used to check whether or not all the roots of a polynomial have negative real parts. We study in Section 8-4 the stability of a system in terms of the state-variable description. We introduce there the concepts of equilibrium state, stability in the sense of Lyapunov, asymptotic stability and total stability. The relationships between these concepts are established. In the last section, a theorem of Lyapunov is introduced and the Routh–Hurwitz criterion is proved.

The references for this chapter are [58], [59], [65], [76], [90], [102], and [116].

## 8-2 Stability Criteria in Terms of the Input–Output Description

**Time-varying case.** We consider first single-variable systems—that is, systems with only one input terminal and only one output terminal. It was shown in Chapter 3 that if the inputs and outputs of an initially relaxed system satisfy the homogeneity and additivity properties, and if the system is causal, then the input $u$ and the output $y$ of the system can be related by

$$y(t) = \int_{-\infty}^{t} g(t, \tau)u(\tau)\, d\tau \qquad \text{for all } t \text{ in } (-\infty, \infty) \tag{8-1}$$

where $g(t, \tau)$ is the impulse response of the system and is, by definition, the output measured at time $t$ due to an impulse function input applied at time $\tau$.

In the qualitative study of a system from the input and output terminals, perhaps the only question which can be asked is that if the input has certain properties, under what condition will the output have the same properties? For example, if the input $u$ is bounded, that is,

$$|u(t)| \leq k_1 < \infty \qquad \text{for all } t \text{ in } (-\infty, \infty) \tag{8-2}$$

under what condition on the system does there exist a constant $k_2$ such that the output $y$ satisfies

$$|y(t)| \leq k_2 < \infty$$

for all $t$ in $(-\infty, \infty)$? If the input is of finite energy, that is,

$$\left(\int_{-\infty}^{\infty} |u(t)|^2\, dt\right)^{1/2} \leq k_3 < \infty \tag{8-3}$$

does there exist a constant $k_4$ such that

$$\left(\int_{-\infty}^{\infty} |y(t)|^2\, dt\right)^{1/2} \leq k_4 < \infty ?$$

If the input approaches a periodic function, under what condition will the output approach another periodic function with the same period? If the input approaches a constant, will the output approach some constant? According to these various properties, we may introduce different stability definitions for the system. We shall introduce here only the one which is most commonly used in linear systems—the bounded-input bounded-output stability.

Recall that the input–output description of a system is applicable only when the system is initially relaxed. A system that is initially relaxed is called a relaxed system. Hence the stability that is defined in terms of the input–output description is applicable only to relaxed systems.

### Definition 8-1

A relaxed system is said to be BIBO (bounded-input bounded-output) stable if and only if for any bounded input, the output is bounded. ▮

We illustrate the importance of the qualification "relaxedness" by showing that a system, being BIBO stable under the relaxedness assumption, might not be BIBO stable if it is not initially relaxed.

### Example 1

Consider the network shown in Figure 8-1. If the system is initially relaxed, that is, the initial voltage across the capacitor is zero, then $y(t) =$

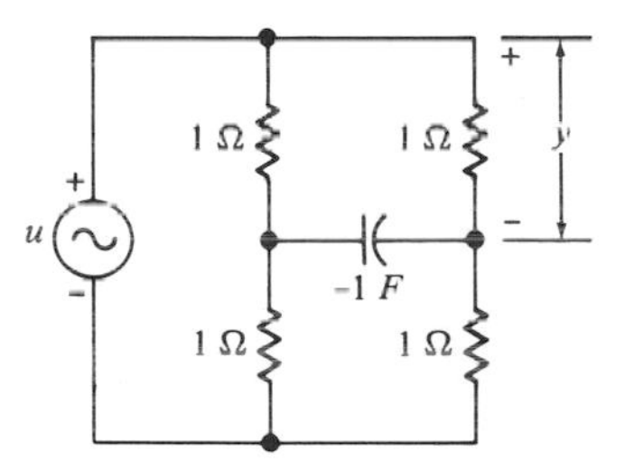


**Figure 8-1** A system whose output is bounded for any bounded input if the initial voltage across the capacitor is zero.

$u(t)/2$ for all $t$. Therefore, for any bounded input, the output is also bounded. However, if the initial voltage across the capacitor is not zero, because of the negative capacitance, the output will increase to infinity even if no input is applied. ▮

### Theorem 8-1

A relaxed single-variable system that is described by

$$y(t) = \int_{-\infty}^{t} g(t, \tau)u(\tau)\, d\tau$$

is BIBO stable if and only if there exists a finite number $k$ such that

$$\int_{-\infty}^{t} |g(t, \tau)|\, d\tau \leq k < \infty$$

for all $t$ in $(-\infty, \infty)$.

Proof

*Sufficiency:* Let $u$ be an arbitrary input and let $|u(t)| \leq k_1$ for all $t$ in $(-\infty, \infty)$. Then

$$|y(t)| = \left| \int_{-\infty}^{t} g(t, \tau)u(\tau)\, d\tau \right| \leq \int_{-\infty}^{t} |g(t, \tau)|\, |u(\tau)|\, d\tau \leq k_1 \int_{-\infty}^{t} |g(t, \tau)|\, d\tau \leq kk_1$$

for all $t$ in $(-\infty, \infty)$. *Necessity:* A rigorous proof of this part is rather involved. We shall exhibit the basic idea by showing that if

$$\int_{-\infty}^{t} |g(t, \tau)|\, d\tau = \infty$$

for some $t$, say $t_1$, then we can find a bounded input that excites an unbounded output. Let us choose

$$u(t) = \operatorname{sgn}\, [g(t_1, t)] \tag{8-4}$$

where

$$\operatorname{sgn} x = \begin{cases} 0 & \text{if } x = 0 \\ 1 & \text{if } x > 0 \\ -1 & \text{if } x < 0 \end{cases}$$

Clearly $u$ is bounded. However, the output excited by this input,

$$y(t_1) = \int_{-\infty}^{t_1} g(t_1, \tau)u(\tau)\, d\tau = \int_{-\infty}^{t_1} |g(t_1, \tau)|\, d\tau = \infty$$

is not bounded. Q.E.D.

We consider now multivariable systems. Consider a system with $p$ input terminals and $q$ output terminals, which is described by

$$\mathbf{y}(t) = \int_{-\infty}^{t} \mathbf{G}(t, \tau)\mathbf{u}(\tau)\, d\tau \tag{8-5}$$

where $\mathbf{u}$ is the $p \times 1$ input vector, $\mathbf{y}$ is the $q \times 1$ output vector, and $\mathbf{G}(t, \tau)$ is the $q \times p$ impulse-response matrix of the system. Let

$$\mathbf{G}(t, \tau) = \begin{bmatrix} g_{11}(t, \tau) & g_{12}(t, \tau) & \cdots & g_{1p}(t, \tau) \\ g_{21}(t, \tau) & g_{22}(t, \tau) & \cdots & g_{2p}(t, \tau) \\ \vdots & \vdots & & \vdots \\ g_{q1}(t, \tau) & g_{q2}(t, \tau) & \cdots & g_{qp}(t, \tau) \end{bmatrix} \tag{8-6}$$

Then $g_{ij}$ is the impulse response between the $j$th input terminal and the $i$th output terminal. Similar to single-variable systems, a relaxed multivariable system is defined to be BIBO stable if and only if for any bounded-input vector, the output vector is bounded. By a bounded vector, we mean that every component of the vector is bounded. Applying Theorem

8-1 to every possible pair of input and output terminals, and using the fact that the sum of a finite number of bounded functions is bounded, we have immediately the following theorem.

### Theorem 8-2

A relaxed multivariable system that is described by

$$\mathbf{y}(t) = \int_{-\infty}^{t} \mathbf{G}(t, \tau)\mathbf{u}(\tau)\, d\tau$$

is BIBO stable if and only if there exists a finite number $k$ such that, for every entry of $\mathbf{G}$,

$$\int_{-\infty}^{t} |g_{ij}(t, \tau)|\, d\tau \leq k < \infty$$

for all $t$ in $(-\infty, \infty)$. ∎

**Time-invariant case.** Consider a single-variable system with the following input–output description:

$$y(t) = \int_0^t g(t - \tau)u(\tau)\, d\tau = \int_0^t g(\tau)u(t - \tau)\, d\tau \tag{8-7}$$

where $g(t)$ is the impulse response of the system. Recall that in order to have a description of the form (8-7), the input–output pairs of the system must satisfy linearity, causality, and time-invariance properties. In addition, the system is assumed to be relaxed at $t = 0$.

### Corollary 8-1

A relaxed single-variable system which is described by

$$y(t) = \int_0^t g(t - \tau)u(\tau)\, d\tau$$

is BIBO stable if and only if

$$\int_0^\infty |g(t)|\, dt \leq k < \infty$$

for some constant $k$.

### Proof

This follows directly from Theorem 8-1 by observing that

$$\int_0^t |g(t, \tau)|\, d\tau = \int_0^t |g(t - \tau)|\, d\tau = \int_0^t |g(\alpha)|\, d\alpha \leq \int_0^\infty |g(\alpha)|\, d\alpha \tag{8-8}$$

For the time-invariant case, the initial time is chosen at $t = 0$; hence the integration in (8-8) starts from 0 instead of $-\infty$. Q.E.D.

A function $g$ is said to be absolutely integrable on $[0, \infty)$ if

$$\int_0^\infty |g(t)|\, dt \leq k < \infty$$

Graphically, it says that the total area under $|g|$ is finite. The fact that $g$ is absolutely integrable does not imply that $g$ is bounded on $[0, \infty)$ nor that $g(t)$ approaches zero as $t \to \infty$. Indeed, consider the function defined by

$$f(t-n) = \begin{cases} n + (t-n)n^4 & \text{for } \dfrac{-1}{n^3} < (t-n) \leq 0 \\ n - (t-n)n^4 & \text{for } 0 \leq (t-n) \leq \dfrac{1}{n^3} \end{cases}$$

for $n = 2, 3, 4, \cdots$. The function $f$ is depicted in Figure 8-2. It is easy to verify that $f$ is absolutely integrable; however, $f$ is neither bounded on $[0, \infty)$ nor approaching zero as $t \to \infty$.

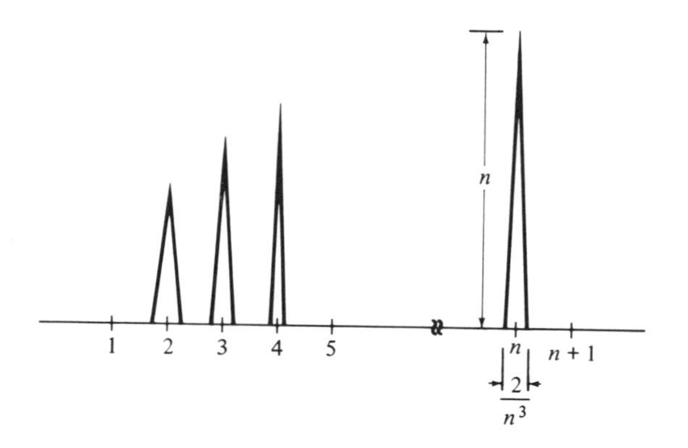


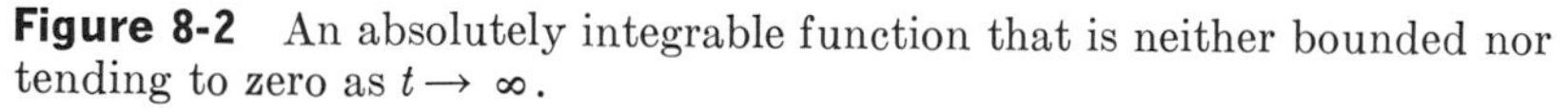
**Figure 8-2** An absolutely integrable function that is neither bounded nor tending to zero as $t \to \infty$.

If $g$ is absolutely integrable on $[0, \infty)$, then[1]

$$\int_{\alpha}^{\infty} |g(t)|\, dt \to 0 \tag{8-9}$$

as $\alpha \to \infty$. We shall use this fact in the proof of the following theorem.

---

[1] Since

$$\int_0^{\infty} |g(t)|\, dt = \int_0^{\alpha} |g(t)|\, dt + \int_{\alpha}^{\infty} |g(t)|\, dt$$

and since

$$\int_0^{\alpha} |g(t)|\, dt$$

is nondecreasing as $\alpha$ increases, hence

$$\lim_{\alpha \to \infty} \int_0^{\alpha} |g(t)|\, dt \to \int_0^{\infty} |g(t)|\, dt$$

Consequently,

$$\lim_{\alpha \to \infty} \int_{\alpha}^{\infty} |g(t)|\, dt \to 0$$

### Theorem 8-3

Consider a relaxed single-variable system whose input $u$ and output $y$ are related by

$$y(t) = \int_0^t g(t - \tau)u(\tau)\, d\tau$$

If

$$\int_0^\infty |g(t)|\, dt \leq k < \infty$$

for some constant $k$, then we have the following:

1. If $u$ is a periodic function with period $T$—that is, $u(t) = u(t + T)$ for all $t \geq 0$—then the output $y$ tends to a periodic function with the same period (not necessarily of the same waveform).
2. If $u$ is bounded and tends to a constant, then the output will tend to a constant.
3.[2] If $u$ is of finite energy, that is,

$$\left(\int_0^\infty |u(t)|^2\, dt\right)^{1/2} \leq k_1 < \infty$$

then the output is also of finite energy; that is, there exists a finite $k_2$ that depends on $k_1$ such that

$$\left(\int_0^\infty |y(t)|^2\, dt\right)^{1/2} \leq k_2 < \infty$$

### Proof

1. We shall show that if $u(t) = u(t + T)$ for all $t \geq 0$, then $y(t) \to y(t + T)$ as $t \to \infty$. It is clear that

$$y(t) = \int_0^t g(\tau)u(t - \tau)\, d\tau \tag{8-10}$$

and

$$y(t + T) = \int_0^{t+T} g(\tau)u(t + T - \tau)\, d\tau = \int_0^{t+T} g(\tau)u(t - \tau)\, d\tau \tag{8-11}$$

---

[2] This can be extended to as follows: For any real number $p$ in $[1, \infty]$, if

$$\left(\int_0^\infty |u(t)|^p\, dt\right)^{1/p} \leq k_1 < \infty$$

then there exists a finite $k_2$ such that

$$\left(\int_0^\infty |y(t)|^p\, dt\right)^{1/p} \leq k_2 < \infty$$

and the system is said to be *$L_p$-stable.*

Subtracting (8-10) from (8-11), we obtain

$$|y(t+T) - y(t)| = \left| \int_t^{t+T} g(\tau)u(t-\tau)\, d\tau \right|$$
$$\leq \int_t^{t+T} |g(\tau)|\, |u(t-\tau)|\, d\tau \leq u_M \int_t^{t+T} |g(\tau)|\, d\tau$$

where $u_M \triangleq \max_{0 \leq t \leq T} |u(t)|$. It follows from (8-9) that $|y(t) - y(t + T)| \to 0$ as $t \to \infty$, or $y(t) \to y(t+T)$ as $t \to \infty$.

**2.** We prove this part by an intuitive argument. Consider

$$y(t) = \int_0^{t_1} g(\tau)u(t-\tau)\, d\tau + \int_{t_1}^{t} g(\tau)u(t-\tau)\, d\tau \qquad \textbf{(8-12)}$$

Let $u_M \triangleq \max_t |u(t)|$. Then we have, with $t > t_1$,

$$\left| \int_{t_1}^{t} g(\tau)u(t-\tau)\, d\tau \right| \leq \int_{t_1}^{t} |g(\tau)|\, |u(t-\tau)|\, d\tau \leq u_M \int_{t_1}^{t} |g(\tau)|\, d\tau$$

which approaches zero as $t_1 \to \infty$, following from (8-9). Hence if $t_1$ is sufficiently large, (8-12) can be approximated by

$$y(t) \doteqdot \int_0^{t_1} g(\tau)u(t-\tau)\, d\tau \qquad \textbf{(8-13)}$$

for all $t \geq t_1$. As can be seen from Figure 8-3, if $t$ is much larger

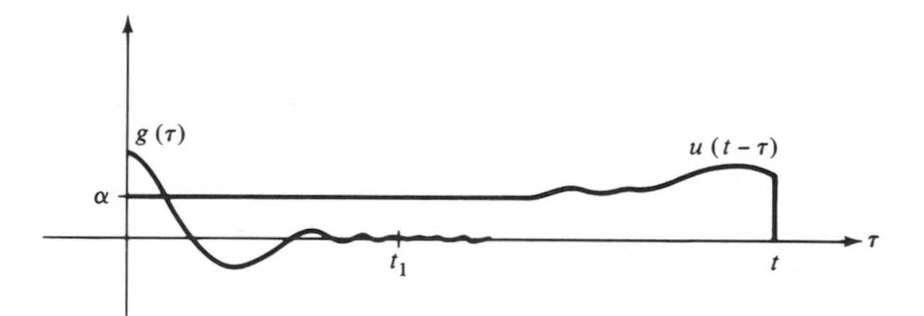


**Figure 8-3** The convolution of $g$ and $u$.

than $t_1$, $u(t-\tau)$ is approximately equal to a constant, say $\alpha$, for all $\tau$ in $[0, t_1]$. Hence (8-13) becomes, for all $t \gg t_1 \gg 0$,

$$y(t) \doteqdot \alpha \int_0^{t_1} g(\tau)\, d\tau$$

which is independent of $t$. This completes the proof.

**3.** It can be shown that the set of all real-valued functions defined over $[0, \infty)$ with the property

$$\left( \int_0^\infty |f(t)|^2\, dt \right)^{1/2} \leq k < \infty$$

for some constant $k$ forms a linear space over $\mathbb{R}$. It is called the $L_2$ function space. In this space we may define a norm and an inner product as follows (see Problem 2-37):

$$\|f\| \triangleq \left(\int_0^\infty |f(t)|^2\,dt\right)^{1/2} \qquad \langle f, g\rangle \triangleq \int_0^\infty f(t)g(t)\,dt$$

And the Schwartz inequality (Theorem 2-15) reads as

$$\left|\int_0^\infty f(t)g(t)\,dt\right| \le \left(\int_0^\infty |f(t)|^2\,dt\right)^{1/2}\left(\int_0^\infty |g(t)|^2\,dt\right)^{1/2}$$

With these preliminaries, we are ready to proceed with the proof. Consider

$$|y(t)| = \left|\int_0^t g(\tau)u(t-\tau)\,d\tau\right| \le \int_0^\infty |g(\tau)|\,|u(t-\tau)|\,d\tau$$

which can be written as

$$|y(t)| \le \int_0^\infty (|g(\tau)|^{1/2})(|g(\tau)|^{1/2}|u(t-\tau)|)\,d\tau \tag{8-14}$$

Applying the Schwartz inequality, (8-14) becomes

$$|y(t)|^2 \le \left(\int_0^\infty |g(\tau)|\,d\tau\right)\left(\int_0^\infty |g(\tau)|\,|u(t-\tau)|^2\,d\tau\right)$$
$$\le k\int_0^\infty |g(\tau)|\,|u(t-\tau)|^2\,d\tau$$

Consider now

$$\int_0^\infty |y(t)|^2\,dt \le k\int_0^\infty \int_0^\infty |g(\tau)|\,|u(t-\tau)|^2\,d\tau\,dt$$
$$= k\int_0^\infty \left(\int_0^\infty |u(t-\tau)|^2\,dt\right)|g(\tau)|\,d\tau \tag{8-15}$$

where we have changed the order of integration. By assumption, $u$ is of finite energy, hence there exists a finite $k_1$ such that

$$\int_0^\infty |u(t-\tau)|^2\,dt \le k_1$$

Hence (8-15) implies that

$$\int_0^\infty |y(t)|^2\,dt \le kk_1\int_0^\infty |g(\tau)|\,d\tau = k^2k_1$$

In other words, if $g$ is absolutely integrable on $[0, \infty)$ and if $u$ is of finite energy, then the output $y$ is also of finite energy. Q.E.D.

We see from Corollary 8-1 and Theorem 8-3 that if a system that is describable by

$$y(t) = \int_0^t g(t-\tau)u(\tau)\,d\tau$$

is BIBO stable, then if the input has certain property, the output will have the same property. For the time-varying case, this is not necessarily true. The following is a very important corollary of Theorem 8-3.

### Corollary 8-3

Consider a relaxed single-variable system whose input $u$ and output $y$ are related by

$$y(t) = \int_0^t g(t-\tau)u(\tau)\,d\tau$$

If

$$\int_0^\infty |g(t)|\,dt \le k < \infty$$

for some $k$ and if $u(t) = \sin \omega t$, for $t \ge 0$, then

$$y(t) \to |\hat{g}(i\omega)| \sin(\omega t + \theta) \qquad \text{as } t \to \infty \tag{8-16}$$

where $\theta = \tan^{-1}(\operatorname{Im}\hat{g}(i\omega)/\operatorname{Re}\hat{g}(i\omega))$ and $\hat{g}(s)$ is the Laplace transform of $g(\cdot)$; Re and Im denote the real part and the imaginary part, respectively.

### Proof

Since $\sin(t-\tau) = \sin t \cos \tau - \cos t \sin \tau$, we have

$$\begin{aligned} y(t) &= \int_0^t g(\tau)u(t-\tau)\,d\tau = \int_0^t g(\tau)[\sin \omega t \cos \omega\tau - \cos \omega t \sin \omega\tau]\,d\tau \\ &= \sin \omega t \int_0^\infty g(\tau) \cos \omega\tau\,d\tau - \cos \omega t \int_0^\infty g(\tau) \sin \omega\tau\,d\tau \\ &\qquad - \int_t^\infty g(\tau) \sin \omega(t-\tau)\,d\tau \end{aligned}$$

It is clear that

$$\left| \int_t^\infty g(\tau) \sin \omega(t-\tau)\,d\tau \right| \le \int_t^\infty |g(\tau)|\,|\sin \omega(t-\tau)|\,d\tau \le \int_t^\infty |g(\tau)|\,d\tau$$

Hence, from (8-9) we conclude that as $t \to \infty$, we obtain

$$y(t) \to \sin \omega t \int_0^\infty g(\tau) \cos \omega\tau\,d\tau - \cos \omega t \int_0^\infty g(\tau) \sin \omega\tau\,d\tau \tag{8-17}$$

By definition, $\hat{g}(s)$ is the Laplace transform of $g(\cdot)$; that is,

$$\hat{g}(s) = \int_0^\infty g(t)e^{-st}\,dt$$

which implies that

$$\hat{g}(j\omega) = \int_0^\infty g(t)e^{-i\omega t}\,dt = \int_0^\infty g(t) \cos \omega t\,dt - i \int_0^\infty g(t) \sin \omega t\,dt$$

Since $g(t)$ is of real-valued, we have

$$\operatorname{Re}\hat{g}(i\omega) = \int_0^\infty g(t) \cos \omega t\,dt \tag{8-18a}$$

$$\operatorname{Im}\hat{g}(i\omega) = -\int_0^\infty g(t) \sin \omega t\,dt \tag{8-18b}$$

Hence (8-17) becomes

$$y(t) \to \sin \omega t\,(\operatorname{Re}\hat{g}(i\omega)) + \cos \omega t\,(\operatorname{Im}\hat{g}(i\omega)) = |\hat{g}(i\omega)| \sin(\omega t + \theta)$$

where $\theta = \tan^{-1}(\operatorname{Im}\hat{g}(i\omega)/\operatorname{Re}\hat{g}(i\omega))$. Q.E.D.

This corollary shows that for a BIBO-stable, linear, time-invariant, relaxed system, if the input is a sinusoidal function, after the transient dies out, the output is also a sinusoidal function. Furthermore, from this sinusoidal output, the magnitude and phase of the transfer function at that frequency can be read out directly. This fact is often used in practice to measure the transfer function of a linear, time-invariant, relaxed system.

Linear time-invariant systems are often described by transfer functions, hence it is useful to study the stability conditions in terms of transfer functions. If $\hat{g}(s)$ is not a rational function of $s$, the stability condition cannot be easily stated in terms of $\hat{g}(s)$ (see Problem 8-3). However if $\hat{g}(s)$ is a proper rational function, then we have the following theorem.

### Theorem 8-4

A relaxed single-variable system that is described by a proper rational function $\hat{g}(s)$ is BIBO stable if and only if all the poles of $\hat{g}(s)$ are in the open left half $s$-plane or, equivalently, all the poles of $\hat{g}(s)$ have negative real parts. ∎

By the open left half $s$-plane, we mean the left half $s$-plane excluding the imaginary axis. On the other hand, the closed left half $s$-plane is the left half $s$-plane including the imaginary axis. Note that stability of a system has nothing to do with the zeros of $\hat{g}(s)$.

### Proof of Theorem 8-4

If $\hat{g}(s)$ is a proper rational function, it can be expanded by partial fraction expansion into a sum of finite number of terms of the form

$$\frac{\beta}{(s - \lambda_i)^k}$$

and possibly of a constant, where $\lambda_i$ is a pole of $\hat{g}(s)$. Consequently, $g(t)$ is a sum of finite number of the term $t^{k-1}e^{\lambda_i t}$ and possibly of a $\delta$-function. It is easy to show that $t^{k-1}e^{\lambda_i t}$ is absolutely integrable if and only if $\lambda_i$ has a negative real part. Hence we conclude that the relaxed system is BIBO stable if and only if all the poles of $\hat{g}(s)$ have negative real part.

Q.E.D.

### Example 2

Consider the system with a transfer function $\hat{g}(s) = 1/s$. The pole of $\hat{g}(s)$ is on the imaginary axis. Hence the system is not BIBO stable. This can also be shown from the definition. Let the input be a unit step function, then $\hat{u}(s) = 1/s$. Corresponding to this bounded input, the output is $\mathcal{L}^{-1}[\hat{g}(s)\hat{u}(s)] = \mathcal{L}^{-1}[1/s^2] = t$, which is not bounded. Hence the system is not BIBO stable. ∎

For multivariable systems, we have the following results.

### Corollary 8-2

A relaxed multivariable system that is described by

$$\mathbf{y}(t) = \int_0^t \mathbf{G}(t-\tau)\mathbf{u}(\tau)\,d\tau$$

is BIBO stable if and only if there exists a finite number $k$ such that, for every entry of $\mathbf{G}$,

$$\int_0^\infty |g_{ij}(t)|\,dt \le k < \infty$$ ∎

If the Laplace transform of $\mathbf{G}(t)$, denoted by $\hat{\mathbf{G}}(s)$, is a proper rational-function matrix, then the BIBO stability condition can also be stated in terms of $\hat{\mathbf{G}}(s)$.

### Theorem 8-5

A relaxed multivariable system that is described by $\hat{\mathbf{y}}(s) = \hat{\mathbf{G}}(s)\hat{\mathbf{u}}(s)$, where $\hat{\mathbf{G}}(s)$ is a proper rational-function matrix, is BIBO stable if and only if all the poles of every entry of $\hat{\mathbf{G}}(s)$ have negative real parts. ∎

Corollary 8-2 and Theorem 8-5 follow immediately from Theorems 8-2 and 8-4 if we consider every entry of $\mathbf{G}$ as the impulse response of a certain input–output pair.

## 8-3 Routh–Hurwitz Criterion and Liénard–Chipart Criterion

If the transfer function of a system is a rational function of $s$, then the BIBO stability of the system is completely determined by the poles of $\hat{g}(s)$ or, equivalently, by the roots of the denominator of $\hat{g}(s)$. We have shown that if all the roots of the denominator of $\hat{g}(s)$, a polynomial, have negative real parts, then the system is BIBO stable. A polynomial is called a *Hurwitz polynomial* if all the roots of the polynomial have negative real parts. Hence the stability problem reduces to determining whether or not a polynomial is a Hurwitz polynomial. A direct method is to solve for all the roots of the polynomial. If the degree of the polynomial is large, solving for the roots is a very tedious job. Furthermore, the knowledge of the exact locations of the roots is not needed as far as the stability is concerned. Hence it is desirable to have some method to determine a Hurwitz polynomial without solving for its roots. In this section, we shall introduce two such methods: the Routh–Hurwitz criterion and the Liénard–Chipart criterion. Consider the polynomial

$$D(s) = a_0 s^n + a_1 s^{n-1} + a_2 s^{n-2} + \cdots + a_{n-1}s + a_n \qquad a_0 > 0 \tag{8-19}$$

where the $a_i$'s are real numbers. Before developing the Routh–Hurwitz criterion, we shall give some necessary condition for $D(s)$ to be Hurwitz. If $D(s)$ is a Hurwitz polynomial—that is, if all the roots of $D(s)$ have negative real parts—then $D(s)$ can be factored as

$$\begin{aligned} D(s) &= a_0 \prod_k (s + \alpha_k) \prod_j (s + \beta_j + i\gamma_j)(s + \beta_j - i\gamma_j) \\ &= a_0 \prod_k (s + \alpha_k) \prod_j (s^2 + 2\beta_j s + \beta_j^2 + \gamma_j^2) \end{aligned} \qquad \text{(8-20)}$$

where $\alpha_k > 0$, $\beta_j > 0$, and $i = \sqrt{-1}$. Since all the coefficients of the factors in the right-hand side of (8-20) are positive, we conclude that if $D(s)$ is a Hurwitz polynomial, its coefficients $a_i$, $i = 1, 2, \cdots, n$ must be all positive. Hence *given a polynomial with a positive leading coefficient, if some of its coefficients are negative or zero then the polynomial is not a Hurwitz polynomial.* The condition that all coefficients of a polynomial be positive is only a necessary condition for the polynomial to be Hurwitz. A polynomial with positive coefficients may still not be a Hurwitz polynomial; for example, the polynomial with positive coefficients

$$s^3 + s^2 + 11s + 51 = (s + 3)(s - 1 + 4i)(s - 1 - 4i)$$

is not a Hurwitz polynomial.

Consider the polynomial $D(s)$ given in (8-19). We form the following polynomials

$$D_1(s) = a_0 s^n + a_2 s^{n-2} + \cdots \qquad \text{(8-21a)}$$
$$D_2(s) = a_1 s^{n-1} + a_3 s^{n-3} + \cdots \qquad \text{(8-21b)}$$

that is, if $n$ is even, $D_1(s)$ consists of the even part of $D(s)$ and $D_2(s)$ consists of the odd part of $D(s)$; if $n$ is odd, $D_1(s)$ consists of the odd part of $D(s)$ and $D_2(s)$ consists of the even part of $D(s)$. Observe that the degree of $D_1(s)$ is always one degree higher than that of $D_2(s)$. Now we expand $D_1(s)/D_2(s)$ in the following Stieljes continued fraction expansion:

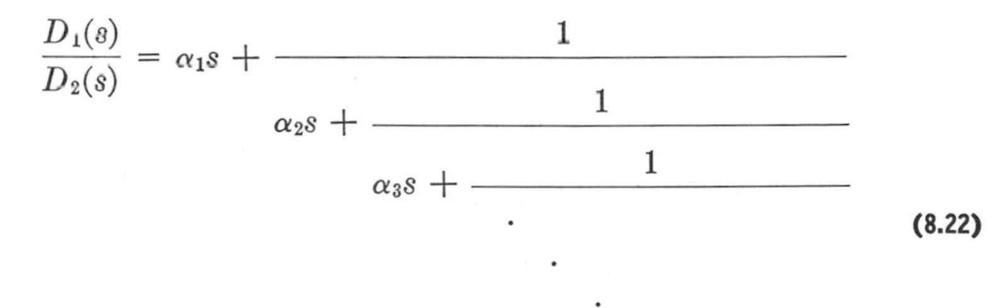

$$\frac{D_1(s)}{D_2(s)} = \alpha_1 s + \cfrac{1}{\alpha_2 s + \cfrac{1}{\alpha_3 s + \cfrac{1}{\ddots + \cfrac{1}{\alpha_{n-1} s + \cfrac{1}{\alpha_n s}}}}} \qquad \text{(8.22)}$$

**Theorem 8-6**

The polynomial $D(s)$ in (8-19) is a Hurwitz polynomial if and only if the $n$ numbers $\alpha_1, \alpha_2, \cdots, \alpha_n$ in (8-22) are all positive. ∎

This theorem will be proved in Section 8-5. Here we give an example to illustrate its application.

### Example 1

Consider $D(s) = s^4 + 2s^3 + 6s^2 + 4s + 1$. Now $D_1(s) = s^4 + 6s^2 + 1$, $D_2(s) = 2s^3 + 4s$ and

$$\frac{D_1(s)}{D_2(s)} = \frac{1}{2}s + \cfrac{1}{\cfrac{2s^3 + 4s}{4s^2 + 1}} = \frac{1}{2}s + \cfrac{1}{\cfrac{1}{2}s + \cfrac{1}{\cfrac{8}{7}s + \cfrac{1}{\frac{7}{2}s}}}$$

Since the four numbers $\alpha_1 = \frac{1}{2}$, $\alpha_2 = \frac{1}{2}$, $\alpha_3 = \frac{8}{7}$, and $\alpha_4 = \frac{7}{2}$ are all positive, we conclude from Theorem 8-6 that the polynomial $D(s)$ is a Hurwitz polynomial. ∎

In order to obtain $\alpha_1, \alpha_2, \cdots, \alpha_n$, a series of long division must be performed. These processes can be carried out in tabular form. First observe that the first long division in the continued fraction expansion is given by

$$\begin{array}{rlll} & \frac{a_0}{a_1}s & & \qquad \alpha_1 \triangleq \frac{a_0}{a_1} \\ a_1s^{n-1} + a_3s^{n-3} + \cdots \,\big) & a_0s^n + & a_2s^{n-2} + & a_4s^{n-4} + \cdots \\ & a_0s^n + & \alpha_1a_3s^{n-2} + & \alpha_1a_5s^{n-4} + \cdots \\ \hline & & b_1s^{n-2} + & b_2s^{n-4} + \cdots \end{array}$$

where $b_1 \triangleq a_2 - \alpha_1a_3 = (a_1a_2 - a_3a_0)/a_1$, $b_2 \triangleq a_4 - \alpha_1a_5 = (a_1a_4 - a_5a_0)/a_1, \cdots$. The second long division is given by

$$\begin{array}{rlll} & \frac{a_1}{b_1}s & & \qquad \alpha_2 \triangleq \frac{a_1}{b_1} \\ b_1s^{n-2} + b_2s^{n-4} + \cdots \,\big) & a_1s^{n-1} + & a_3s^{n-3} + & a_5s^{n-5} + \cdots \\ & a_1s^{n-1} + & \alpha_2b_2s^{n-3} + & \alpha_2b_3s^{n-5} + \cdots \\ \hline & & c_1s^{n-3} + & c_2s^{n-5} + \cdots \end{array}$$

where $c_1 \triangleq a_3 - \alpha_2b_2 = (b_1a_3 - a_1b_2)/b_1$, $c_2 \triangleq a_5 - \alpha_2b_3 = (b_1a_5 - a_1b_3)/b_1, \cdots$. The third long division is given by

$$\begin{array}{rlll} & \frac{b_1}{c_1}s & & \qquad \alpha_3 \triangleq \frac{b_1}{c_1} \\ c_1s^{n-3} + c_2s^{n-5} + \cdots \,\big) & b_1s^{n-2} + & b_2s^{n-4} + \cdots & \\ & b_1s^{n-2} + & \alpha_3c_2s^{n-4} + \cdots & \\ \hline & & d_1s^{n-4} + & d_2s^{n-6} + \cdots \end{array}$$

where $\alpha_1 \triangleq b_2 - \alpha_3c_2 = (c_1b_2 - b_1c_2)/c_1, \cdots$.

These processes can be summarized in tabular form, as shown in Table 8-1. The first two rows are the coefficients of $D_1(s)$ and $D_2(s)$, respectively.

**Table 8-1** STABILITY TEST

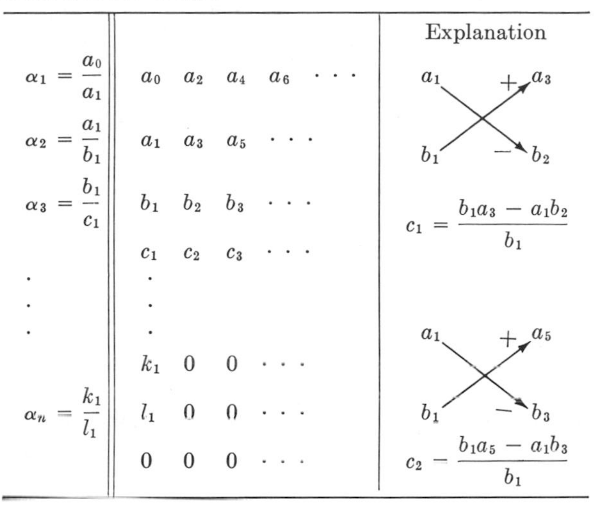

| | | Explanation |
|---|---|---|
| $\alpha_1 = \frac{a_0}{a_1}$ | $a_0 \quad a_2 \quad a_4 \quad a_6 \quad \cdots$ | $a_1$ $+$ $a_3$ |
| $\alpha_2 = \frac{a_1}{b_1}$ | $a_1 \quad a_3 \quad a_5 \quad \cdots$ | $b_1$ $-$ $b_2$ |
| $\alpha_3 = \frac{b_1}{c_1}$ | $b_1 \quad b_2 \quad b_3 \quad \cdots$ | $c_1 = \frac{b_1 a_3 - a_1 b_2}{b_1}$ |
| | $c_1 \quad c_2 \quad c_3 \quad \cdots$ | |
| $\cdot$ | $\cdot$ | |
| $\cdot$ | $\cdot$ | |
| $\cdot$ | $\cdot$ | $a_1$ $+$ $a_5$ |
| | $k_1 \quad 0 \quad 0 \quad \cdots$ | |
| $\alpha_n = \frac{k_1}{l_1}$ | $l_1 \quad 0 \quad 0 \quad \cdots$ | $b_1$ $-$ $b_3$ |
| | $0 \quad 0 \quad 0 \quad \cdots$ | $c_2 = \frac{b_1 a_5 - a_1 b_3}{b_1}$ |

The coefficients in the $i$th row $(3 \leq i \leq n)$ is obtained from the coefficients in the $(i-2)$th and the $(i-1)$th row as shown in the right-hand side of Table 8-1. In this table, we may obtain at most $n$ nonzero numbers $a_1, b_1, \cdots, l_1$ in the first column. From the relation between the $\alpha_i$'s and $a_1, b_1, \cdots$, which is shown in the table, we see immediately that the $n$ numbers $\alpha_1, \alpha_2, \cdots, \alpha_n$ are positive if and only if the $n$ numbers $a_1, b_1, \cdots, l_1$ are positive. Consequently we have the following corollary of Theorem 8-6.

### Corollary 8-6 (Routh–Hurwitz criterion)

A polynomial $D(s)$ of degree $n$ given in (8-19) is a Hurwitz polynomial if and only if the $n$ numbers $a_1, b_1, \cdots, l_1$ defined in Table 8-1 are all positive.[3]

### Example 2

Consider the same polynomial as in Example 1. We have

$$\begin{array}{ccc} 1 & 6 & 1 \\ 2 & 4 & 0 \\ 4 & 1 & 0 \\ 3.5 & 0 & \\ 1 & 0 & \end{array}$$

[3] If the $n$ numbers $a_1, b_1, \ldots, l_1$ are all different from zero, then the number of changes of sign in $\{a_1, b_1, \ldots, l_1\}$ is equal to the number of roots of $D(s)$ with positive real parts. However, if some of them are equal to zero, then it is *not* always possible to tell the number of roots with nonnegative real parts from the signs of these numbers (after certain modifications). The interested reader is referred to Reference [39]—in particular, the example on p. 219 of vol. II.

There are four positive numbers in the first column (excluding the first component); hence the polynomial is a Hurwitz polynomial. ▮

In applying Corollary 8-6, if a zero or a negative number appears in the first column before we obtain $n$ positive numbers, we may stop there and conclude that the polynomial is not a Hurwitz polynomial.

### Example 3

Consider $2s^4 + 2s^3 + s^2 + 3s + 2$. We have

$$\begin{array}{rrr} 2 & 1 & 2 \\ 2 & 3 & \\ -2 & & \end{array}$$

A negative number appears in the first column; hence the polynomial is not a Hurwitz polynomial.

### Example 4

Consider $D(s) = s^3 + s^2 + s + 1$. We have

$$\begin{array}{rr} 1 & 1 \\ 1 & 1 \\ 0 & \end{array}$$

$D(s)$ is a Hurwitz polynomial if and only if there are three positive numbers in the first column excluding the first component. Since this is not so, hence $D(s)$ is not a Hurwitz polynomial. ▮

We note that in computing Table 8-1, the signs of the numbers in the table are not affected if a row is multiplied by a positive number. By using this fact, the computation of Table 8-1 may often be simplified.

### Example 5

Consider $D(s) = 2s^4 + 5s^3 + 5s^2 + 2s + 1$. We form the following table:

$$\begin{array}{rrrl} 2 & 5 & 1 & \\ 5 & 2 & & \\ 21 & 5 & & \text{(after the multiplication of 5)} \\ 17 & 0 & & \text{(after the multiplication of 21)} \\ 5 & & & \end{array}$$

There are four positive numbers in the first column (excluding the first number); hence $D(s)$ is a Hurwitz polynomial. ▮

A remark is in order concerning the Routh–Hurwitz criterion. The algorithm in Table 8-1 was actually discovered by Routh alone. The stability test discovered by Hurwitz is as follows: Given a polynomial, say,

$$D(s) = a_0s^7 + a_1s^6 + a_2s^5 + \cdots + a_6s + a_7 \qquad a_0 > 0$$

form the so-called *Hurwitz matrix*

$$\mathbf{H} \triangleq \begin{bmatrix} a_1 & a_3 & a_5 & a_7 & 0 & 0 & 0 \\ a_0 & a_2 & a_4 & a_6 & 0 & 0 & 0 \\ 0 & a_1 & a_3 & a_5 & a_7 & 0 & 0 \\ 0 & a_0 & a_2 & a_4 & a_6 & 0 & 0 \\ 0 & 0 & a_1 & a_3 & a_5 & a_7 & 0 \\ 0 & 0 & a_0 & a_2 & a_4 & a_6 & 0 \\ 0 & 0 & 0 & a_1 & a_3 & a_5 & a_7 \end{bmatrix}$$

Note that the elements on the diagonal of $\mathbf{H}$ are $a_1, a_2, \cdots, a_7$. Then the polynomial $D(s)$ is a Hurwitz polynomial if and only if the following leading principal minors of $\mathbf{H}$

$$\Delta_1 = a_1$$

$$\Delta_2 = \det \begin{bmatrix} a_1 & a_3 \\ a_0 & a_2 \end{bmatrix}$$

$$\Delta_3 = \det \begin{bmatrix} a_1 & a_3 & a_5 \\ a_0 & a_2 & a_4 \\ 0 & a_1 & a_3 \end{bmatrix}$$

$$\vdots$$

$$\Delta_7 = \det \mathbf{H}$$

are all positive. By purely algebraic manipulation, it can be shown that $\Delta_1, \Delta_2, \cdots$ and the set of numbers $a_1, b_1, \cdots$ in Table 8-1 are related by $\Delta_1 = a_1$, $\Delta_2 = a_1b_1$, $\Delta_3 = a_1b_1c_1, \cdots$. Hence we see that these two tests are actually equivalent. The determinants $\Delta_1, \Delta_2, \cdots$ are called the *Hurwitz determinants*. It was proved by Liénard and Chipart that under the condition that all the coefficients of a polynomial be positive, if all the Hurwitz determinants of odd order are positive, then all the Hurwitz determinants of even order are also positive and vice versa. We state this as a theorem; its proof can be found in Reference [39].

**Theorem 8-7 (the Liénard–Chipart criterion)**

A polynomial with real coefficients is a Hurwitz polynomial if and only if either all the coefficients are positive and all the Hurwitz determinants of even order are positive, or all the coefficients are positive and all the Hurwitz determinants of odd order are positive. ∎

Example 6

Find the range of $k$ so that $D(s) = s^4 + 2s^3 + ks^2 + s + 2$ is a Hurwitz polynomial.

1. *The Routh–Hurwitz criterion:* We form the following table

$$\begin{array}{ccl} 1 & k \quad 2 & \\ 2 & 1 & \\ 2k-1 & 4 & \text{(after the multiplication of 2)} \\ 2k-9 & 0 & \text{(after the multiplication of } (2k-1)) \\ 4 & & \end{array}$$

Clearly if $k > 4.5$, then the numbers in the first column are all positive, and consequently $D(s)$ is a Hurwitz polynomial.

2. *The Liénard–Chipart criterion:* The Hurwitz matrix is

$$\begin{bmatrix} 2 & 1 & 0 & 0 \\ 1 & k & 2 & 0 \\ 0 & 2 & 1 & 0 \\ 0 & 1 & k & 2 \end{bmatrix}$$

The Hurwitz determinants are

$$\Delta_1 = 2 \qquad \Delta_2 = 2k - 1 \qquad \Delta_3 = 2k - 9 \qquad \Delta_4 = 2(2k - 9)$$

From either $\Delta_1 > 0$, $\Delta_3 > 0$ or $\Delta_2 > 0$, $\Delta_4 > 0$, we conclude that if $k > 4.5$, then $D(s)$ is a Hurwitz polynomial. ∎

We see that Theorem 8-7 and Corollary 8-6 give the same result. For high-order polynomials, it seems that the computation required in using the Routh–Hurwitz criterion is slightly less involved than that in using the Liénard–Chipart criterion. This might be the reason why the Routh–Hurwitz criterion is more commonly used.

## 8-4 Stability of Linear Dynamical Equations

**Time-varying case.** Consider the $n$-dimensional linear time-varying dynamical equation

$$E: \quad \dot{\mathbf{x}} = \mathbf{A}(t)\mathbf{x} + \mathbf{B}(t)\mathbf{u} \qquad \textbf{(8-23a)}$$

$$\mathbf{y} = \mathbf{C}(t)\mathbf{x} \qquad \textbf{(8-23b)}$$

where $\mathbf{x}$ is the $n \times 1$ state vector, $\mathbf{u}$ is the $p \times 1$ input vector, $\mathbf{y}$ is the $q \times 1$ output vector; and $\mathbf{A}$, $\mathbf{B}$, and $\mathbf{C}$ are $n \times n$, $n \times p$, and $q \times n$ matrices, respectively. It is assumed that every entry of $\mathbf{A}$, $\mathbf{B}$, and $\mathbf{C}$ are continuous functions of $t$ in $(-\infty, \infty)$. No assumption as to boundedness of these entries is made. We have assumed that there is no direct transmission part in (8-23) (that is, $\mathbf{D} = \mathbf{0}$) because it does not play any role in the stability study.

The response of the state equation (8-23a) can always be decomposed into the zero-input response and the zero-state response as

$$\mathbf{x}(t) = \boldsymbol{\phi}(t; t_0, \mathbf{x}^0, \mathbf{u}) = \boldsymbol{\phi}(t; t_0, \mathbf{x}^0, \mathbf{0}) + \boldsymbol{\phi}(t; t_0, \mathbf{0}, \mathbf{u}) \tag{8-24}$$

Hence it is very convenient to study the stabilities of the zero-input response and of the zero-state response separately. Combining these results, we shall immediately obtain the stability properties of the entire dynamical equation.

First we consider the stability of the zero-state response. The response of $E$ with $\mathbf{0}$ as the initial state at time $t_0$ is given by

$$\begin{aligned} \mathbf{y}(t) &= \int_{t_0}^{t} \mathbf{C}(t)\boldsymbol{\Phi}(t, \tau)\mathbf{B}(\tau)\,\mathbf{u}(\tau)\,d\tau \\ &= \int_{t_0}^{t} \mathbf{G}(t, \tau)\mathbf{u}(\tau)\,d\tau \end{aligned} \tag{8-25}$$

where $\mathbf{G}(t, \tau) \triangleq \mathbf{C}(t)\boldsymbol{\Phi}(t, \tau)\mathbf{B}(\tau)$ is, by definition, the impulse-response matrix of $E$. What we have discussed in Section 8-2 (in particular, Theorem 8-2) can be applied here. We shall rephrase Theorem 8-2 by using the notion of norm (see Section 2-8).

### Theorem 8-8

The zero-state response of the dynamical equation $E$ is BIBO stable if and only if there exists a finite number $k$ such that

$$\int_{t_0}^{t} \|\mathbf{C}(t)\boldsymbol{\Phi}(t, \tau)\mathbf{B}(\tau)\|\,d\tau \le k < \infty \tag{8-26}$$

for any $t_0$ and for all $t \ge t_0$. ∎

The norm used in (8-26) is defined in terms of the norm of $\mathbf{u}$. At any instant of time, $\|\mathbf{u}(t)\|$ can be chosen as $\sum_i |u_i(t)|$, $\max_i |u_i(t)|$, or $\left(\sum_i |u_i(t)|^2\right)^{1/2}$. Corresponding to these different norms of $\mathbf{u}$,

$$\|\mathbf{C}(t)\boldsymbol{\Phi}(t, \tau)\mathbf{B}(\tau)\|$$

has different values. However, as far as stability is concerned, any norm can be used.

Next we study the stability of the zero-input response. More specifically we study the response of

$$\dot{\mathbf{x}} = \mathbf{A}(t)\mathbf{x} \tag{8-27}$$

due to any initial state. The response of $\dot{\mathbf{x}} = \mathbf{A}(t)\mathbf{x}(t)$ due to the initial state $\mathbf{x}(t_0) = \mathbf{x}^0$ is given by

$$\mathbf{x}(t) \triangleq \boldsymbol{\phi}(t; t_0, \mathbf{x}^0, \mathbf{0}) = \boldsymbol{\Phi}(t, t_0)\mathbf{x}^0 \tag{8-28}$$

It is clear that the bounded-input bounded-output stability can no longer be applied here. Hence we must introduce different kinds of stability. Before doing so we need the concept of equilibrium state.

### Definition 8-2

A state $\mathbf{x}^e$ of a dynamical equation is said to be an *equilibrium state* if and only if

$$\mathbf{x}^e = \boldsymbol{\phi}(t; t_0, \mathbf{x}^e, \mathbf{0}) \tag{8-29}$$

for any $t_0$ and for all $t \geq t_0$. ∎

We see from this definition that if a trajectory reaches an equilibrium state and if no input is applied, the trajectory will stay at the equilibrium state forever. Hence at any equilibrium state, $\dot{\mathbf{x}}^e(t) = \mathbf{0}$ for all $t$. Consequently, an equilibrium state of $\dot{\mathbf{x}} = \mathbf{A}(t)\mathbf{x}$ is a solution of $\mathbf{A}(t)\mathbf{x} = \mathbf{0}$ for all $t$. Or from definition, an equilibrium state of $\dot{\mathbf{x}} = \mathbf{A}(t)\mathbf{x}$ is a solution of

$$\mathbf{x}^e = \mathbf{\Phi}(t, t_0)\mathbf{x}^e \qquad \text{or} \qquad (\mathbf{\Phi}(t, t_0) - \mathbf{I})\mathbf{x}^e = \mathbf{0}$$

for any $t_0$ and all $t \geq t_0$. Clearly, the zero state, $\mathbf{0}$, is always an equilibrium state of $\dot{\mathbf{x}} = \mathbf{A}(t)\mathbf{x}$.

We shall now define the stability of an equilibrium state in terms of the zero-input response.

### Definition 8-3

An equilibrium state $\mathbf{x}^e$ is said to be *stable in the sense of Lyapunov*[4] if and only if for any positive $\epsilon$, there exists a positive number $\delta(\epsilon)$ such that $\|\mathbf{x}^0 - \mathbf{x}^e\| \leq \delta$ implies that

$$\|\boldsymbol{\phi}(t; t_0, \mathbf{x}^0, \mathbf{0}) - \mathbf{x}^e\| \leq \epsilon$$

for any $t_0$ and for all $t \geq t_0$. ∎

Roughly speaking, an equilibrium state $\mathbf{x}^e$ is stable i.s.L. (in the sense of Lyapunov) if the response due to any initial state that is sufficiently near to $\mathbf{x}^e$ will not move far away from $\mathbf{x}^e$. If the response will, in addition, go back to $\mathbf{x}^e$, then $\mathbf{x}^e$ is said to be asymptotically stable.

### Definition 8-4

An equilibrium state $\mathbf{x}^e$ is said to be *asymptotically stable* if it is stable i.s.L. and if every motion starting sufficiently near $\mathbf{x}^e$ converges to $\mathbf{x}^e$ as $t \to \infty$. More precisely, there is some $\gamma > 0$, and for every $\epsilon > 0$ there corresponds

[4] To be more precise, the state $\mathbf{x}^e$ is said to be *uniformly* stable in the sense of Lyapunov.

a positive $T(\epsilon, \gamma)$ (independent of $t_0$ and $\mathbf{x}^0$), such that $\|\mathbf{x}^0 - \mathbf{x}^e\| \leq \gamma$ implies that

$$\|\boldsymbol{\phi}(t; t_0, \mathbf{x}^0, \mathbf{0}) - \mathbf{x}^e\| \leq \epsilon$$

for all $t \geq t_0 + T$, and for any $t_0$. ∎

The concept of asymptotic stability is illustrated in Figure 8-4. We see that the stabilities defined in Definitions 8-3 and 8-4 are local proper-

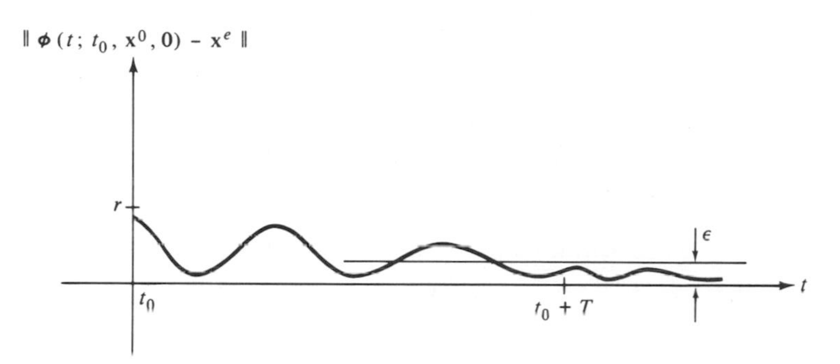


**Figure 8-4** Asymptotic stability.

ties, because we do not know how small $\delta$ in Definition 8-3 and $\gamma$ in Definition 8-4 should be chosen. However, for linear systems, because of the homogeneity property, $\delta$ and $\gamma$ can be extended to the entire state space. Hence if an equilibrium state of a linear equation is stable at all, it will be *globally stable* or *stable in the large*.

### Theorem 8-9

Every equilibrium state of $\dot{\mathbf{x}} = \mathbf{A}(t)\mathbf{x}$ is stable in the sense of Lyapunov if and only if there exists some constant $k$ such that

$$\|\boldsymbol{\Phi}(t, t_0)\| \leq k < \infty$$

for any $t_0$ and for all $t \geq t_0$.

### Proof

*Sufficiency:* Let $\mathbf{x}^e$ be an equilibrium state of $\dot{\mathbf{x}} = \mathbf{A}(t)\mathbf{x}$; that is

$$\mathbf{x}^e = \boldsymbol{\Phi}(t, t_0)\mathbf{x}^e \qquad \text{for all } t \geq t_0$$

Then we have

$$\mathbf{x}(t) - \mathbf{x}^e = \boldsymbol{\Phi}(t, t_0)\mathbf{x}^0 - \mathbf{x}^e = \boldsymbol{\Phi}(t, t_0)(\mathbf{x}^0 - \mathbf{x}^e) \tag{8-30}$$

It follows from (2-90) that

$$\|\mathbf{x}(t) - \mathbf{x}^e\| \leq \|\boldsymbol{\Phi}(t, t_0)\| \, \|\mathbf{x}^0 - \mathbf{x}^e\| \leq k\|\mathbf{x}^0 - \mathbf{x}^e\|$$

for all $t$. Hence, for any $\epsilon$, if we choose $\delta = \epsilon/k$, then $\|\mathbf{x}^0 - \mathbf{x}^e\| \leq \delta$ implies that

$$\|\mathbf{x}(t) - \mathbf{x}^e\| \leq \epsilon \qquad \text{for all } t \geq t_0$$

*Necessity:* If $\mathbf{x}^e$ is stable i.s.L., then $\mathbf{\Phi}(t, t_0)$ is uniformly bounded. We prove this by contradiction. Suppose that $\mathbf{x}^e$ is stable i.s.L. and that $\mathbf{\Phi}(t, t_0)$ is not uniformly bounded, then for some $t_0$, say $\bar{t}_0$, at least one element of $\mathbf{\Phi}(t, \bar{t}_0)$, say $\phi_{ij}(t, \bar{t}_0)$, becomes arbitrarily large as $t \to \infty$. Let us choose $\mathbf{x}^0$ at time $\bar{t}_0$ such that all the components of $(\mathbf{x}^0 - \mathbf{x}^e)$ are zero except the $j$th component, which is equal to $\alpha$. Then the $i$th component of $(\mathbf{x}(t) - \mathbf{x}^e)$ is equal to $\phi_{ij}(t, \bar{t}_0) \cdot \alpha$, which becomes arbitrarily large as $t \to \infty$, no matter how small $\alpha$ is. Hence $\mathbf{x}^e$ is not stable i.s.L. This is a contradiction; hence, if $\mathbf{x}^e$ is stable i.s.L., then $\mathbf{\Phi}(t, t_0)$ is uniformly bounded. Q.E.D.

### Theorem 8-10

The zero state of $\dot{\mathbf{x}} = \mathbf{A}(t)\mathbf{x}$ is asymptotically stable if and only if there exist positive numbers $k_1$ and $k_2$ such that

$$\|\mathbf{\Phi}(t, t_0)\| \leq k_1 e^{-k_2(t-t_0)} \tag{8-31}$$

for any $t_0$ and for all $t \geq t_0$. ∎

The zero state is, as mentioned earlier, an equilibrium state. If *the zero state is asymptotically stable, then the zero state is the only equilibrium state of* $\dot{\mathbf{x}} = \mathbf{A}(t)\mathbf{x}$. Indeed, if there were another equilibrium state different from the zero state, then by choosing that equilibrium state as an initial state, the response would not approach the zero state. Hence if $\dot{\mathbf{x}} = \mathbf{A}(t)\mathbf{x}$ is asymptotically stable, the zero state is the only equilibrium state. This can also be seen from the fact that an equilibrium state is a solution of $(\mathbf{\Phi}(t, t_0) - \mathbf{I})\mathbf{x}^e = \mathbf{0}$ for all $t \geq t_0$. Now if the zero state is asymptotically stable, Theorem 8-10 implies that $\mathbf{\Phi}(t, t_0) \to \mathbf{0}$ as $t \to \infty$. Consequently, the zero state is the only solution satisfying $(\mathbf{\Phi}(t, t_0) - \mathbf{I})\mathbf{x}^e = \mathbf{0}$ for all $t \geq t_0$.

### Proof of Theorem 8-10

*Sufficiency:* If

$$\|\mathbf{\Phi}(t, t_0)\| \leq k_1 e^{-k_2(t-t_0)}$$

then

$$\|\mathbf{x}(t)\| = \|\mathbf{\Phi}(t, t_0)\mathbf{x}(t_0)\| \leq \|\mathbf{\Phi}(t, t_0)\| \, \|\mathbf{x}^0\| \leq k_1 e^{-k_2(t-t_0)} \|\mathbf{x}^0\|$$

which implies that $\|\mathbf{x}(t)\| \to 0$ as $t \to \infty$. *Necessity:* If the zero state is asymptotically stable, then by definition it is stable in the sense of

Lyapunov. Consequently, there exists a finite number $k_3$ such that $\|\mathbf{\Phi}(t, t_0)\| \leq k_3$ for any $t_0$ and all $t \geq t_0$ (Theorem 8-9). From Definition 8-4, there is some $\gamma > 0$, and for every $\epsilon > 0$ there exists a positive $T$ such that

$$\|\mathbf{x}(t_0 + T)\| = \|\mathbf{\Phi}(t_0 + T, t_0)\mathbf{x}^0\| \leq \epsilon \tag{8-32}$$

for all $\|\mathbf{x}^0\| \leq \gamma$, and for any $t_0$. Now choose an $\mathbf{x}^0$ such that $\|\mathbf{x}^0\| = \gamma$ and $\|\mathbf{\Phi}(t_0 + T, t_0)\mathbf{x}^0\| = \|\mathbf{\Phi}(t_0 + T, t_0)\| \, \|\mathbf{x}^0\|$ and choose $\epsilon = \gamma/2$; then (8-32) implies that

$$\|\mathbf{\Phi}(t_0 + T, t_0)\| \leq \tfrac{1}{2} \qquad \text{for any } t_0$$

This condition and $\|\mathbf{\Phi}(t, t_0)\| \leq k_3$ imply that

$$\|\mathbf{\Phi}(t, t_0)\| \leq k_3 \qquad \text{for all } t \text{ in } [t_0, t_0 + T)$$

$$\|\mathbf{\Phi}(t, t_0)\| = \|\mathbf{\Phi}(t, t_0 + T)\mathbf{\Phi}(t_0 + T, t_0)\| \leq \|\mathbf{\Phi}(t, t_0 + T)\| \, \|\mathbf{\Phi}(t_0 + T, t_0)\| \leq k_3/2 \qquad \text{for all } t \text{ in } [t_0 + T, t_0 + 2T)$$

$$\|\mathbf{\Phi}(t, t_0)\| \leq \|\mathbf{\Phi}(t, t_0 + 2T)\| \, \|\mathbf{\Phi}(t_0 + 2T, t_0 + T)\| \, \|\mathbf{\Phi}(t_0 + T, t_0)\| \leq k_3/2^2 \qquad \text{for all } t \text{ in } [t_0 + 2T, t_0 + 3T)$$

and so forth, as shown in Figure 8-5. Let us choose $k_2$ such that $e^{-k_2T} = \tfrac{1}{2}$ and let $k_1 = 2k_3$. Then from Figure 8-5, we see immediately that

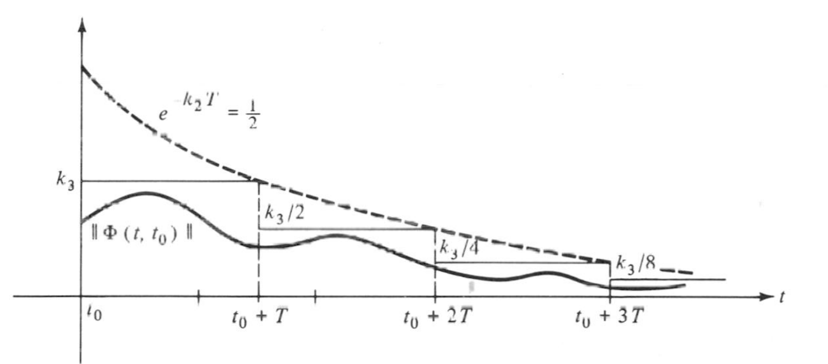


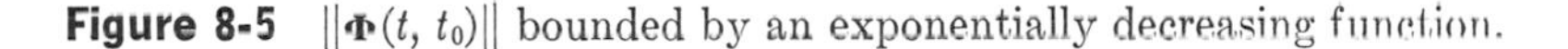

**Figure 8-5** $\|\mathbf{\Phi}(t, t_0)\|$ bounded by an exponentially decreasing function.

$$\|\mathbf{\Phi}(t, t_0)\| \leq k_1 e^{-k_2(t - t_0)}$$

for any $t_0$ and for all $t \geq t_0$. Q.E.D

If the zero state of $\dot{\mathbf{x}} = \mathbf{A}(t)\mathbf{x}$ is asymptotically stable, then for any initial state the zero-input response will tend to zero exponentially. Hence, for linear equations, if the zero state is asymptotically stable it is also said to be *exponentially stable*.

We have discussed separately the stability of the zero-input response and the stability of the zero-state response. By combining these results, we may give various definitions and theorems for the stability of the entire dynamical equation $E$. Before doing so, we shall discuss the relation between the stability of the zero-state response and the stability of the zero-input response.

The necessary and sufficient condition for the zero-state response of $E$ to be bounded-input bounded-output stable is that, for some finite $k$,

$$\int_{t_0}^{t} \|\mathbf{C}(t)\mathbf{\Phi}(t, \tau)\mathbf{B}(\tau)\| \, d\tau \leq k < \infty$$

for any $t_0$ and for all $t \geq t_0$. It has been shown by the function given in Figure 8-2 that an absolutely integrable function is not necessarily bounded. Conversely, a bounded function need not be absolutely integrable. Hence, the stability i.s.L. of an equilibrium state, in general, does not imply nor is implied by the BIBO stability of the zero-state response. However, with some condition on the matrices **B** and **C**, the asymptotic stability of the zero state does imply the BIBO stability of the zero-state response.

### Theorem 8-11

Consider the dynamical equation $E$ given in Equation (8-23). If the matrices **B** and **C** are bounded on $(-\infty, \infty)$, then the asymptotic stability of the zero state implies the BIBO stability of the zero-state response. ▮

### Proof

This theorem follows directly from the fact that

$$\int \|\mathbf{C}(t)\mathbf{\Phi}(t, \tau)\mathbf{B}(\tau)\| \, d\tau \leq \int \|\mathbf{C}(t)\| \, \|\mathbf{\Phi}(t, \tau)\| \, \|\mathbf{B}(\tau)\| \, d\tau \leq k_1 k_2 \int \|\mathbf{\Phi}(t, \tau)\| \, d\tau$$

where $\|\mathbf{B}(t)\| \leq k_1$, $\|\mathbf{C}(t)\| \leq k_2$ for all $t$. Q.E.D.

The converse problem of Theorem 8-11—that of determining the conditions under which the BIBO stability of the zero-state response implies the asymptotic stability of the zero state—is much more difficult. In order to solve this problem, the concepts of uniform controllability and uniform observability (Definitions 5-4 and 5-8) are needed. We state only the result; its proof can be found in Reference [102].

### *Theorem 8-12

Consider the dynamical equation $E$ given in (8-23). If the matrices **A**, **B**, and **C** are bounded on $(-\infty, \infty)$ and if $E$ is uniformly controllable and uniformly observable, then the zero state of $E$ is asymptotically stable (under the zero-input response) if and only if its zero-state response is bounded-input bounded-output stable. ▮

We have studied the stability of the zero-input response and the stability of the zero-state response. Their relations are also established. We shall now study the entire response.

### Definition 8-5

A linear dynamical equation is said to be *totally stable,* or *T-stable* for short, if and only if for any initial state and for any bounded input, the output as well as all the state variables are bounded. ∎

We see that the conditions of T-stability are more stringent than those of BIBO stability; they require not only the boundedness of the output but also of all state variables; the boundedness must hold not only for the zero state but also for any initial state. A system that is BIBO stable sometimes cannot function properly, because some of the state variables might increase with time, and the system will burn out or at least be saturated. Therefore, in practice every system is required to be T-stable.

### Theorem 8-13

A system that is described by the linear dynamical equation $E$ in (8-23) is T-stable if the matrices $\mathbf{B}$ and $\mathbf{C}$ are bounded on $(-\infty, \infty)$ and the zero state of $\dot{\mathbf{x}} = \mathbf{A}\mathbf{x}$ is asymptotically stable. ∎

This theorem can be proved by using Theorems 8-10 and 8-11 and is left as an exercise. Note that the conditions in Theorem 8-13 are sufficient, but not necessary, for a system to be T-stable. In applying this theorem, it is not necessary to check the controllability and observability of the dynamical equation $E$.

**Time-invariant case.** Although various stability conditions have been obtained for linear, time-varying dynamical equations, they can hardly be used, because all the conditions are stated in terms of state transition matrices, which are very difficult, if not impossible, to obtain. In the stability study of linear, time-invariant dynamical equations, the knowledge of the state transition matrix is, however, not needed. The stability can be determined directly from the matrix $\mathbf{A}$.

Consider the $n$-dimensional, linear, time-invariant dynamical equation

$$FE: \quad \dot{\mathbf{x}} = \mathbf{A}\mathbf{x} + \mathbf{B}\mathbf{u} \tag{8-33a}$$

$$\mathbf{y} = \mathbf{C}\mathbf{x} \tag{8-33b}$$

where $\mathbf{A}$, $\mathbf{B}$, $\mathbf{C}$ are $n \times n$, $n \times p$, $q \times n$ real constant matrices, respectively. As in the time-varying case, we study first the zero-state response and the zero-input response and then the entire response. The zero-state response of $FE$ is characterized by

$$\hat{\mathbf{G}}(s) = \mathbf{C}(s\mathbf{I} - \mathbf{A})^{-1}\mathbf{B} \tag{8-34}$$

From Theorem 8-5, *the zero-state response of $FE$ is BIBO stable if and only if all the poles of every entry of $\hat{\mathbf{G}}(s)$ have negative real parts.* The zero-input response of $FE$ is governed by $\dot{\mathbf{x}} = \mathbf{A}\mathbf{x}$ or $\mathbf{x}(t) = e^{\mathbf{A}t}\mathbf{x}^0$. Recall that an equilibrium state of $\dot{\mathbf{x}} = \mathbf{A}\mathbf{x}$ is a solution of $\dot{\mathbf{x}} = \mathbf{0}$ or $\mathbf{A}\mathbf{x} = \mathbf{0}$, and has the property $\mathbf{x}^e = e^{\mathbf{A}t}\mathbf{x}^e$ for all $t \geq 0$.

### Theorem 8-14

Every equilibrium state of $\dot{\mathbf{x}} = \mathbf{A}\mathbf{x}$ is stable in the sense of Lyapunov if and only if all the eigenvalues of $\mathbf{A}$ have nonpositive (negative or zero) real parts and those with zero real parts are simple zeros of the minimal polynomial of $\mathbf{A}$.[5]

### Proof

Let $\mathbf{x}^e$ be an equilibrium state of $\dot{\mathbf{x}} = \mathbf{A}\mathbf{x}$. Then $\mathbf{x}(t) - \mathbf{x}^e = e^{\mathbf{A}t}(\mathbf{x}^0 - \mathbf{x}^e)$. Hence every equilibrium state is stable i.s.L. if and only if there is a constant $k$ such that $\|e^{\mathbf{A}t}\| \leq k < \infty$ for all $t \geq 0$. Let $\mathbf{P}$ be the nonsingular matrix such that $\hat{\mathbf{A}} = \mathbf{P}\mathbf{A}\mathbf{P}^{-1}$ and $\hat{\mathbf{A}}$ is in the Jordan form. Since $e^{\hat{\mathbf{A}}t} = \mathbf{P}e^{\mathbf{A}t}\mathbf{P}^{-1}$, then

$$\|e^{\hat{\mathbf{A}}t}\| \leq \|\mathbf{P}\| \, \|e^{\mathbf{A}t}\| \, \|\mathbf{P}^{-1}\|$$

Consequently, if $\|e^{\mathbf{A}t}\|$ is bounded, so is $\|e^{\hat{\mathbf{A}}t}\|$. Conversely, from the equation $e^{\mathbf{A}t} = \mathbf{P}^{-1}e^{\hat{\mathbf{A}}t}\mathbf{P}$, we see that if $\|e^{\hat{\mathbf{A}}t}\|$ is bounded, so is $\|e^{\mathbf{A}t}\|$. Hence we conclude that every equilibrium state is stable i.s.L. if and only if $\|e^{\hat{\mathbf{A}}t}\|$ is bounded on $[0, \infty)$. Now $\|e^{\hat{\mathbf{A}}t}\|$ is bounded if and only if every entry of $e^{\hat{\mathbf{A}}t}$ is bounded. Since $\hat{\mathbf{A}}$ is in the Jordan form, every entry of $e^{\hat{\mathbf{A}}t}$ is of the form $t^k e^{\alpha_j t + i\omega_j t}$, where $\alpha_j + i\omega_j$ is an eigenvalue of $\hat{\mathbf{A}}$ (see Section 2-7). If $\alpha_j$ is negative, it is easy to see that $t^k e^{\alpha_j t + i_j \omega t}$ is bounded on $[0, \infty)$ for any integer $k$. If $\alpha_j = 0$, the function $t^k e^{i\omega_j t}$ is bounded if and only if $k = 0$—that is, the order of the Jordan block associated with the eigenvalue with $\alpha_j = 0$ is 1. Q.E.D.

### Theorem 8-15

The zero state of $\dot{\mathbf{x}} = \mathbf{A}\mathbf{x}$ is asymptotically stable if and only if all the eigenvalues of $\mathbf{A}$ have negative real parts.

### Proof

In order for the zero state to be asymptotically stable, in addition to the boundedness of $\|e^{\mathbf{A}t}\|$, it is required that $\|e^{\mathbf{A}t}\|$ tend to zero as $t \to \infty$, or

[5] Equivalently, if $\mathbf{A}$ is transformed into the Jordan form, the order of every Jordan blocks associated with the eigenvalue with zero real part is 1.

equivalently, that $\|e^{\bar{\mathbf{A}}t}\| \to 0$ as $t \to \infty$. Since every entry of $e^{\bar{\mathbf{A}}t}$ is of the form $t^k e^{\alpha_i t + i\omega_i t}$, we conclude that $\|e^{\bar{\mathbf{A}}t}\| \to 0$ as $t \to \infty$ if and only if all the eigenvalues of $\bar{\mathbf{A}}$, and consequently of $\mathbf{A}$ have negative real parts. Q.E.D.

The eigenvalues of $\mathbf{A}$ are the roots of the characteristic equation of $\mathbf{A}$, $\det(s\mathbf{I} - \mathbf{A}) = 0$. We have introduced in the previous section the Routh-Hurwitz criterion to check whether or not all the roots of a polynomial have negative real parts. Hence the asymptotic stability of the zero-input response of $\dot{\mathbf{x}} = \mathbf{A}\mathbf{x}$ can be easily determined by first forming the characteristic polynomial of $\mathbf{A}$ and then applying the Routh–Hurwitz criterion.

The BIBO stability of the linear, time-invariant dynamical *FE* is determined by the poles of $\hat{\mathbf{G}}(s)$. Since

$$\hat{\mathbf{G}}(s) = \mathbf{C}(s\mathbf{I} - \mathbf{A})^{-1}\mathbf{B} = \frac{1}{\det(s\mathbf{I} - \mathbf{A})} \mathbf{C} \cdot [\operatorname{Adj}(s\mathbf{I} - \mathbf{A})]\mathbf{B}$$

every pole of $\hat{\mathbf{G}}(s)$ is an eigenvalue of $\mathbf{A}$ (the converse is not true). Consequently, if the zero state of *FE* is asymptotically stable, the zero-state response of *FE* will also be BIBO stable. (This fact can also be deduced directly from Theorem 8-11.) Conversely, the BIBO stability of the zero-state response in general does not imply the asymptotic stability of the zero state, because the zero-state response is determined by the transfer function, which, however, describes only the controllable and observable part of a dynamical equation.

### Example 1

Consider a system with the following dynamical equation description:

$$\dot{\mathbf{x}} = \begin{bmatrix} 1 & 0 \\ 1 & -1 \end{bmatrix} \mathbf{x} + \begin{bmatrix} 0 \\ 1 \end{bmatrix} u$$
$$\mathbf{y} = [1 \quad 1]\mathbf{x}$$

Its transfer function is

$$\hat{g}(s) = [1 \quad 1] \begin{bmatrix} s-1 & 0 \\ -1 & s+1 \end{bmatrix}^{-1} \begin{bmatrix} 0 \\ 1 \end{bmatrix} = \frac{1}{s+1}$$

Hence the zero-state response of the dynamical equation is BIBO stable; however, the zero state is not asymptotically stable, because there is a positive eigenvalue. ∎

If a linear, time-invariant dynamical equation is controllable and observable, then the characteristic polynomial of $\mathbf{A}$ is equal to the characteristic polynomial of $\hat{\mathbf{G}}(s)$ (Theorem 6-2). This implies that every eigenvalue of $\mathbf{A}$ is a pole of $\hat{\mathbf{G}}(s)$, and every pole of $\hat{\mathbf{G}}(s)$ is an eigenvalue of $\mathbf{A}$. Consequently, we have the following theorem.

### Theorem 8-16

If a linear, time-invariant dynamical equation *FE* is controllable and observable, then the following statements are equivalent:

1. The dynamical equation is totally stable.
2. The zero-state response of *FE* is BIBO stable.
3. The zero state of *FE* is asymptotically stable (under the zero-input response).
4. All the poles of the transfer function matrix of *FE* have negative real parts.
5. All the eigenvalues of the matrix **A** of *FE* have negative real parts. ▮

### Proof

The equivalences of statements 2 through 5 follow directly from Theorems 6-2, 8-5, and 8-15. Statement 1 implies statement 2 following from Definition 8-5. It is easy to show that statement 3 implies statement 1. Q.E.D.

### Corollary 8-16

A system that is described by a linear, time-invariant dynamical equation *FE* is T-stable if all the eigenvalues of the matrix **A** of *FE* have negative real parts. ▮

The dynamical equation *FE* in this corollary is not required to be controllable and observable. The condition in this corollary is only sufficient as can be seen from the following example.

### Example 2

Consider a system with the dynamical-equation description

$$FE: \quad \begin{aligned} \dot{\mathbf{x}} &= \begin{bmatrix} -1 & 0 \\ 0 & 0 \end{bmatrix} \mathbf{x} + \begin{bmatrix} 1 \\ 0 \end{bmatrix} u \\ y &= [1 \quad 1]\mathbf{x} \end{aligned}$$

Not all eigenvalues of *FE* have negative real parts; however, the output as well as all the state variables are bounded for any bounded input and for any initial state. Hence, *FE* is T-stable. Note that the mode associated with the eigenvalue 0 is not controllable. ▮

A system is said to be completely or faithfully characterized by its transfer-function matrix if the dynamical-equation description of the system is controllable and observable. We see from Theorem 8-16 that *if a system is completely characterized by its transfer-function matrix, then the total stability of the system can be determined from its transfer-function matrix alone* with no need of considering the dynamical-equation description of the system.

A remark is in order concerning the stability of $\dot{\mathbf{x}} = \mathbf{A}(t)\mathbf{x}$. If the matrix $\mathbf{A}$ is independent of $t$ and if all the eigenvalues of $\mathbf{A}$ have negative real parts, then the zero state is asymptotically stable. Hence one might be tempted to suggest that if for each $t$ all the eigenvalues of $\mathbf{A}(t)$ have negative real parts, then the zero state of $\dot{\mathbf{x}} = \mathbf{A}(t)\mathbf{x}$ is asymptotically stable. This is not so, as can be seen from the following example.

### Example 3

Consider the linear time-varying equation

$$\dot{\mathbf{x}} = \begin{bmatrix} -1 & e^{2t} \\ 0 & -1 \end{bmatrix} \mathbf{x}$$

The characteristic polynomial of the matrix $\mathbf{A}$ at each $t$ is given by

$$\det [\lambda \mathbf{I} - \mathbf{A}] = \det \begin{bmatrix} \lambda + 1 & -e^{2t} \\ 0 & \lambda + 1 \end{bmatrix} = (\lambda + 1)^2$$

Hence the eigenvalues of $\mathbf{A}$ are $-1$ and $-1$ for all $t$. However, the zero state of the equation is not asymptotically stable or stable i.s.L., because the state transition matrix of the equation is

$$\mathbf{\Phi}(t, 0) = \begin{bmatrix} e^{-t} & (e^{t} - e^{-t})/2 \\ 0 & e^{-t} \end{bmatrix}$$

(as in Problem 4-1 or by direct verification), whose norm tends to infinity as $t \to \infty$.

## *8-5 A Proof of the Routh–Hurwitz Criterion

In this section we shall use the results obtained in the previous section to prove Theorem 8-6, the Routh–Hurwitz criterion. Before proceeding, we need the concept of positive definite matrix. An $n \times n$ matrix $\mathbf{M}$ with elements in the field of complex numbers is said to be a *hermitian matrix* if $\mathbf{M}^* = \mathbf{M}$, where $\mathbf{M}^*$ is the complex conjugate transpose of $\mathbf{M}$. If $\mathbf{M}$ is a real matrix, $\mathbf{M}$ is said to be *symmetric*. The matrix $\mathbf{M}$ can be considered as an operator that maps $(\mathbb{C}^n, \mathbb{C})$ into itself. It is shown in Appendix D that all the eigenvalues of a hermitian matrix are real, and that there exists a nonsingular matrix $\mathbf{P}$ such that $\mathbf{P}^{-1} = \mathbf{P}^*$ and $\hat{\mathbf{M}} =$

$\mathbf{PMP}^*$, where $\hat{\mathbf{M}}$ is a diagonal matrix with eigenvalues on the diagonal (Theorem D-4). We shall use this fact to establish the following theorem.

### Theorem 8-17

Let $\mathbf{M}$ be a hermitian matrix and let $\lambda_{\min}$ and $\lambda_{\max}$ be the smallest eigenvalue and the largest eigenvalue of $M$, respectively. Then

$$\lambda_{\min}\|\mathbf{x}\|^2 \leq \mathbf{x}^*\mathbf{M}\mathbf{x} \leq \lambda_{\max}\|\mathbf{x}\|^2 \tag{8-35}$$

for any $\mathbf{x}$ in the $n$-dimensional complex vector space $\mathbb{C}^n$, where

$$\|\mathbf{x}\|^2 \triangleq \langle \mathbf{x}, \mathbf{x}\rangle \triangleq \mathbf{x}^*\mathbf{x} = \sum_{i=1}^{n} |x_i|^2$$

and $x_i$ is the $i$th component of $\mathbf{x}$.

### Proof

Note that $\mathbf{x}^*\mathbf{M}\mathbf{x}$ is a real number for any $\mathbf{x}$ in $\mathbb{C}^n$. Let $\mathbf{P}$ be the nonsingular matrix such that $\mathbf{P}^{-1} = \mathbf{P}^*$ and $\hat{\mathbf{M}} = \mathbf{PMP}^*$, where $\hat{\mathbf{M}}$ is a diagonal matrix with eigenvalues of $\mathbf{M}$ on the diagonal. Let $\hat{\mathbf{x}} = \mathbf{Px}$ or $\mathbf{x} = \mathbf{P}^{-1}\hat{\mathbf{x}} = \mathbf{P}^*\hat{\mathbf{x}}$, then

$$\mathbf{x}^*\mathbf{M}\mathbf{x} = \hat{\mathbf{x}}^*\mathbf{PMP}^*\hat{\mathbf{x}} = \hat{\mathbf{x}}^*\hat{\mathbf{M}}\hat{\mathbf{x}} = \sum_{i=1}^{n} \lambda_i|\hat{x}_i|^2$$

where the $\lambda_i$'s are the eigenvalues of $\mathbf{M}$. It follows that

$$\lambda_{\min}\sum_{i=1}^{n} |\hat{x}_i|^2 \leq \mathbf{x}^*\mathbf{M}\mathbf{x} = \hat{\mathbf{x}}^*\hat{\mathbf{M}}\hat{\mathbf{x}} = \sum_{i=1}^{n} \lambda_i|\hat{x}_i|^2 \leq \lambda_{\max}\sum_{i=1}^{n} |\hat{x}_i|^2 \tag{8-36}$$

The fact that $\mathbf{P}^{-1} = \mathbf{P}^*$ implies that

$$\|\mathbf{x}\|^2 = \mathbf{x}^*\mathbf{x} = \hat{\mathbf{x}}^*\hat{\mathbf{x}} = \sum_{i=1}^{n} |\hat{x}_i|^2$$

Hence, the inequality (8-36) implies (8-35).

### Definition 8-6

A hermitian matrix $\mathbf{M}$ is said to be *positive definite* if and only if $\mathbf{x}^*\mathbf{M}\mathbf{x} > 0$ for all nonzero $\mathbf{x}$ in $\mathbb{C}^n$. A hermitian matrix $\mathbf{M}$ is said to be *positive semidefinite* or *nonnegative definite* if and only if $\mathbf{x}^*\mathbf{M}\mathbf{x} \geq 0$ for all $\mathbf{x}$ in $\mathbb{C}^n$, and the equality holds for some nonzero $\mathbf{x}$ in $\mathbb{C}^n$. ∎

### Theorem 8-18

A hermitian matrix $\mathbf{M}$ is positive definite (positive semidefinite) if and only if any one of the following conditions holds:

1. All the eigenvalues of $\mathbf{M}$ are positive (nonnegative).
2. All the leading principal minors[6] of $\mathbf{M}$ are positive (all the principal minors of $\mathbf{M}$ are nonnegative.)[7]
3. There exists a nonsingular matrix $\mathbf{N}$ (a singular matrix $\mathbf{N}$) such that $\mathbf{M} = \mathbf{N}^*\mathbf{N}$.

### Proof

Condition 1 follows directly from Theorem 8-17. The proof of condition 2 can be found, for example, in References [5] and [39]. For the proof of condition 3, see Problem 8-24. Q.E.D

With these preliminaries, we are ready to introduce the Lyapunov theorem and its extension. They will be used to prove the Routh–Hurwitz criterion.

### Theorem 8-19 (Lyapunov theorem)

The zero state of the linear, time-invariant equation $\dot{\mathbf{x}} = \mathbf{A}\mathbf{x}$ is asymptotically stable if and only if for any given positive definite hermitian matrix $\mathbf{N}$, there exists a positive definite hermitian matrix $\mathbf{M}$ that satisfies the matrix equation

$$\mathbf{A}^*\mathbf{M} + \mathbf{M}\mathbf{A} = -\mathbf{N} \tag{8-37}$$

∎

---

[6] The *principal minors* of the matrix

$$\mathbf{M} = \begin{bmatrix} m_{11} & m_{12} & m_{13} \\ m_{21} & m_{22} & m_{23} \\ m_{31} & m_{32} & m_{33} \end{bmatrix}$$

arc

$$m_{11},\ m_{22},\ m_{23},\ \det\begin{bmatrix} m_{11} & m_{12} \\ m_{21} & m_{22} \end{bmatrix},\ \det\begin{bmatrix} m_{11} & m_{13} \\ m_{31} & m_{33} \end{bmatrix},\ \det\begin{bmatrix} m_{22} & m_{23} \\ m_{32} & m_{33} \end{bmatrix},\ \text{and } \det \mathbf{M}$$

that is, the minors whose diagonal elements are also diagonal elements of the matrix. The *leading principal minors* of $\mathbf{M}$ are

$$m_{11},\ \det\begin{bmatrix} m_{11} & m_{12} \\ m_{21} & m_{22} \end{bmatrix},\ \text{and } \det \mathbf{M}$$

that is, the minors obtained by deleting the last $k$ columns and the last $k$ rows, for $k = 2, 1,$ and $0$.

[7] It is shown in Reference [39] that if all the leading principal minors of a matrix are positive, then all the principal minors are positive. However, it is *not* true that if all the leading principal minors are nonnegative, then all the principal minors are nonnegative. For a counter example, try Problem 8-32 b.

### Corollary 8-19

The zero state of the linear, time-invariant equation $\dot{\mathbf{x}} = \mathbf{A}\mathbf{x}$ is asymptotically stable if and only if, for any given positive semidefinite hermitian matrix $\mathbf{N}$ that has the property that $\mathbf{x}^*\mathbf{N}\mathbf{x}$ is not identically zero along any nontrivial solution of $\dot{\mathbf{x}} = \mathbf{A}\mathbf{x}$, there exists a positive definite hermitian matrix $\mathbf{M}$ that satisfies the matrix equation

$$\mathbf{A}^*\mathbf{M} + \mathbf{M}\mathbf{A} = -\mathbf{N}$$ ∎

The implication of Theorem 8-19 and Corollary 8-19 is that if $\mathbf{A}$ is asymptotically stable and if $\mathbf{N}$ is positive definite or positive semidefinite, then the solution $\mathbf{M}$ of (8-37) must be positive definite. However, it does *not* say that if $\mathbf{A}$ is asymptotically stable and if $\mathbf{M}$ is positive definite, then the matrix $\mathbf{N}$ computed from (8-37) is positive definite or positive semidefinite.

Before proving the Lyapunov theorem, let us make a few comments. Since Theorem 8-19 holds for any positive definite hermitian matrix $\mathbf{N}$, the matrix $\mathbf{N}$ in (8-37) is often chosen to be a unit matrix. Since $\mathbf{M}$ is a hermitian matrix, there are $n^2$ unknown numbers in $\mathbf{M}$ to be solved. If $\mathbf{M}$ is a real symmetric matrix there are $n(n+1)/2$ unknown numbers in $\mathbf{M}$ to be solved. Hence the matrix equation (8-37) actually consists of $n^2$ linear algebraic equations. To apply Theorem 8-19, first we have to solve these $n^2$ equations for $\mathbf{M}$, and then check whether or not $\mathbf{M}$ is positive definite. This is not an easy task. Hence Theorem 8-19 and its corollary are generally not used in determining the stability of $\dot{\mathbf{x}} = \mathbf{A}\mathbf{x}$. However, Theorem 8-19 is very important in the stability study of nonlinear, time-varying systems by using the so-called second method of Lyapunov. Furthermore, we shall use it to prove the Routh–Hurwitz criterion.

We give now a physical interpretation of the Lyapunov theorem. If the hermitian matrix $\mathbf{M}$ is positive definite, the plot of $V(\mathbf{x})$

$$V(\mathbf{x}) \triangleq \mathbf{x}^*\mathbf{M}\mathbf{x} \tag{8-38}$$

will be bowl shaped, as shown in Figure 8-6. Consider now the successive values taken by $V$ along a trajectory of $\dot{\mathbf{x}} = \mathbf{A}\mathbf{x}$. We like to know whether the value of $V$ will increase or decrease with time as the state moving along the trajectory. Taking the derivative of $V$ with respect to $t$ along any trajectory of $\dot{\mathbf{x}} = \mathbf{A}\mathbf{x}$, we obtain

$$\begin{aligned}\frac{d}{dt}V(\mathbf{x}(t)) &= \frac{d}{dt}(\mathbf{x}^*(t)\mathbf{M}\mathbf{x}(t)) = \left(\frac{d}{dt}\mathbf{x}^*(t)\right)\mathbf{M}\mathbf{x}(t) + \mathbf{x}^*(t)\mathbf{M}\left(\frac{d}{dt}\mathbf{x}(t)\right)\\ &= \mathbf{x}^*(t)\mathbf{A}^*\mathbf{M}\mathbf{x}(t) + \mathbf{x}^*(t)\mathbf{M}\mathbf{A}\mathbf{x}(t) = \mathbf{x}^*(t)(\mathbf{A}^*\mathbf{M} + \mathbf{M}\mathbf{A})\mathbf{x}(t)\\ &= -\mathbf{x}^*(t)\mathbf{N}\mathbf{x}(t)\end{aligned} \tag{8-39}$$

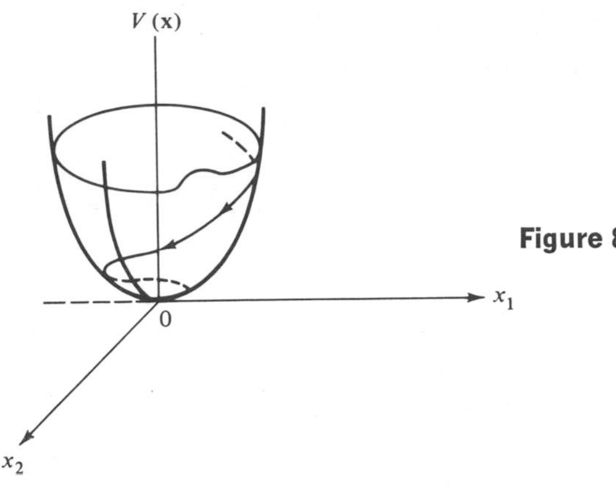


**Figure 8-6** A Lyapunov function $V(\mathbf{x})$.

where $\mathbf{N} \triangleq -(\mathbf{A}^*\mathbf{M} + \mathbf{MA})$. This equation gives the rate of change of $V(\mathbf{x})$ along any trajectory of $\dot{\mathbf{x}} = \mathbf{Ax}$. Now if $\mathbf{N}$ is positive definite, the function $-\mathbf{x}^*(t)\mathbf{N}\mathbf{x}(t)$ is always negative. This implies that $V(\mathbf{x}(t))$ decreases with time along any trajectory of $\dot{\mathbf{x}} = \mathbf{Ax}$; hence $V(\mathbf{x}(t))$ will eventually approach zero as $t \to \infty$. Now since $V(\mathbf{x})$ is positive definite, we have $V(\mathbf{x}) = 0$ only at $\mathbf{x} = \mathbf{0}$; hence we conclude that if we can find positive definite matrices $\mathbf{M}$ and $\mathbf{N}$ that are related by (8-37), then every trajectory of $\dot{\mathbf{x}} = \mathbf{Ax}$ will approach the zero vector as $t \to \infty$. The function $V(\mathbf{x})$ is called a *Lyapunov function* of $\dot{\mathbf{x}} = \mathbf{Ax}$. A Lyapunov function can be considered as a generalization of the concept of distance or energy. If the "distance" of the state along any trajectory of $\dot{\mathbf{x}} = \mathbf{Ax}$ decreases with time, then $\mathbf{x}(t)$ must tend to $\mathbf{0}$ as $t \to \infty$.

## Proof of Theorem 8-19

*Sufficiency:* Consider $V(\mathbf{x}) = \mathbf{x}^*\mathbf{M}\mathbf{x}$. Then we have

$$\dot{V}(\mathbf{x}) \triangleq \frac{d}{dt} V(\mathbf{x}) = -\mathbf{x}^*\mathbf{N}\mathbf{x}$$

along any trajectory of $\dot{\mathbf{x}} = \mathbf{Ax}$. From Theorem 8-17, we have

$$\frac{\dot{V}}{V} = -\frac{\mathbf{x}^*\mathbf{N}\mathbf{x}}{\mathbf{x}^*\mathbf{M}\mathbf{x}} \leq -\frac{(\lambda_{\mathbf{N}})_{\min}}{(\lambda_{\mathbf{M}})_{\max}} \tag{8-40}$$

where $(\lambda_{\mathbf{N}})_{\min}$ is the smallest eigenvalue of $\mathbf{N}$ and $(\lambda_{\mathbf{M}})_{\max}$ is the largest eigenvalue of $\mathbf{M}$. From Theorem 8-18 and from the assumption that the matrices $\mathbf{M}$ and $\mathbf{N}$ are positive definite, we have $(\lambda_{\mathbf{N}})_{\min} > 0$ and $(\lambda_{\mathbf{M}})_{\max} > 0$. If we define

$$\alpha \triangleq \frac{(\lambda_{\mathbf{N}})_{\min}}{(\lambda_{\mathbf{M}})_{\max}}$$

then inequality (8-40) becomes $\dot{V} \leq -\alpha V$, which implies that $V(t) \leq e^{-\alpha t}V(0)$. It is clear that $\alpha > 0$, hence $V$ decreases exponentially to zero on every trajectory of $\dot{\mathbf{x}} = \mathbf{A}\mathbf{x}$. Now $V(t) = 0$ only at $\mathbf{x} = \mathbf{0}$; hence we conclude that the response of $\dot{\mathbf{x}} = \mathbf{A}\mathbf{x}$ due to any intial state $\mathbf{x}^0$ tends to $\mathbf{0}$ as $t \to \infty$. This proves that the zero state of $\dot{\mathbf{x}} = \mathbf{A}\mathbf{x}$ is asymptotically stable. *Necessity:* If the zero state of $\dot{\mathbf{x}} = \mathbf{A}\mathbf{x}$ is asymptotically stable, then all the eigenvalues of $\mathbf{A}$ have negative real parts. Consequently, for any $\mathbf{N}$, there exists a matrix $\mathbf{M}$ satisfying

$$\mathbf{A}^*\mathbf{M} + \mathbf{M}\mathbf{A} = -\mathbf{N}$$

(see Appendix E). Now we show that if $\mathbf{N}$ is positive definite, $\mathbf{M}$ is also positive definite. Suppose that this is not the case, then there exists an $\mathbf{x}^0 \neq \mathbf{0}$ such that $V(\mathbf{x}^0) = (\mathbf{x}^0)^*\mathbf{M}\mathbf{x}^0 \leq 0$. Now since $\dot{V}$ is always negative, the value of $V(\mathbf{x})$ decreases monotonically to $-\infty$ as $t \to \infty$. This contradicts the fact that $V(\mathbf{x}(t)) \to 0$ as $t \to \infty$, which is implied by the assumption of the asymptotic stability of the zero state. Q.E.D.

With a slight modification, Corollary 8-19 can be similarly proved.

We are now ready to prove the Routh–Hurwitz criterion. Consider the polynomial

$$D(s) = a_0 s^n + a_1 s^{n-1} + a_2 s^{n-2} + \cdots + a_n \qquad a_0 > 0 \tag{8-41}$$

with real coefficients $a_i$, $i = 0, 1, 2, \cdots, n$. We form the polynomials

$$D_1(s) = a_0 s^n + a_2 s^{n-2} + \cdots$$
$$D_2(s) = a_1 s^{n-1} + a_3 s^{n-3} + \cdots$$

and compute

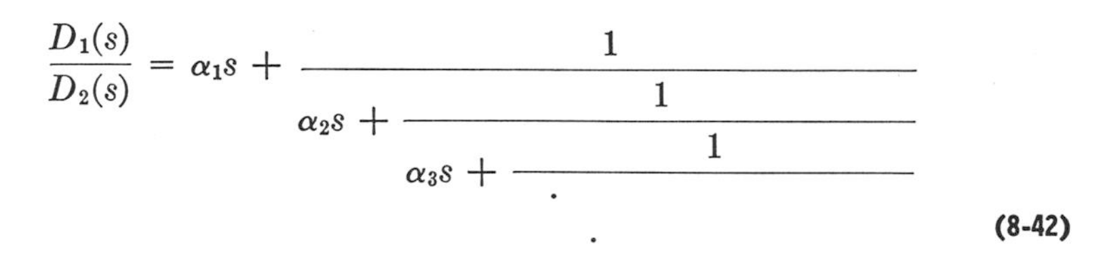

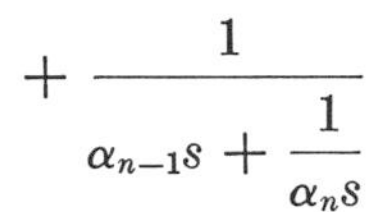

$$\frac{D_1(s)}{D_2(s)} = \alpha_1 s + \cfrac{1}{\alpha_2 s + \cfrac{1}{\alpha_3 s + \cfrac{1}{\ddots + \cfrac{1}{\alpha_{n-1}s + \cfrac{1}{\alpha_n s}}}}} \tag{8-42}$$

For convenience, we shall restate the theorem here.

### Theorem 8-6

The polynomial $D(s)$ is a Hurwitz polynomial if and only if all the $n$ numbers $\alpha_1, \alpha_2, \cdots, \alpha_n$ are positive.

### Proof

First we assume that all the $\alpha_i$'s are different from zero. Consider the rational function

$$\hat{g}(s) \triangleq \frac{D_2(s)}{D(s)} = \frac{D_2(s)}{D_2(s) + D_1(s)} = \frac{1}{1 + \dfrac{D_1(s)}{D_2(s)}}$$

The assumption $\alpha_i \neq 0$, for $i = 1, 2, \cdots, n$, implies that there is no common factor between $D_2(s)$ and $D_1(s)$. Consequently, there is no common factor between $D_2(s)$ and $D(s)$; in other words, $\hat{g}(s)$ is irreducible.

Consider the block diagram shown in Figure 8-7. We show that the transfer function from $u$ to $y$ is $\hat{g}(s)$. Let $\hat{h}_1(s)$ be the transfer function

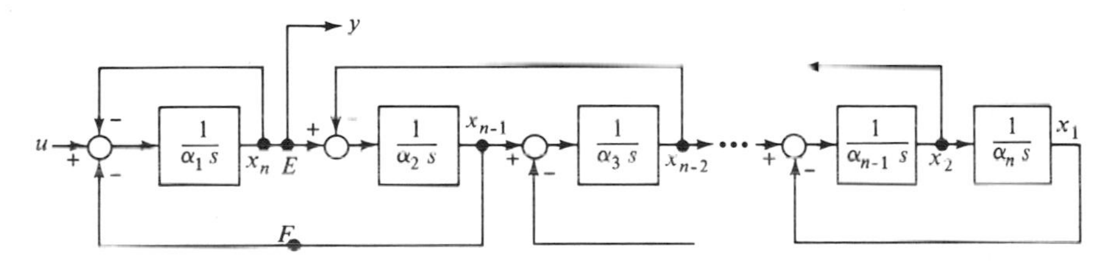


**Figure 8-7**

from $x_n$ to $x_{n-1}$ or, equivalently, from the terminal $E$ to the terminal $F$, as shown in Figure 8-7. Then

$$\frac{\hat{y}(s)}{\hat{u}(s)} = \frac{\dfrac{1}{\alpha_1 s}}{1 + \dfrac{1 + \hat{h}_1(s)}{\alpha_1 s}} = \frac{1}{1 + \alpha_1 s + \hat{h}_1(s)}$$

(see Problem 3-32). Let $\hat{h}_2(s)$ be the transfer function from $x_{n-1}$ to $x_{n-2}$, then $\hat{h}_1(s)$ can be written as

$$\hat{h}_1(s) = \frac{\dfrac{1}{\alpha_2 s}}{1 + \dfrac{\hat{h}_2}{\alpha_2 s}} = \frac{1}{\alpha_2 s + \hat{h}_2(s)}$$

Proceeding forward, we can show easily that the transfer function from $u$ to $y$ is indeed $\hat{g}(s)$. With the state variable chosen as shown, we can readily write the dynamical equation of the block diagram as

$$\begin{bmatrix} \dot{x}_1 \\ \dot{x}_2 \\ \dot{x}_3 \\ \cdot \\ \cdot \\ \cdot \\ \dot{x}_{n-1} \\ \dot{x}_n \end{bmatrix} = \begin{bmatrix} 0 & \frac{1}{\alpha_n} & 0 & \cdots & 0 & 0 & 0 \\ \frac{-1}{\alpha_{n-1}} & 0 & \frac{1}{\alpha_{n-1}} & \cdots & 0 & 0 & 0 \\ 0 & \frac{-1}{\alpha_{n-2}} & 0 & & 0 & 0 & 0 \\ \cdot & \cdot & \cdot & & \cdot & \cdot & \cdot \\ \cdot & \cdot & \cdot & & \cdot & \cdot & \cdot \\ \cdot & \cdot & \cdot & & \cdot & \cdot & \cdot \\ 0 & 0 & 0 & \cdots & \frac{-1}{\alpha_2} & 0 & \frac{1}{\alpha_2} \\ 0 & 0 & 0 & \cdots & 0 & \frac{-1}{\alpha_1} & \frac{-1}{\alpha_1} \end{bmatrix} \begin{bmatrix} x_1 \\ x_2 \\ x_3 \\ \cdot \\ \cdot \\ \cdot \\ x_{n-1} \\ x_n \end{bmatrix} + \begin{bmatrix} 0 \\ 0 \\ 0 \\ \cdot \\ \cdot \\ \cdot \\ 0 \\ 1 \end{bmatrix} u \tag{8-43}$$

$$y = [0 \quad 0 \quad 0 \quad \cdots \quad 0 \quad 0 \quad 1]\mathbf{x}$$

Irreducibility of (8-43) can be verified either by showing that it is controllable and observable or by the fact that its dimension is equal to the degree of the denominator of $\hat{g}(s)$. Consequently, the characteristic polynomial of the matrix $\mathbf{A}$ in (8-43) is equal to the denominator of $\hat{g}(s)$ (Theorem 6-2). Now we shall derive the condition for the zero state of (8-43) to be asymptotically stable. Let the $\mathbf{M}$ matrix in Corollary 8-19 be chosen as

$$\mathbf{M} = \begin{bmatrix} \alpha_n & 0 & \cdots & 0 & 0 \\ 0 & \alpha_{n-1} & \cdots & 0 & 0 \\ \cdot & \cdot & & \cdot & \cdot \\ \cdot & \cdot & & \cdot & \cdot \\ \cdot & \cdot & & \cdot & \cdot \\ 0 & 0 & \cdots & \alpha_2 & 0 \\ 0 & 0 & \cdots & 0 & \alpha_1 \end{bmatrix} \tag{8-44}$$

Then it is easy to verify that

$$\mathbf{A}^*\mathbf{M} + \mathbf{M}\mathbf{A} = -\begin{bmatrix} 0 & 0 & 0 & \cdots & 0 & 0 \\ 0 & 0 & 0 & \cdots & 0 & 0 \\ 0 & 0 & 0 & \cdots & 0 & 0 \\ \cdot & \cdot & \cdot & & \cdot & \cdot \\ \cdot & \cdot & \cdot & & \cdot & \cdot \\ \cdot & \cdot & \cdot & & \cdot & \cdot \\ 0 & 0 & 0 & \cdots & 0 & 0 \\ 0 & 0 & 0 & \cdots & 0 & 2 \end{bmatrix} \triangleq -\mathbf{N} \tag{8-45}$$

It is clear that $\mathbf{N}$ is a positive semidefinite matrix. We show that $\mathbf{x}^*\mathbf{N}\mathbf{x}$ cannot be identically zero along any solution of $\dot{\mathbf{x}} = \mathbf{A}\mathbf{x}$ except $\mathbf{x} \equiv \mathbf{0}$. Indeed, if $\mathbf{x}^*\mathbf{N}\mathbf{x} = -2x_n^2 \equiv 0$, then $x_n \equiv 0$ and $\dot{x}_n \equiv 0$, which, from (8-43) implies that $x_{n-1} \equiv 0$ and consequently $\dot{x}_{n-1} \equiv 0$. Proceeding forward, we conclude that $\mathbf{x}^*\mathbf{N}\mathbf{x} \equiv 0$ implies $\mathbf{x} \equiv \mathbf{0}$. Now the matrix $\mathbf{M}$ in (8-44) is the solution of (8-45). Hence from Corollary 8-19 we conclude that the zero state of $\dot{\mathbf{x}} = \mathbf{A}\mathbf{x}$ is asymptotically stable if and only if the matrix $\mathbf{M}$ is positive definite, or equivalently, the $n$ numbers $\alpha_1, \alpha_2, \cdots, \alpha_n$ are positive. Now the zero state of $\dot{\mathbf{x}} = \mathbf{A}\mathbf{x}$ is asymptotically stable if and only if all the eigenvalues of $\mathbf{A}$, or equivalently all the roots of $D(s)$, have negative real parts. In other words, $D(s)$ is a Hurwitz polynomial if and only if all the $n$ numbers $\alpha_1, \alpha_2, \cdots, \alpha_n$ are positive.

Consider now the case in which not all the $\alpha_i$'s are different from zero. In other words, some of the coefficients in the first column of Table 8-1 are equal to zero. Suppose $c_1 = 0$. If all the coefficients in the $c$ row of Table 8-1 are equal to zero, it implies that $D_1(s)$ and $D_2(s)$ have at least one common factor. The common factor is clearly either an even or an odd function of $s$, say $f(s)$. Then $D(s)$ can be factored as $f(s)\bar{D}(s)$. Since not all the roots of an even function or an odd function can have negative real parts, hence $D(s)$ is not a Hurwitz polynomial. If not all the coefficients in the $c$ row are equal to zero, we may replace $c_1$ by a very small positive number $\epsilon$ and continue to complete Table 8-1. In this case it can be seen that some $\alpha_i$ will be negative. If some $\alpha_i$ is negative, at least one root of the modified $D(s)$ (since $c_1$ is replaced by $\epsilon$) has a positive real part. Now the roots of a polynomial are continuous functions of its coefficients. Hence, as $\epsilon \to 0$, at least one root of $D(s)$ has a positive or zero real part. Q.E.D.

## 8-6 Concluding Remarks

In this chapter we studied various stability concepts of linear systems. For the input–output description, we have the BIBO stability; for the zero-input response of a dynamical equation, we have stability in the sense of Lyapunov and asymptotic stability; for an entire dynamical equation, we have total stability. Various theorems for these stabilities were derived. The relations among them are listed in Table 8-2. The theorems in parentheses are for single-variable systems; those without parentheses are for multivariable systems. The BIBO stability of the zero-state response and the asymptotic stability of the zero-input response of a linear, time-invariant dynamical equation are determined by the roots of some polynomial (Theorems 8-4, 8-5, and 8-15). Solving for the roots of a polynomial is a difficult task. Theorems 8-6, 8-7, and 8-19 provide three methods of testing these stabilities without solving the roots.

**Table 8-2** THEOREMS FOR STABILITY

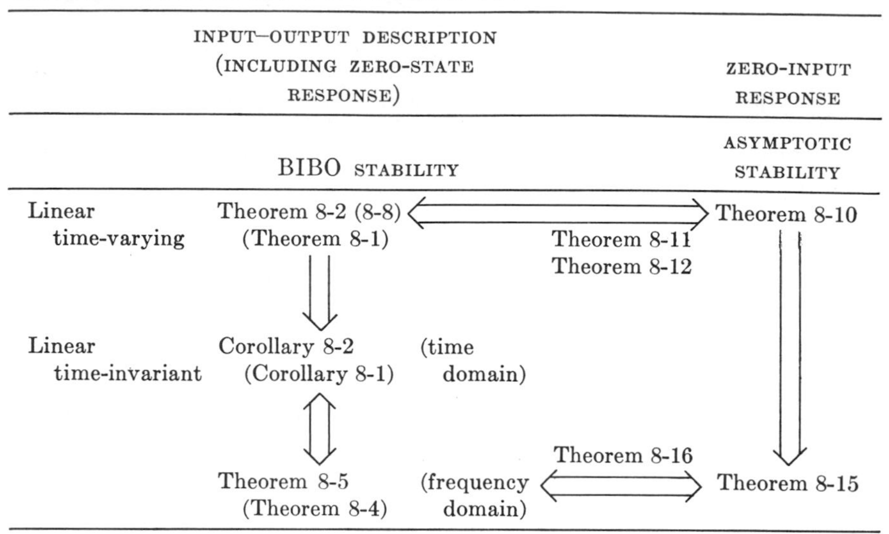

| | INPUT–OUTPUT DESCRIPTION (INCLUDING ZERO-STATE RESPONSE) | | | ZERO-INPUT RESPONSE |
|---|---|---|---|---|
| | BIBO STABILITY | | | ASYMPTOTIC STABILITY |
| Linear time-varying | Theorem 8-2 (8-8) (Theorem 8-1) | | Theorem 8-11 Theorem 8-12 | Theorem 8-10 |
| Linear time-invariant | Corollary 8-2 (Corollary 8-1) | (time domain) | | |
| | Theorem 8-5 (Theorem 8-4) | (frequency domain) | Theorem 8-16 | Theorem 8-15 |

## PROBLEMS

**8-1** Is a system with the impulse responses $g(t, \tau) = e^{-2|t|-|\tau|}$, for $t \geq \tau$, BIBO stable? How about $g(t, \tau) = \sin t \, e^{-(t-\tau)} \cos \tau$?

**8-2** Is the network shown in Figure P8-2 BIBO stable? If not, find a bounded input that will excite an unbounded output.

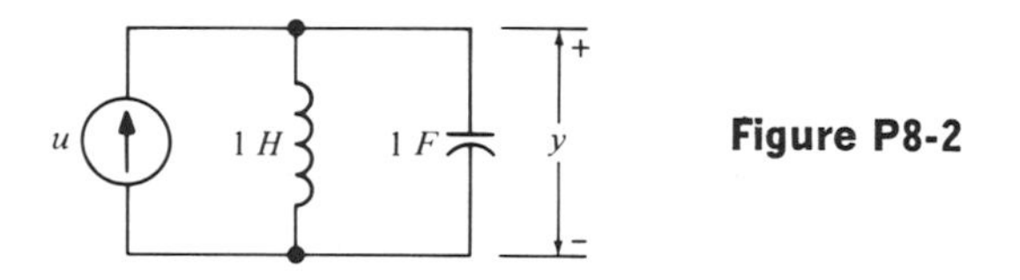


**Figure P8-2**

**8-3** Consider a system with the transfer function $\hat{g}(s)$ that is not necessarily a rational function of $s$. Show that a necessary condition for the system to be BIBO stable is that $|\hat{g}(s)|$ is finite for all Re $s \geq 0$.

**8-4** Consider a system with the impulse response shown in Figure P8-4. If the input $u(t) = \sin 2\pi t$, for $t \geq 0$, is applied, what is the waveform of

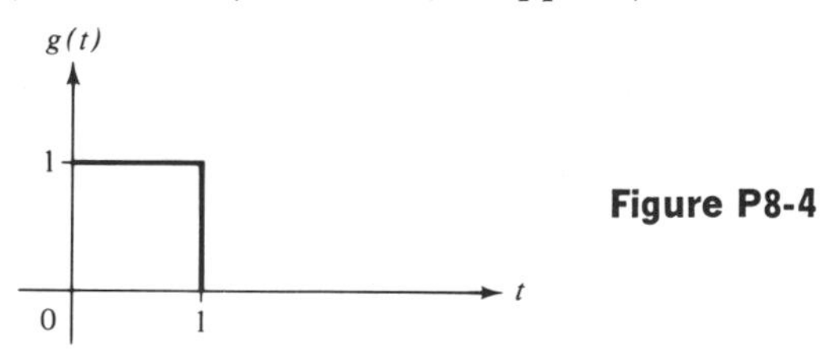


**Figure P8-4**

the output? After how many seconds will the output reach its steady state?

**8-5** Is a system with the impulse response $g(t) = \dfrac{1}{1+t}$ BIBO stable?

**8-6** Is a system with the transfer function $\hat{g}(s) = \dfrac{e^{-s}}{s+1}$ BIBO stable?

**8-7** Use the Routh–Hurwitz criterion and the Liénard–Chipart criterion to determine which of the following polynomials are Hurwitz polynomials. Which criterion do you prefer?

**a.** $s^5 + 4s^4 + 10s^3 + 2s^2 + 5s + 6$
**b.** $s^5 + s^4 + 2s^3 + 2s^2 + 5s + 5$
**c.** $-2s^4 - 7s^3 - 4s^2 - 5s - 10$

**8-8** Can you determine without solving the roots that the real parts of all the roots of $s^4 + 14s^3 + 71s^2 + 154s + 120$ are smaller than $-1$? *Hint:* Let $s = s' - 1$.

**8-9** Give the necessary and sufficient conditions for the following polynomials to be Hurwitz polynomials:

**a.** $a_0s^2 + a_1s + a_2$
**b.** $a_0s^3 + a_1s^2 + a_2s + a_3$

**8-10** Find the dynamical-equation description of the network shown in Problem 8-2. Find the equilibrium states of the equation. Is the equilibrium state stable in the sense of Lyapunov? Is it asymptotically stable?

**8-11** Prove parts 1 and 2 of Theorem 8-3 by using the Laplace transform for the class of systems describable by rational transfer functions.

**8-12** Consider the dynamical equation

$$\dot{\mathbf{x}} = \begin{bmatrix} 0 & 1 & 0 & 0 & 0 \\ 0 & 0 & 1 & 0 & 0 \\ 0 & 0 & 0 & 1 & 0 \\ 0 & 0 & 0 & 0 & 1 \\ -1 & -1 & -2 & -10 & -4 \end{bmatrix} \mathbf{x} + \begin{bmatrix} 0 \\ 0 \\ 0 \\ 0 \\ 1 \end{bmatrix} u$$

$$y = [0 \quad 0 \quad 0 \quad 0 \quad 0]\mathbf{x}$$

Is the zero state asymptotically stable (for the case $u \equiv 0$)? Is its zero-state response BIBO stable? Is the equation totally stable?

**8-13** Consider

$$\begin{bmatrix} \dot{x}_1 \\ \dot{x}_2 \\ \dot{x}_3 \end{bmatrix} = \begin{bmatrix} 0 & 0 & 0 \\ 0 & 0 & 0 \\ 0 & 0 & 0 \end{bmatrix} \begin{bmatrix} x_1 \\ x_2 \\ x_3 \end{bmatrix} + \begin{bmatrix} 1 \\ 0 \\ 0 \end{bmatrix} u$$
$$y = [1 \quad 1 \quad 1]\mathbf{x}$$

Find all the equilibrium states of the equation. Is every equilibrium state stable i.s.L.? Is it asymptotically stable? Is its zero-state response BIBO stable? Is the equation totally stable?

**8-14** Check the BIBO stability of a system with the transfer function

$$\frac{2s^2 - 1}{s^5 + 2s^4 + 3s^3 + 5s^2 + 2s + 2}$$

**8-15** Find the ranges of $k_1$ and $k_2$ such that the system with the transfer function

$$\frac{s + k_1}{s^3 + 2s^2 + k_2 s + 4}$$

is BIBO stable.

**8-16** It is known that the dynamical equation

$$\dot{\mathbf{x}} = \begin{bmatrix} \lambda & 1 & 0 \\ 0 & \lambda & 0 \\ 0 & 0 & \lambda \end{bmatrix} \mathbf{x} + \begin{bmatrix} 0 & 0 \\ 1 & 0 \\ 0 & 1 \end{bmatrix} \mathbf{u}$$
$$\mathbf{y} = \begin{bmatrix} 1 & 2 & 0 \\ 0 & 1 & 1 \end{bmatrix} \mathbf{x}$$

is BIBO stable. Can we conclude that the real part of $\lambda$ is negative? Why?

**8-17** Show that $\mathbf{x} = \mathbf{0}$ is the only solution satisfying $(\mathbf{\Phi}(t, t_0) - \mathbf{I})\mathbf{x} \equiv \mathbf{0}$ if $\|\mathbf{\Phi}(t, t_0)\| \to 0$ as $t \to \infty$.

**8-18** Consider the following linear, time-varying dynamical equation:

$$E: \qquad \dot{x} = 2tx + u$$
$$y = x$$

Show that the zero state of $E$ is not stable i.s.L. (under the zero-input response).

**8-19** Consider the equivalent equation of $E$ in Problem 8-18 obtained by the equivalence transformation $\bar{x} = P(t)x$, where $P(t) = e^{-t^2}$:

$$\bar{E}: \qquad \dot{\bar{x}} = (2te^{-t^2} - 2te^{-t^2})e^{t^2}\bar{x} + e^{-t^2}u = 0 + e^{-t^2}u$$
$$y = e^{t^2}\bar{x}$$

Show that the zero state of $\bar{E}$ is stable i.s.L. (under the zero-input response). From Problems 8-18 and 8-19, we conclude that *an equivalence transformation need not preserve the stability of the zero state.* Does an equivalence transformation preserve the BIBO stability of the zero-state response?

**8-20** An equivalence transformation $\mathbf{P}(t)$ (see Definition 4-5) is called a *Lyapunov transformation* if, in addition, $\|\mathbf{P}(t)\| < k_1$ and $\|\mathbf{P}^{-1}(t)\| < k_2$ for all $t$. Show that *the stability* (stability i.s.L. or asymptotic stability) *of the zero state is preserved under any Lyapunov transformation.* Check whether or not the transformation $P(t) = e^{-t^2}$ in Problem 8-19 is a Lyapunov transformation.

**8-21** Is the transformation in Theorem 4-8 a Lyapunov transformation?

**8-22** Consider a system with the following dynamical-equation description:

$$\dot{\mathbf{x}} = \begin{bmatrix} -1 & 1 & 0 \\ 0 & -1 & 0 \\ 0 & 0 & 0 \end{bmatrix} \mathbf{x} + \begin{bmatrix} 1 \\ 2 \\ 0 \end{bmatrix} u$$

$$y = [2 \quad 3 \quad 1]\mathbf{x}$$

Is the zero state asymptotically stable? Is the zero-state response BIBO stable? Is the system T-stable?

**8-23** Prove Theorem 8-13.

**8-24** Prove condition 3 of Theorem 8-18. *Hint:* Use $\mathbf{M} = \mathbf{P}\hat{\mathbf{M}}\mathbf{P}^*$ and $\hat{\mathbf{M}} = (\hat{\mathbf{M}})^{1/2}(\hat{\mathbf{M}})^{1/2}$.

**8-25** Prove Corollary 8-19.

**8-26** Are the networks shown in Figure P8-26 totally stable?
*Answers:* no; yes.

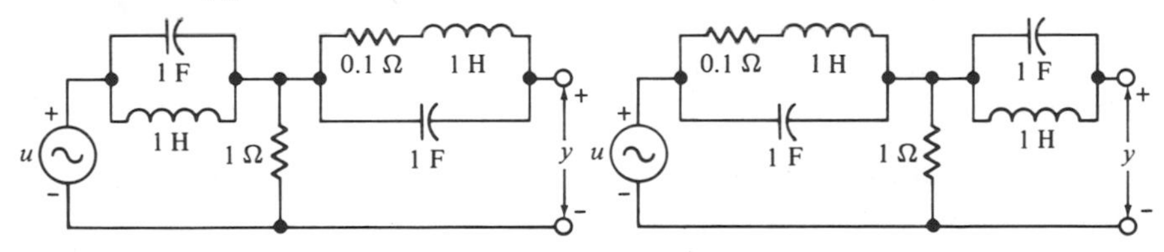


**Figure P8-26**

**8-27** Consider a discrete-time system that is described by

$$y(n) = \sum_{m=-\infty}^{n} g(n, m)u(m)$$

Show that any bounded-input sequence $\{u(n)\}$ excites a bounded-output sequence $\{y(n)\}$ if and only if

$$\sum_{m=-\infty}^{n} |g(n, m)| \leq k \leq \infty \qquad \text{for all } n$$

**8-28** Consider a discrete-time system that is described by

$$y(n) = \sum_{m=0}^{n} g(n - m)u(m)$$

Show that any bounded-input sequence $\{u(n)\}$ excites a bounded-output sequence $\{y(n)\}$ if and only if

$$\sum_{m=0}^{\infty} |g(m)| \leq k < \infty$$

**8-29** Consider a system with the impulse response

$$g(t) = g_1(t) + \sum_{i=0}^{\infty} a_i\delta(t - \tau_i)$$

Show that the system is BIBO stable if and only if

$$\int_0^\infty |g_1(t)|\, dt \leq k_1 < \infty$$

and

$$\sum_{i=0}^{\infty} |a_i| \leq k_2 < \infty$$

**8-30** Prove Corollary 8-3 by using partial fraction expansion for the class of systems that have proper rational transfer-function descriptions.

**8-31** Is the function

$$[x_1 \quad x_2 \quad x_3] \begin{bmatrix} 1 & 2 & 2 \\ 0 & 1 & 1 \\ 2 & 2 & 6 \end{bmatrix} \begin{bmatrix} x_1 \\ x_2 \\ x_3 \end{bmatrix}$$

positive definite or semidefinite? *Hint:* Use Equation (D-2) in Appendix D.

**8-32** Which of the following hermitian (symmetric) matrices are positive definite or positive semidefinite?

**a.** $\begin{bmatrix} 2 & 3 & 2 \\ 3 & 1 & 0 \\ 2 & 0 & 2 \end{bmatrix}$

**b.** $\begin{bmatrix} 0 & 0 & 1 \\ 0 & 0 & 0 \\ 1 & 0 & 2 \end{bmatrix}$

**c.** $\begin{bmatrix} 0 & 0 & 0 \\ 0 & 1 & 0 \\ 0 & 0 & 0 \end{bmatrix}$

**d.** $\begin{bmatrix} a_1a_1 & a_1a_2 & a_1a_3 \\ a_1a_2 & a_2a_2 & a_2a_3 \\ a_1a_3 & a_2a_3 & a_3a_3 \end{bmatrix}$

**8-33** Let $\lambda_1 = -1$, $\lambda_2 = -2$, $\lambda_3 = -3$ and let $a_1$, $a_2$, $a_3$ be arbitrary real numbers, not all zero. Prove by using Corollary 8-19 that the matrix

$$\mathbf{M} = \begin{bmatrix} -\dfrac{a_1^2}{2\lambda_1} & -\dfrac{a_1a_2}{\lambda_1+\lambda_2} & -\dfrac{a_1a_3}{\lambda_1+\lambda_3} \\ -\dfrac{a_1a_2}{\lambda_2+\lambda_1} & -\dfrac{a_2^2}{2\lambda_2} & -\dfrac{a_2a_3}{\lambda_2+\lambda_3} \\ -\dfrac{a_3a_1}{\lambda_1+\lambda_3} & -\dfrac{a_3a_2}{\lambda_2+\lambda_3} & -\dfrac{a_3^2}{2\lambda_3} \end{bmatrix}$$

is a positive definite matrix. *Hint:* Let $\mathbf{A} = \text{diag}(\lambda_1, \lambda_2, \lambda_3)$.

**8-34** A real matrix $\mathbf{M}$ (not necessarily symmetric) is defined to be, as in Definition 8-6, positive definite if $\mathbf{x}'\mathbf{M}\mathbf{x} > 0$ for all nonzero $\mathbf{x}$ in $\mathbb{R}^n$. Is it true that the matrix $\mathbf{M}$ is positive definite if all the eigenvalues of $\mathbf{M}$ are positive real or if all the leading principal minors are positive? If not, how do you check its positive definiteness? *Hint:* Try

$$\begin{bmatrix} 0 & 1 \\ -2 & 3 \end{bmatrix}, \begin{bmatrix} 2 & 1 \\ 1.9 & 1 \end{bmatrix}$$

CHAPTER

# 9

# Linear, Time-Invariant Composite Systems

## 9-1 Introduction

In this chapter[1] we shall study various problems associated with composite systems. By a composite system we mean that the system consists of a collection of subsystems. Although there are many forms of composite systems, they are mainly built up from three basic connections: parallel, tandem, and feedback connections. Hence we shall restrict ourselves to the studies of these three connections. If all the subsystems that form a composite system are all linear and time-invariant, it is easy to show that the composite system is again a linear, time-invariant system. For linear composite systems, all the results in the previous chapters can be directly applied. For example, consider the feedback connection of two linear, time-invariant systems with a transfer function $\hat{g}_1(s)$ in the forward path and a transfer function $\hat{g}_2(s)$ in the feedback path. If the transfer function of the feedback connection, $\hat{g}_f(s) = (1 + \hat{g}_1(s)\hat{g}_2(s))^{-1}\hat{g}_1(s)$, is computed, then the input–output stability of the feedback system can be determined

[1] The time-invariant part of Section 3-6 is a prerequisite of this chapter.

from $\hat{g}_f(s)$. However, many questions can still be raised regarding the composite system. For example, what is the implication if there are pole–zero cancellations between $\hat{g}_1(s)$ and $\hat{g}_2(s)$? Is it possible to determine the stability of the feedback system from $\hat{g}_1$ and $\hat{g}_2$ without computing $\hat{g}_f$? In this chapter we shall study these and other related problems.

Before proceeding, we introduce a definition.

### Definition 9-1

A system is said to be *completely characterized* by its rational transfer-function matrix if and only if the dynamical-equation description of the system is controllable and observable. ∎

The motivation of this definition is as follows. If the dynamical-equation description of a system is uncontrollable and/or unobservable, then the transfer-function matrix of the system describes only part of the system (only those modes that are connected to the input and output terminals); hence the transfer function does not describe the system fully. On the other hand, if the dynamical-equation description of a system is controllable and observable, then the information obtained from the dynamical equation and the one from the transfer function of the system will be essentially the same. Hence, in this case the system is said to be completely characterized by its transfer-function matrix.

All the subsystems that form a composite system will be assumed to be completely characterized by their transfer-function matrices. This assumption, however, does not imply that a composite system is completely characterized by its composite transfer-function matrix. In Section 9-2 we study the conditions of complete characterization of composite systems. For single-variable systems, the conditions are very simple; if there is no common pole in the parallel connection, or no pole–zero cancellation in the tandem and the feedback connections, then the composite system is completely characterized by its transfer function. In Section 9-3 we study the controllability and observability of composite dynamical equations. The stabilities of single-variable and multivariable feedback systems are studied from their open-loop transfer-function matrices in Section 9-4. In Section 9-5 we study the design of compensators to achieve pole placement for single-input multiple-output systems.

All the rational functions are assumed to be irreducible; that is, there is no common factor between the denominator and the numerator of each rational function. It is also assumed that there is no loading effect in any connection of two systems; that is, the transfer-function matrix of each system remains unchanged after connection.

The references for this chapter are [7], [15], [16], [18], [19], [20], [22], [28], [30], [33], [40], [49], [94], and [116].

## 9-2 Transfer-Function Descriptions of Composite Systems

Consider two systems $S_i$, for $i = 1, 2$, with the dynamical-equation descriptions

$$FE^i: \quad \dot{\mathbf{x}}_i = \mathbf{A}_i\mathbf{x}_i + \mathbf{B}_i\mathbf{u}_i \tag{9-1a}$$
$$\mathbf{y}_i = \mathbf{C}_i\mathbf{x}_i + \mathbf{D}_i\mathbf{u}_i \tag{9-1b}$$

where $\mathbf{x}_i$, $\mathbf{u}_i$ and $\mathbf{y}_i$ are, respectively, the state, the input and the output of the system $S_i$. $\mathbf{A}_i$, $\mathbf{B}_i$, $\mathbf{C}_i$ and $\mathbf{D}_i$ are real constant matrices. The transfer-function matrix of $S_i$ is

$$\hat{\mathbf{G}}_i(s) = \mathbf{C}_i(s\mathbf{I} - \mathbf{A}_i)^{-1}\mathbf{B}_i + \mathbf{D}_i \tag{9-2}$$

It is assumed that the systems $S_1$ and $S_2$ are completely characterized by their transfer-function matrices $\hat{\mathbf{G}}_1(s)$ and $\hat{\mathbf{G}}_2(s)$; or, equivalently, the dynamical equations (9-1) are controllable and observable. It was shown in Section 3-6 that the transfer-function matrix of the parallel connection of $S_1$ and $S_2$ is $\hat{\mathbf{G}}_1(s) + \hat{\mathbf{G}}_2(s)$; the transfer-function matrix of the tandem connection of $S_1$ followed by $S_2$ is $\hat{\mathbf{G}}_2(s)\hat{\mathbf{G}}_1(s)$; the transfer-function matrix of the feedback connection of $S_1$ with $S_2$ in the feedback path is $\hat{\mathbf{G}}_1(s)(\mathbf{I} + \hat{\mathbf{G}}_2(s)\hat{\mathbf{G}}_1(s))^{-1} = (\mathbf{I} + \hat{\mathbf{G}}_1(s)\hat{\mathbf{G}}_2(s))^{-1}\hat{\mathbf{G}}_1(s)$. Although $\hat{\mathbf{G}}_1(s)$ and $\hat{\mathbf{G}}_2(s)$ completely characterize the systems $S_1$ and $S_2$, respectively, it does not follow that a composite transfer function $\hat{\mathbf{G}}(s)$ completely characterizes a composite system.

### Example 1

Consider the parallel connection of two single-variable systems $S_1$ and $S_2$ whose dynamical-equation descriptions are, respectively,

$$FE_1{}^1: \quad \dot{x}_1 = x_1 + u_1$$
$$y_1 = x_1 + u_1$$

and

$$FE_1{}^2: \quad \dot{x}_2 = x_2 - u_2$$
$$y_2 = x_2$$

Their transfer functions are

$$\hat{g}_1(s) = \frac{s}{s-1} \qquad \text{and} \qquad \hat{g}_2(s) = \frac{-1}{s-1}$$

The composite transfer function of the parallel connection of $S_1$ and $S_2$ is

$$g(s) = \hat{g}_1(s) + \hat{g}_2(s) = \frac{s}{s-1} + \frac{-1}{s-1} = 1$$

It is clear that $\hat{g}(s) = 1$ does not characterize completely the composite system, because $\hat{g}(s)$ does not reveal the unstable mode $e^t$ in the system.

This can also be checked from the composite dynamical equation. In the parallel connection, we have $u_1 = u_2 = u$ and $y = y_1 + y_2$; hence the composite dynamical equation is

$$\begin{bmatrix} \dot{x}_1 \\ \dot{x}_2 \end{bmatrix} = \begin{bmatrix} 1 & 0 \\ 0 & 1 \end{bmatrix} \begin{bmatrix} x_1 \\ x_2 \end{bmatrix} + \begin{bmatrix} 1 \\ -1 \end{bmatrix} u$$
$$y = [1 \quad 1]\mathbf{x} + u$$

It is easy to check that the composite equation is not controllable and not observable; hence from Definition 9-1, the composite system is not completely characterized by its composite transfer function. ∎

In the following we shall study the conditions under which composite transfer functions completely describe composite systems. We study this problem directly from transfer-function matrices without looking into dynamical equations. In the next section, we shall study the same problem from dynamical equations.

Recall from Definition 6-1 that the characteristic polynomial of a proper rational matrix $\hat{\mathbf{G}}(s)$ is the least common denominator of all minors of $\hat{\mathbf{G}}(s)$. The degree of $\hat{\mathbf{G}}(s)$, denoted by $\delta\hat{\mathbf{G}}(s)$, is the degree of the characteristic polynomial of $\hat{\mathbf{G}}(s)$. If a system is completely characterized by its transfer-function matrix, then the dimension of the dynamical-equation description of the system is equal to the degree of its transfer-function matrix (Theorem 6-2). Therefore, whether or not a system is completely characterized by its transfer-function matrix can be checked from the number of state variables of the system. If a system is an $RLC$ network,[2] then the number of state variables is equal to the number of energy-storage elements (inductors and capacitors), hence *an $RLC$ network[2] is completely characterized by its transfer-function matrix if and only if the number of energy storage elements is equal to the degree of its transfer-function matrix.* Consider now two $RLC$ networks $S_1$ and $S_2$, which are completely characterized by their transfer-function matrices $\hat{\mathbf{G}}_1(s)$ and $\hat{\mathbf{G}}_2(s)$, respectively. The number of energy-storage elements in any composite connection of $S_1$ and $S_2$ is clearly equal to $\delta\hat{\mathbf{G}}_1(s) + \delta\hat{\mathbf{G}}_2(s)$. Let $\hat{\mathbf{G}}(s)$ be the transfer-function matrix of the composite connection of $S_1$ and $S_2$. Now the composite system consists of $(\delta\hat{\mathbf{G}}_1(s) + \delta\hat{\mathbf{G}}_2(s))$ energy-storage elements; hence, in order for $\hat{\mathbf{G}}(s)$ to characterize the composite system completely, it is necessary and sufficient to have $\delta\hat{\mathbf{G}}(s) = \delta\hat{\mathbf{G}}_1(s) + \delta\hat{\mathbf{G}}_2(s)$. This is stated as a theorem.

### Theorem 9-1

Consider two systems $S_1$ and $S_2$, which are completely characterized by their proper transfer-function matrices $\hat{\mathbf{G}}_1(s)$ and $\hat{\mathbf{G}}_2(s)$, respectively. Any

[2] We assume that there are no capacitors-only loops and inductors-only cutsets in the network.

composite connection of $S_1$ and $S_2$ is completely characterized by its composite transfer-function matrix $\hat{\mathbf{G}}(s)$ if and only if

$$\delta\hat{\mathbf{G}}(s) = \delta\hat{\mathbf{G}}_1(s) + \delta\hat{\mathbf{G}}_2(s) \qquad \blacksquare$$

This theorem can also be verified from the dynamical-equation descriptions of systems. Recall from Section 3-6 that the state space of any composite connection of $S_1$ and $S_2$ is chosen to be the direct sum of the state spaces of $S_1$ and $S_2$; consequently, the dimension of the composite dynamical equation is the sum of the dimensions of the dynamical-equation descriptions of $S_1$ and $S_2$. Hence Theorem 9-1 follows directly from Definition 9-1 and Theorem 6-2.

In order to apply Theorem 9-1 we must first compute the transfer-function matrix of a composite system. This is not desirable, particularly in the synthesis of feedback control systems. Hence the conditions in terms of $\hat{\mathbf{G}}_1$ and $\hat{\mathbf{G}}_2$ for $\hat{\mathbf{G}}$ to characterize completely the composite connections of $S_1$ and $S_2$ will be studied. We study in this section only single-variable systems. The multivariable systems will be studied in the next section.

The transfer function of a single-variable system is a scalar, and its degree is just the degree of its denominator. Note that every rational function is assumed to be irreducible.

### Theorem 9-2

Consider two single-variable systems $S_1$ and $S_2$, which are completely characterized by their proper rational transfer functions $\hat{g}_1(s)$ and $\hat{g}_2(s)$.

1. The parallel connection of $S_1$ and $S_2$ is completely characterized by $\hat{g}(s) = \hat{g}_1(s) + \hat{g}_2(s)$ if and only if $\hat{g}_1(s)$ and $\hat{g}_2(s)$ do not have any pole in common.
2. The tandem connection of $S_1$ and $S_2$ is completely characterized by $\hat{g}(s) = \hat{g}_2(s)\hat{g}_1(s)$ if and only if there is no pole–zero cancellation between $\hat{g}_1$ and $\hat{g}_2$.
3. The feedback connection of $S_1$ and $S_2$ shown in Figure 9-1 is completely characterized by $\hat{g}(s) = (1 + \hat{g}_1\hat{g}_2)^{-1}\hat{g}_1$ if and only if there is no pole of $\hat{g}_2(s)$ canceled by any zero of $\hat{g}_1(s)$.

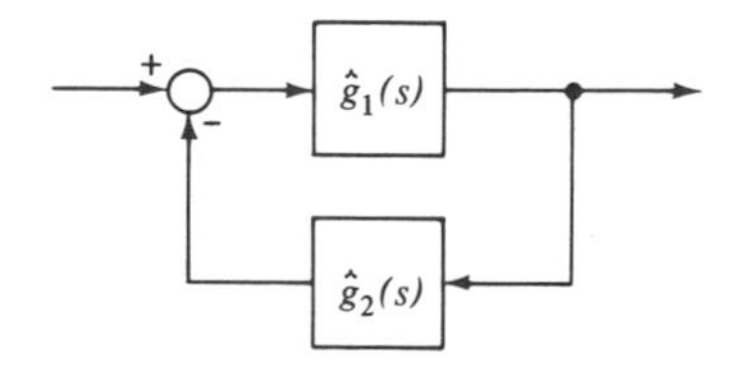


**Figure 9-1** A single-variable feedback system.

### Proof

**1.** It is obvious that if $\hat{g}_1$ and $\hat{g}_2$ have at least one pole in common, then $\delta\hat{g} < \delta\hat{g}_1 + \delta\hat{g}_2$. Let $\hat{g}_i = N_i/D_i$, for $i = 1, 2$; then

$$\hat{g} = \hat{g}_1 + \hat{g}_2 = \frac{N_1D_2 + N_2D_1}{D_1D_2}$$

We show now that if $\hat{g}_1$ and $\hat{g}_2$ do not have any pole in common, then $\delta\hat{g} = \delta\hat{g}_1 + \delta\hat{g}_2$. We prove this by contradiction. Suppose $\delta\hat{g} < \delta\hat{g}_1 + \delta\hat{g}_2$, then there is at least one common factor between $(N_1D_2 + N_2D_1)$ and $(D_1D_2)$. If there is a common factor, say, between $N_1D_2 + N_2D_1$ and $D_1$, then the assumption that there is no common factor between $D_1$ and $D_2$ implies that there is a common factor between $N_1$ and $D_1$. This contradicts the assumption that $\hat{g}_1$ is irreducible. Hence we conclude that if $\hat{g}_1$ and $\hat{g}_2$ have no pole in common, then $\hat{g}(s) = \hat{g}_1(s) + \hat{g}_2(s)$ characterizes completely the parallel connection of $S_1$ and $S_2$.
**2.** The proof of this part is obvious and is omitted.
**3.** The transfer function of the feedback system shown in Figure 9-1 is

$$\hat{g}(s) = \frac{\hat{g}_1(s)}{1 + \hat{g}_1(s)\hat{g}_2(s)} = \frac{D_2N_1}{D_1D_2 + N_1N_2}$$

By the irreducibility assumption, $D_1$ and $N_1$ have no common factor, nor have $D_2$ and $N_2$. Hence $D_2N_1$ and $(D_1D_2 + N_1N_2)$ have common factors if and only if $D_2$ and $N_1$ have common factors.
Q.E.D.

### Example 2

Consider the tandem connection of $S_1$ and $S_2$ with transfer functions

$$\frac{1}{s-1} \quad \text{and} \quad \frac{s-1}{s+1}$$

as shown in Figure 9-2(a). There is a pole–zero cancellation between $\hat{g}_1$ and $\hat{g}_2$. Hence the composite transfer function

$$\hat{g}(s) = \hat{g}_2(s)\hat{g}_1(s) = \frac{1}{s+1}$$

does not completely characterize the tandem connection. This can be seen by applying a unit step input to the composite system; although the output of the tandem connection is bounded, the output of $S_1$ increases exponentially with time, as shown in Figure 9-2(b).

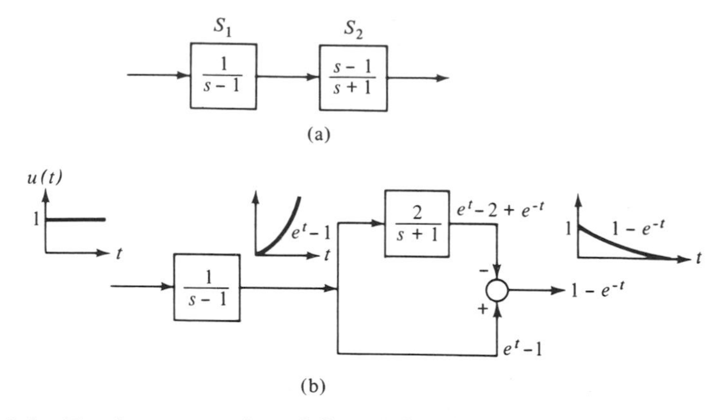


**Figure 9-2** Tandem connection of $S_1$ and $S_2$, which is not characterized completely by $\hat{g}(s) = \hat{g}_2\hat{g}_1 = 1/(s+1)$.

### Example 3

Consider the feedback connections shown in Figure 9-3. In Figure 9-3(a), the pole of the transfer function in the feedback path is canceled by the zero of the transfer function in the forward path. Hence the transfer function of the feedback system does not completely describe the feedback system. Indeed, its transfer function is

$$\hat{g}(s) = \frac{\dfrac{s-1}{s+1}}{1 + \dfrac{s-1}{s+1} \cdot \dfrac{1}{s-1}} = \frac{s-1}{s+2}$$

the degree of which is smaller than 2.

On the other hand, although the pole of the transfer function in the forward path of Figure 9-3(b) is canceled by the zero of the transfer function in the feedback path, the transfer function of the feedback system still completely characterizes the feedback system. Its transfer function is

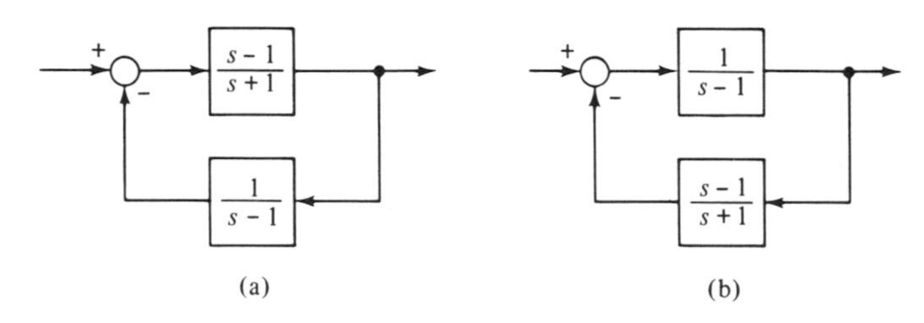


**Figure 9-3** Feedback systems.

$$\hat{g}(s) = \frac{s+1}{(s-1)(s+2)}$$ ∎

Consider a special case of the feedback system shown in Figure 9-1 with $\hat{g}_2(s) = k$, where $k$ is a real constant. Since there is no pole in $\hat{g}_2(s)$ to be canceled by $\hat{g}_1(s)$, the transfer function

$$\hat{g}(s) = (1 + k\hat{g}_1(s))^{-1}\hat{g}_1(s) = \frac{N_1}{D_1 + kN_1}$$

always characterizes the feedback system completely.

## *9-3 Controllability and Observability of Composite Systems

The necessary and sufficient conditions in terms of $\hat{g}_1(s)$ and $\hat{g}_2(s)$ are given in the preceding section for composite transfer functions to characterize single-variable composite systems completely. For multivariable systems, the situation is much more complicated. Although whether or not a composite transfer-function matrix completely characterizes a composite system still depends on the transfer matrices $\hat{\mathbf{G}}_1(s)$ and $\hat{\mathbf{G}}_2(s)$, the condition cannot be stated explicitly in terms of poles and zeros as in the single-variable case. The conditions can be stated only in terms of irreducible realizations of $\hat{\mathbf{G}}_1(s)$ and $\hat{\mathbf{G}}_2(s)$. Furthermore, we have to study first the controllability and observability of composite dynamical equations and then combine these to give the conditions under which a composite transfer-function matrix will completely characterize a composite connection.

We study first the controllability and observability of the parallel connection, and then the tandem connection. Finally we show that the controllability and observability of the feedback system can be reduced to those of a tandem connection.

**Parallel connection.** Consider systems $S_i$, for $i = 1, 2$, with the dynamical-equation descriptions

$$FE^i: \quad \begin{aligned} \dot{\mathbf{x}}_i &= \mathbf{A}_i\mathbf{x}_i + \mathbf{B}_i\mathbf{u}_i &\quad \text{(9-3a)} \\ \mathbf{y}_i &= \mathbf{C}_i\mathbf{x}_i + \mathbf{D}_i\mathbf{u}_i &\quad \text{(9-3b)} \end{aligned}$$

Their transfer-function descriptions are $\hat{\mathbf{G}}_i = \mathbf{C}_i(s\mathbf{I} - \mathbf{A}_i)^{-1}\mathbf{B}_i + \mathbf{D}_i$. We show now that in the study of controllability and observability of the dynamical-equation description of the parallel connection of $S_1$ and $S_2$, there is no loss of generality in assuming that the matrices $\mathbf{A}_i$ in (9-3) are in the Jordan form. The dynamical equation of the parallel connection of $S_1$ and $S_2$ shown in Figure 9-4 is

$$\begin{bmatrix} \dot{\mathbf{x}}_1 \\ \dot{\mathbf{x}}_2 \end{bmatrix} = \begin{bmatrix} \mathbf{A}_1 & \mathbf{0} \\ \mathbf{0} & \mathbf{A}_2 \end{bmatrix} \begin{bmatrix} \mathbf{x}_1 \\ \mathbf{x}_2 \end{bmatrix} + \begin{bmatrix} \mathbf{B}_1 \\ \mathbf{B}_2 \end{bmatrix} \mathbf{u} \tag{9-4a}$$

$$\mathbf{y} = [\mathbf{C}_1 \quad \mathbf{C}_2] \begin{bmatrix} \mathbf{x}_1 \\ \mathbf{x}_2 \end{bmatrix} + (\mathbf{D}_1 + \mathbf{D}_2)\mathbf{u} \tag{9-4b}$$

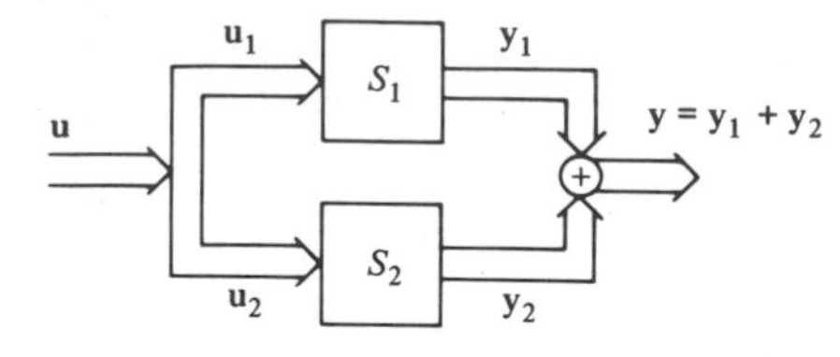


**Figure 9-4** The parallel connection of $S_1$ and $S_2$.

Suppose that $\mathbf{A}_1$ and $\mathbf{A}_2$ are not in the Jordan form. Let $\mathbf{P}_i$ be the nonsingular matrix that transforms $FE^i$ into the Jordan-canonical-form dynamical equation $F\tilde{E}^i$. Then the equivalence transformation

$$\tilde{\mathbf{x}} = \begin{bmatrix} \mathbf{P}_1 & \mathbf{0} \\ \mathbf{0} & \mathbf{P}_2 \end{bmatrix} \begin{bmatrix} \mathbf{x}_1 \\ \mathbf{x}_2 \end{bmatrix} \triangleq \mathbf{P} \begin{bmatrix} \mathbf{x}_1 \\ \mathbf{x}_2 \end{bmatrix}$$

will transform (9-4) into

$$\dot{\tilde{\mathbf{x}}} = \begin{bmatrix} \mathbf{P}_1\mathbf{A}_1\mathbf{P}_1^{-1} & \mathbf{0} \\ \mathbf{0} & \mathbf{P}_2\mathbf{A}_2\mathbf{P}_2^{-1} \end{bmatrix} \tilde{\mathbf{x}} + \begin{bmatrix} \mathbf{P}_1\mathbf{B}_1 \\ \mathbf{P}_2\mathbf{B}_2 \end{bmatrix} \mathbf{u} \tag{9-5a}$$

$$\mathbf{y} = [\mathbf{C}_1\mathbf{P}_1^{-1} \quad \mathbf{C}_2\mathbf{P}_2^{-1}]\tilde{\mathbf{x}} + (\mathbf{D}_1 + \mathbf{D}_2)\mathbf{u} \tag{9-5b}$$

Clearly this is the dynamical-equation description of the parallel connection of $F\tilde{E}^1$ and $F\tilde{E}^2$. Since the controllability of a dynamical equation is invariant under any nonsingular transformation (Theorem 5-15), Equation (9-4) is controllable if and only if Equation (9-5) is controllable. Hence we conclude that there is no loss of generality in assuming that the dynamical equations $FE^i$ are in the Jordan form. At this point, the reader is advised to review Theorem 5-16, the controllability and observability conditions in Jordan-form dynamical equations.

### Theorem 9-3

Assume that $S_1$ and $S_2$ are completely characterized by $\hat{\mathbf{G}}_1(s)$ and $\hat{\mathbf{G}}_2(s)$. Then the parallel connection of $S_1$ and $S_2$ is controllable (observable) if and only if either there are no common eigenvalues between $S_1$ and $S_2$; or in case there are common eigenvalues, each pair of common eigenvalues, say $\lambda_\alpha{}^1 = \lambda_\beta{}^2$, is such that all the rows of $\mathbf{B}_1$ and $\mathbf{B}_2$ corresponding to the last row of the Jordan blocks in $\mathbf{A}_1$ and $\mathbf{A}_2$ associated with the eigenvalue $\lambda_\alpha{}^1$ or $\lambda_\beta{}^2$ are linearly independent. (All the columns of $\mathbf{C}_1$ and $\mathbf{C}_2$ corresponding to the first column of the Jordan blocks in $\mathbf{A}_1$ and $\mathbf{A}_2$ associated with $\lambda_\alpha{}^1$ or $\lambda_\beta{}^2$ are linearly independent.)

### Proof

The dynamical-equation description of the parallel connection of $S_1$ and $S_2$ is given in (9-4). The matrices $\mathbf{A}_1$ and $\mathbf{A}_2$ are in the Jordan form by assumption. Hence this theorem follows directly from Theorem 5-16.

Q.E.D.

It is shown in Reference [15] that the conditions of Theorem 9-3 are uniquely determinable from $\hat{\mathbf{G}}_1(s)$ and $\hat{\mathbf{G}}_2(s)$. By combining the controllability and observability parts of Theorem 9-3, we have the necessary and sufficient conditions for $\hat{\mathbf{G}}_1(s) + \hat{\mathbf{G}}_2(s)$ to characterize completely the parallel connection of $S_1$ and $S_2$.

**Tandem connection.** We consider now the tandem connection of $S_1$ followed by $S_2$. As in the parallel connection, in order to state the necessary and sufficient conditions in terms of the subsystems for $\hat{\mathbf{G}}_2(s)\hat{\mathbf{G}}_1(s)$ to characterize completely the tandem connection, the irreducible Jordan-form realizations of $\hat{\mathbf{G}}_1(s)$ and $\hat{\mathbf{G}}_2(s)$ must be used. Instead of doing so, we introduce one sufficient condition that can be stated in terms of $\hat{\mathbf{G}}_1(s)$ and $\hat{\mathbf{G}}_2(s)$ alone. Let $\hat{\mathbf{G}}_i(s)$ be a $q_i \times p_i$ rational transfer-function matrix. In the tandem connection shown in Figure 9-5, we have $q_1 = p_2$; that is,

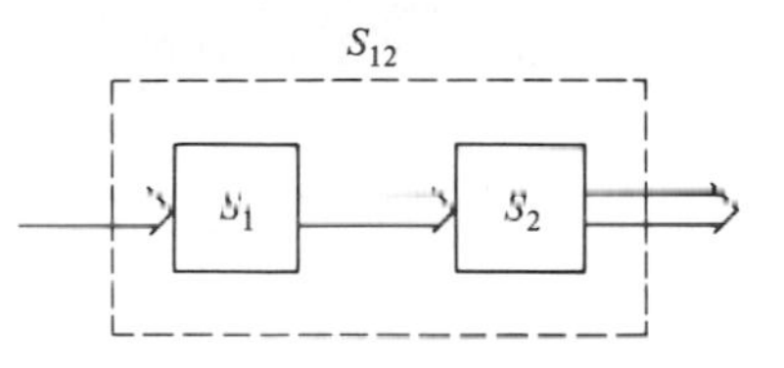


**Figure 9-5** $S_{12}$, the tandem connection of $S_1$ followed by $S_2$.

the number of the output terminals of $S_1$ is equal to the number of the input terminals of $S_2$.

### Theorem 9-4

Assume that $S_1$ and $S_2$ are completely characterized by $\hat{\mathbf{G}}_1(s)$ and $\hat{\mathbf{G}}_2(s)$, respectively. Then the composite dynamical equation of $S_1$ followed by $S_2$ is controllable (observable) if, for each pole $\lambda_i^2$ of $\hat{\mathbf{G}}_2(s)$,

$$\rho\hat{\mathbf{G}}_1(\lambda_i^2) = q_1$$

(For each pole $\lambda_i^1$ of $\hat{\mathbf{G}}_1(s)$, $\rho\hat{\mathbf{G}}_2(\lambda_i^1) = p_2$.) ∎

Recall that $\lambda_i^2$ is a pole of $\hat{\mathbf{G}}_2(s)$ if $\lambda_i^2$ is a pole of at least one element of $\hat{\mathbf{G}}_2(s)$ (Definition 3-5). The rank of a matrix is denoted by $\rho(\cdot)$. Note that this theorem is applicable only when the set of poles of $\hat{\mathbf{G}}_1(s)$ and the set of poles of $\hat{\mathbf{G}}_2(s)$ are disjoint; otherwise $\rho\hat{\mathbf{G}}_1(\lambda_i^2)$ may not be defined. Furthermore the condition is sufficient but not necessary for the composite dynamical equation of the tandem connection of $S_1$ and $S_2$ to be controllable or observable. The proof and a more general version of Theorem 9-4 can be found in References [15] and [20].

The composite matrix $\hat{\mathbf{G}}_2(s)\hat{\mathbf{G}}_1(s)$ characterizes completely the tandem connection of $S_1$ and $S_2$, denoted by $S_{12}$, if and only if its composite dynamical equation is controllable and observable. Hence, from Theorem 9-4, $\hat{\mathbf{G}}_2(s)\hat{\mathbf{G}}_1(s)$ completely characterizes $S_{12}$ if $\rho\hat{\mathbf{G}}_1(\lambda_i^2) = q_1$ for all the poles $\lambda_i^2$ of $\hat{\mathbf{G}}_2(s)$ and $\rho\hat{\mathbf{G}}_2(\lambda_i^1) = p_2$ for all the poles $\lambda_i^1$ of $\hat{\mathbf{G}}_1(s)$.

It is of interest to give an interpretation of Theorem 9-4 for single-variable systems. For single-variable systems, $q_1 = p_1 = q_2 = p_2 = 1$, and the theorem reduces to a statement that the tandem connection of $S_1$ followed by $S_2$, denoted by $S_{12}$, is controllable if and only if $\rho\hat{g}_1(\lambda_i^2) = 1$. Since $\hat{g}_1(s)$ is a scalar, $\rho\hat{g}_1(\lambda_i^2) = 1$ implies that $\hat{g}_1(\lambda_i^2) \neq 0$; that is, the zeros of $\hat{g}_1(s)$ are different from the poles of $\hat{g}_2(s)$. In other words, if there is no pole of $\hat{g}_2(s)$ canceled by any zero of $\hat{g}_1(s)$, then the composite dynamical equation of the tandem connection of $S_1$ followed by $S_2$ is controllable. Similarly, if there is no pole of $\hat{g}_1(s)$ canceled by any zero of $g_2(s)$, the dynamical equation of $S_{12}$ is observable. For ease of reference, we state these as a theorem.

### Theorem 9-5

Consider two single-variable systems $S_1$ and $S_2$, which are completely characterized by their proper rational functions $\hat{g}_1(s)$ and $\hat{g}_2(s)$, respectively. The dynamical equation of $S_{12}$ is controllable (observable) if and only if there is no pole of $\hat{g}_2(s)$ canceled by any zero of $\hat{g}_1(s)$. (There is no pole of $\hat{g}_1(s)$ canceled by any zero of $\hat{g}_2(s)$.) ∎

A formal proof of Theorem 9-5 can be found in Reference [20]. Combining the controllability and observability part of this theorem, we obtain a different proof of part 2 of Theorem 9-1.

### Corollary 9-5

Consider two single-variable systems $S_1$ and $S_2$, which are completely characterized by $\hat{g}_1(s)$ and $\hat{g}_2(s)$, respectively. Then the composite dynamical equation of $S_{12}$ is controllable if and only if the composite dynamical equation of $S_{21}$ is observable. ∎

$S_{12}$ denotes the tandem connection of $S_1$ followed by $S_2$; $S_{21}$ denotes the tandem connection of $S_2$ followed by $S_1$. Corollary 9-5 follows directly from Theorem 9-5. Note that Corollary 9-5 does not hold for multivariable systems.

**Feedback connection.** We study now the controllability and observability of composite dynamical equations of feedback systems. We shall show that the controllability of a feedback system can be reduced to the controllability of a tandem system. Before considering the general case, we study first the unit feedback system shown in Figure 9-6. Let the

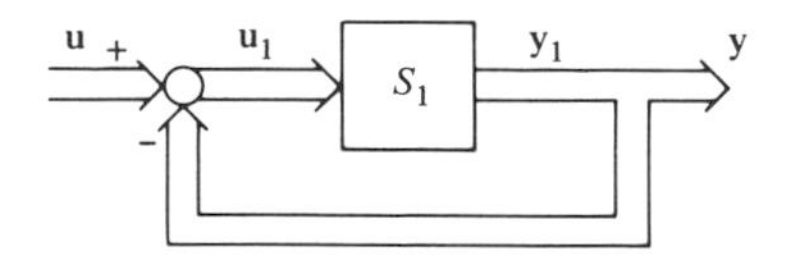


**Figure 9-6** The feedback system $S_f$.

dynamical-equation description of the system $S_1$ be

$$FE^1: \quad \begin{aligned} \dot{\mathbf{x}}_1 &= \mathbf{A}_1\mathbf{x}_1 + \mathbf{B}_1\mathbf{u}_1 && \text{(9-6a)} \\ \mathbf{y}_1 &= \mathbf{C}_1\mathbf{x}_1 + \mathbf{D}_1\mathbf{u}_1 && \text{(9-6b)} \end{aligned}$$

Then the dynamical-equation description of the feedback system $S_f$, denoted by $FE^f$, is

$$FE^f: \quad \dot{\mathbf{x}}_1 = [\mathbf{A}_1 - \mathbf{B}_1(\mathbf{I} + \mathbf{D}_1)^{-1}\mathbf{C}_1]\mathbf{x}_1 + [\mathbf{B}_1 - \mathbf{B}_1(\mathbf{I} + \mathbf{D}_1)^{-1}\mathbf{D}_1]\mathbf{u} \qquad \text{(9-7a)}$$

$$\mathbf{y} = (\mathbf{I} + \mathbf{D}_1)^{-1}\mathbf{C}_1\mathbf{x}_1 + (\mathbf{I} + \mathbf{D}_1)^{-1}\mathbf{D}_1\mathbf{u} \qquad \text{(9-7b)}$$

which is obtained by substituting $\mathbf{u} - \mathbf{y}_1 = \mathbf{u}_1$ and $\mathbf{y} = \mathbf{y}_1$ into (9-6). It is obvious that in order for $FE^f$ to be defined, we need the assumption that $(\mathbf{I} + \mathbf{D}_1)^{-1}$ exists. Because of the identity

$$\mathbf{I} - (\mathbf{I} + \mathbf{D}_1)^{-1}\mathbf{D}_1 = (\mathbf{I} + \mathbf{D}_1)^{-1} \qquad \text{(9-8)}$$

Equation (9-7) can be reduced to

$$FE^f: \quad \begin{aligned} \dot{\mathbf{x}}_1 &= [\mathbf{A}_1 - \mathbf{B}_1(\mathbf{I} + \mathbf{D}_1)^{-1}\mathbf{C}_1]\mathbf{x}_1 + \mathbf{B}_1(\mathbf{I} + \mathbf{D}_1)^{-1}\mathbf{u} && \text{(9-9a)} \\ \mathbf{y} &= (\mathbf{I} + \mathbf{D}_1)^{-1}\mathbf{C}_1\mathbf{x}_1 + (\mathbf{I} + \mathbf{D}_1)^{-1}\mathbf{D}_1\mathbf{u} && \text{(9-9b)} \end{aligned}$$

### Theorem 9-6

Consider the feedback system shown in Figure 9-6. It is assumed that $\det(\mathbf{I} + \mathbf{D}_1) \neq 0$. Then the dynamical-equation description of the feedback system $FE^f$ is controllable (observable) if and only if the dynamical-equation description of the open-loop system $FE^1$ is controllable (observable). ∎

In order to distinguish from the state-feedback systems introduced in Chapter 7, the feedback system shown in Figure 9-6 is also called an *output feedback system*. Recall that although the controllability of a linear, time-invariant dynamical equation is preserved after the introduction of constant state feedback, the observability property is generally not preserved. However, the controllability and observability of a linear, time-invariant dynamical equation are always preserved after the introduction of constant output feedback (see Problem 9-14).

Theorem 9-6 can be proved either by using the criterion $\rho[\mathbf{B} \quad \mathbf{AB} \quad \cdots \quad \mathbf{A}^{n-1}\mathbf{B}] = n$ or directly from the definitions of controllability and observability. We give in the following a proof based on the definitions. Note that the proof of the controllability part is identical to the proof of Theorem 7-3.

### Proof of Theorem 9-6

The assumption $\det(\mathbf{I} + \mathbf{D}_1) \neq 0$ implies that for any input $\mathbf{u}$ of the feedback system, there is a unique output $\mathbf{y}$. If $FE^1$ is controllable, then for any $\mathbf{x}^0$ and any $\mathbf{x}^1$ in the state space, there exists an input $\mathbf{u}_1$ to $S_1$ which will transfer $\mathbf{x}^0$ to $\mathbf{x}^1$ in a finite time. Call $\mathbf{y}_1$ the corresponding output. Now for the feedback system $S_f$, if we choose $\mathbf{u} = \mathbf{u}_1 + \mathbf{y}_1$, then this input will transfer $\mathbf{x}^0$ to $\mathbf{x}^1$. This proves that if $FE^1$ is controllable, so is $FE^f$. Conversely, if $FE^f$ is controllable, corresponding to any $\mathbf{u}$, if we choose $\mathbf{u}_1 = \mathbf{u} - \mathbf{y}$, then $FE^1$ is controllable.

Next we show that the observability of $FE^1$ implies the observability of $FE^f$. Let the zero-input response of $FE^f$ be $\mathbf{y}^0$ (due to some unknown initial state $\mathbf{x}^0$). If $FE^1$ is in the same initial state and if $-\mathbf{y}^0$ is applied to $FE^1$, from Figure 9-6 we see that the output of $FE^1$ is $\mathbf{y}^0$. Since $FE^1$ is observable by assumption, from the knowledge of the input $-\mathbf{y}^0$ and the output $\mathbf{y}^0$, the initial state $\mathbf{x}^0$ can be determined; consequently, $FE^f$ is observable. We prove now the converse. Assume that $FE^f$ is observable. Let the zero-input response of $FE^1$ due to some initial state be denoted by $\mathbf{y}_1{}^0$. Now if we apply $\mathbf{u} = \mathbf{y}_1{}^0$ to $FE^f$ and if $FE^f$ has the same initial state as $FE^1$, then the output of $FE^f$ is $\mathbf{y}_1{}^0$. From the input and the output of $FE^f$, the initial state can be determined. Consequently $FE^1$ is observable. Q.E.D.

Next we consider the feedback connection of two systems $S_1$ and $S_2$, as shown in Figure 9-7. Let $S_{12}$ denote the tandem connection of $S_1$ followed by $S_2$ and let $S_{21}$ denote the tandem connection of $S_2$ followed by $S_1$.

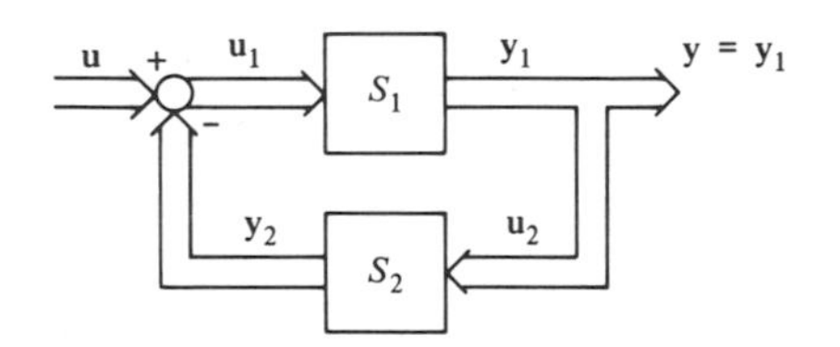


**Figure 9-7** Feedback connection of $S_1$ and $S_2$.

### Theorem 9-7

Consider the system $S_i$, for $i = 1, 2$, with controllable and observable dynamical equations $FE^i$. It is assumed that $\det(\mathbf{I} + \mathbf{D}_1\mathbf{D}_2) \neq 0$. Then the dynamical-equation description of the feedback system $S_f$ shown in Figure 9-7 is controllable (observable) if and only if the dynamical equation of the tandem connection $S_{12}(S_{21})$ is controllable (observable).

### Proof

Since the controllability of $S_f$ does not involve the output, we may just as well consider the system shown in Figure 9-8(a) in the controllability study. Hence from Theorem 9-6 we conclude that $S_f$ is controllable if and only if $S_{12}$ is controllable.

In the study of observability, the input $\mathbf{u}$ can be assumed, without loss of generality, to be zero. The zero-input response of the system shown in Figure 9-7 and the zero-input response of the system shown in Figure

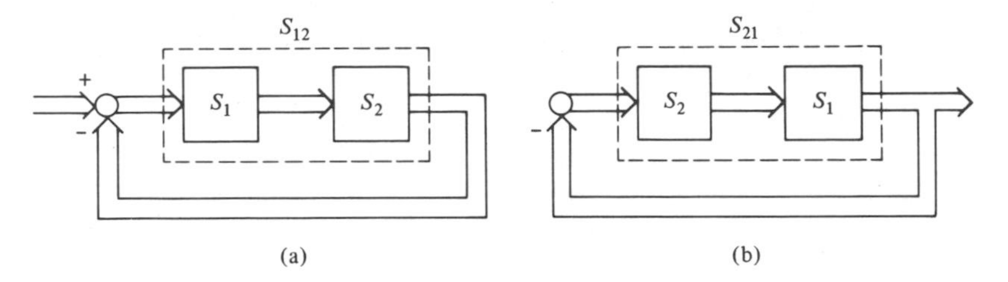


**Figure 9-8** Controllability and observability of feedback system.

9-8(b) are identical. Hence from Theorem 9-6 we conclude that $S_f$ is observable if and only if $S_{21}$ is observable. Q.E.D.

The controllability of $S_{12}$ and the observability of $S_{21}$ are, in general, not related. However, for single-variable systems, $S_{12}$ is controllable if and only if $S_{21}$ is observable (Corollary 9-5). Hence, a single-variable feedback system is controllable and observable if and only if $S_{12}$ is controllable. From Theorem 9-5, $S_{12}$ is controllable if and only if no pole of $\hat{g}_2(s)$ is canceled by any zero of $\hat{g}_1(s)$. Hence we conclude that a single-variable feedback system is controllable and observable, and consequently is completely characterized by its composite transfer function if and only if there is no pole of $\hat{g}_2(s)$ canceled by any zero of $\hat{g}_1(s)$. This checks with what we have in part 3 of Theorem 9-2.

## 9-4 Stability of Linear, Time-Invariant Feedback Systems

In this section we shall study the stability of composite connections of two systems $S_1$ and $S_2$ which are describable by linear, time-invariant dynamical equations. Recall that a system is defined to be totally stable or T-stable if for any initial state and for any bounded input, the output and the state variables are all bounded. It was shown in Theorem 8-16 that if the dynamical-equation description of a system is controllable and observable, or equivalently, if the system is completely characterized by its transfer-function matrix (Definition 9-1), then the T-stability of the system can be determined from its transfer-function matrix alone. Since the stability of a system is, in many cases, easier to determine from the transfer function than from the dynamical equation, we shall study the stability of composite systems from the transfer-function descriptions.

Consider two linear, time-invariant systems $S_1$ and $S_2$. It is assumed that $S_1$ and $S_2$ are completely characterized by their transfer-function matrices $\hat{\mathbf{G}}_1(s)$ and $\hat{\mathbf{G}}_2(s)$, respectively. Since any connection of $S_1$ and $S_2$ is again a linear, time-invariant system, if the composite-transfer-function description is obtained and if we are certain that the composite system is completely characterized by the composite transfer function, then the stability of the composite system can be determined. However, it is not a simple task to compute the composite transfer function, especially in the feedback connection. Even if the composite transfer function is obtained, it is still quite tedious to check whether or not the composite system is completely characterized. Therefore, it is desirable to be able to determine the stability of composite systems from $\hat{\mathbf{G}}_1(s)$ and $\hat{\mathbf{G}}_2(s)$ without computing composite-transfer-function matrices. We shall study this problem in this section.

For the parallel and tandem connections of $S_1$ and $S_2$, the problem is very simple. *The parallel or the tandem connections of $S_1$ and $S_2$ is T-stable if and only if $S_1$ and $S_2$ are T-stable.* This assertion is quite obvious because the parallel and tandem connections do not affect the behavior of the state variables of $S_1$ and $S_2$. For the feedback connection, the situation is much more complicated. A feedback system might be stable with unstable subsystems; conversely, a feedback system might be unstable with stable subsystems. We shall discuss first single-variable feedback systems and then multivariable systems.

**Single-variable feedback systems.** Consider the single-variable feedback system shown in Figure 9-9. The linear, time-invariant system $S_1$ is

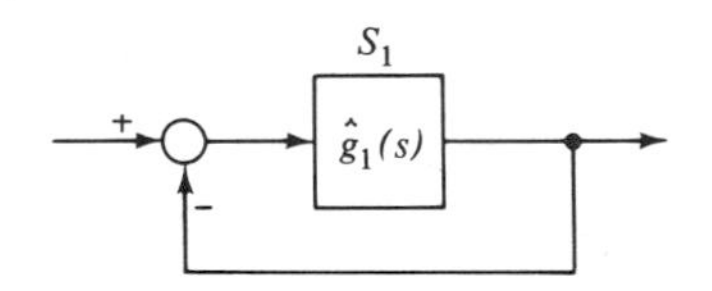


**Figure 9-9** Single-variable feedback system.

assumed to be completely characterized by its transfer function $\hat{g}_1(s)$. Let $\hat{g}_f(s)$ be the transfer function of the entire feedback system. Then we have

$$\hat{g}_f(s) = \frac{\hat{g}_1(s)}{1 + \hat{g}_1(s)} = \frac{N_1(s)}{D_1(s) + N_1(s)} \tag{9-10}$$

where

$$\hat{g}_1(s) \triangleq \frac{N_1(s)}{D_1(s)}$$

It can be deduced either from part 3 of Theorem 9-2 or from Theorem 9-6 that the feedback system shown in Figure 9-9 is completely characterized by the transfer function $\hat{g}_f(s)$. Hence the feedback system is T-stable if and only if all the poles of $\hat{g}_f(s)$ have negative real parts; or correspondingly, all the zeros of the rational function $1 + \hat{g}_1(s)$ or the roots of the polynomial $D_1(s) + N_1(s)$ have negative real parts; or correspondingly, the polynomial $D_1(s) + N_1(s)$ is a Hurwitz polynomial.

The Routh–Hurwitz and the Liénard–Chipart criteria introduced in the previous chapter can be employed to test the stability of the feedback system. Their applications to determine whether or not $D_1(s) + N_1(s)$ is a Hurwitz polynomial are straightforward. We introduce in the following the Nyquist criterion,[3] which can also be used to check the stability of a system, or more precisely, to check whether or not a polynomial is a Hurwitz polynomial. The Nyquist criterion is based on the following principle, whose proof can be found, for example, in Reference [113]:

### Principle of the argument

Let $C_1$ be a simple closed curve[4] in the $s$-plane. Let $\hat{f}$ be a rational function of $s$ that has neither pole nor zero on $C_1$. Let $Z$ and $P$ be the numbers of zeros and poles of $\hat{f}$ (counting the multiplicities) encircled by $C_1$ and let $C_2$ be the mapping of $C_1$ on the $\hat{f}$-plane. Then $C_2$ will encircle the origin of the $\hat{f}$-plane $(Z - P)$ times in the same direction as $C_1$. ∎

If $(Z - P)$ is a negative number, then $C_2$ will encircle the origin in the opposite direction to $C_1$. For example, consider the mapping shown in Figure 9-10. There are five poles and three zeros of a rational function

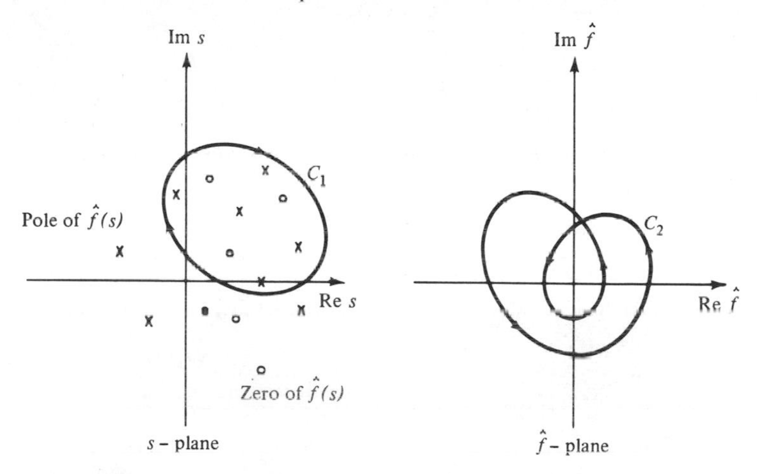


**Figure 9-10** Mapping of the contour $C_1$ by $\hat{f}(s)$.

$\hat{f}(s)$ encircled by $C_1$, which is chosen in the clockwise direction. Hence if we plot $\hat{f}(s)$ along the contour $C_1$, the resulting curve $C_2$ will encircle the origin of the $\hat{f}$-plane $Z - P = -2$ times, that is, encircle the origin twice in the counterclockwise direction. If $C_2$ does not encircle the origin $-2$ times, there must be an error in the plotting according to the principle of the argument.

In the application of the principle of the argument to the stability study, the problem is usually posed as follows: Given a rational function

[3] In addition to the Routh-Hurwitz criterion and the Nyquist criterion, there is another technique called the root-locus method, which can be used to determine the stability of feedback systems. This method is very important in the design of feedback systems; its discussion can be found in any control book.

[4] A simple curve is one that does not cross itself.

$\hat{f}(s)$, we know the poles of $\hat{f}(s)$, the question is whether or not the numerator of $\hat{f}(s)$ is a Hurwitz polynomial. To answer this question, the contour $C_1$ is chosen as the one shown in Figure 9-11(a) or the one in Figure 9-11(b) if

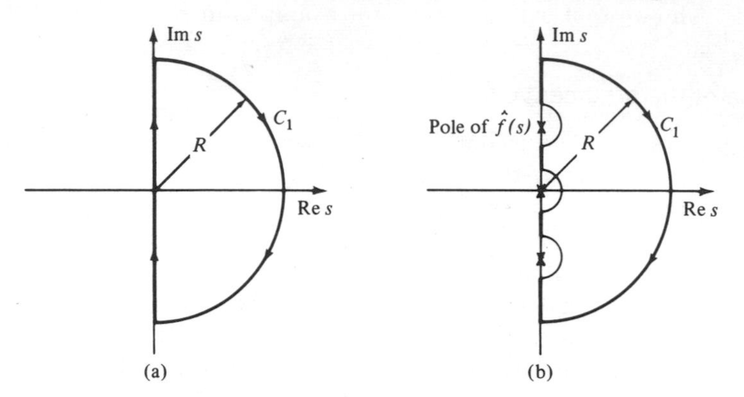


**Figure 9-11** The contour $C_1$.

$\hat{f}(s)$ has poles on the $i\omega$-axis. Theoretically, the radius of the semicircle $R$ should be infinity so that the contour $C_1$ encloses the entire right half $s$-plane.

### Definition 9-2

The mapping of the contour $C_1$ shown in Figure 9-11 by a rational function $\hat{f}$ is called the *Nyquist plot* of $\hat{f}$. ∎

In determining whether or not the numerator of $\hat{f}$ is a Hurwitz polynomial, we plot the Nyquist plot of $\hat{f}$ and count the number of encirclements around the origin, say $N$ (including positive or negative sign). Since we have assumed that the poles of $\hat{f}$ are known, we can count the poles of $\hat{f}$ encircled by $C_1$, or equivalently, the number of poles of $\hat{f}$ in the open right half $s$-plane. Now the principle of the argument implies that the number of the roots of the numerator of $\hat{f}(s)$ encircled by $C_1$ is equal to $P + N$. Hence we conclude that the numerator of $\hat{f}(s)$ has no roots in the open right half $s$-plane if and only if $P + N = 0$. If the Nyquist plot of $\hat{f}(s)$ goes through the origin, then the numerator of $\hat{f}(s)$ has at least one root on the $i\omega$-axis. We summarize what we have in the following: The numerator of $\hat{f}(s)$ is a Hurwitz polynomial (has no roots in the closed right half $s$-plane) if and only if the Nyquist plot of $\hat{f}(s)$ does not go through the origin and encircle the origin of the $\hat{f}$-plane $N = -P$ times.

Now we shall apply this fact to the stability study of the system shown in Figure 9-9. The feedback system is T-stable if and only if the numerator of $(1 + \hat{g}_1(s))$ is a Hurwitz polynomial. Hence we conclude that *the feedback system is T-stable if and only if the Nyquist plot of* $(1 + \hat{g}(s))$ *encircles the origin of the* $(1 + \hat{g}_1(s))$*-plane* $-P$ *times—that is,* $P$ *times in* *the counterclockwise direction,*[5] *where P is the number of the open right-half-plane poles of* $\hat{g}_1(s)$.

Observe that the Nyquist plot of $(1 + \hat{g}_1(s))$ is equal to the Nyquist plot of $\hat{g}_1(s)$ shifted to the right by one. Consequently, if we plot the Nyquist plot of $\hat{g}_1(s)$ and then shift the origin to $(-1, 0)$, we will obtain the Nyquist plot of $(1 + \hat{g}_1(s))$. Hence we have the following *Nyquist stability criterion.*

### Theorem 9-8

Assume that the system $S_1$ in Figure 9-9 is completely characterized by its proper rational transfer function $\hat{g}_1(s)$ and that $1 + \hat{g}_1(\infty) \neq 0$. Then the feedback system is T-stable if and only if the numerator of $1 + \hat{g}_1(s)$ is a Hurwitz polynomial or if and only if the Nyquist plot of $\hat{g}_1(s)$ does not go through the $(-1, 0)$ point and encircles the $(-1, 0)$ point $P$ times in the counterclockwise direction, where $P$ is the number of the open right-half-plane poles of $\hat{g}_1(s)$. ∎

We give a remark concerning the condition $1 + \hat{g}_1(\infty) \neq 0$. It was pointed out in Section 3-6 that in order for the existence of the transfer-function description of a feedback system, we need the condition $1 + \hat{g}_1(s) \neq 0$, for some $s$. (This is reduced from $\det(\mathbf{I} + \hat{\mathbf{G}}_1(s)\hat{\mathbf{G}}_2(s)) \neq 0$ for some $s$ for the multivariable system.) Clearly, the condition $1 + \hat{g}_1(\infty) \neq 0$ implies $1 + \hat{g}_1(s) \neq 0$ for some $s$; but not conversely. Hence the condition $1 + \hat{g}_1(\infty) \neq 0$ is more than necessary to insure the existence of the transfer function for a feedback system. The reason for introducing this stronger condition is to insure that if $\hat{g}_1(s)$ is a proper rational function, then the feedback transfer function $\hat{g}_f(s)$ is again a *proper* rational function. This can be seen from the dynamical-equation description of the feedback system. Since $\hat{g}_1(\infty) = d_1$, the condition $1 + \hat{g}_1(\infty) = 1 + d_1 \neq 0$ implies the existence of the dynamical-equation description of the feedback system. Hence the feedback transfer function must be proper. It is easy to show by example that if $1 + \hat{g}_1(\infty) = 0$, and $1 + \hat{g}_1(s) \neq 0$ for some $s$, then $\hat{g}_f(s) = (1 + \hat{g}_1(s))^{-1}\hat{g}_1(s)$ will not be a proper rational function (Try $\hat{g}_1(s) = (-s + 1)/(s + 2)$). If $\hat{g}_f(s)$ is not proper, then the stability definitions we have introduced cannot be applied directly, because if the input is a discontinuous function, the output will contain $\delta$-functions which have infinite magnitude.

If $\hat{g}_1(s)$ does not have open right-half-plane poles, then the theorem reduces to a statement that the system is T-stable if and only if the Nyquist plot of $\hat{g}_1(s)$ does not encircle or go through the point $(-1, 0)$.

In practical application of the Nyquist criterion, some simplifications can be made. If $\hat{g}_1(s)$ is a strictly proper rational function, the value of $\hat{g}_1(s)$ along the right half semicircle of the contour $C_1$ is zero. Hence

[5] Since $C_1$ in Figure 9-10 is chosen to be in the clockwise direction.

only $\hat{g}_1(s)$ along the $i\omega$-axis has to be plotted. Since $\hat{g}_1(s)$, in general, has real coefficients, the plot of $\hat{g}_1(i\omega)$ from $\omega = -\infty$ to 0 is just the mirror image, with respect to the real axis, of the plot of $\hat{g}_1(i\omega)$ from $\omega = 0$ to $\infty$. (See Problem 9-13.) Hence if we plot $\hat{g}_1(i\omega)$ from $\omega = 0$ to $\infty$, the remainder of the Nyquist plot can be obtained immediately.

Example 1

Consider the feedback system shown in Figure 9-9, with

$$\hat{g}_1(s) = \frac{8}{(s+1)(s^2+2s+2)}$$

Is the feedback system T-stable?

**1.** *Using the Routh–Hurwitz criterion:*

$$D(s) + N(s) = (s+1)(s^2+2s+2) + 8 = s^3 + 3s^2 + 4s + 10$$

$$\begin{matrix} 1 & 4 \\ 3 & 10 \\ \frac{2}{3} & 0 \\ 10 & \end{matrix}$$

Since we have three positive numbers (except the first element) in the first column, the numerator of $1 + \hat{g}_1(s)$—namely, $D(s) + N(s)$—is a Hurwitz polynomial; hence the feedback system is T-stable.

**2.** *Using the Nyquist stability criterion:*
The Nyquist plot of

$$\hat{g}_1(s) = \frac{8}{(s+1)(s^2+2s+2)}$$

is shown in Figure 9-12. Since $\hat{g}_1(s)$ has no right-half-plane pole, and since the Nyquist plot of $\hat{g}_1(s)$ does not encircle nor go through the point $(-1, 0)$, the feedback system is T-stable. ▮

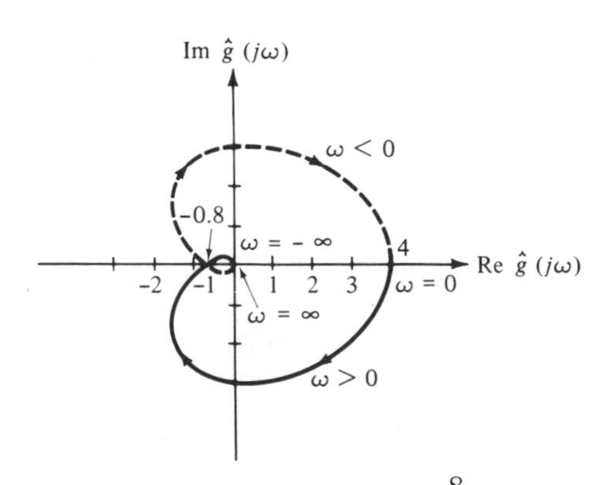


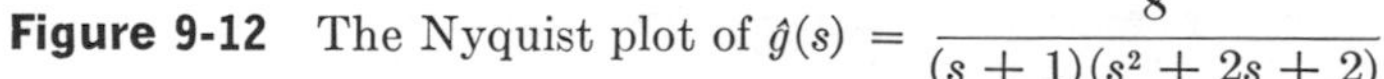
**Figure 9-12** The Nyquist plot of $\hat{g}(s) = \dfrac{8}{(s+1)(s^2+2s+2)}$.

In the study or design of feedback systems, we often encounter the following problem: Given the feedback system shown in Figure 9-13, find

**Figure 9-13** A feedback system.

the range of the constant gain $k$ such that the feedback system is T-stable. The transfer function of the feedback system is

$$\hat{g}_f(s) = \frac{k\hat{g}_1(s)}{1 + k\hat{g}_1(s)} = \frac{kN_1(s)}{D_1(s) + kN_1(s)} = \frac{\hat{g}_1(s)}{\frac{1}{k} + \hat{g}_1(s)} \tag{9-11}$$

We can apply the Routh–Hurwitz criterion or the Liénard–Chipart criterion to $D_1(s) + kN_1(s)$ or apply the Nyquist criterion to $(1/k) + \hat{g}_1(s)$ to find the stability range of $k$. This is illustrated by the following example:

### Example 2

Consider the feedback system shown in Figure 9-13 with

$$\hat{g}_1(s) = \frac{8}{(s+1)(s^2+2s+2)}$$

Find the stability range of $k$.

**1.** *Using the Routh–Hurwitz criterion.*

$$D(s) + kN(s) = s^3 + 3s^2 + 4s + 2 + 8k$$

$$\begin{array}{cc} 1 & 4 \\ 3 & 2+8k \\ \dfrac{12-2-8k}{3} & 0 \\ 2+8k & \end{array}$$

The conditions for the feedback system to be stable are $(10 - 8k)/3 > 0$ and $2 + 8k > 0$, which give $-\frac{1}{4} < k < \frac{5}{4}$.

**2.** *Using the Nyquist criterion.* The Nyquist plot of $\hat{g}_1(s)$ is given in Figure 9-12. Since $\hat{g}_1(s)$ has no right-half-plane pole, in order for the feedback system to be stable the Nyquist plot of $\hat{g}_1(s)$ should not encircle the point $(-1/k, 0)$. Consequently, from Figure 9-12, we have $-\infty < -1/k < -0.8$ and $\infty > -1/k > 4$. The inequality $-\infty < -1/k < -0.8$ implies that $0 < k < \frac{5}{4}$. The inequality $\infty > -1/k > 4$ implies that $0 > k > -\frac{1}{4}$. Hence, the stability range of $k$ is $-\frac{1}{4} < k < \frac{5}{4}$. ∎

A feedback system without any dynamic in its feedback path is always completely characterized by its feedback transfer function. If there is a dynamic system in its feedback path, then this is not always the case. Consider the feedback system shown in Figure 9-14. The transfer function of the feedback system is

$$\hat{g}_f(s) = \frac{k\hat{g}_1(s)}{1 + k\hat{g}_1(s)\hat{g}_2(s)} = \frac{\hat{g}_1(s)}{\dfrac{1}{k} + \hat{g}_1(s)\hat{g}_2(s)} = \frac{N_1(s)D_2(s)}{D_1(s)D_2(s) + kN_1(s)N_2(s)}$$

$$\triangleq \frac{N_f(s)}{D_f(s)} \tag{9-12}$$

where $\hat{g}_i(s) \triangleq N_i(s)/D_i(s)$, for $i = 1, 2$, and $N_f$ and $D_f$ are assumed to have no common factor. Recall that every transfer function is assumed to be irreducible, therefore when we speak of $\hat{g}_f(s)$, we mean $\hat{g}_f(s) = N_f(s)/D_f(s)$. It was shown in Theorem 9-2 that if $N_1$ and $D_2$ have no common factor, then $D_f = D_1D_2 + kN_1N_2$ and the system is completely characterized by $\hat{g}_f(s)$. If there are common factors between $N_1$ and $D_2$, then $D_f$ consists of only one part of $D_1D_2 + kN_1N_2$, and the system is not completely characterized by $\hat{g}_f(s)$. In either case, the T-stability of the feedback system can be determined by using the following theorem *without* checking whether or not the feedback system is characterized completely by $\hat{g}_f(s)$.

### Theorem 9-9

Assume that the systems $S_1$ and $S_2$ in the feedback connection shown in Figure 9-14 are completely characterized by $\hat{g}_1(s)$ and $\hat{g}_2(s)$ and that

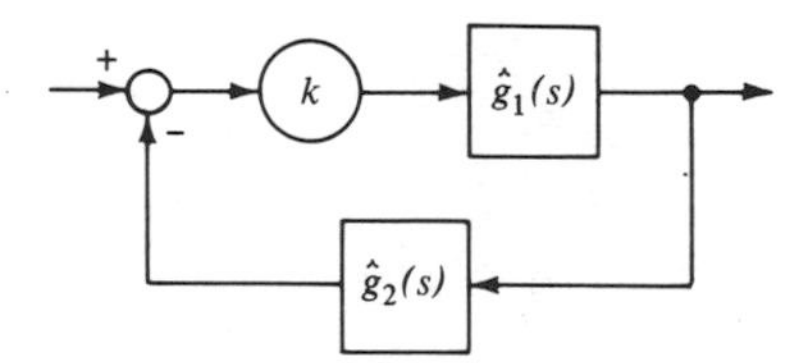


**Figure 9-14** A feedback system.

$1 + \hat{g}_1(\infty)\hat{g}_2(\infty) \neq 0$. Then the feedback system is T-stable if all the zeros of $D_1(s)D_2(s) + kN_1(s)N_2(s)$ have negative real parts or if all the zeros of the largest common factor of $N_1(s)N_2(s)$ and $D_1(s)D_2(s)$ have negative real parts and the Nyquist plot of $\hat{g}_1(s)\hat{g}_2(s)$ does not go through the point $(-1/k, 0)$ and encircles the point $(-1/k, 0)$ $P$ times in the counterclockwise direction, where $P$ is the number of the open right-half-plane poles of $\hat{g}_1(s)$ and $\hat{g}_2(s)$.

### Proof

If $N_1(s)$ and $D_2(s)$ have no common factor, the feedback system is completely characterized by $\hat{g}_f$. If $N_1(s)$ and $D_2(s)$ have common factors, say

$\hat{D}_2(s)$, then $D_1D_2 + kN_1N_2 = \hat{D}_2D_f$. The zeros of $D_f$ are the new poles that are introduced by feedback; the zeros of $\hat{D}_2$ are poles of the subsystem that are unobservable and/or uncontrollable from the input and output of the feedback system. Hence if all the zeros of $D_f$ and $\hat{D}_2$ have negative real parts, then the system is T-stable. This proves the first part of the theorem.

Let the largest common factor of $N_1N_2$ and $D_1D_2$ be $\hat{D}$, and let

$$\hat{g}_1(s)\hat{g}_2(s) \triangleq \frac{\hat{D}\bar{N}}{\hat{D}\bar{D}}$$

Then

$$D_1D_2 + kN_1N_2 = \hat{D}(\bar{D} + k\bar{N})$$

Observe that when we plot the Nyquist plot of $\hat{g}_1(s)\hat{g}_2(s)$, we are in fact plotting the Nyquist plot of $\bar{N}/\bar{D}$. Hence the Nyquist criterion studies only the zeros of $\bar{D} + k\bar{N}$. The zeros of $\hat{D}$ must be checked separately.

Q.E.D.

This theorem gives only the sufficient condition. In order to state it as necessary and sufficient, the condition on $\hat{D}_2$, the common factors of $N_1$ and $D_2$, must be relaxed. This will not be discussed here. The interested reader is referred to Example 2 in Section 8-4 and Problems 9-8 and 9-9.

If $\hat{g}_1(s)$ and $\hat{g}_2(s)$ have no closed right-half-plane poles, the T-stability of the feedback system can be determined from the Nyquist plot of $\hat{g}_1(s)\hat{g}_2(s)$ alone, because the conditions on the common factors of $N_1N_2$ and $D_1D_2$ are always satisfied.

**Multivariable feedback systems.** Consider the multivariable feedback system shown in Figure 9-15. The system $S_1$ is assumed to be completely

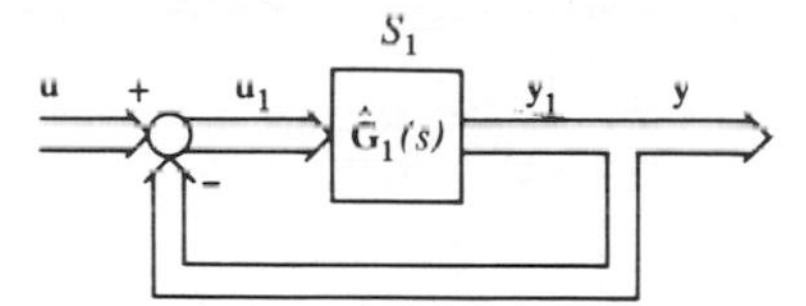


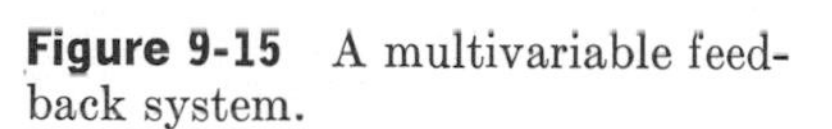

**Figure 9-15** A multivariable feedback system.

characterized by its $p \times p$ proper rational-function matrix $\hat{\mathbf{G}}_1(s)$. The assumption that the number of the input terminals of $S_1$ is equal to the number of its output terminals is a necessity, otherwise the feedback path cannot be properly connected to the input. Let $\hat{\mathbf{G}}_f(s)$ be the transfer-function matrix of the feedback system; then

$$\hat{\mathbf{G}}_f(s) = (\mathbf{I} + \hat{\mathbf{G}}_1(s))^{-1}\hat{\mathbf{G}}_1(s) = \frac{1}{\det(\mathbf{I} + \hat{\mathbf{G}}_1(s))}[\text{Adj}(\mathbf{I} + \hat{\mathbf{G}}_1(s))]\hat{\mathbf{G}}_1(s) \tag{9-13}$$

where Adj $(\cdot)$ denotes the adjoint of a matrix. Since $S_1$ is completely characterized by $\hat{\mathbf{G}}_1(s)$, it follows from Theorem 9-6 that the feedback system in Figure 9-15 is completely characterized by $\hat{\mathbf{G}}_f(s)$. If the dimension of $\hat{\mathbf{G}}_1(s)$ reduces to 1, then $\det(\mathbf{I} + \hat{\mathbf{G}}_1(s))$ becomes $1 + \hat{g}_1(s)$ and the stability depends only on the zeros of $1 + \hat{g}_1(s)$. Hence, one might be tempted to reach the *false* conclusion that the stability of multivariable feedback systems depends only on the zeros of $\det(\mathbf{I} + \hat{\mathbf{G}}_1(s))$.

### Example 3

Consider the multivariable feedback system shown in Figure 9-15, with

$$\hat{\mathbf{G}}_1(s) = \begin{bmatrix} \dfrac{-s}{s-1} & \dfrac{s}{s+1} \\ 1 & \dfrac{-2}{s+1} \end{bmatrix}$$

It can be easily verified that

$$\hat{\mathbf{G}}_f(s) = \begin{bmatrix} \dfrac{2s}{s+1} & \dfrac{-s}{s+1} \\ -1 & \dfrac{s-2}{s-1} \end{bmatrix}$$

which is unstable. However, we have $\det(\mathbf{I} + \hat{\mathbf{G}}_1(s)) = -1$, which does not have any right-half-plane zeros. ▮

In order to study the stability of multivariable feedback systems, the concept of the characteristic polynomial of a proper rational matrix is needed. Recall from Definition 6-1 that the characteristic polynomial of $\hat{\mathbf{G}}_1(s)$, denoted by $\Delta_1(s)$, is the least common denominator of all the minors of $\hat{\mathbf{G}}_1(s)$.

### Theorem 9-10

Consider the feedback system shown in Figure 9-15. Assume that the system $S_1$ is completely characterized by $\hat{\mathbf{G}}_1(s)$ and assume[6] that $\det(\mathbf{I} + \hat{\mathbf{G}}_1(\infty)) \neq 0$. Then the feedback system is totally stable (T-stable) if and only if all the zeros of the polynomial $\Delta_1(s)\det(\mathbf{I} + \hat{\mathbf{G}}_1(s))$ have negative real parts, where $\Delta_1(s)$ is the characteristic polynomial of $\hat{\mathbf{G}}_1(s)$. ▮

Before proving the theorem, we first show that $\Delta_1(s)\det(\mathbf{I} + \hat{\mathbf{G}}_1(s))$ is a polynomial. Define

$$\det(\mathbf{I} + \hat{\mathbf{G}}_1(s)) \triangleq \frac{E(s)}{F(s)} \tag{9-14}$$

[6] This condition is to insure that the dynamical equation of the feedback system is well-defined. See the remark following Theorem 9-8.

where $E(s)$ and $F(s)$ are assumed to have no common factor, then

$$\Delta_1(s) \det (\mathbf{I} + \hat{\mathbf{G}}_1(s)) = \frac{\Delta_1(s)E(s)}{F(s)}$$

To show that $\Delta_1(s) \det (\mathbf{I} + \hat{\mathbf{G}}_1(s))$ is a polynomial, it is the same to show that $\Delta_1(s)$ is divisible without remainder by $F(s)$. This will be shown by using the following identity:

$$\det (\mathbf{I} + \hat{\mathbf{G}}_1(s)) = \sum_{i=1}^{p} \alpha_i(s) + 1 \tag{9-15}$$

where $p$ is the order of $\hat{\mathbf{G}}_1(s)$ and $\alpha_i$ is the sum of all the principal minors of $\hat{\mathbf{G}}_1(s)$ of order $i$; for example, $\alpha_p(s) = \det \hat{\mathbf{G}}_1(s)$, $\alpha_1(s) =$ sum of the diagonal elements of $\hat{\mathbf{G}}_1(s)$. The verification of (9-15) is straightforward and is omitted. Now $\Delta_1(s)$ is the least common denominator of all the minors of $\hat{\mathbf{G}}_1(s)$, whereas $F(s)$ is at most equal to the least common denominator of all the principal minors of $\hat{\mathbf{G}}_1(s)$ following from (9-14) and (9-15); hence we conclude that $\Delta_1(s)$ is divisible without remainder by $F(s)$ and that $\Delta_1(s) \det (\mathbf{I} + \hat{\mathbf{G}}_1(s))$ is a polynomial.

### Proof of Theorem 9-10

We shall use the dynamical-equation description of the system to prove the theorem. Let

$$FE^1\text{:} \qquad \dot{\mathbf{x}}_1 = \mathbf{A}_1\mathbf{x}_1 + \mathbf{B}_1\mathbf{u}_1 \tag{9-16a}$$

$$\mathbf{y}_1 = \mathbf{C}_1\mathbf{x}_1 + \mathbf{D}_1\mathbf{u}_1 \tag{9-16b}$$

be the dynamical-equation description of the system $S_1$. It is clear that

$$\hat{\mathbf{G}}_1(s) = \mathbf{C}_1(s\mathbf{I} - \mathbf{A}_1)^{-1}\mathbf{B}_1 + \mathbf{D}_1 \tag{9-17}$$

Note that the dynamical equation $FE^1$ is used only implicitly in the proof; its knowledge is not required in the application of the theorem. Now the assumption that $S_1$ is completely characterized by $\hat{\mathbf{G}}_1(s)$ implies that $FE^1$ is controllable and observable. Consequently, the characteristic polynomial, $\Delta_1(s)$, of $\hat{\mathbf{G}}_1(s)$ is equal to the characteristic polynomial of $\mathbf{A}_1$; that is,

$$\Delta_1(s) = \det (s\mathbf{I} - \mathbf{A}_1) \tag{9-18}$$

(Theorem 6-2). By the substitution $\mathbf{u}_1 = \mathbf{u} - \mathbf{y}_1$, $\mathbf{y} = \mathbf{y}_1$, the dynamical-equation description of the feedback system is

$$FE^f\text{:} \qquad \dot{\mathbf{x}}_1 = [\mathbf{A}_1 - \mathbf{B}_1(\mathbf{I} + \mathbf{D}_1)^{-1}\mathbf{C}_1]\mathbf{x}_1 + \mathbf{B}_1(\mathbf{I} + \mathbf{D}_1)^{-1}\mathbf{u} \tag{9-19a}$$

$$\mathbf{y} = (\mathbf{I} + \mathbf{D}_1)\mathbf{C}_1\mathbf{x}_1 + (\mathbf{I} + \mathbf{D}_1)^{-1}\mathbf{D}_1\mathbf{u} \tag{9-19b}$$

The dynamical equation $FE^f$ is controllable and observable following from Theorem 9-6. Hence we conclude that the feedback system is totally stable if and only if all the eigenvalues of $[\mathbf{A}_1 - \mathbf{B}_1(\mathbf{I} + \mathbf{D}_1)^{-1}\mathbf{C}_1]$ have negative real parts (Theorem 8-16). We show in the following that the set of the eigenvalues of $[\mathbf{A}_1 - \mathbf{B}_1(\mathbf{I} + \mathbf{D}_1)^{-1}\mathbf{C}_1]$ is equal to the set of the zeros of $\Delta_1(s) \det (\mathbf{I} + \hat{\mathbf{G}}(s))$. Consider

$$\begin{aligned} J \triangleq \det [s\mathbf{I} - \mathbf{A}_1 + \mathbf{B}_1(\mathbf{I} + \mathbf{D}_1)^{-1}\mathbf{C}_1] &= \det [(s\mathbf{I} - \mathbf{A}_1) + \mathbf{B}_1(\mathbf{I} + \mathbf{D}_1)^{-1}\mathbf{C}_1] \\ &= \det \{(s\mathbf{I} - \mathbf{A}_1)[\mathbf{I} + (s\mathbf{I} - \mathbf{A}_1)^{-1}\mathbf{B}_1(\mathbf{I} + \mathbf{D}_1)^{-1}\mathbf{C}_1]\} \end{aligned} \tag{9-20}$$

It is well known that $\det \mathbf{AB} = \det \mathbf{A} \det \mathbf{B}$; hence (9-20) becomes

$$J = \det (s\mathbf{I} - \mathbf{A}_1) \det [\mathbf{I} + (s\mathbf{I} - \mathbf{A}_1)^{-1}\mathbf{B}_1(\mathbf{I} + \mathbf{D}_1)^{-1}\mathbf{C}_1] \tag{9-21}$$

It is shown in Theorem 3-2 that $\det (\mathbf{I} + \hat{\mathbf{G}}_1(s)\hat{\mathbf{G}}_2(s)) = \det (\mathbf{I} + \hat{\mathbf{G}}_2(s)\hat{\mathbf{G}}_1(s))$ and, since $\Delta_1(s) = \det (s\mathbf{I} - \mathbf{A}_1)$, Equation (9-21) implies that

$$\begin{aligned} J &= \Delta_1(s) \det [\mathbf{I} + \mathbf{C}_1(s\mathbf{I} - \mathbf{A}_1)^{-1}\mathbf{B}_1(\mathbf{I} + \mathbf{D}_1)^{-1}] \\ &= \Delta_1(s) \det \{[(\mathbf{I} + \mathbf{D}_1) + \mathbf{C}_1(s\mathbf{I} - \mathbf{A}_1)^{-1}\mathbf{B}_1](\mathbf{I} + \mathbf{D}_1)^{-1}\} \\ &= \Delta_1(s) \det [\mathbf{I} + \hat{\mathbf{G}}_1(s)] \det (\mathbf{I} + \mathbf{D}_1)^{-1} \end{aligned} \tag{9-22}$$

Since $\det (\mathbf{I} + \mathbf{D}_1)^{-1}$ is a real number, we conclude from (9-22) that the set of the eigenvalues of $[\mathbf{A}_1 - \mathbf{B}_1(\mathbf{I} + \mathbf{D}_1)^{-1}\mathbf{C}_1]$ is equal to the set of the zeros of $\Delta_1(s) \det [\mathbf{I} + \hat{\mathbf{G}}_1(s)]$. Q.E.D.

### Example 4

Determine the T-stability of the feedback system in Figure 9-15, with

$$\hat{\mathbf{G}}_1(s) = \begin{bmatrix} \dfrac{-s}{s-1} & \dfrac{s}{s+1} \\ 1 & \dfrac{-2}{s+1} \end{bmatrix}$$

The characteristic polynomial of $\hat{\mathbf{G}}_1(s)$ is $\Delta_1(s) = (s-1)(s+1)$. It is easy to verify that $\det (\mathbf{I} + \hat{\mathbf{G}}_1(s)) = -1$. Hence the feedback system is totally stable if and only if all the zeros of $\Delta_1(s) \det (\mathbf{I} + \hat{\mathbf{G}}_1(s)) = -(s-1)(s+1)$ have negative real parts. This is not the case, hence the feedback system is not T-stable. ∎

It is of interest to give some interpretation of the zeros of the polynomial

$$\Delta_1(s) \det (\mathbf{I} + \hat{\mathbf{G}}_1(s)) = \Delta_1(s) \frac{E(s)}{F(s)}$$

where $E(s)$ and $F(s)$ have no common factor. The zeros of $E(s)$ are poles of the feedback system introduced by feedback, whereas the zeros of $\Delta_1(s)/F(s)$ are poles which are possessed by the open-loop system $S_1$ as well as by the feedback system. For example, in Example 3 we have $E(s) = -1$, $F(s) = 1$, $\Delta_1(s)/F(s) = (s+1)(s-1)$; hence no new pole is introduced in the feedback system $\hat{\mathbf{G}}_f(s)$; $\hat{\mathbf{G}}_1(s)$ and $\hat{\mathbf{G}}_f(s)$ have the same poles, $(s+1)$ and $(s-1)$.

Roughly speaking, the polynomial $\Delta_1(s)/F(s)$ takes care of the missing zeros of det $(\mathbf{I} + \hat{\mathbf{G}}_1(s))$. Therefore, one might suggest that if we do not cancel out any common factors in det $(\mathbf{I} + \hat{\mathbf{G}}_1(s))$, then the stability of the feedback system might be determinable from the zeros of det $(\mathbf{I} + \hat{\mathbf{G}}_1(s))$. This is not necessarily correct, as can be seen from the following example.

### Example 5

Consider the feedback system shown in Figure 9-15, with

$$\hat{\mathbf{G}}_1(s) = \begin{bmatrix} \dfrac{-s^2+s+1}{(s+1)(s-1)} & \dfrac{1}{s-1} \\ \dfrac{1}{(s+1)(s-1)} & \dfrac{1}{s-1} \end{bmatrix}$$

It is easy to verify that

$$\det(\mathbf{I} + \hat{\mathbf{G}}_1(s)) = \frac{s^2}{(s+1)(s-1)^2} - \frac{1}{(s+1)(s-1)^2} = \frac{(s+1)(s-1)}{(s+1)(s-1)^2}$$

which has a right-half-plane zero. However, by applying Theorem 9-10, we have

$$\Delta_1(s) = (s+1)(s-1) \qquad \text{and} \qquad \frac{E(s)}{F(s)} = \frac{1}{s-1}$$

and the feedback system is totally stable. ∎

### Corollary 9-10

Under the assumption of Theorem 9-10 and if, in addition, $\hat{\mathbf{G}}_1(s)$ has no closed right-half-plane poles, then the feedback system is T-stable if and only if the Nyquist plot of det $(\mathbf{I} + \hat{\mathbf{G}}_1(s))$ does not encircle or go through the origin, or equivalently, the Nyquist plot[6a] of [det $(\mathbf{I} + \hat{\mathbf{G}}_1(s)) - 1$] does not encircle or go through the point $(-1, 0)$. ∎

This follows directly from Theorem 9-10. Indeed if $\hat{\mathbf{G}}_1(s)$ has no closed right-half-plane poles, all the zeros of $\Delta_1(s)$, and consequently of $\Delta_1(s)/F(s)$ have negative real parts. This corollary holds even if $\hat{\mathbf{G}}_1(s)$ is not a rational-function matrix. The interested reader is referred to References [28] and [34].

---

[6a] We note that the Nyquist plot of [det $(\mathbf{I} + \hat{\mathbf{G}}_1(s)) - 1$] is equal to, following from Equation (9-15), the Nyquist plot of $\left(\sum_{i=1}^{p} \alpha_i(s)\right)$, where $\alpha_i(s)$ is the sum of all the principal minors of $\hat{\mathbf{G}}_1(s)$ of order $i$.

### Theorem 9-11

Consider the feedback system shown in Figure 9-16. It is assumed that $S_1$ and $S_2$ are completely characterized by $\hat{\mathbf{G}}_1(s)$ and $\hat{\mathbf{G}}_2(s)$, respectively, and that $\det(\mathbf{I} + \hat{\mathbf{G}}_1(\infty)\hat{\mathbf{G}}_2(\infty)) \neq 0$. Then the feedback system is totally stable (T-stable) if all the zeros of the polynomial $\Delta_1\Delta_2 \det(\mathbf{I} + \hat{\mathbf{G}}_1(s)\hat{\mathbf{G}}_2(s))$ have negative real parts, where $\Delta_1$ and $\Delta_2$ are the characteristic polynomials of $\hat{\mathbf{G}}_1(s)$ and $\hat{\mathbf{G}}_2(s)$, respectively. ∎

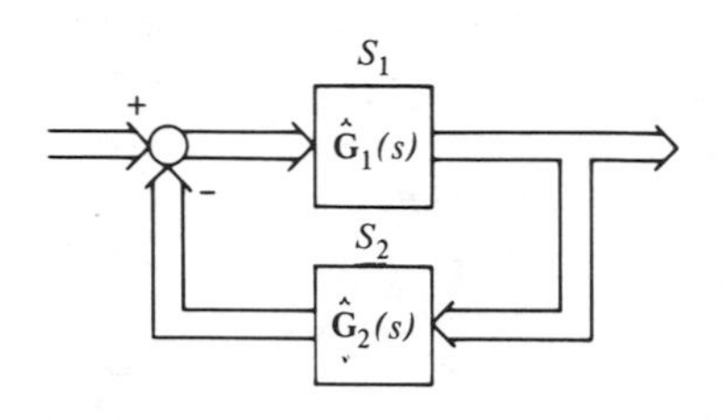


**Figure 9-16** A multivariable feedback system.

This theorem can be proved by using the dynamical-equation descriptions of $S_1$ and $S_2$ and using Corollary 8-16. The procedure of the proof is the same as that of Theorem 9-10. It involves showing that the characteristic polynomial $J(s)$ of the composite dynamical-equation description of the feedback system is equal to

$$J(s) = \Delta_1(s)\Delta_2(s) \det[\mathbf{I} + \hat{\mathbf{G}}_1(s)\hat{\mathbf{G}}_2(s)]/\det[\mathbf{I} + \hat{\mathbf{G}}_1(\infty)\hat{\mathbf{G}}_2(\infty)] \quad \textbf{(9-23)}$$

The proof of (9-23) is left as an exercise (Problem 9-10). In employing Theorem 9-11, it is not required to check the controllability and observability of the feedback system. The conditions insure not only the boundedness of the output but also of all the state variables. This theorem, however, gives only sufficient conditions. In order to state it as necessary and sufficient, the condition on the uncontrollable poles must be relaxed. This will not be discussed here.

We give one example to illustrate how to apply Theorem 9-11 to obtain the stability range of a multivariable feedback system.

### Example 6

Consider the multivariable feedback system shown in Figure 9-17. It can be written in the form of Figure 9-15 with

$$\hat{\mathbf{G}}_1(s) = \begin{bmatrix} \dfrac{k_1}{s-1} & k_2 \\ k_1 & \dfrac{k_2}{s+2} \end{bmatrix}$$

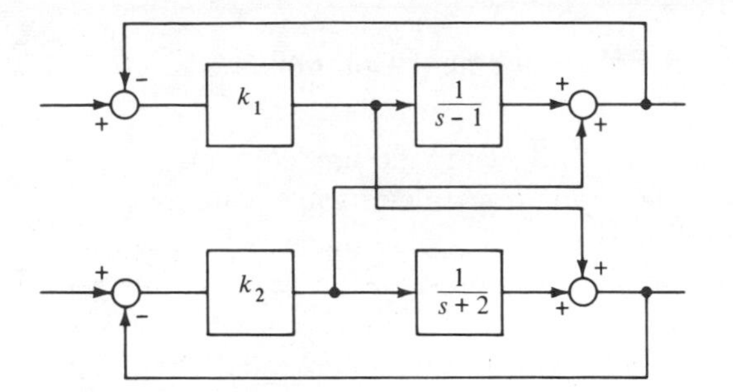


**Figure 9-17** A multivariable feedback system.

If $k_1 \neq 0$, $k_2 \neq 0$, the characteristic polynomial of $\hat{\mathbf{G}}_1(s)$ is $(s-1)(s+2)$ and

$$\det(\mathbf{I} + \hat{\mathbf{G}}_1(s)) = \frac{(1 - k_1k_2)s^2 + (k_1 + k_2 + 1 - k_1k_2)s + 3k_1k_2 - k_2 + 2k_1 - 2}{(s-1)(s+2)}$$

Hence if $k_1 \neq 0$ and $k_2 \neq 0$, the stability range of $k_1$ and $k_2$ is determined by the zeros of $\Delta_1(s) \det(\mathbf{I} + \hat{\mathbf{G}}_1(s))$ which is equal to

$$(1 - k_1k_2)s^2 + (k_1 + k_2 + 1 - k_1k_2)s + 3k_1k_2 - k_2 + 2k_1 - 2 = 0$$

For this simple case, we may apply the Routh–Hurwitz criterion to obtain the stability range. For the case in which either $k_1 = 0$ or $k_2 = 0$, the system reduces to a single-loop system and the stability range can be found easily. The stability range of the feedback system is shown in the shaded area of Figure 9-18. For a more complicated system, the stability range can still be easily obtained with the aid of a digital computer.

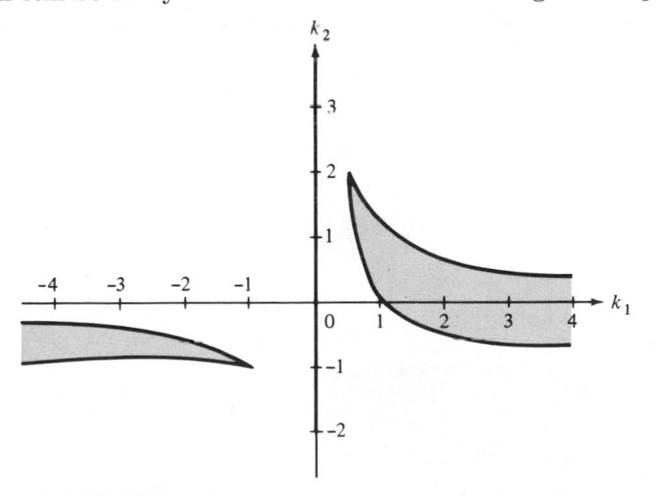


**Figure 9-18** The stability range of the system in Figure 9-17.

## 9-5 Design of Pole-Placement Compensators

In this section we shall study design of compensators to achieve pole placement for one-input multiple-output systems that are describable by proper rational matrices. By pole placement we mean that, given a system and a set of desired poles, we design a compensator such that the resulting system has the set of desired poles. The compensator designed is required—for realizability by operational amplifier circuits—to have a proper rational transfer-function matrix. It is also required, for economic reasons, that the degree of the transfer-function matrix of the compensator be as small as possible.

The systems studied are assumed to be describable by rational matrices; therefore this design problem can also be studied by means of dynamical-equation descriptions. If an irreducible (controllable and observable) dynamical-equation realization is found for a system with a proper rational matrix $\hat{\mathbf{G}}(s)$, then the results in Section 7-3 and 7-4 can be applied to accomplish this design task. We recall that if an $n$-dimensional, linear, time-invariant dynamical equation is controllable, by introducing the state feedback $\mathbf{u} = \mathbf{v} + \mathbf{Kx}$, where $\mathbf{K}$ is the feedback gain matrix, we can arbitrarily assign the $n$ eigenvalues of the resulting equation (Theorem 7-6). If an $n$-dimensional, linear, time-invariant dynamical equation is observable, a set of estimators with arbitrary eigenvalues can be constructed to generate all the state variables (Theorem 7-9). Now because of the separation property (see Section 7-4), if the feedback gain matrix $\mathbf{K}$ is applied to the output of the estimators, the resulting system will have a set of desired poles. The state estimators, together with the gain matrix $\mathbf{K}$, will be called a *compensator*, as shown in Figure 9-19. We see that in

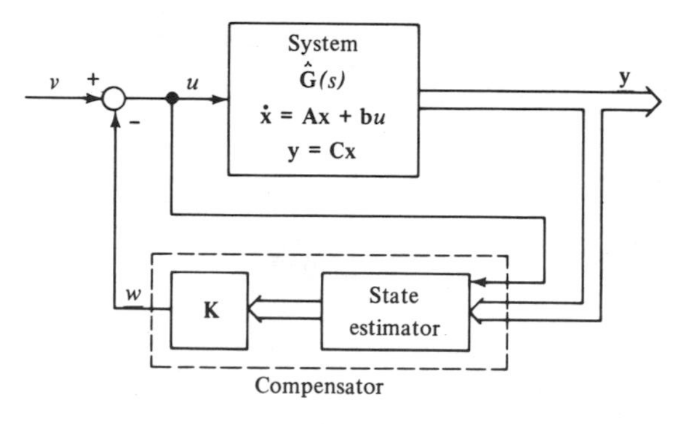


**Figure 9-19** Design of compensator for a multivariable system.

designing a compensator to achieve pole placement, it is not necessary to generate all the state variables; what is needed is to generate $\mathbf{Kx}$. Hence from the discussion following Theorem 7-9, the dimension of the compensator can be designed of dimension $\mu - 1$, where $\mu$ is the observability index of an irreducible realization of $\hat{\mathbf{G}}(s)$.

Although this design can be achieved by using the dynamical-equation description, we would like to study it directly from the transfer-function matrix. The reason for doing this is twofold: first, to develop a design scheme that is simpler conceptually and computationally than the one by using dynamical equations; second, to establish some links between the dynamical equation and the transfer-function matrix. However, the design procedure developed in this section is applicable only to single-input systems. On the other hand, the state-variable approach is applicable to multiple-input multiple-output systems and, moreover, can be extended to time-varying dynamical equations.

This section will be divided into three subsections. In the first subsection we shall establish two algebraic theorems. In the second subsection we apply these theorems to the design of compensators to achieve pole placement for single-input two-output systems. Two examples are given to illustrate how the design scheme may be applied. In the last subsection we give some extension and discussion.

**Two algebraic theorems.** In this subsection we shall introduce two algebraic theorems. These theorems will form the basis for the design of compensators to achieve pole placement for multivariable systems.

### Theorem 9-12

Given two arbitrary polynomials $P_1(s)$ and $P_2(s)$ of degrees $n_1$ and $n_2$, respectively. For any polynomial $\Delta(s)$ of degree $n_1 + n_2 - 1$, there exist two polynomials $C_1(s)$ and $C_2(s)$, respectively, of degree $n_2 - 1$ and $n_1 - 1$ or less such that

$$\Delta(s) = P_1(s)C_1(s) + P_2(s)C_2(s) \tag{9-24}$$

if and only if $P_1(s)$ and $P_2(s)$ have no nontrivial common factor.

### Proof

This theorem will be proved for $n_1 = 4$ and $n_2 = 2$; the general case can be similarly proved. Let

$$P_1(s) = p_{14}s^4 + p_{13}s^3 + p_{12}s^2 + p_{11}s + p_{10} \tag{9-25}$$
$$P_2(s) = p_{22}s^2 + p_{21}s + p_{20} \tag{9-26}$$

From the statement of the theorem, $\Delta(s)$, $C_1(s)$ and $C_2(s)$ are, respectively, of degree 5, 1, and 3; hence they may be assumed to be of the forms

$$\Delta(s) = d_5s^5 + d_4s^4 + d_3s^3 + d_2s^2 + d_1s + d_0 \tag{9-27}$$
$$C_1(s) = c_{11}s + c_{10} \tag{9-28}$$
$$C_2(s) = c_{23}s^3 + c_{22}s^2 + c_{21}s + c_{20} \tag{9-29}$$

Equating the coefficients of the same power of (9-24) yields

$$\mathbf{A}\mathbf{c} \triangleq \left[\begin{array}{cc:cccc} p_{14} & 0 & p_{22} & 0 & 0 & 0 \\ p_{13} & p_{14} & p_{21} & p_{22} & 0 & 0 \\ p_{12} & p_{13} & p_{20} & p_{21} & p_{22} & 0 \\ p_{11} & p_{12} & 0 & p_{20} & p_{21} & p_{22} \\ p_{10} & p_{11} & 0 & 0 & p_{20} & p_{21} \\ 0 & p_{10} & 0 & 0 & 0 & p_{20} \end{array}\right] \begin{bmatrix} c_{11} \\ c_{10} \\ c_{23} \\ c_{22} \\ c_{21} \\ c_{20} \end{bmatrix} = \begin{bmatrix} d_5 \\ d_4 \\ d_3 \\ d_2 \\ d_1 \\ d_0 \end{bmatrix} \tag{9-30}$$

(the first $n_2$ columns and the last $n_1$ columns are marked by braces: "$n_2$ columns", "$n_1$ columns")

Hence the problem reduces to the following: Given a set of arbitrary $d_i$'s under what conditions on $P_1(s)$ and $P_2(s)$ do there exist $c_{ij}$'s satisfying (9-30). It is clear that for any $d_i$, Equation (9-30) has a solution if and only if the matrix $\mathbf{A}$ is nonsingular. Hence if we show that the matrix $\mathbf{A}$ is nonsingular if and only if $P_1(s)$ and $P_2(s)$ have no common factors,[6b] then the theorem is proved.

Consider the following division algorithm

$$P_1(s) = P_2(s)q_1(s) + r_1(s) \qquad \delta r_1(s) < \delta P_2(s) \tag{9-31}$$

$$P_2(s) = r_1(s)q_2(s) + r_2(s) \qquad \delta r_2(s) < \delta r_1(s) \tag{9-32}$$

In the present case, $\delta P_1(s) = 4$, $\delta P_2(s) = 2$, and the algorithm may stop here; otherwise, the algorithm should proceed until the last remainder $r_i(s)$ is zero or a nonzero constant. It is well known that $P_1(s)$ and $P_2(s)$ have no common factor if and only if the last remainder is a nonzero constant.

In the following we shall show that the division algorithm can be carried out in the matrix $\mathbf{A}$ by applying a sequence of elementary transformations.[7] Now if the third and fourth columns of $\mathbf{A}$ are multiplied by $-p_{14}/p_{22}$ and added, respectively, to the first and second columns of $\mathbf{A}$, the matrix becomes

$$\mathbf{A}^{(1)} = \begin{bmatrix} 0 & 0 & p_{22} & 0 & 0 & 0 \\ p'_{13} & 0 & p_{21} & p_{22} & 0 & 0 \\ p'_{12} & p'_{13} & p_{20} & p_{21} & p_{22} & 0 \\ p_{11} & p'_{12} & 0 & p_{20} & p_{21} & p_{22} \\ p_{10} & p_{11} & 0 & 0 & p_{20} & p_{21} \\ 0 & p_{10} & 0 & 0 & 0 & p_{20} \end{bmatrix}$$

---

[6b] This statement can be deduced from a theorem in modern algebra, for example, from Theorem XVI, p. 131, *Modern Algebra*, by F. Ayres, Jr., Schaum Publishing Co., New York, 1965. However, in order to be self-contained, we shall prove it directly by using the division algorithm.

[7] Elementary transformations are the transformations that (1) interchange two rows or columns, (2) multiply a row or column by a nonzero constant, or (3) add one row or column to another. These transformations can be achieved by multiplications of nonsingular matrices of special forms. See References [39] and [86].

By adding the fourth and fifth columns (after proper multiplications) to the first and second columns, $\mathbf{A}^{(1)}$ may become

$$\mathbf{A}^{(2)} = \begin{bmatrix} 0 & 0 & p_{22} & 0 & 0 & 0 \\ 0 & 0 & p_{21} & p_{22} & 0 & 0 \\ 0 & 0 & p_{20} & p_{21} & p_{22} & 0 \\ p'_{11} & 0 & 0 & p_{20} & p_{21} & p_{22} \\ p'_{10} & p'_{11} & 0 & 0 & p_{20} & p_{21} \\ 0 & p'_{10} & 0 & 0 & 0 & p_{20} \end{bmatrix}$$

Note that $p'_{11}s + p'_{10}$ is the remainder $r_1(s)$ defined in (9-31). Now if $p'_{11} = 0$ and $p'_{10} \neq 0$, then $P_1(s)$ and $P_2(s)$ have no common factor. In this case the determinant of $\mathbf{A}^{(2)}$ can be shown to be $p_{22}{}^4(p'_{10})^2$. Consequently, from the fact that the rank of a matrix is invariant under elementary transformations (Theorem 2-7), we conclude that the original matrix $\mathbf{A}$ is nonsingular. If $p'_{11} = 0$ and $p'_{10} = 0$, then $P_1(s)$ and $P_2(s)$ have common factor and $\mathbf{A}$ is clearly singular. Now if $p'_{11} \neq 0$ and if we add the first and second columns (after proper multiplications) to the last column, $\mathbf{A}^{(2)}$ becomes

$$\mathbf{A}^{(3)} = \begin{bmatrix} 0 & 0 & p_{22} & 0 & 0 & 0 \\ 0 & 0 & p_{21} & p_{22} & 0 & 0 \\ 0 & 0 & p_{20} & p_{21} & p_{22} & 0 \\ p'_{11} & 0 & 0 & p_{20} & p_{21} & 0 \\ p'_{10} & p'_{11} & 0 & 0 & p_{20} & 0 \\ 0 & p'_{10} & 0 & 0 & 0 & p'_{20} \end{bmatrix}$$

If $p'_{20} = 0$, $P_1(s)$ and $P_2(s)$ have common factor and $\mathbf{A}$ is singular. If $p'_{20} \neq 0$, $P_1(s)$ and $P_2(s)$ have no common factor and $\mathbf{A}$ is nonsingular. Hence we conclude that the matrix $\mathbf{A}$ is nonsingular if and only if $P_1(s)$ and $P_2(s)$ have no common factor. Q.E.D.

Observe that if $P_1(s)$ and $P_2(s)$ have no common factor, then all the rows of a matrix of the form

$$\begin{array}{c} (n_1 + n_2 + m) \\ \text{rows} \end{array} \left[\begin{array}{ccccc|ccccc} p_{14} & 0 & \cdots & 0 & 0 & p_{22} & 0 & \cdots & 0 & 0 \\ p_{13} & p_{14} & \cdots & 0 & 0 & p_{21} & p_{22} & \cdots & 0 & 0 \\ p_{12} & p_{13} & \cdots & 0 & 0 & p_{20} & p_{21} & \cdots & 0 & 0 \\ p_{11} & p_{12} & \cdots & & \cdot & 0 & p_{20} & \cdots & 0 & 0 \\ p_{10} & p_{11} & \cdots & & \cdot & 0 & 0 & \cdots & \cdot & \cdot \\ 0 & p_{10} & \cdots & & \cdot & 0 & 0 & \cdots & \cdot & \cdot \\ \cdot & \cdot & & & \cdot & \cdot & \cdot & & & \cdot \\ \cdot & \cdot & & & \cdot & \cdot & \cdot & & & \cdot \\ \cdot & \cdot & & & \cdot & \cdot & \cdot & & & \cdot \\ 0 & 0 & \cdots & p_{10} & p_{11} & 0 & 0 & \cdots & p_{20} & p_{21} \\ 0 & 0 & \cdots & 0 & p_{10} & 0 & 0 & \cdots & 0 & p_{20} \end{array}\right]$$

$$\underbrace{\hphantom{xxxxxxxxxxxx}}_{(n_2 + m) \text{ columns}} \quad \underbrace{\hphantom{xxxxxxxxxxxx}}_{(n_1 + m) \text{ columns, with } m \geq 0} \tag{9-33}$$

will be linearly independent. This assertion can be proved by using the argument given in the proof of Theorem 9-12. Because of this fact, Theorem 9-12 can be extended as follows.

### Theorem 9-13

Given two polynomials $P_1(s)$ and $P_2(s)$ of degree $n_1$ and $n_2$, respectively. It is assumed, without loss of generality, that $n_1 \geq n_2$. Let $\Delta(s)$, $C_1(s)$, and $C_2(s)$ be polynomials, respectively, of degree $n_1 + m$, $m$, and $l + m$, where $0 \leq l \leq n_1 - n_2$. If $P_1(s)$ and $P_2(s)$ have no common factor and if

$$n_1 - l - 1 \leq m \tag{9-34}$$

then for any given $\Delta(s)$, there exist $C_1(s)$ and $C_2(s)$ which satisfy the equation

$$\Delta(s) = P_1(s)C_1(s) + P_2(s)C_2(s) \tag{9-35}$$

### Proof

If the polynomials in (9-35) are chosen to be of the degrees specified, then by equating the coefficients of the same powers of (9-35), a set of $n_1 + m + 1$ linear algebraic equations can be obtained. The number of unknowns in $C_1(s)$ and $C_2(s)$ is $(m + 1) + (l + m + 1)$. Now in order to have a solution in a set of algebraic equations, it is necessary to have the number of unknowns greater than the number of equations, that is

$$n_1 + m + 1 \leq m + 1 + l + m + 1$$

or

$$n_1 - l - 1 \leq m$$

This proves the necessity of inequality (9-34). Let

$$P_i(s) = \sum_{j=0}^{n_i} p_{ij}s^j \qquad C_1(s) = \sum_{i=0}^{m} c_{1i}s^i$$

$$C_2(s) = \sum_{i=0}^{m+l} c_{2i}s^i \qquad \Delta(s) = \sum_{i=0}^{n_1+m} d_is^i$$

Then equating the coefficients of the same power of (9-35) yields an equation of the form

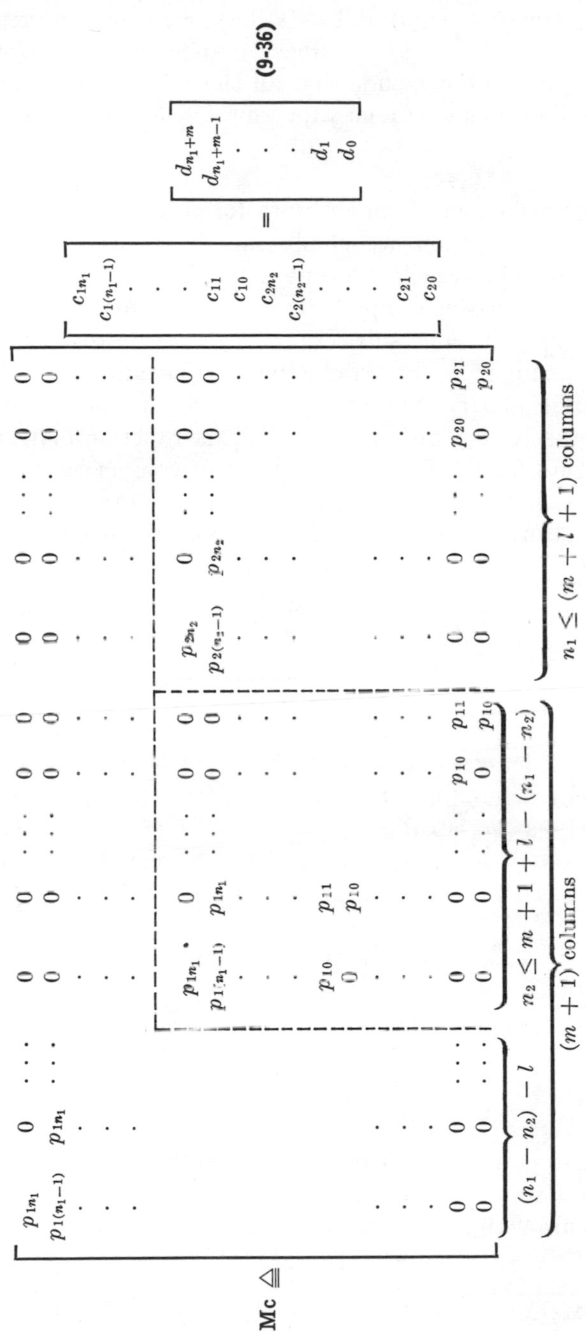
Mc ≜
(9-36)
(n1 − n2) − l
n2 ≤ m + 1 + l − (n1 − n2)
(m + 1) columns
n1 ≤ (m + l + 1) columns

We note that in the matrix **M** defined in (9-36) there is a submatrix of the form (9-33). Hence, if $P_1(s)$ and $P_2(s)$ have no common factor, then all the rows of the submatrix are linearly independent. Consequently, because of $p_{1n_1} \neq 0$, we conclude that all the rows of **M** defined in (9-36) are linearly independent. Hence, for any $d_i$'s, there exist $c_{ij}$'s satisfying (9-36). Q.E.D.

**Design of pole-placement compensators for single-input two-output systems.** In this subsection we shall employ Theorem 9-13 to design compensators to achieve pole placement for single-input two-output systems. Extension to single-input three-or-more-output systems will be discussed in the following subsection. We recall that every rational function are assumed to be irreducible. It is also assumed that the transfer-function matrix of a given system is a proper rational matrix.

Consider the system with a 2 × 1 transfer-function matrix $\hat{\mathbf{G}}(s)$ illustrated in Figure 9-19. The configuration of compensators is chosen as shown.[8] This configuration of compensators is motivated by the design of state estimator. First we write the transfer-function matrix of the given system as

$$\hat{\mathbf{G}}(s) \triangleq \begin{bmatrix} \hat{g}_1(s) \\ \hat{g}_2(s) \end{bmatrix} \triangleq \begin{bmatrix} \dfrac{N_1'(s)}{D_1(s)} \\ \dfrac{N_2'(s)}{D_2(s)} \end{bmatrix} = \frac{1}{\Delta(s)} \begin{bmatrix} N_1(s) \\ N_2(s) \end{bmatrix} \qquad \textbf{(9-37)}$$

where $\Delta(s)$ is the least common denominator of $\hat{g}_1(s)$ and $\hat{g}_2(s)$ and, in this case, is the characteristic polynomial of $\hat{\mathbf{G}}(s)$; see Definition 6-1. The degree of $\Delta(s)$, say $n$, is called the degree of $\hat{\mathbf{G}}(s)$. That $\hat{\mathbf{G}}(s)$ is proper implies $\delta N_1 \triangleq n_1 \leq n$ and $\delta N_2 \triangleq n_2 \leq n$. As shown in Figure 9-19, the compensator has three inputs and one output. Let

$$\hat{\mathbf{G}}_c(s) = \frac{1}{\Delta_c(s)} [C_1(s) \quad C_2(s) \quad C_0(s)] \qquad \textbf{(9-38)}$$

be the transfer-function matrix of the compensator, where

$$\frac{1}{\Delta_c(s)} [C_1(s) \quad C_2(s)]$$

is the transfer-function matrix from $\mathbf{y}$ to $w$, and $C_0(s)/\Delta_c(s)$ is the transfer function from $u$ to $w$. It can be shown that the overall transfer-function matrix of Figure 9-19 is (see Problem 9-15)

$$\hat{\mathbf{G}}_f = \frac{\Delta(s)\Delta_c(s)}{\Delta(s)\Delta_c(s) + N_1(s)C_1(s) + N_2(s)C_2(s) + \Delta(s)C_0(s)} \hat{\mathbf{G}}(s) \qquad \textbf{(9-39)}$$

[8] The subsequent development is also applicable to other configurations of compensators.

Consequently, the product of all the poles of the overall system, $\Delta_f$, is equal to

$$\begin{aligned}\Delta_f(s) &= \Delta(s)(\Delta_c(s) + C_0(s)) + N_1(s)C_1(s) + N_2(s)C_2(s) \\ &= \Delta(s)\Delta_c(s)\left(1 + \hat{\mathbf{G}}_c(s)\begin{bmatrix}\hat{\mathbf{G}}(s)\\ 1\end{bmatrix}\right)\end{aligned} \tag{9-40}$$

Hence the problem of pole placement reduces to the following: Given $\Delta_f$, $\Delta$, $N_1$, and $N_2$, find compensators $\Delta_c$, $C_1$, $C_2$, and $C_0$ that satisfy (9-40); or equivalently, given $\hat{\mathbf{G}}(s)$, find $\hat{\mathbf{G}}_c(s)$ that satisfies (9-40). For realizability, the compensator $\hat{\mathbf{G}}_c(s)$ is required to be a proper rational matrix; that is, $\delta C_i(s) \leq \delta\Delta_c(s)$, for $i = 0, 1, 2$. For economical reasons, it is desirable to have $\delta\Delta_c$ as small as possible.

Let $N_{12}(s)$ be the greatest common divisor of $N_1(s)$ and $N_2(s)$. Define $\bar{N}_1 \triangleq N_1/N_{12}$, $\bar{N}_2 \triangleq N_2/N_{12}$. Clearly, $\bar{N}_1$ and $\bar{N}_2$ have no common factor. Let $\bar{\Delta}_c(s) \triangleq \Delta_c(s) + C_0(s)$. Then (9-40) can be written as

$$\Delta_f(s) = \bar{\Delta}_c(s)\Delta(s) + [\bar{N}_1(s)C_1(s) + \bar{N}_2(s)C_2(s)]N_{12}(s) \tag{9-41}$$

Since $\hat{g}_1(s)$ and $\hat{g}_2(s)$ are assumed to be irreducible (that is, there is no common factor between $N_i'$ and $D_i$, for $i = 1, 2$), we can show that the three polynomials $\Delta(s)$, $N_1(s)$ and $N_2(s)$ defined in (9-37) have no common factor (Problem 9-16). Consequently, the two polynomials $\Delta(s)$ and $N_{12}(s)$ have no common factor. Let the degrees of $\bar{\Delta}_c(s)$, $C_1(s)$, and $C_2(s)$ be $m$ and let the degree of $N_{12}$ be $n_{12}$. First consider $\bar{N}_1(s)C_1(s) + \bar{N}_2(s)C_2(s)$. Since $\bar{N}_1(s)$ and $\bar{N}_2(s)$ have no common factor, and since $C_1(s)$ and $C_2(s)$ are chosen of degree $m$, Theorem 9-13 implies that if

$$m \geq \max(n_1 - n_{12}, n_2 - n_{12}) - 1 \tag{9-42}$$

then given any polynomial, $F(s)$, of degree $\max(n_1 - n_{12}, n_2 - n_{12}) + m$, $C_1(s)$ and $C_2(s)$ can be found such that $F(s) = \bar{N}_1(s)C_1(s) + \bar{N}_2C_2(s)$. Next consider $\Delta_f(s) = \bar{\Delta}_c(s)\Delta(s) + F(s)N_{12}(s)$. Since $\Delta(s)$ and $N_{12}(s)$ have no common factor, and since $F(s)$ is of degree $\max(n_1 - n_{12}, n_2 - n_{12}) + m$, Theorem 9-13 implies that if

$$m \geq n - \max(n_1 - n_{12}, n_2 - n_{12}) - 1 \tag{9-43}$$

then for any $\Delta_f(s)$ of degree $n + m$, $\bar{\Delta}_c(s)$ and $F(s)$ can be found such that $\Delta_f(s) = \bar{\Delta}_c(s)\Delta(s) + F(s)N_{12}(s)$. Since the degree of the compensator is required to satisfy (9-42) and (9-43), and since it is desirable to have it as small as possible, the degree of the compensator $m$ will be chosen as

$$m = \max[\max(n_1 - n_{12}, n_2 - n_{12}) - 1, n - \max(n_1 - n_{12}, n_2 - n_{12}) - 1] \tag{9-44}$$

After $m$ is chosen, the design can be easily carried out. We recapitulate the design procedure as follows:

**1.** Given

$$\hat{\mathbf{G}}(s) = \begin{bmatrix} \dfrac{N_1'(s)}{D_1(s)} \\ \dfrac{N_2'(s)}{D_2(s)} \end{bmatrix}$$

Compute

$$\hat{\mathbf{G}}(s) = \frac{1}{\Delta(s)} \begin{bmatrix} N_1(s) \\ N_2(s) \end{bmatrix}$$

and compute $N_{12}(s)$, the greatest common divisor of $N_1'$ and $N_2'$ [or equivalently, of $N_1(s)$ and $N_2(s)$]. Let $n \triangleq \delta\Delta(s)$, $n_{12} \triangleq \delta N_{12}(s)$, $n_1 \triangleq \delta N_1(s)$, $n_2 \triangleq \delta N_2(s)$.

**2.** Compute

$$m = \max\,[\max\,(n_1 - n_{12}, n_2 - n_{12}) - 1, n - \max\,(n_1 - n_{12}, n_2 - n_{12}) - 1]$$

**3.** Choose a set of $n + m$ desired poles, or equivalently, a polynomial, $\Delta_f(s)$, of degree $n + m$.

**4.** Let $\bar{\Delta}_c(s)$, $F(s)$ be unknown polynomials, respectively, of degree $m$ and $\max\,(n_1 - n_{12}, n_2 - n_{12}) + m$. Solve for $\bar{\Delta}_c(s)$ and $F(s)$ from $\Delta_f(s) = \bar{\Delta}_c(s)\Delta(s) + F(s)N_{12}(s)$ by using Equation (9-36).

**5.** Let $C_1(s)$ and $C_2(s)$ be unknown polynomials of degree $m$. Solve for $C_1(s)$ and $C_2(s)$ from $F(s) = \bar{N}_1(s)C_1(s) + \bar{N}_2(s)C_2(s)$ by using (9-36), where $\bar{N}_1(s) = N_1(s)/N_{12}(s)$, $\bar{N}_2(s) = N_2(s)/N_{12}(s)$.

**6.** The polynomial $\Delta_c(s)$ may be chosen arbitrarily. After $\Delta_c(s)$ is chosen, compute $C_0(s) = \bar{\Delta}_c(s) - \Delta_c(s)$. Then the compensator is

$$\hat{\mathbf{G}}_c(s) = \frac{1}{\Delta_c(s)} [C_1(s) \quad C_2(s) \quad C_0(s)]$$

Example 1

Consider

$$\hat{\mathbf{G}}(s) = \begin{bmatrix} \dfrac{(s+3)(s-2)}{(s+1)^2(s-1)} \\ \dfrac{s+3}{s(s-1)(s+2)} \end{bmatrix} = \frac{1}{s(s-1)(s+1)^2(s+2)} \begin{bmatrix} s(s+2)(s+3)(s-2) \\ (s+1)^2(s+3) \end{bmatrix}$$

It is clear that $n = 5$, $N_{12}(s) = s + 3$, $n_{12} = 1$, $n_1 = 4$, $n_2 = 3$. The degree of the compensator is $m = \max[3 - 1, 5 - 3 - 1] = 2$. Let

$$\bar{\Delta}_c(s) = \sum_{i=0}^{2} e_i s^i \qquad \text{and} \qquad F(s) = \sum_{i=0}^{5} f_i s^i$$

Since $\Delta(s) = s^5 + 3s^4 + s^3 - 3s^2 - 2s$, and $N_{12}(s) = s + 3$, $f_i$ and $e_i$ can be computed, by the use of (9-36), from

$$\left[\begin{array}{ccc|cccccc} 1 & 0 & 0 & 0 & 0 & 0 & 0 & 0 & 0 \\ 3 & 1 & 0 & 1 & 0 & 0 & 0 & 0 & 0 \\ 1 & 3 & 1 & 3 & 1 & 0 & 0 & 0 & 0 \\ -3 & 1 & 3 & 0 & 3 & 1 & 0 & 0 & 0 \\ -2 & -3 & 1 & 0 & 0 & 3 & 1 & 0 & 0 \\ 0 & -2 & -3 & 0 & 0 & 0 & 3 & 1 & 0 \\ 0 & 0 & -2 & 0 & 0 & 0 & 0 & 3 & 1 \\ 0 & 0 & 0 & 0 & 0 & 0 & 0 & 0 & 3 \end{array}\right] \begin{bmatrix} e_2 \\ e_1 \\ e_0 \\ f_5 \\ f_4 \\ f_3 \\ f_2 \\ f_1 \\ f_0 \end{bmatrix} = \begin{bmatrix} d_7 \\ d_6 \\ d_5 \\ d_4 \\ d_3 \\ d_2 \\ d_1 \\ d_0 \end{bmatrix} \tag{9-45}$$

Let

$$C_i(s) = \sum_{j=0}^{2} c_{ij} s^j \qquad i = 1, 2$$

Since $\bar{N}_1(s) = s(s+2)(s-2) = s^3 - 4s$, $\bar{N}_2(s) = (s+1)^2 = s^2 + 2s + 1$, the $c_{ij}$'s can be computed from

$$\left[\begin{array}{ccc|ccc} 1 & 0 & 0 & 0 & 0 & 0 \\ 0 & 1 & 0 & 1 & 0 & 0 \\ -4 & 0 & 1 & 2 & 1 & 0 \\ 0 & -4 & 0 & 1 & 2 & 1 \\ 0 & 0 & -4 & 0 & 1 & 2 \\ 0 & 0 & 0 & 0 & 0 & 1 \end{array}\right] \begin{bmatrix} c_{12} \\ c_{11} \\ c_{10} \\ c_{22} \\ c_{21} \\ c_{20} \end{bmatrix} = \begin{bmatrix} f_5 \\ f_4 \\ f_3 \\ f_2 \\ f_1 \\ f_0 \end{bmatrix} \tag{9-46}$$

The design can be achieved by solving these two sets of algebraic equations. As an example, if $\Delta_f(s)$ is chosen as

$$s^7 + 7s^6 + 22s^5 + 40s^4 + 45s^3 + 31s^2 + 12s + 2$$

or, equivalently, if the desired poles are chosen as $-1 + i$, $-1 - i$, and $-1$ with multiplicity 5, then solving (9-45), by taking $f_5 = 0$, yields $e_2 = 1$, $e_1 = 4$, $e_0 = 6.34$, $f_5 = 0$, $f_4 = 2.66$, $f_3 = 12$, $f_2 = 16.66$, $f_1 = 8$, and $f_0 = 0.66$. Substituting the $f_i$'s into (9-46) and solving for the $c_{ij}$'s, we obtain $c_{10} = 0$, $c_{11} = 0$, $c_{22} = 2.65$, $c_{21} = 6.68$, $c_{20} = 0.66$. If the poles of the compensator are chosen as $-2$ and $-3$, then

$$\begin{aligned} \Delta_c(s) &= s^2 + 5s + 6 \\ C_0(s) &= \bar{\Delta}_c - \Delta_c = s^2 + 4s + 6.34 - s^2 - 5s - 6 \\ &= -s + 0.34 \end{aligned}$$

Hence the compensator is

$$\hat{\mathbf{G}}_c(s) = \frac{1}{s^2 + 5s + 6}[0 \quad 2.65s^2 + 6.68s + 0.66 \quad -s + 0.34] \quad \blacksquare$$

### Example 2

We study the design problem in Section 7-5 directly from the transfer function. For convenience, the problem is restated here. Given a system with the transfer function

$$\hat{g}(s) = \frac{N(s)}{D(s)} = \frac{1}{s(s+1)}$$

design a compensator such that the resulting system has poles $-1 + i$ and $-1 - i$. In order for the result to be compared with that in Section 7-5, the configuration of compensator is chosen as shown in Figure 9-20.

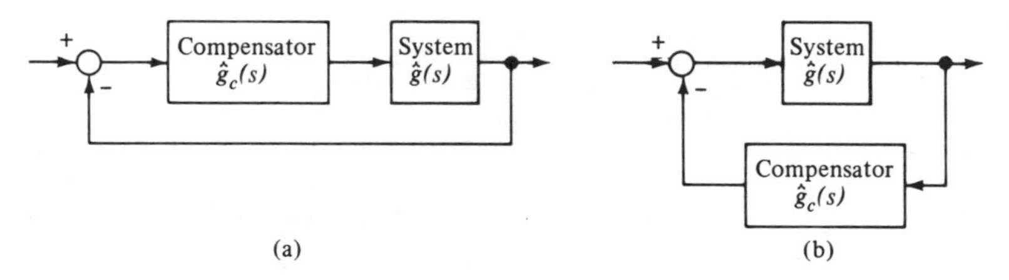


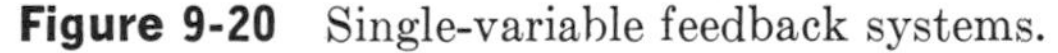

**Figure 9-20** Single-variable feedback systems.

Let $\hat{g}_c(s) = N_c(s)/D_c(s)$ and $\delta N_c = \delta D_c = m$ be the transfer function of the compensator. It is clear that the poles of the feedback system is given by

$$\Delta_f(s) = D(s)D_c(s) + N(s)N_c(s) \tag{9-47}$$

From (9-34), we know that the degree of the compensator $m$ be at least 1. Let $m = 1$ and let $N_c(s) = c_1 s + c_0$, $D_c(s) = a_1 s + a_0$. The desired poles of the feedback system are chosen as $-1 + i$, $-1 - i$, and $-2$. Note that the pole $-2$ is added because the compensator introduces one additional degree on $\Delta_f(s)$. Hence we have

$$\Delta_f(s) = (s + 1 + i)(s + 1 - i)(s + 2) = s^3 + 4s^2 + 6s + 4$$

Equating the coefficients of the same power of (9-47), we obtain an equation of the form of (9-36) as follows:

$$\left[\begin{array}{cc:cc} 0 & 0 & 1 & 0 \\ 0 & 0 & 1 & 1 \\ 1 & 0 & 0 & 1 \\ 0 & 1 & 0 & 0 \end{array}\right]\begin{bmatrix} c_1 \\ c_0 \\ a_1 \\ a_0 \end{bmatrix} = \begin{bmatrix} 1 \\ 4 \\ 6 \\ 4 \end{bmatrix} \tag{9-48}$$

The solution of (9-48) can be easily found as $c_1 = 3$, $c_0 = 4$, $a_1 = 1$, and $a_0 = 3$. Hence the required compensator is

$$\hat{g}_c(s) = \frac{3s + 4}{s + 3}$$

which is the same as the one obtained in Section 7-5.

**Discussion and extension.** At this point we offer a number of comments concerning the design introduced in the previous subsection.

**1.** Compared with the state-variable approach, the design of compensators for achieving pole placement is simpler by using transfer functions, not only because the concepts of controllability and observability are not required, but also because the amount of computation required seems much less. However the transfer-function approach is, at present, applicable only to linear, time-invariant single-input systems and a small class of linear, time-invariant multivariable systems; see Reference [19]. Whereas, the state-variable approach is applicable to any linear, time-invariant multivariable systems; it can also be extended to instantaneously controllable and observable linear, time-varying multivariable systems.
**2.** How we choose the desired poles for a system is not discussed. This belongs to the domain of control theory.
**3.** The design of compensator is not unique. We see from Equation (9-45) that the number of unknowns is one more than the number of equations; hence there are many different solutions of (9-45). Consequently, there are many different compensators to achieve the same pole placement.
**4.** In the previous subsection, the equation

$$\Delta_f(s) = \bar{\Delta}_c(s)\Delta(s) + N_1(s)C_1(s) + N_2(s)C_2(s)$$

is grouped as

$$\Delta_f(s) = \bar{\Delta}_c(s)\Delta(s) + (\bar{N}_1(s)C_1(s) + \bar{N}_2(s)C_2(s))N_{12}(s)$$

By a different grouping such as

$$\Delta_f(s) = (\bar{\Delta}_c(s)\bar{\Delta}(s) + \bar{N}_1(s)C_1(s))N_{1\Delta}(s) + N_2(s)C_2(s)$$

where $N_{1\Delta}(s)$ is the greatest common divisor of $N_1$ and $\Delta(s)$, a different design scheme can be obtained. It is found by many examples that the degree of the compensator by using this scheme is the same as the $m$ defined in (9-44).
**5.** The smallest possible $m$ defined in (9-44) occurs when $\max (n_1 - n_{12}, n_2 - n_{12}) - 1 = n - \max (n_1 - n_{12}, n_2 - n_{12}) - 1$, or $\max (n_1 - n_{12}, n_2 - n_{12}) = [n/2]$ where $[\cdot]$ denotes the integer

greater than or equal to $[n/2]$. Hence in any case, $m$ is not less than $[n/2] - 1$.

**6.** In using the configuration of compensators shown in Figure 9-19, the poles of the compensator $\hat{\mathbf{G}}_c(s)$ [the roots of $\Delta_c(s)$] can be arbitrarily chosen. This will not be the case if other configuration of compensators is used.

**7.** We essentially proved in the previous subsection the combination of Theorems 7-6 and 7-9 for the class of single-input, linear, time-invariant systems. That is, if every element of a $q \times 1$ proper rational matrix is irreducible, then a compensator with arbitrary poles can be designed such that the resulting system has any desired poles.

In the remainder of this subsection, we shall extended the design procedure to one-input three-output systems. The configuration of compensators is again chosen as shown in Figure 9-19. Let the transfer-function matrix of a given system be

$$\hat{\mathbf{G}}(s) = \begin{bmatrix} \hat{g}_1(s) \\ \hat{g}_2(s) \\ \hat{g}_3(s) \end{bmatrix} = \frac{1}{\Delta(s)} \begin{bmatrix} N_1(s) \\ N_2(s) \\ N_2(s) \end{bmatrix}$$

The transfer-function matrix of the compensator is assumed to be

$$\hat{\mathbf{G}}_c(s) = \frac{1}{\Delta_c(s)} [C_1(s) \quad C_2(s) \quad C_3(s) \quad C_0(s)]$$

Then we can show that the product of all poles of the overall system is

$$\Delta_f(s) = \Delta(s)[\Delta_c(s) + C_0(s)] + N_1(s)C_1(s) + N_2(s)C_2(s) + N_3(s)C_3(s) \quad \textbf{(9-49)}$$

Assume, without loss of generality, that $\delta N_1(s) \geq \delta N_2(s) \geq \delta N_3(s)$. Then Equation (9-49) can be grouped as

$$\Delta_f(s) = \Delta(s)\bar{\Delta}_c(s) + N_{123}(s)[\bar{N}_1(s)C_1(s) + N_{23}(s)(\bar{N}_2(s)C_2(s) + \bar{N}_3(s)C_3(s))] \quad \textbf{(9-50)}$$

where $\bar{\Delta}_c(s) = \Delta_c(s) + C_0(s)$; $N_{123}$ is the greatest common divisor of $N_1$, $N_2$, and $N_3$; $N_{23}$ is the greatest common divisor of $N_2/N_{123}$ and $N_3/N_{123}$. Now by applying Theorem 9-13 three times to (9-50), the design can be achieved. Design of compensators for one-input four-or-more-output systems can be similarly carried out and its discussion will not be repeated here.

## 9-6 Concluding Remarks

The characterization of linear, time-invariant composite systems by transfer-function matrices was studied in this chapter. For single-variable systems, if there is no common pole in the parallel connection, and if there is no pole–zero cancellation in the tandem or feedback connection, then composite transfer functions will characterize completely composite systems. For multivariable systems, in order to give a complete solution, irreducible Jordan-form realizations of the transfer-function matrices of subsystems have to be used.

We also studied stabilities of single-variable and multivariable feedback systems. The stability conditions are stated in terms of the characteristic polynomials of the transfer-function matrices of the subsystems. The Nyquist criterion was also discussed.

The design of compensators to achieve pole placement was studied. We saw that this is purely an algebraic problem. The compensators can be obtained by solving sets of linear algebraic equations which can be read off directly from the transfer-function matrix of the system. The design procedure as presented is applicable only to single-input multiple-output systems. It can, however, be extended to a class of multiple-input systems. The interested reader is referred to Reference [19].

## PROBLEMS

**9-1** Consider systems with the transfer functions

$$\frac{1}{(s+1)(s+2)} \quad \text{and} \quad \frac{s+2}{s+3}$$

It is shown in Theorem 9-2 that the tandem connection of $1/(s+1)(s+2)$ followed by $(s+2)/(s+3)$ is not completely characterized by the composite transfer function. Verify this by showing that its composite dynamical equation is unobservable (but controllable).

**9-2** In Problem 9-1 the pole which does not appear in the composite transfer function is $1/(s+2)$. Does this pole have any serious effect on the response or design of the system?

**9-3** In the design of feedback systems, it is often suggested that we employ a tachometer, whose transfer function is $ks$, in the feedback path to stabilize a system. Can the system in Figure P9-3 be stable for some $k$?

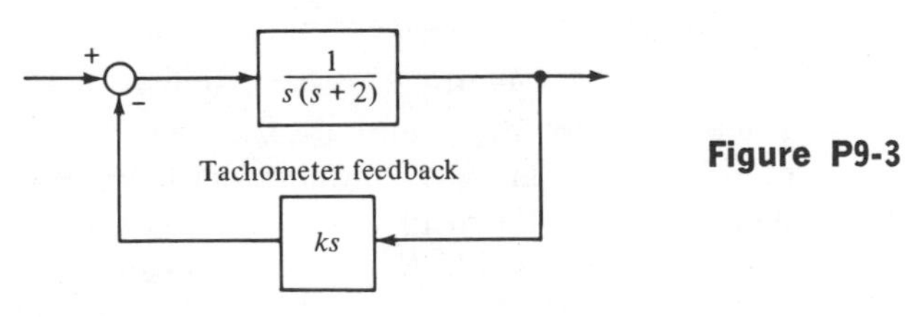


**Figure P9-3**

**9-4** Find the stability range of $k$ by using the Routh–Hurwitz criterion for the system shown in Figure P9-4.

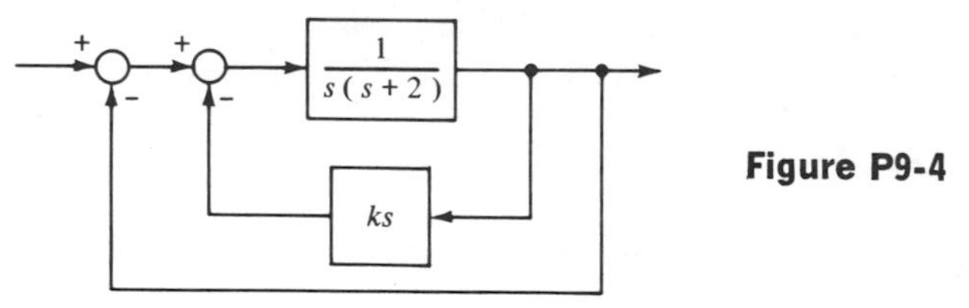


**Figure P9-4**

**9-5** Find the stability range of $k$ by using the Nyquist plot for the system shown in Figure P9-5.

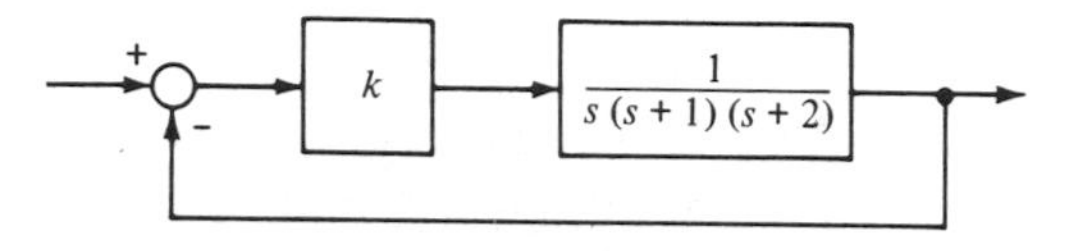


**Figure P9-5**

**9-6** Find the range of $k$ by using the Nyquist plot and the Routh–Hurwitz criterion for the polynomial $s^3 + 2s^2 + s + k$ to be a Hurwitz polynomial. *Hint:* Write the polynomial $s^3 + 2s^2 + s + k$ as

$$(s^3 + 2s^2 + s)\left(1 + \frac{k}{s^3 + 2s^2 + s}\right) \text{ and plot } \frac{1}{s^3 + 2s^2 + s}$$

**9-7** Use the Nyquist plot to determine whether or not the polynomial $3s^3 + 2s^2 + s + 4$ is a Hurwitz polynomial. *Note:* We may plot the Nyquist plot of the polynomial itself or plot

$$\left(1 + \frac{s + 4}{s^2(3s + 2)}\right) \quad \text{or} \quad \left(1 + \frac{3s^3 + 2s^2}{s + 4}\right)$$

or some others.

**9-8** Verify that the feedback system in Figure P9-8 is totally stable.

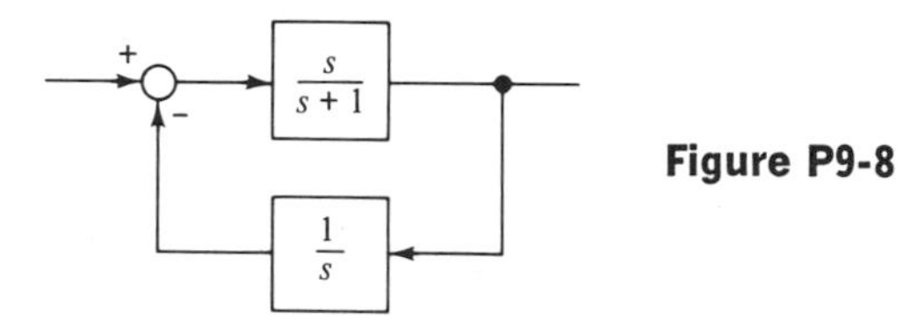


**Figure P9-8**

**9-9** Prove by using physical argument the following statement: Assume that the systems $S_1$ and $S_2$ in the feedback system shown in Figure 9-14 are completely characterized by $\hat{g}_1(s)$ and $\hat{g}_2(s)$. Let $\hat{g}_1(s) = N_1/D_1$, $\hat{g}_2(s) = N_2/D_2$, and let $\hat{D}_2$ be the largest common factor between $N_1$ and $D_2$. Then the feedback system shown in Figure 9-14 is totally stable if and only if all the roots of $(D_1D_2 + kN_1N_2)/\hat{D}_2$ have negative real parts, and all the roots of $\hat{D}_2$ have nonpositive real parts and the multiplicity of the root with zero real part is 1.

**9-10** Prove Theorem 9-11.

**9-11** Is the multivariable feedback system shown in Figure 9-16 with

$$\hat{G}_1(s) = \begin{bmatrix} \dfrac{1}{(s+1)(s-1)} & \dfrac{1}{s+1} \\ \dfrac{1}{(s-1)} & 1 \end{bmatrix} \qquad \hat{G}_2(s) = \begin{bmatrix} \dfrac{1}{s+2} & \dfrac{1}{s+3} \\ \dfrac{1}{s+1} & \dfrac{1}{s+3} \end{bmatrix}$$

totally stable?

**9-12** Find the stability range of $k_1$ and $k_2$ for the multivariable feedback shown in Figure P9-12 to be totally stable.

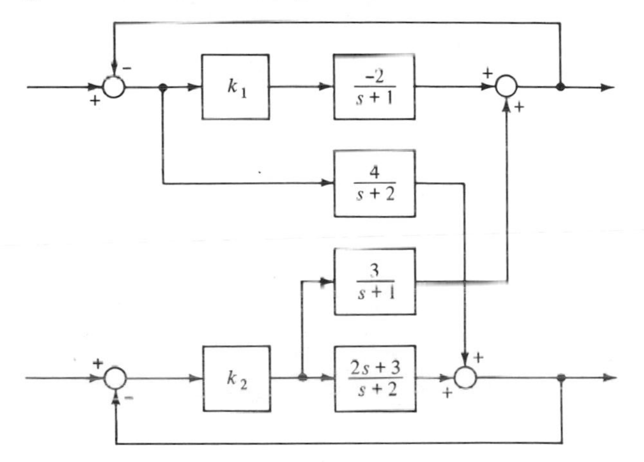


**Figure P9-12**

**9-13** Show that if the coefficients of a rational transfer function $\hat{g}(s)$ are all real, then the real part of $\hat{g}(i\omega)$ is an even function of $\omega$ and the imaginary part of $\hat{g}(i\omega)$ is an odd function of $\omega$.

**9-14** Consider the output feedback system shown in Figure P9-14. Show that the feedback system is controllable (observable) for any

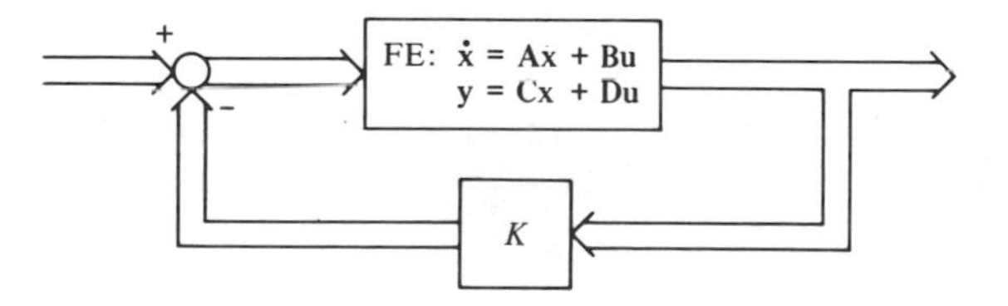

**Figure P9-14**

constant $\mathbf{K}$ if and only if $FE$ is controllable (observable). In the state feedback, we can control all the eigenvalues. Do we have the same situation in the output feedback?

**9-15** Verify that the transfer-function matrix of the feedback system shown in Figure 9-19 is equal to the one given in Equation (9-39).

**9-16** Show that if $N_i'$ and $D_i$, for $i = 1, 2$, have no common factor, then the polynomials $\Delta(s)$, $N_1(s)$ and $N_2(s)$ defined in Equation (9-37) have no common factor. Note that any two can have common factors, but not all three. *Hint:* $\Delta(s) = D_1\bar{D}_2 = D_2\bar{D}_1$ where $\bar{D}_1/\bar{D}_2 = D_1/D_2$, and $\bar{D}_1$ and $\bar{D}_2$ have no common factor.

**9-17** Given a system with the transfer-function matrix

$$\hat{\mathbf{G}}(s) = \begin{bmatrix} \dfrac{(s+3)(s+4)}{s(s-1)(s+2)^2} \\ \dfrac{(s+3)(s-1)}{s(s+1)(s+2)} \end{bmatrix}$$

design a compensator such that the resulting system has a pole $-1$ with a multiplicity 5. The additional poles introduced by the compensator may be chosen to be $-2$ with proper multiplicity. Do you have freedom in choosing the poles of the compensator?

APPENDIX

# A

# Analytic Functions of a Real Variable

Let $D$ be an open interval in the real line $\mathbb{R}$ and let $f(\cdot)$ be a function defined on $D$; that is, to each point in $D$, a unique number is assigned to $f$. The function $f(\cdot)$ may be real-valued or complex-valued.

A function $f(\cdot)$ of a real variable is said to be an element of class $C^n$ on $D$ if its $n$th derivative, $f^{(n)}(\cdot)$, exists and is continuous for all $t$ in $D$. $C^\infty$ is the class of functions having derivatives of all orders.

A function of a real variable, $f(\cdot)$, is said to be *analytic* on $D$ if $f$ is an element of $C^\infty$ *and* if for each $t_0$ in $D$ there exists a positive real number $\epsilon_0$ such that, for all $t$ in $(t_0 - \epsilon_0, t_0 + \epsilon_0)$, $f(t)$ is representable by a Taylor series about the point $t_0$

$$f(t) = \sum_{n=0}^{\infty} \frac{(t - t_0)^n}{n!} f^{(n)}(t_0) \tag{A-1}$$

A remark is in order at this point. If $f$ is a function of a complex variable and if $f$ has continuous derivative, then it can be shown that $f$ has continuous second derivative, third derivative, . . . , and in fact a Taylor-series expansion. Therefore, a function of a *complex* variable may be defined as analytic if it has continuous derivative. However, for functions of a *real* variable, even a function possesses derivatives of all order, it may still not be analytic. For example, the function

$$f(t) = \begin{cases} e^{1/t^2} & \text{for } t \neq 0 \\ 0 & \text{for } t = 0 \end{cases}$$

is not analytic at $t = 0$, even though it is infinitely differentiable at $t = 0$; see Reference [9].

The sum, product, or quotient (provided the denominator is not equal to 0 at any point) of analytic functions of a real variable is analytic. All polynomials, exponential functions, sinusoidal functions are analytic in the entire real line.

If a function is known to be analytic in $D$, then the function is completely determinable from an arbitrary point in $D$ if all the derivatives at that point are known. This can be argued as follows: Suppose that the values of $f$ and its derivatives at $t_0$ are known, then by using (A-1), we can compute $f$ over $(t_0 - \epsilon_0, t_0 + \epsilon_0)$. Next we choose a point $t_1$ that is almost equal to $t_0 + \epsilon_0$; again, from (A-1), we can compute $f$ over $(t_0 + \epsilon_0, t_0 + \epsilon_0 + \epsilon_1)$. Proceeding in both directions, we can compute the function $f$ over $D$. This process is called *analytic continuation.*

### Theorem A-1

If a function $f$ is analytic on $D$ and if $f$ is known to be identically zero on an arbitrarily small nonzero interval in $D$, then the function $f$ is identically zero on $D$.

### Proof

If the function is identically zero on an arbitrarily small nonzero interval, say $(t_0, t_1)$, then the function and its derivatives are all equal to zero on $(t_0, t_1)$. By analytic continuation, the function can be shown to be identically zero. Q.E.D.

APPENDIX

# B

# Minimum-Energy Control[1]

Consider the n-dimensional linear, time-varying state equation

$$E: \qquad \dot{\mathbf{x}} = \mathbf{A}(t)\mathbf{x} + \mathbf{B}(t)\mathbf{u}$$

where $\mathbf{x}$ is the $n \times 1$ state vector, $\mathbf{u}$ is the $p \times 1$ input vector, and $\mathbf{A}$ and $\mathbf{B}$ are, respectively, $n \times n$ and $n \times p$ matrices whose entries are continuous functions of $t$ defined over $(-\infty, \infty)$. If the state equation is controllable at time $t_0$, then given any initial state $\mathbf{x}^0$ at time $t_0$ and any desired final state $\mathbf{x}^1$, there exist a finite $t_1 > t_0$ and an input $\mathbf{u}_{[t_0,t_1]}$ that will transfer the state $\mathbf{x}^0$ at time $t_0$ to $\mathbf{x}^1$ at time $t_1$, denoted as $(\mathbf{x}^0, t_0)$ to $(\mathbf{x}^1, t_1)$. Define

$$\mathbf{W}(t_0, t_1) \triangleq \int_{t_0}^{t_1} \mathbf{\Phi}(t_0, \tau)\mathbf{B}(\tau)\mathbf{B}^*(\tau)\mathbf{\Phi}^*(t_0, \tau)\, d\tau$$

Then the input $\mathbf{u}^0_{[t_0,t_1]}$ defined by

$$\mathbf{u}^0(t) = (\mathbf{\Phi}(t_0, t)\mathbf{B}(t))^*\mathbf{W}^{-1}(t_0, t_1)[\mathbf{\Phi}(t_0, t_1)\mathbf{x}^1 - \mathbf{x}^0] \qquad \text{for all } t \text{ in } [t_0, t_1] \tag{B-1}$$

will transfer $(\mathbf{x}^0, t_0)$ to $(\mathbf{x}^1, t_1)$. This is proved in Theorem 5-4. Now we show that the input $\mathbf{u}^0_{[t_0,t_1]}$ consumes the minimal amount of energy, among all the $\mathbf{u}$'s that can transfer $(\mathbf{x}^0, t_0)$ to $(\mathbf{x}^1, t_1)$.

### Theorem B-1

Let $\mathbf{u}^1_{[t_0,t_1]}$ be any control that transfers $(\mathbf{x}^0, t_0)$ to $(\mathbf{x}^1, t_1)$, and let $\mathbf{u}^0$ be the control defined in (B-1) that accomplishes the same transfer; then

$$\int_{t_0}^{t_1} \|\mathbf{u}^1(t)\|^2\, dt \geq \int_{t_0}^{t_1} \|\mathbf{u}^0(t)\|^2\, dt$$

where $\|\mathbf{u}(t)\| \triangleq (\langle \mathbf{u}(t), \mathbf{u}(t)\rangle)^{1/2} \triangleq (\mathbf{u}^*(t)\mathbf{u}(t))^{1/2}$, the Euclidean norm of $\mathbf{u}(t)$. (See Section 2-8.)

### Proof

The solution of the state equation $E$ is

$$\mathbf{x}(t_1) = \mathbf{\Phi}(t_1, t_0)[\mathbf{x}(t_0) + \int_{t_0}^{t_1} \mathbf{\Phi}(t_0, \tau)\mathbf{B}(\tau)\mathbf{u}(\tau)\, d\tau] \tag{B-2}$$

[1] This appendix closely follows Reference [69].

Define

$$\bar{\mathbf{x}} \triangleq \mathbf{\Phi}^{-1}(t_1, t_0)\mathbf{x}(t_1) - \mathbf{x}(t_0) = \mathbf{\Phi}(t_0, t_1)\mathbf{x}^1 - \mathbf{x}^0$$

Then the assumptions that $\mathbf{u}^1$ and $\mathbf{u}^0$ transfer $(\mathbf{x}^0, t_0)$ to $(\mathbf{x}^1, t_1)$ imply that

$$\bar{\mathbf{x}} = \int_{t_0}^{t_1} \mathbf{\Phi}(t_0, \tau)\mathbf{B}(\tau)\mathbf{u}^1(\tau)\, d\tau = \int_{t_0}^{t_1} \mathbf{\Phi}(t_0, \tau)\mathbf{B}(\tau)\mathbf{u}^0(\tau)\, d\tau$$

Subtracting both sides, we obtain

$$\int_{t_0}^{t_1} \mathbf{\Phi}(t_0, \tau)\mathbf{B}(\tau)(\mathbf{u}^1(\tau) - \mathbf{u}^0(\tau))\, d\tau = \mathbf{0}$$

which implies that

$$\left\langle \int_{t_0}^{t_1} \mathbf{\Phi}(t_0, \tau)\mathbf{B}(\tau)(\mathbf{u}^1(\tau) - \mathbf{u}^0(\tau))\, d\tau, \mathbf{W}^{-1}(t_0, t_1)\bar{\mathbf{x}} \right\rangle = 0$$

By using (2-94), we can write this equation as

$$\int_{t_0}^{t_1} \langle \mathbf{u}^1(\tau) - \mathbf{u}^0(\tau), (\mathbf{\Phi}(t_0, \tau)\mathbf{B}(\tau))^*\mathbf{W}^{-1}(t_0, t_1)\bar{\mathbf{x}} \rangle\, d\tau = 0 \tag{B-3}$$

With the use of (B-1), Equation (B-3) becomes

$$\int_{t_0}^{t_1} \langle \mathbf{u}^1(\tau) - \mathbf{u}^0(\tau), \mathbf{u}^0(\tau) \rangle\, d\tau = 0 \tag{B-4}$$

Consider now

$$\int_{t_0}^{t_1} \|\mathbf{u}^1\|^2\, d\tau$$

By some manipulation and using (B-4), we obtain

$$\begin{aligned}
\int_{t_0}^{t_1} \|\mathbf{u}^1(\tau)\|^2\, d\tau &= \int_{t_0}^{t_1} \|\mathbf{u}^1(\tau) - \mathbf{u}^0(\tau) + \mathbf{u}^0(\tau)\|^2\, d\tau \\
&= \int_{t_0}^{t_1} \|\mathbf{u}^1(\tau) - \mathbf{u}^0(\tau)\|^2\, d\tau + \int_{t_0}^{t_1} \|\mathbf{u}^0(\tau)\|^2\, d\tau \\
&\qquad + 2\int_{t_0}^{t} \langle \mathbf{u}^1(\tau) - \mathbf{u}^0(\tau), \mathbf{u}^0(\tau) \rangle\, d\tau \\
&= \int_{t_0}^{t_1} \|\mathbf{u}^1(\tau) - \mathbf{u}^0(\tau)\|^2\, d\tau + \int_{t_0}^{t_1} \|\mathbf{u}^0(\tau)\|^2\, d\tau
\end{aligned}$$

Since

$$\int_{t_0}^{t_1} \|\mathbf{u}^1(\tau) - \mathbf{u}^0(\tau)\|^2\, d\tau$$

is always nonnegative, we conclude that

$$\int_{t_0}^{t_1} \|\mathbf{u}^1(\tau)\|^2\, d\tau \geq \int_{t_0}^{t_1} \|\mathbf{u}^0(\tau)\|^2\, d\tau \qquad \text{Q.E.D.}$$

We see that in transferring $(\mathbf{x}^0, t_0)$ to $(\mathbf{x}^1, t_1)$, if minimizing

$$\int_{t_0}^{t_1} \|\mathbf{u}(\tau)\|^2\, d\tau$$

is used as a criterion, the control defined in (B-1) is optimal. Since, in many instances,

$$\int_{t_0}^{t_1} \|\mathbf{u}(\tau)\|^2\, d\tau$$

is related to energy, the control $\mathbf{u}^0$ is called the *minimum-energy control.*

APPENDIX

# C

# Controllability after the Introduction of Sampling[1]

Consider the linear, time-invariant dynamical equation

$$FE: \quad \dot{\mathbf{x}} = \mathbf{A}\mathbf{x} + \mathbf{B}\mathbf{u} \tag{C-1a}$$
$$\mathbf{y} = \mathbf{C}\mathbf{x} \tag{C-1b}$$

where $\mathbf{x}$ is the $n \times 1$ state vector, $\mathbf{u}$ is the $p \times 1$ input vector, $\mathbf{y}$ is the $q \times 1$ output vector; $\mathbf{A}$, $\mathbf{B}$, and $\mathbf{C}$ are $n \times n$, $n \times p$, and $q \times n$ constant matrices, respectively. The response of $FE$ is given by

$$\mathbf{x}(t) = e^{\mathbf{A}t}\mathbf{x}^0 + \int_0^t e^{\mathbf{A}(t-\tau)}\mathbf{B}\mathbf{u}(\tau)\,d\tau \tag{C-2a}$$
$$\mathbf{y}(t) = \mathbf{C}\mathbf{x}(t) \tag{C-2b}$$

where $\mathbf{x}^0$ is the initial state at $t = 0$.

We consider now the case in which the input $\mathbf{u}$ is piecewise constant; that is, the input $\mathbf{u}$ changes value only at a discrete instant of time. Inputs of this type occur in sampled-data systems or in systems in which digital computers are used to generate $\mathbf{u}$. A piecewise-constant function is often generated by a sampler and a filter, called zero-order hold, as shown in Figure C-1. Let

$$\mathbf{u}(t) = \mathbf{u}(n) \qquad \text{for } nT \leq t < (n+1)T;\ n = 0, 1, 2, \ldots \tag{C-3}$$

where $T$ is a positive constant, called the *sampling period*. The discrete times $0, T, 2T$, are called *sampling instants*. The behavior of $FE$ with the peicewise-constant inputs given in (C-3) can be computed from (C-2). However, if only the behavior at sampling instants $0, T, 2T, \cdots$ is of interest, a discrete-time dynamical equation can be written to give the response of $\mathbf{x}(n) \triangleq \mathbf{x}(nT)$ at $n = 0, 1, 2, \cdots$. From (C-2a), we have

$$\begin{aligned}\mathbf{x}(n+1) &= e^{\mathbf{A}(n+1)T}\mathbf{x}^0 + \int_0^{(n+1)T} e^{\mathbf{A}[nT+T-\tau]}\mathbf{B}\mathbf{u}(\tau)\,d\tau \\ &= e^{\mathbf{A}T}\left[e^{\mathbf{A}nT}\mathbf{x}^0 + \int_0^{nT} e^{\mathbf{A}(nT-\tau)}\mathbf{B}\mathbf{u}(\tau)\,d\tau\right] \\ &\qquad + \int_{nT}^{(n+1)T} e^{\mathbf{A}(nT+T-\tau)}\mathbf{B}\mathbf{u}(\tau)\,d\tau\end{aligned} \tag{C-4}$$

[1] See Reference [69].

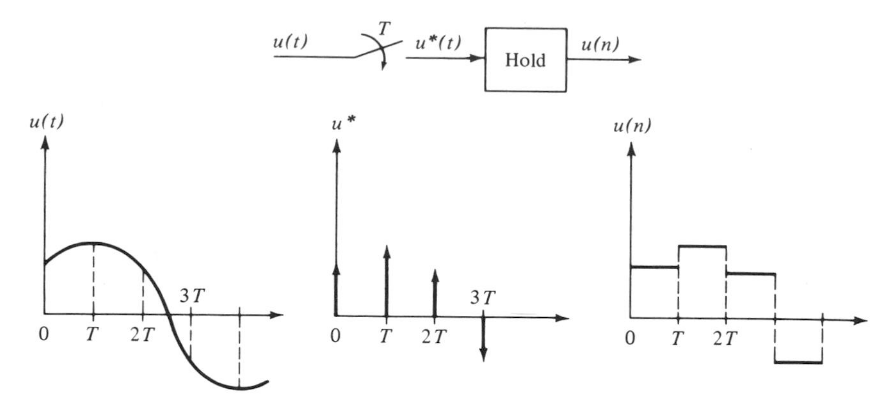


**Figure C-1** A piecewise-constant function, which can be generated by a sampler and a hold.

The term in the brackets of (C-4) is equal to $\mathbf{x}(n)$; the input $\mathbf{u}(\tau)$ is constant in the interval $[nT, nT + T)$ and is equal to $\mathbf{u}(n)$; hence (C-4) becomes, after the change of variable $\alpha = nT + T - \tau$,

$$\mathbf{x}(n+1) = e^{\mathbf{A}T}\mathbf{x}(n) + \left(\int_0^T e^{\mathbf{A}\alpha}\, d\alpha\right)\mathbf{B}\mathbf{u}(n)$$

which is a discrete-time state equation. Therefore, if the input is piecewise constant over the same interval $T$, and if only the response at the sampling instants is of interest, the dynamical equation $FE$ in (C-1) can be replaced by the following discrete-time, linear, time-invariant dynamical equation

$$DFE: \qquad \mathbf{x}(n+1) = \tilde{\mathbf{A}}\mathbf{x}(n) + \tilde{\mathbf{B}}\mathbf{u}(n) \tag{C-5a}$$
$$\mathbf{y}(n) = \tilde{\mathbf{C}}\mathbf{x}(n) \tag{C-5b}$$

where

$$\tilde{\mathbf{A}} = e^{\mathbf{A}T} \tag{C-6}$$
$$\tilde{\mathbf{B}} = \left(\int_0^T e^{\mathbf{A}\tau}\, d\tau\right)\mathbf{B} \triangleq \mathbf{MB} \tag{C-7}$$
$$\tilde{\mathbf{C}} = \mathbf{C} \tag{C-8}$$

If the dynamical equation $FE$ is controllable, it is of interest to study whether the system remains controllable after the introduction of sampling or, correspondingly, whether the discrete-time dynamical equation $DFE$ is controllable. This problem is important in the design of dead-beat sampled-data systems.

A discrete-time dynamical equation $DFE$ is defined to be controllable if for any initial state $\mathbf{x}^0$ and any $\mathbf{x}^1$ in the state space, there exists an input sequence $\{\mathbf{u}(n)\}$ of finite length that transfers $\mathbf{x}^0$ to $\mathbf{x}^1$. Similar to the continuous case, we have the following theorem.

### Theorem C-1

The discrete-time dynamical equation *DFE* given in (C-5) is controllable if and only if

$$\rho[\tilde{\mathbf{B}} \quad \tilde{\mathbf{A}}\tilde{\mathbf{B}} \quad \cdots \quad \tilde{\mathbf{A}}^{n-1}\tilde{\mathbf{B}}] = n$$

or if and only if the rows of $(z\mathbf{I} - \tilde{\mathbf{A}})^{-1}\tilde{\mathbf{B}}$ are linearly independent over the field of complex numbers.

### Proof

**1.** By successive substituting, from (C-5a) we have

$$\mathbf{x}(n) = \tilde{\mathbf{A}}^n\mathbf{x}(0) + \tilde{\mathbf{A}}^{n-1}\tilde{\mathbf{B}}\mathbf{u}(0) + \tilde{\mathbf{A}}^{n-2}\tilde{\mathbf{B}}\mathbf{u}(1) + \cdots + \tilde{\mathbf{A}}\tilde{\mathbf{B}}\mathbf{u}(n-2) + \tilde{\mathbf{B}}\mathbf{u}(n-1)$$

which can be written as

$$\mathbf{x}(n) - \tilde{\mathbf{A}}^n\mathbf{x}(0) = [\tilde{\mathbf{B}} \quad \tilde{\mathbf{A}}\tilde{\mathbf{B}} \quad \cdots \quad \tilde{\mathbf{A}}^{n-1}\tilde{\mathbf{B}}]\begin{bmatrix} \mathbf{u}(n-1) \\ \mathbf{u}(n-2) \\ \cdot \\ \cdot \\ \cdot \\ \mathbf{u}(0) \end{bmatrix} \qquad \text{(C-9)}$$

[illegible]ws from Theorem 2-4 that for any $\mathbf{x}(0)$ and any $\mathbf{x}^1$, there [illegible] solution $\mathbf{u}(0), \mathbf{u}(1), \cdots, \mathbf{u}(n-1)$ satisfying (C-9) if and [illegible] $\rho[\tilde{\mathbf{B}} \quad \tilde{\mathbf{A}}\tilde{\mathbf{B}} \cdots \tilde{\mathbf{A}}^{n-1}\tilde{\mathbf{B}}] = n$. We see that if an $n$-dimensional, [illegible]te-time, time-invariant dynamical equation is controllable, then any initial state can be transferred to any other state within $n$ steps. If a state cannot be transferred to some other state in $n$ steps, no matter how long the input sequence $\{\mathbf{u}(n)\}$ is, it still cannot be achieved. (Why?)

**2.** We show that $\rho[\tilde{\mathbf{B}} \quad \tilde{\mathbf{A}}\tilde{\mathbf{B}} \quad \cdots \quad \tilde{\mathbf{A}}^{n-1}\tilde{\mathbf{B}}] = n$ if and only if the $n$ rows of $(z\mathbf{I} - \tilde{\mathbf{A}})^{-1}\mathbf{B}$ are linearly independent over the field of complex numbers. From (2-85), we have

$$(z\mathbf{I} - \tilde{\mathbf{A}})^{-1}\tilde{\mathbf{B}} = z^{-1}(\mathbf{I} - z^{-1}\tilde{\mathbf{A}})^{-1}\tilde{\mathbf{B}} = z^{-1}\tilde{\mathbf{B}} + z^{-2}\tilde{\mathbf{A}}\tilde{\mathbf{B}} + z^{-3}\tilde{\mathbf{A}}^2\tilde{\mathbf{B}} + \cdots \qquad \text{(C-10)}$$

It is clear from (C-10) that if all the rows of $[\tilde{\mathbf{B}} \quad \tilde{\mathbf{A}}\tilde{\mathbf{B}} \quad \cdots \quad \tilde{\mathbf{A}}^{n-1}\tilde{\mathbf{B}}]$ are linearly independent over the field of complex numbers, so are the rows of $(z\mathbf{I} - \tilde{\mathbf{A}})^{-1}\tilde{\mathbf{B}}$. Now if $\rho[\tilde{\mathbf{B}} \quad \tilde{\mathbf{A}}\tilde{\mathbf{B}} \quad \cdots \quad \tilde{\mathbf{A}}^{n-1}\tilde{\mathbf{B}}] < n$, by definition there exists a nonzero constant vector $\boldsymbol{\alpha}$ such that $\boldsymbol{\alpha}\tilde{\mathbf{B}} = \mathbf{0}$, $\boldsymbol{\alpha}\tilde{\mathbf{A}}\tilde{\mathbf{B}} = \mathbf{0}$, $\cdots$, $\boldsymbol{\alpha}\tilde{\mathbf{A}}^{n-1}\tilde{\mathbf{B}} = \mathbf{0}$. It follows from the

Cayley–Hamilton theorem that $\boldsymbol{\alpha}\tilde{\mathbf{A}}^k\tilde{\mathbf{B}} = \mathbf{0}$ for $k = n, n+1, \cdots$. Hence, from (C-10), we have $\boldsymbol{\alpha}(z\mathbf{I} - \tilde{\mathbf{A}})^{-1}\tilde{\mathbf{B}} = \mathbf{0}$. In other words, if $\rho[\tilde{\mathbf{B}} \quad \tilde{\mathbf{A}}\tilde{\mathbf{B}} \quad \cdots \quad \tilde{\mathbf{A}}^{n-1}\tilde{\mathbf{B}}] < n$, then the rows of $(z\mathbf{I} - \tilde{\mathbf{A}})^{-1}\tilde{\mathbf{B}}$ are not linearly independent over the field of complex numbers.

Q.E.D.

Let $\lambda_i(\mathbf{A})$ denote an eigenvalue of $\mathbf{A}$, and let "Im" and "Re" stand for the imaginary part and the real part of an eigenvalue, respectively.

### Theorem C-2

Assume that the dynamical equation *FE* given in (C-1) is controllable. A sufficient condition for the discrete-time dynamical equation *DFE* given in (C-5) to be controllable is that $\operatorname{Im}[\lambda_i(\mathbf{A}) - \lambda_j(\mathbf{A})] \neq 2\pi\alpha/T$ for $\alpha = \pm 1, \pm 2, \cdots$ whenever $\operatorname{Re}[\lambda_i(\mathbf{A}) - \lambda_j(\mathbf{A})] = 0$.

For the single-input case ($p = 1$), the condition is necessary as well.

### Proof

We assume, without loss of generality, that $\mathbf{A}$ is in the Jordan canonical form shown in Table 5-1. Then, from (C-6), (2-69), and (2-71), we have

$$\tilde{\mathbf{A}} = e^{\mathbf{A}T} = \begin{bmatrix} e^{\mathbf{A}_{11}T} & \mathbf{0} & \cdots & \mathbf{0} \\ \mathbf{0} & e^{\mathbf{A}_{12}T} & \cdots & \mathbf{0} \\ \cdot & \cdot & & \cdot \\ \cdot & \cdot & & \cdot \\ \cdot & \cdot & & \cdot \\ \mathbf{0} & \mathbf{0} & \cdots & e^{\mathbf{A}_{mr(m)}T} \end{bmatrix} \tag{C-11}$$

where

$$e^{\mathbf{A}_{ij}T} = e^{\lambda_i T} \begin{bmatrix} 1 & T & \cdots & T^{(n_{ij}-1)}/(n_{ij}-1)! \\ 0 & 1 & \cdots & T^{(n_{ij}-2)}/(n_{ij}-2)! \\ 0 & 0 & \cdots & T^{(n_{ij}-3)}/(n_{ij}-3)! \\ \cdot & \cdot & & \cdot \\ \cdot & \cdot & & \cdot \\ \cdot & \cdot & & \cdot \\ 0 & 0 & \cdots & 1 \end{bmatrix} \tag{C-12}$$

Consider now

$$(z\mathbf{I} - \tilde{\mathbf{A}})^{-1}\tilde{\mathbf{B}} = (z\mathbf{I} - \mathbf{e}^{AT})^{-1}\mathbf{M}\mathbf{B} \tag{C-13}$$

where

$$\mathbf{M} \triangleq \int_0^T e^{\mathbf{A}\tau}\, d\tau$$

Since the matrices $(z\mathbf{I} - e^{\mathbf{A}T})^{-1}$ and $\mathbf{M}$ are functions of the same matrix $\mathbf{A}$, we have $(z\mathbf{I} - e^{\mathbf{A}T})^{-1}\mathbf{M} = \mathbf{M}(z\mathbf{I} - e^{\mathbf{A}T})^{-1}$ (see Problem 2-33), and (C-13) becomes

$$(z\mathbf{I} - \tilde{\mathbf{A}})^{-1}\tilde{\mathbf{B}} = \mathbf{M}(z\mathbf{I} - e^{\mathbf{A}T})^{-1}\mathbf{B} \tag{C-14}$$

From (C-11) and (C-12) it is easy to verify that

$$\det \mathbf{M} = \prod_{i=1}^{m} \rho_i^{n_i}$$

where

$$\rho_i = \begin{cases} \dfrac{1 - e^{-\lambda_i T}}{\lambda_i} & \text{if } \lambda_i \neq 0 \\ T & \text{if } \lambda_i = 0 \end{cases}$$

which is nonzero, following from the assumption that $T > 0$. Hence the matrix $\mathbf{M}$ is nonsingular. Consequently, if we show that the rows of $(z\mathbf{I} - e^{\mathbf{A}T})^{-1}\mathbf{B}$ are linearly independent over the field of complex numbers, so are the rows of $(z\mathbf{I} - \tilde{\mathbf{A}})^{-1}\tilde{\mathbf{B}}$. Since $e^{\mathbf{A}_{ij}T}$ is a triangular matrix, it is easy to verify that

$$(z\mathbf{I} - e^{\mathbf{A}_{ij}T})^{-1} = \begin{bmatrix} (z - \tilde{\lambda}_i)^{-1} & T\tilde{\lambda}_i(z - \tilde{\lambda}_i)^{-2} & \cdots & [(T\tilde{\lambda}_i)^{-n_{ij}+1}(z - \tilde{\lambda}_i)^{-n_{ij}} + \cdots] \\ 0 & (z - \tilde{\lambda}_i)^{-1} & \cdots & [(T\tilde{\lambda}_i)^{-n_{ij}+2}(z - \tilde{\lambda}_i)^{-n_{ij}+1} + \cdots] \\ \vdots & \vdots & & \vdots \\ 0 & 0 & \cdots & (z - \tilde{\lambda}_i)^{-1} \end{bmatrix} \tag{C-15}$$

where $\tilde{\lambda}_i \triangleq e^{\lambda_i T}$. The fact we need in the proof is that the first component of the last column of $(z\mathbf{I} - e^{\mathbf{A}_{ij}T})^{-1}$ contains a factor $(z - \tilde{\lambda}_i)^{-n_{ij}}$; the second component contains a factor $(z - \lambda_i)^{-n_{ij}+1}$, and so forth. Now by assumption, the dynamical equation $FE$ is in the Jordan form and is controllable; hence $\mathbf{b}_{lij}$ (see Table 5-1) is different from the zero vector. Consequently, all the rows of $(z\mathbf{I} - e^{\mathbf{A}_{ij}T})^{-1}\mathbf{B}_{ij}$ are linearly independent. The controllability assumption of $FE$ implies that the set $\{\mathbf{b}_{lij}, \text{ for } j = 1, 2, \cdots, r(i)\}$ is a linearly independent set (see Theorem 5-16); hence all the rows of $(z\mathbf{I} - e^{\mathbf{A}_iT})^{-1}\mathbf{B}_i$ are linearly independent for each $i$. Now the assumption that $\text{Im}\,[\lambda_i(\mathbf{A}) - \lambda_j(\mathbf{A})] \neq 2\pi\alpha/T$ implies that if $\lambda_i$ and $\lambda_j$ are distinct, so are $\tilde{\lambda}_i = e^{\lambda_i T}$ and $\tilde{\lambda}_j = e^{\lambda_j T}$. Consequently, we conclude that all the rows of $(z\mathbf{I} - e^{\mathbf{A}T})^{-1}\mathbf{B}$ are linearly independent, and thus the discrete-time dynamical equation $DFE$ is controllable.

Now we show that for the single-input case, if the condition is not met, then (C-5) is not controllable. If $\text{Im}\,[\lambda_i(\mathbf{A}) - \lambda_j(\mathbf{A})] = 2\pi\alpha/T$, for $\text{Re}\,[\lambda_i(\mathbf{A}) - \lambda_j(\mathbf{A})] = 0$, then $\tilde{\lambda}_i = \tilde{\lambda}_j$ and the last row of $(z\mathbf{I} - e^{\mathbf{A}_iT})^{-1}\mathbf{B}_i$ (a scalar) and the last row of $(z\mathbf{I} - e^{\mathbf{A}_jT})^{-1}\mathbf{B}_j$ are linearly dependent. Hence the single-input, discrete-time dynamical equation is not controllable. Q.E.D.

### Example

Consider the sampled-data system shown in Figure C-2. By partial fraction expansion, we have

$$\frac{8}{(s+1)(s+1+2i)(s+1-2i)} = \frac{2}{s+1} - \frac{1}{(s+1+2i)} - \frac{1}{(s+1-2i)}$$

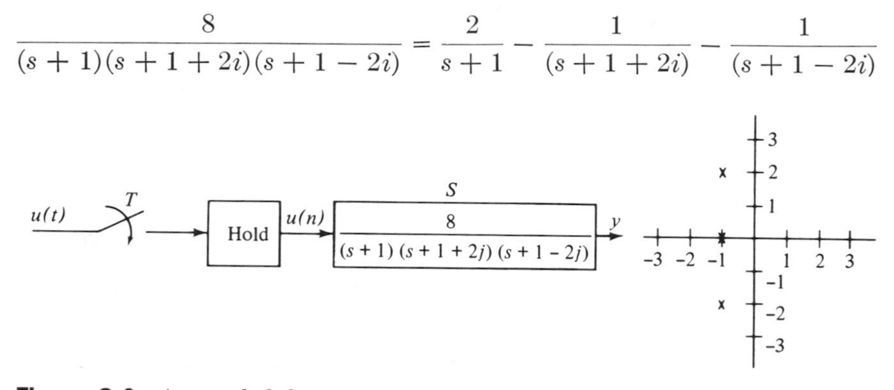


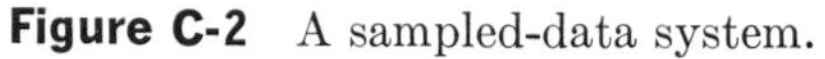

**Figure C-2** A sampled-data system.

Consequently, an irreducible Jordan-form realization of the system $S$ can be found as

$$\begin{bmatrix} \dot{x}_1 \\ \dot{x}_2 \\ \dot{x}_3 \end{bmatrix} = \begin{bmatrix} -1 & 0 & 0 \\ 0 & -1-2i & 0 \\ 0 & 0 & -1+2i \end{bmatrix} \begin{bmatrix} x_1 \\ x_2 \\ x_3 \end{bmatrix} + \begin{bmatrix} 1 \\ 1 \\ 1 \end{bmatrix} u$$

$$y = [2 \quad -1 \quad -1]\mathbf{x}$$

By using (C-5) through (C-8), a discrete-time state equation can be computed as

$$\mathbf{x}(n+1) = \begin{bmatrix} e^{-T} & 0 & 0 \\ 0 & e^{-T-2iT} & 0 \\ 0 & 0 & e^{-T+2iT} \end{bmatrix} \mathbf{x}(n) + \begin{bmatrix} 1 - e^{-T} \\ \dfrac{1}{1+2i}(1 - e^{-T-2iT}) \\ \dfrac{1}{1-2i}(1 - e^{-T+2iT}) \end{bmatrix} u(n) \tag{C-16}$$

We conclude from Theorem C-2 that the discrete-time state equation (C-16) is controllable if and only if

$$T \neq \frac{2\pi\alpha}{2} = \pi\alpha$$

and

$$T \neq \frac{2\pi\alpha}{4} = \frac{\pi}{2}\alpha \qquad \alpha = \pm 1, \pm 2, \cdots$$

This fact can also be verified directly from (C-16) by using either the criterion $\rho[\tilde{\mathbf{B}} \quad \tilde{\mathbf{A}}\tilde{\mathbf{B}} \quad \tilde{\mathbf{A}}^2\tilde{\mathbf{B}}] = 3$ or Corollary 5-16.

APPENDIX

# D

# Hermitian Forms

A *hermitian form*[1] of $n$ complex variables $x_1, x_2, \cdots, x_n$ is a *real-valued* homogeneous polynomial of the form

$$\sum_{i,j=1}^{n} m_{ij}\bar{x}_i x_j$$

or, in matrix form,

$$[\bar{x}_1 \quad \bar{x}_2 \quad \cdots \quad \bar{x}_n]\begin{bmatrix} m_{11} & m_{12} & \cdots & m_{1n} \\ m_{21} & m_{22} & \cdots & m_{2n} \\ \cdot & \cdot & & \cdot \\ \cdot & \cdot & & \cdot \\ \cdot & \cdot & & \cdot \\ m_{n1} & m_{n2} & \cdots & m_{nn} \end{bmatrix}\begin{bmatrix} x_1 \\ x_2 \\ \cdot \\ \cdot \\ \cdot \\ x_n \end{bmatrix} \triangleq \mathbf{x}^*\mathbf{M}_1\mathbf{x} \qquad \text{(D-1)}$$

where the $m_{ij}$'s are any complex numbers and $\bar{x}_i$ is the complex conjugate of $x_i$. Since every hermitian form is assumed to be real-valued, we have

$$\mathbf{x}^*\mathbf{M}_1\mathbf{x} = (\mathbf{x}^*\mathbf{M}_1\mathbf{x})^* = \mathbf{x}^*\mathbf{M}_1^*\mathbf{x}$$

where $\mathbf{M}_1^*$ is the complex conjugate transpose of $\mathbf{M}_1$; hence

$$\mathbf{x}^*\mathbf{M}_1\mathbf{x} = \mathbf{x}^*(\tfrac{1}{2}(\mathbf{M}_1 + \mathbf{M}_1^*))\mathbf{x} \triangleq \mathbf{x}^*\mathbf{M}\mathbf{x} \qquad \text{(D-2)}$$

where $\mathbf{M} = \frac{1}{2}(\mathbf{M}_1 + \mathbf{M}_1^*)$. It is clear that $\mathbf{M} = \mathbf{M}^*$. Thus *every hermitian form can be written as* $\mathbf{x}^*\mathbf{M}\mathbf{x}$ *with* $\mathbf{M} = \mathbf{M}^*$. A matrix $\mathbf{M}$ with the property $\mathbf{M} = \mathbf{M}^*$ is called a hermitian matrix.

In the study of hermitian forms, it is convenient to use the notation of inner product. Observe that the hermitian matrix $\mathbf{M}$ can be considered as a linear operator that maps the $n$-dimensional complex vector space $(\mathbb{C}^n, \mathbb{C})$ into itself. If the inner product of $(\mathbb{C}^n, \mathbb{C})$ is chosen as

$$\langle \mathbf{x}, \mathbf{y} \rangle \triangleq \mathbf{x}^*\mathbf{y} \qquad \text{(D-3)}$$

where $\mathbf{x}$ and $\mathbf{y}$ are any vectors in $(\mathbb{C}^n, \mathbb{C})$, then the hermitian form can be written as

$$\mathbf{x}^*\mathbf{M}\mathbf{x} = \langle \mathbf{x}, \mathbf{M}\mathbf{x} \rangle = \langle \mathbf{M}^*\mathbf{x}, \mathbf{x} \rangle = \langle \mathbf{M}\mathbf{x}, \mathbf{x} \rangle \qquad \text{(D-4)}$$

where, in the last step, we have used the fact that $\mathbf{M}^* = \mathbf{M}$.

---

[1] See References [38] and [39].

### Theorem D-1

All the eigenvalues of a hermitian matrix $\mathbf{M}$ are real.

### Proof

Let $\lambda$ be any eigenvalue of $\mathbf{M}$ and let $\mathbf{e}$ be an eigenvector of $\mathbf{M}$ associated with $\lambda$; that is, $\mathbf{Me} = \lambda\mathbf{e}$. Consider

$$\langle \mathbf{e}, \mathbf{Me} \rangle = \langle \mathbf{e}, \lambda\mathbf{e} \rangle = \lambda \langle \mathbf{e}, \mathbf{e} \rangle \tag{D-5}$$

Since $\langle \mathbf{e}, \mathbf{Me} \rangle$ is a real number and since $\langle \mathbf{e}, \mathbf{e} \rangle$ is a real positive number, from (D-5) we conclude that $\lambda$ is a real number. Q.E.D.

### Theorem D-2

The Jordan-form representation of a hermitian matrix $\mathbf{M}$ is a diagonal matrix.

### Proof

Recall from Section 2-6 that every square matrix which maps $(\mathbb{C}^n, \mathbb{C})$ into itself has a Jordan-form representation. The basis vectors that give a Jordan-form representation consist of eigenvectors and generalized eigenvectors of the matrix. We show that if a matrix is hermitian, then there is no generalized eigenvector of rank $k \geq 2$; we show this by contradiction. Suppose there exists a vector $\mathbf{e}$ such that $(\mathbf{M} - \lambda_i\mathbf{I})^k\mathbf{e} = \mathbf{0}$ and $(\mathbf{M} - \lambda_i\mathbf{I})^{k-1}\mathbf{e} \neq \mathbf{0}$ for some eigenvalue $\lambda_i$ of $\mathbf{M}$. Consider now, for $k \geq 2$,

$$\begin{aligned} 0 &= \langle (\mathbf{M} - \lambda_i\mathbf{I})^{k-2}\mathbf{e}, (\mathbf{M} - \lambda_i\mathbf{I})^k\mathbf{e} \rangle = \langle (\mathbf{M} - \lambda_i\mathbf{I})^{k-1}\mathbf{e}, (\mathbf{M} - \lambda_i\mathbf{I})^{k-1}\mathbf{e} \rangle \\ &= \|(\mathbf{M} - \lambda_i\mathbf{I})^{k-1}\mathbf{e}\|^2 \end{aligned}$$

which implies $(\mathbf{M} - \lambda_i\mathbf{I})^{k-1}\mathbf{e} = \mathbf{0}$. This is a contradiction. Hence there is no generalized eigenvector of rank $k \geq 2$. Consequently, there is no Jordan block whose order is greater than one. Hence the Jordan-form representation of a hermitian matrix is a diagonal matrix. In other words, there exists a nonsingular matrix $\mathbf{P}$ such that $\mathbf{PMP}^{-1} = \hat{\mathbf{M}}$ and $\hat{\mathbf{M}}$ is a diagonal matrix with eigenvalues on the diagonal. Q.E.D.

Two vectors $\mathbf{x}, \mathbf{y}$ are said to be *orthogonal* if and only if $\langle \mathbf{x}, \mathbf{y} \rangle = 0$. A vector $\mathbf{x}$ is said to be *normalized* if and only if $\langle \mathbf{x}, \mathbf{x} \rangle \triangleq \|\mathbf{x}\|^2 = 1$. It is clear that every vector $\mathbf{x}$ can be normalized by choosing $\hat{\mathbf{x}} = (1/\|\mathbf{x}\|)\mathbf{x}$. A set of basis vectors $\{\mathbf{q}^1, \mathbf{q}^2, \cdots, \mathbf{q}^n\}$ is said to be an *orthonormal basis* if and only if

$$\begin{aligned} \langle \mathbf{q}^i, \mathbf{q}^j \rangle &= 0 \qquad i \neq j \\ &= 1 \qquad i = j \end{aligned} \tag{D-6}$$

Now we show that the basis of a Jordan-form representation of a hermitian matrix can be chosen as an orthonormal basis. This is derived in part from the following theorem.

### Theorem D-3

The eigenvectors of a hermitian matrix $\mathbf{M}$ corresponding to different eigenvalues are orthogonal.

### Proof

Let $\mathbf{e}^i$ and $\mathbf{e}^j$ be the eigenvectors of $\mathbf{M}$ corresponding to the distinct eigenvalues $\lambda_i$ and $\lambda_j$, respectively; that is, $\mathbf{Me}^i = \lambda_i\mathbf{e}^i$ and $\mathbf{Me}^j = \lambda_j\mathbf{e}^j$. Consider

$$\langle \mathbf{e}^j, \mathbf{Me}^i \rangle = \langle \mathbf{e}^j, \lambda_i\mathbf{e}^i \rangle = \lambda_i\langle \mathbf{e}^j, \mathbf{e}^i \rangle \tag{D-7}$$

and

$$\langle \mathbf{e}^j, \mathbf{Me}^i \rangle = \langle \mathbf{Me}^j, \mathbf{e}^i \rangle = \langle \lambda_j\mathbf{e}^j, \mathbf{e}^i \rangle = \lambda_j\langle \mathbf{e}^j, \mathbf{e}^i \rangle \tag{D-8}$$

where we have used the fact that the eigenvalues are real. Subtracting (D-8) from (D-7), we obtain $(\lambda_i - \lambda_j)\langle \mathbf{e}^j, \mathbf{e}^i \rangle = 0$. Since $\lambda_i \neq \lambda_j$, we conclude that $\langle \mathbf{e}^i, \mathbf{e}^j \rangle = 0$. Q.E.D.

Since every eigenvector can be normalized and since eigenvectors of a hermitian matrix $\mathbf{M}$ associated with distinct eigenvalues are orthogonal, hence the eigenvectors associated with different eigenvalues can be made to be orthonormal. We consider now the linearly independent eigenvectors associated with the same eigenvalue. Let $\{\mathbf{e}^1, \mathbf{e}^2, \cdots, \mathbf{e}^m\}$ be a set of linearly independent eigenvectors associated with the same eigenvalue. Now we shall obtain a set of orthonormal vectors from the set $\{\mathbf{e}^1, \mathbf{e}^2, \cdots, \mathbf{e}^m\}$. Let

$$\begin{aligned}
\mathbf{u}^1 &= \mathbf{e}^1 & \mathbf{q}^1 &= \mathbf{u}^1/\|\mathbf{u}^1\| \\
\mathbf{u}^2 &= \mathbf{e}^2 - \langle \mathbf{q}^1, \mathbf{e}^2 \rangle \mathbf{q}^1 & \mathbf{q}^2 &= \mathbf{u}^2/\|\mathbf{u}^2\| \\
\mathbf{u}^3 &= \mathbf{e}^3 - \langle \mathbf{q}^1, \mathbf{e}^3 \rangle \mathbf{q}^1 - \langle \mathbf{q}^2, \mathbf{e}^3 \rangle \mathbf{q}^2 & \mathbf{q}^3 &= \mathbf{u}^3/\|\mathbf{u}^3\| \\
&\vdots & &\vdots \\
\mathbf{u}^m &= \mathbf{e}^m - \sum_{k=1}^{m-1} \langle \mathbf{q}^k, \mathbf{u}^m \rangle \mathbf{q}^k & \mathbf{q}^m &= \mathbf{u}^m/\|\mathbf{u}^m\|
\end{aligned}$$

The procedure for defining $\mathbf{q}^i$ is illustrated in Figure D-1. It is called the *Schmidt orthonormalization procedure*. By direct verification, it can be shown that $\langle \mathbf{q}^i, \mathbf{q}^j \rangle = 0$ for $i \neq j$.

From Theorem D-3 and the Schmidt orthonormalization procedure we conclude that for any hermitian matrix there exists a set of orthonormal vectors with respect to which the hermitian matrix has a diagonal-form

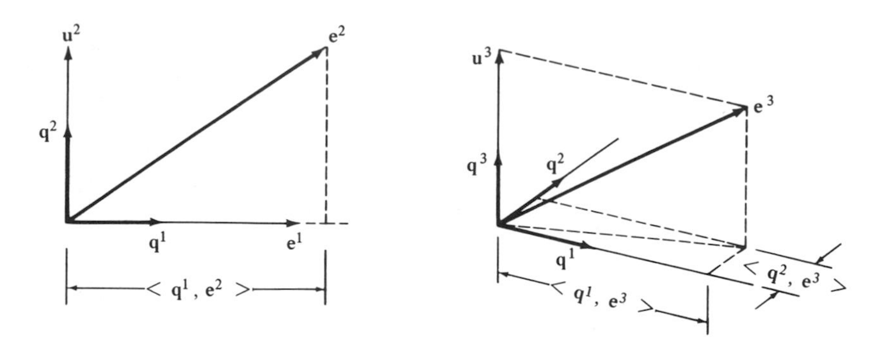


**Figure D-1** The Schmidt orthonormalization procedure.

representation; or equivalently, for any hermitian matrix $\mathbf{M}$, there exists a nonsingular matrix $\mathbf{Q}$ whose columns are orthonormal, such that

$$\hat{\mathbf{M}} = \mathbf{Q}^{-1}\mathbf{M}\mathbf{Q} \triangleq \mathbf{P}\mathbf{M}\mathbf{P}^{-1}$$

where $\hat{\mathbf{M}}$ is a diagonal matrix and $\mathbf{P} \triangleq \mathbf{Q}^{-1}$. Let $\mathbf{Q} = [\mathbf{q}^1 \quad \mathbf{q}^2 \quad \cdots \quad \mathbf{q}^n]$. Because of the orthonormal assumption, we have

$$\mathbf{Q}^*\mathbf{Q} = \begin{bmatrix} (\mathbf{q}^1)^* \\ (\mathbf{q}^2)^* \\ \cdot \\ \cdot \\ \cdot \\ (\mathbf{q}^n)^* \end{bmatrix} [\mathbf{q}^1 \quad \mathbf{q}^2 \quad \cdots \quad \mathbf{q}^n] = \mathbf{I}$$

hence $\mathbf{Q}^{-1} = \mathbf{Q}^*$, and $\mathbf{P}^{-1} = \mathbf{P}^*$. A matrix $\mathbf{Q}$ with the property $\mathbf{Q}^* = \mathbf{Q}^{-1}$ is called a *unitary matrix*. We summarize what we have achieved in this appendix in the following theorem.

### Theorem D-4

A hermitian matrix $\mathbf{M}$ can be transformed by a unitary matrix into a diagonal matrix with real elements; or, equivalently, for any hermitian matrix $\mathbf{M}$ there exists a nonsingular matrix $\mathbf{P}$ with the property $\mathbf{P}^{-1} = \mathbf{P}^*$ such that

$$\hat{\mathbf{M}} = \mathbf{P}\mathbf{M}\mathbf{P}^*$$

where $\hat{\mathbf{M}}$ is a diagonal matrix with real eigenvalues of $\mathbf{M}$ on the diagonal. ∎

A *quadratic form* of $n$ real variables $x_1, x_2, \cdots, x_n$ is a real-valued homogeneous polynomial of the form

$$\sum_{i,j=1}^{n} m_{ij}x_ix_j$$

where the $m_{ij}$'s are any *real* numbers. Similar to the hermitian form, a quadratic form can be written as $\mathbf{x}'\mathbf{M}\mathbf{x}$ with $\mathbf{M}' = \mathbf{M}$, where the prime denotes the transpose. A real matrix $\mathbf{M}$ with the property $\mathbf{M}' = \mathbf{M}$ is called a symmetric matrix. Since the class of real symmetric matrices is a subset of the class of hermitian matrices, all the results in this appendix can be applied without any modification to any real symmetric matrix. In this case the complex conjugate transpose reduces to simply the transpose. For example, for the class of real symmetric matrices Theorem D-4 reads as: for any real symmetric matrix $\mathbf{M}$ there exists a nonsingular matrix $\mathbf{P}$ with the property $\mathbf{P}^{-1} = \mathbf{P}'$ such that $\hat{\mathbf{M}} = \mathbf{PMP}'$, where $\hat{\mathbf{M}}$ is a diagonal matrix with the real eigenvalues of $\mathbf{M}$ on the diagonal.

## PROBLEM

**D-1** If $\mathbf{M}$ is an $n \times n$ matrix with complex coefficients, verify that $\mathbf{x}^*\mathbf{M}\mathbf{x}$ is a real number for any $\mathbf{x}$ in $\mathbb{C}^n$ if and only if $\mathbf{M}^* = \mathbf{M}$. If $\mathbf{M}$ is an $n \times n$ matrix with real coefficients, is it true that $\mathbf{x}'\mathbf{M}\mathbf{x}$ is a real number for any $\mathbf{x}$ in $\mathbb{R}^n$ if and only if $\mathbf{M}' = \mathbf{M}$?

APPENDIX

# E

# On the Matrix Equation AM + MB = N

In this appendix[1] we shall study the matrix equation $\mathbf{AM} + \mathbf{MB} = \mathbf{N}$, where $\mathbf{A}$, $\mathbf{B}$, $\mathbf{M}$, and $\mathbf{N}$ are $n \times n$ complex-valued matrices. We observe that all the $n \times n$ complex-valued matrices, with the usual rules of multiplication and addition, form a linear space (Problem 2-10). Let us denote this space by $(\chi, \mathbb{C})$. The dimension of $(\chi, \mathbb{C})$ is $n^2$. Consider the operator $\mathcal{A}$ defined by

$$\mathcal{A}(\mathbf{M}) \triangleq \mathbf{AM} + \mathbf{MB} \qquad \text{for all } \mathbf{M} \text{ in } \chi$$

It is clear that the operator $\mathcal{A}$ maps $(\chi, \mathbb{C})$ into itself, and is a linear operator.

### Theorem E-1

Let $\mathcal{A}$: $(\chi, \mathbb{C}) \rightarrow (\chi, \mathbb{C})$ be the operator defined by $\mathcal{A}(\mathbf{M}) = \mathbf{AM} + \mathbf{MB}$ for all $\mathbf{M}$ in $\chi$. Let $\lambda_i$, for $i = 1, 2, \cdots, l \leq n$, be the distinct eigenvalues of $\mathbf{A}$ and let $\mu_j$, for $j = 1, 2, \cdots, m \leq n$, be the distinct eigenvalues of $\mathbf{B}$. Then $(\lambda_i + \mu_j)$ is an eigenvalue of $\mathcal{A}$. Conversely, let $\eta_k$, $k = 1, 2, \cdots, p \leq n^2$ be the distinct eigenvalues of $\mathcal{A}$, then for each $k$,

$$\eta_k = \lambda_i + \mu_j$$

for some $i$ and some $j$.

### Proof

We prove, first, that $\lambda_i + \mu_j$ is an eigenvalue of $\mathcal{A}$. Let $\mathbf{x}$ and $\mathbf{y}$ be nonzero $n \times 1$ vectors such that

$$\mathbf{Ax} = \lambda_i \mathbf{x}$$
$$\mathbf{y}'\mathbf{B} = \mu_j \mathbf{y}'$$

where $\mathbf{y}'$ is the transpose of $\mathbf{y}$, $\mathbf{y}'$ may be called the left eigenvector of $\mathbf{B}$. Applying the operator $\mathcal{A}$ on the $n \times n$ matrix $(\mathbf{xy}')$ that is clearly a nonzero matrix, we obtain

$$\mathcal{A}(\mathbf{xy}') = \mathbf{Axy}' + \mathbf{xy}'\mathbf{B} = \lambda_i\mathbf{xy}' + \mu_j\mathbf{xy}' = (\lambda_i + \mu_j)\mathbf{xy}'$$

[1] This appendix closely follows Reference [50].

Hence $(\lambda_i + \mu_j)$ is an eigenvalue of $\mathcal{A}$ for $i = 1, 2, \cdots, l$, for $j = 1, 2, \cdots, m$. See Definition 2-12.

Next we prove that all eigenvalues of $\mathcal{A}$ are of the form $(\lambda_i + \mu_j)$. Let us suppose that $\eta_k$ is an eigenvalue of $\mathcal{A}$. Then, by definition, there exists a $\mathbf{M} \neq \mathbf{0}$ such that

$$\mathcal{A}(\mathbf{M}) = \mathbf{AM} + \mathbf{MB} = \eta_k\mathbf{M}$$

or

$$(\eta_k\mathbf{I} - \mathbf{A})\mathbf{M} = \mathbf{MB} \tag{E-1}$$

We now show that the matrices $\eta_k\mathbf{I} - \mathbf{A}$ and $\mathbf{B}$ have at least one eigenvalue in common. We prove this by contradiction. Suppose $\eta_k\mathbf{I} - \mathbf{A}$ and $\mathbf{B}$ have no common eigenvalue and let

$$g(\lambda) = \prod_{i=1}^{q} (\lambda - \xi_i)^{s_i}$$

be the minimal polynomial of $\eta_k\mathbf{I} - \mathbf{A}$. Then

$$g(\eta_k\mathbf{I} - \mathbf{A}) = \mathbf{0} \tag{E-2}$$

and

$$g(\mathbf{B}) = \prod_{i=1}^{q} (\mathbf{B} - \xi_i\mathbf{I})^{s_i} \tag{E-3}$$

From (E-1), we have

$$(\eta_k\mathbf{I} - \mathbf{A})^2\mathbf{M} = (\eta_k\mathbf{I} - \mathbf{A})\mathbf{MB} = \mathbf{MB}^2$$

$$\vdots$$

$$(\eta_k\mathbf{I} - \mathbf{A})^n\mathbf{M} = \mathbf{MB}^n$$

Consequently, we have

$$g(\eta_k\mathbf{I} - \mathbf{A})\mathbf{M} = \mathbf{M}g(\mathbf{B})$$

which, together with (E-2), implies that

$$\mathbf{0} = \mathbf{M}g(\mathbf{B}) \tag{E-4}$$

From (E-3), we have

$$|g(\mathbf{B})| = \prod_{i=1}^{q} |(\mathbf{B} - \xi_i\mathbf{I})^{s_i}|$$

where $|\cdot|$ denotes the determinant of a matrix. By assumption, $(\eta_k\mathbf{I} - \mathbf{A})$ and $\mathbf{B}$ have no eigenvalue in common, hence $|\mathbf{B} - \xi_i\mathbf{I}| \neq 0$ for all $i$. It implies that $g(\mathbf{B})$ is nonsingular; this fact, together with (E-4), implies that $\mathbf{M} = \mathbf{0}$, which contradicts the hypothesis that $\mathbf{M} \neq \mathbf{0}$. Therefore, the matrices $\eta_k\mathbf{I} - \mathbf{A}$ and $\mathbf{B}$ have at least one common eigenvalue. Now the eigenvalue of $\eta_k\mathbf{I} - \mathbf{A}$ is of the form[2] $\eta_k - \lambda_i$. Consequently, for some $i$ and for some $j$,

$$\eta_k - \lambda_i = \mu_j \qquad \text{or} \qquad \eta_k = \lambda_i + \mu_j$$ Q.E.D.

### Corollary E-1a

Any matrix representation of the operator $\mathcal{A}$ is nonsingular if and only if $\lambda_i + \mu_j \neq 0$ for all $i, j$.

### Proof

Since the linear operator $\mathcal{A}$ maps $n^2$ dimensional linear space into itself, it has a matrix representation (Theorem 2-3). A matrix representation can be easily obtained by writing the $n^2$ equations $\mathbf{AM} + \mathbf{MB} = \mathbf{C}$ in the form of $\bar{\mathbf{A}}\bar{\mathbf{m}} = \bar{\mathbf{c}}$, where $\bar{\mathbf{m}}$ is an $n^2 \times 1$ column vector consisting of all the $n^2$ elements of $\mathbf{M}$. The corollary follows directly from the fact that the determinant of $\bar{\mathbf{A}}$ is the product of its eigenvalues (Problem 2-22).

Q.E.D.

### Corollary E-1b

If all the eigenvalues of $\mathbf{A}$ have negative real parts, then for any $\mathbf{N}$ there exists a unique $\mathbf{M}$ that satisfies the matrix equation

$$\mathbf{A}^*\mathbf{M} + \mathbf{MA} = -\mathbf{N}$$

### Proof

Since the eigenvalues of $\mathbf{A}^*$ are the complex conjugate of the eigenvalues of $\mathbf{A}$, hence all the eigenvalues of $\mathbf{A}$ and $\mathbf{A}^*$ have negative real parts. Consequently, $\lambda_i + \mu_j \neq 0$ for all $i, j$, which implies that the matrix representation of $\mathcal{A}(\mathbf{M}) = \mathbf{A}^*\mathbf{M} + \mathbf{MA}$ is nonsingular. Hence, for any $\mathbf{N}$ there exists a unique $\mathbf{M}$ satisfying $\mathcal{A}(\mathbf{M}) = -\mathbf{N}$. Q.E.D.

---

[2] If $\lambda_i$ is an eigenvalue of $\mathbf{A}$, then $f(\lambda_i)$ is an eigenvalue of $f(\mathbf{A})$, where $f(\lambda) = \eta_k - \lambda$. (See Problem 2-30.)

## References


[1] Anderson, B. D. O., and D. G. Luenberger, "Design of multivariable feedback systems," *Proc. IEE (London)*, vol. 114, 1967, pp. 295–399, 1967.

[2] ———, R. W. Newcomb, R. E. Kalman, and D. C. Youla, "Equivalence of linear time-invariant dynamical systems," *J. Franklin Inst.*, vol. 281, pp. 371–378, 1966.

[3] Athans, M., and P. L. Falb, *Optimal Control*. New York: McGraw-Hill, 1966.

[4] Bass, R. W., and I. Gura, "High order system design via state-space considerations," *Preprints 1965 JACC*, pp. 311–318.

[5] Bellman, R., *Introduction to Matrix Analysis*. New York: McGraw-Hill, 1960.

[6] Brand, L., "The companion matrix and its properties," *Am. Math. Monthly*, vol. 71, pp. 629–634, 1964.

[7] Brockett, R. W., "Poles, zeros, and feedback: State space interpretation," *IEEE Trans. Automatic Control*, vol. AC-10, pp. 129–135, 1965.

[8] ———, and M. Mesarovic, "The reproducibility of multivariable systems," *J. Math. Anal. Appl.*, vol. 11, pp. 548–563, 1965.

[9] Buck, R. C., *Advanced Calculus*. New York: McGraw-Hill, 1956.

[10] Bucy, R. S., "Canonical forms for multivariable systems," *IEEE Trans. Automatic Control*, vol. AC-13, pp. 567–569, 1968.

[11] Chang, A., "An algebraic characterization of controllability," *IEEE Trans. Automatic Control*, vol. AC-10, pp. 112–113, 1965.

[12] Chang, S. S. L., *Synthesis of Optimal Control Systems*. New York: McGraw-Hill, 1961.

[13] Chen, C. T., "Output controllability of composite systems," *IEEE Trans. Automatic Control*, vol. AC-12, p. 201, 1967.

[14] ———, "Linear independence of analytic functions," *SIAM J. Appl. Math.*, vol. 15, pp. 1272–1274, 1967.

[15] ———, "Representation of linear time-invariant composite systems," *IEEE Trans. Automatic Control*, vol. AC-13, pp. 277–283, 1968.

[16] ———, "Stability of linear multivariable feedback systems," *Proc. IEEE*, vol. 56, pp. 821–828, 1968.

[17] ———, "A note on pole assignment," *IEEE Trans. Automatic Control*, vol. AC-13, pp. 597–598, 1968.

[18] ———, "Design of feedback control systems," *Proc. Natl. Electron. Conf.* vol. 57, pp. 46–51, 1969.

[19] ———, "Design of pole-placement compensators for multivariable systems," *Preprints 1970 JACC.*

[20] ———, and Desoer, C. A., "Controllability and observability of composite systems," *IEEE Trans. Automatic Control,* vol. AC-12, pp. 402–409, 1967.

[21] ———, and ———, "A proof of controllability of Jordan form state equations," *IEEE Trans. Automatic Control,* vol. AC-13, pp. 195–196, 1968.

[22] ———, ———, and A. Niederlinski, "Simplified condition for controllability and observability of linear time-invariant systems," *IEEE Trans. Automatic Control,* vol. AC-11, pp. 613–614, 1966.

[23] Chi, H. H., and C. T. Chen, "A sensitivity study of analog computer simulation," *Proc. Allerton Conf.*, pp. 845–854, 1969.

[24] Coddington, E. A., and N. Levinson, *Theory of Ordinary Differential Equations.* New York: McGraw-Hill, 1955.

[25] Dellon, F., and P. E. Sarachik, "Optimal control of unstable linear plants with inaccessible states," *IEEE Trans. Automatic Control,* vol. AC-13, pp. 491–495, 1968.

[26] Dertouzos, M. L., M. E. Kaliski, and K. P. Polzen, "On-line simulation of block-diagram systems," *IEEE Trans. Computers,* vol. C-18, pp. 333–342, 1969.

[27] DeRusso, P. M., R. J. Roy, and C. M. Close, *State Variables for Engineers.* New York: Wiley, 1965.

[28] Desoer, C. A., "A general formulation of the Nyquist criterion," *IEEE Trans. Circuit Theory,* vol. CT-12, pp. 230–234, 1965.

[29] ———, EECS 222 Lecture Notes, University of California, Berkeley, Fall 1968.

[30] ———, and C. T. Chen, "Controllability and observability of feedback systems," *IEEE Trans. Automatic Control,* vol. AC-12, pp. 474–475, 1967.

[31] ———, and E. S. Kuh, *Basic Circuit Theory.* New York: McGraw-Hill, 1969.

[32] ———, and P. Varaiya, "The minimal realization of a nonanticipative impulse response matrix," *SIAM J. Appl. Math.,* vol. 15, pp. 754–763, 1967.

[33] ———, and M. Y. Wu, "Stability of linear time-invariant systems," *IEEE Trans. Circuit Theory,* vol. CT-15, pp. 245–250, 1968.

[34] ———, and ———, "Stability of multiple-loop feedback linear time-invariant systems," *J. Math. Anal. Appl.,* vol. 23, pp. 121–129, 1968.

[35] Duffin, R. J., and D. Hazony, "The degree of a rational matrix function," *J. SIAM,* vol. 11, pp. 645–658, 1963.

[36] Falb, P. L., and W. A. Wolovich, "Decoupling in the design of multivariable control systems," *IEEE Trans. Automatic Control,* vol. AC-12, pp. 651–659, 1967.

[37] Ferguson, J. D., and Z. V. Rekasius, "Optimal linear control systems with incomplete state measurements," *IEEE Trans. Automatic Control,* vol. AC-14, pp. 135–140, 1969.

[38] Friedman, B., *Principles and Techniques of Applied Mathematics.* New York: Wiley, 1956.

[39] Gantmacher, F. R., *The Theory of Matrices,* vols. 1 and 2. New York: Chelsea, 1959.

[40] Gilbert, E. G., "Controllability and observability in multivariable control systems," *SIAM J. Control,* vol. 1, pp. 128–151, 1963.

[41] ———, "The decoupling of multivariable systems by state feedback," *SIAM J. Control,* vol. 7, pp. 50–63, 1969.

[42] Gueguen, C. J., and E. Toumire, "Comments on 'Irreducible Jordan form realization of a rational matrix,'" *IEEE Trans. Automatic Control,* vol. AC-15, 1970.

[43] Hadley, G., *Linear Algebra.* Reading, Mass.: Addison-Wesley, 1961.

[44] Halmos, P. R., *Finite Dimensional Vector Spaces,* 2d ed. Princeton, N.J.: Van Nostrand, 1950.

[45] Herstein, I. N., *Topics in Algebra.* Waltham, Mass.: Blaisdell, 1964.

[46] Heymann, M., "Comments 'On pole assignment in multi-input controllable linear systems,'" *IEEE Trans. Automatic Control,* vol. AC-13, pp. 748–749, 1968.

[47] Ho, B. L., and R. E. Kalman, "Effective construction of linear state variable models from input/output data," *Proc. Third Allerton Conf.,* pp. 449–459, 1965.

[48] Ho, Y. C., "What constitutes a controllable system," *IRE Trans. Automatic Control,* vol. AC-7, p. 76, 1962.

[49] Hsu, C. H., and C. T. Chen, "A proof of stability of multivariable feedback systems," *Proc. IEEE,* vol. 56, pp. 2061–2062, 1968.

[50] Jacob, J. P., and E. Polak, "On the inverse of the operator $(\cdot) = A(\cdot) + B(\cdot)$," *Am. Math. Monthly,* vol. 73, pp. 388–390, 1966.

[51] Johnson, C. D., and W. M. Wonham, "A note on the transformation to canonical (phase-variable) form," *IEEE Trans. Automatic Control,* vol. AC-9, pp. 312–313, 1964.

[52] Joseph, P. D., and J. T. Tou, "On linear control theory," *AIEE Trans. Applications and Industry,* vol. 80, pt. II, pp. 193–196, 1961.

[53] Jury, E. I., *Sampled-Data Control Systems.* New York: Wiley, 1958.

[54] Kalman, R. E., "A new approach to linear filtering and prediction problems," *Trans. ASME*, ser. D, vol. 82, pp. 35–45, 1960.

[55] ———, "On the general theory of control systems," *Proc. First Intern. Congr. Autom. Control*, Butterworth, London, pp. 481–493, 1960.

[56] ———, "Contribution to the theory of optimal control," *Bol. Soc. Mat. Mex.*, vol. 5, pp. 102–119, 1960.

[57] ———, "Canonical structure of linear dynamical systems," *Proc. Natl. Acad. Sci. U.S.*, vol. 48, no. 4, pp. 596–600, 1962.

[58] ———, "On the stability of linear time-varying systems," *IRE Trans. Circuit Theory*, vol. CT-9, pp. 420–423, 1962.

[59] ———, "Lyapunov functions for the problems of Lur'e in automatic control," *Proc. Natl. Acad. Sci. U.S.*, vol. 49, pp. 201–205, 1962.

[60] ———, "Mathematical description of linear dynamical system," *SIAM J. Control*, vol. 1, pp. 152–192, 1963.

[61] ———, "When is a linear control system optimal?" *Trans. ASME*, ser. D, vol. 86, pp. 51–60, 1964.

[62] ———, "Irreducible realizations and the degree of a rational matrix," *SIAM J.*, vol. 13, pp. 520–544, 1965.

[63] ———, "Toward a theory of difficulty of computation in optimal control," *Proc. 4th IBM Sci. Comput. Symp.*, pp. 25–43, 1964.

[64] ———, "On structural properties of linear constant multivariable systems," reprint of paper 6A, Third Congress of the International Federation of Automatic Control, 1966.

[65] ———, and J. E. Bertram, "Control system analysis and design via the 'second method' of Lyapunov," *Trans. ASME*, ser. D, vol. 82, pp. 371–393, 1960.

[66] ———, and R. S. Bucy, "New results in linear filtering and prediction theory," *Trans. ASME*, ser. D, vol. 83, pp. 95–108, 1961.

[67] ———, and T. S. Engler, *A User's Manual for Automatic Synthesis Programs*, NASA, Washington, D.C., 1966.

[68] ———, P. L. Falb, and M. A. Arbib, *Topics in Mathematical System Theory*. New York: McGraw-Hill, 1969.

[69] ———, Y. C. Ho, and K. S. Narendra, "Controllability of linear dynamical systems," *Contrib. Differential Equations*, vol. 1, pp. 189–213, 1961.

[70] Kaplan, W., *Operational Methods for Linear Systems*. Reading, Mass: Addison-Wesley, 1962.

[71] Kreindler, A., and P. E. Sarachik, "On the concepts of controllability and observability of linear systems," *IEEE Trans. Automatic Control*, vol. AC-9, pp. 129–136, 1964.

[72] Kuh, E. S., and R. A. Rohrer, "The state-variable approach to network analysis," *Proc. IEEE*, vol. 53, pp. 672–686, 1965.

[73] ———, and ———, *Theory of Linear Active Networks.* San Francisco: Holden-Day, 1967.

[74] Kumar, S., "Computer-aided design of multivariable systems," M.S. thesis, State University of New York, Stony Brook, 1969.

[75] Kuo, F. F., and J. F. Kaiser, *System Analysis by Digital Computer.* New York: Wiley, 1966.

[76] LaSalle, J., and S. Lefschetz, *Stability of Liapunov's Direct Method.* New York: Academic Press, 1961.

[77] Lefschetz, S., *Differential Equations: Geometric Theory.* New York: Interscience, 1957.

[78] Leunberger, D. G., "Observing the state of a linear system," *IEEE Trans. Military Electronics*, vol. MIL-8, pp. 74–80, 1964.

[79] ———, "Observers for multivariable systems," *IEEE Trans. Automatic Control*, vol. AC 11, pp. 190–197, 1966.

[80] ———, "Canonical forms for linear multivariable systems," *IEEE Trans. Automatic Control*, vol. AC-12, pp. 290–293, 1967.

[81] Mantey, P. E., "Eigenvalue sensitivity and state-variable selection," *IEEE Trans. Automatic Control*, vol. AC-13, pp. 263–269, 1968.

[82] Mayne, D. Q., "Computational procedure for the minimal realization of transfer-function matrices," *Proc. IEE (London)*, vol. 115, pp. 1363–1368, 1968.

[83] McMillan, B., "Introduction to formal realizability theory," *Bell System Tech. J.*, vol. 31, pp. 217–279, 541–600, 1952.

[84] Morgan, B. S., Jr., "The synthesis of linear multivariable systems by state variable feedback," *Proc. 1964 JACC*, pp. 468–472.

[85] Narendra, K. S., and C. P. Neuman, "Stability of a class of differential equations with a single monotone linearity," *SIAM J. Control*, vol. 4, pp. 295–308, 1966.

[86] Nering, E. D., *Linear Algebra and Matrix Theory.* New York: Wiley, 1963.

[87] Newcomb, R. W., *Active Integrated Circuit Synthesis.* Englewood Cliffs, N.J.: Prentice-Hall, 1968.

[88] Ogata, K., *State Space Analysis of Control Systems.* Englewood Cliffs, N.J.: Prentice-Hall, 1967.

[89] Panda, S. P., and C. T. Chen, "Irreducible Jordan form realization of a rational matrix," *IEEE Trans. Automatic Control*, vol. AC-14, pp. 66–69, 1969.

[90] Parks, P. C., "A new proof of the Routh–Hurwitz stability criterion using the 'second method' of Lyapunov," *Proc. Cambridge Phil. Soc.*, vol. 58, pt. 4, pp. 694–720, 1962.

[91] Polak, E., "An algorithm for reducing a linear, time-invariant differential system to state form," *IEEE Trans. Automatic Control*, vol. AC-11, pp. 577–579, 1966.

[92] Pontryagin, L. S., *Ordinary Differential Equations.* Reading, Mass.: Addison-Wesley, 1962.

[93] Rekasisu, Z. V., "Decoupling of multivariable systems by means of state variable feedback," *Proc. Third Allerton Conf.*, pp. 439–447, 1965.

[94] Sandberg, I. W., "Linear multiloop feedback systems," *Bell System Tech. J.*, vol. 42, pp. 355–382, 1963.

[95] ———, "On the $L_2$-boundness of solutions of nonlinear functional equations," *Bell System Tech. J.*, vol. 43, pp. 1581–1599, 1964.

[96] Schwartz, L., *Théorie des distributions.* Paris: Hermann & Cie, 1951, 1957.

[97] Schwarz, R. J., and B. Friedland, *Linear Systems.* New York: McGraw-Hill, 1965.

[98] Silverman, L. M., "Structural properties of time-variable linear systems," Ph.D. dissertation, Dept. of Elec. Eng., Columbia University, 1966.

[99] ———, "Transformation of time-variable systems to canonical (phase-variable) form," *IEEE Trans. Automatic Control*, vol. AC-11, pp. 300–303, 1966.

[100] ———, "Stable realization of impulse response matrices," *1967 IEEE Intern. Conv. Record*, vol. 15, pt. 5, pp. 32–36.

[101] ———, "Synthesis of impulse response matrices by internally stable and passive realizations," *IEEE Trans. Circuit Theory*, vol. CT-15, pp. 238–245, 1968.

[102] ———, and B. D. O. Anderson, "Controllability, observability and stability of linear systems," *SIAM J. Control*, vol. 6, pp. 121–129, 1968.

[103] ———, and H. E. Meadows, "Controllability and observability in time-variable linear systems," *SIAM J. Control*, vol. 5, pp. 64–73, 1967.

[104] Truxall, J. G., *Control System Synthesis.* New York: McGraw-Hill, 1955.

[105] Weiss, L., "The concepts of differential controllability and differential observability," *J. Math. Anal. Appl.*, vol. 10, pp. 442–449, 1965.

[106] ———, "On the structure theory of linear differential systems," *Proc. Second Princeton Conf.*, pp. 243–249, 1968; also *SIAM J. Control*, vol. 6, pp. 659–680, 1968.

[107] ———, "Lectures on controllability and observability," Tech. Note BN-590, University of Maryland, Jan. 1969.

[108] ———, and P. L. Falb, "Dolezal's theorem, linear algebra with continuously parametrized elements, and time-varying systems," *Math. System Theory*, vol. 3, pp. 67–75, 1969.

[109] ———, and R. E. Kalman, "Contributions to linear system theory," *Int. J. Eng. Soc.*, vol. 3, pp. 141–171, 1965.

[110] Wolovich, N. A., "One state estimation of observable systems," *Preprints 1968 JACC*, pp. 210–220.

[111] Wonham, W. M., "On pole assignment in multi-input controllable linear systems," *IEEE Trans. Automatic Control*, vol. AC-12, pp. 660–665, 1967.

[112] ———, and A. S. Morse, "Decoupling and pole-assignment in linear multivariable systems—a geometric approach," *Proc. Third Princeton Conf.*, pp. 114–121, 1969.

[113] Wylie, C. R., Jr., *Advanced Engineering Mathematics.* New York: McGraw-Hill, 1951.

[114] Youla, D. C., "The synthesis of linear dynamical systems from prescribed weighting patterns," *SIAM J. Appl. Math.*, vol. 14, pp. 527–549, 1966.

[115] ———, and Plinio Tissi, "n-port synthesis via reactance extraction—part I," *1966 IEEE Intern. Conv. Record*, vol. 14, pt. 7, pp. 183–208.

[116] Zadeh, L. A., and C. A. Desoer, *Linear System Theory.* New York: McGraw-Hill, 1963.

[117] Zames, G., "On the input-output stability of time-varying nonlinear feedback systems," pts. I and II, *IEEE Trans. Automatic Control*, vol. AC-11, pp. 228–238 and 465–476, 1966.

# Index


**A**

Additivity, 76
Adjoint equation, 156, 306
Analytic continuation, 400
Analytic function, 81, 137, 168, 174, 177, 187, 399
Anderson, B. D. O., 417, 422
Arbib, M. A., 420
Asymptotic stability, 330, 332, 334, 336, 338, 341
Asymptotic state estimator, $n$-dimensional, 281–286
  ($n$-1)-dimensional, 289–291
Athans, M., 417
Ayres, F., 384

**B**

Basis, 16
  change of, 20, 24
  orthonormal, 19, 91, 138, 410
Bass, R. W., 417
Bellman, R., 417
Bertram, J. E., 420
Black box, 74
Bounded-input bounded-output (BIBO) stability, 313, 315, 321, 322
Brand, L., 417
Brockett, R. W., 417
Buck, R. C., 417
Bucy, R. S., 417

**C**

Cancellation, pole-zero, 6, 358, 364, 367
Canonical decomposition theorem, 201
Canonical-form dynamical equation, controllable, 228, 233, 262
  Jordan, 142, 191, 229
  observable, 227, 265
Canonical-form matrix, Jordan, 43

Causality, 79
Cayley–Hamilton theorem, 53, 71
Chang, A., 417
Chang, S. S. L., 417
Characteristic polynomial, of constant matrix, 36, 57, 262, 337
  of rational matrix, 219, 376, 380
Characterization, complete, 338, 355
Chen, C. T., 254, 417, 421
Chi, H. H., 418
Close, C. M., 418
Coddington, E. A., 418
Commutativity, 70, 130
Companion matrix, 69
Compensator, 298
  pole-placement, 382, 388
Control law, 271
Controllability, 169, 179, 191, 197, 203, 209, 365
  differential, 174
  instantaneous, 175
  matrix, 178, 261, 268
  output, 206
  output function, 208
  uniform, 175
Convolution integral, 84
Coupling, 299

## D

Decoupling, 299
Degree of rational matrix, 219
Dellon, F., 418
Delta function, 77
DeRusso, P. M., 418
Desoer, C. A., 418
Diagonalization, of Hermitian matrix, 410
  of matrix, 36
Difference equation, first-order, 115
Differential equation, first-order, 90
Direct sum, 109
Discrete-time, system, 113
  dynamical equation, 115, 160, 214, 404
Discretization of continuous-time dynamical equation, 160, 215, 404
Domain, of function, 21
  of linear operator, 22
Duality, theorem of, 185
Duffin, R. J., 418
Dynamical equation, 89
  computer simulation of, 92
  discrete-time, 115
  equivalent, 138
  Jordan-form, 142, 189, 191, 192
  linear time-invariant, 91
  linear time-varying, 90

## E

Eigenvalue, 35, 414
  index of, 51, 70
  multiplicity of, 44, 70
Eigenvector, 35
  generalized, 40
Elementary operations, 238, 384
Engler, T. S., 420
Equilibrium state, 330
Equivalence, 139, 144
  algebraic, 144
  topological, 144
  zero-input, 140
  zero-state, 140
Equivalence transformation, 139, 144
Estimation of state, 281
exp $\mathbf{A}t$, 57, 59
  computation of, 134
  Laplace transform of, 61
  properties of, 60
Exponential stability, 333
External description of system, 72

## F

Falb, P. L., 253, 281, 417, 419, 420
Feedback connection, 107, 356, 364
Feedback control system, 271
Feedback gain matrix, 271
Ferguson, J. D., 419
Field, 9
  of complex numbers, 10
  of rational functions, 10, 15
  of real numbers, 10, 15

Fixedness (*see* Time-invariance)
Floquet, theory of, 146
Friedland, B., 422
Friedman, B., 419
Function, analytic, 399
  integrable, 315
  linear, 22, 123
  of matrix, 49, 65
  sinusodial, 321
  transfer (*see* Transfer function)
Function reproducibility (*see* Output function controllability)
Function space, 11
Fundamental cutset, 100
Fundamental loop, 100
Fundamental matrix, 128

## G

Gantmacher, F. R., 419
Generalized eigenvector, 40
  chain of, 40
Gilbert, E. G., 419
Gram determinant, 165
Gueguen, C. J., 419
Gura, I., 417

## H

Hadley, G., 419
Halmos, P. R., 419
Hazony, D., 418
Hermitian matrix, 339, 409
  nonnegative definite, 340
  positive definite, 340
  positive semidefinite, 340
Herstein, I. N., 419
Heymann, N., 253, 419
Ho, B. L., 236, 419
Ho, Y. C., 419
Ho and Kalman's algorithm, 236
Homogeneity, 76
Hsu, C. H., 419
Hurwitz determinant, 327
Hurwitz matrix, 327
Hurwitz polynomial, 322
Hurwitz stability test, 327

## I

Impulse function, 77
Impulse response, 78
Impulse-response matrix, 78, 133
Inner product, 61, 62, 319
Input–output description, 74
Input–output pair, 75
Instantaneous controllability, 175
Instantaneous observability, 187
Internal description of system, 72
Inverse of ($s$**I**-**A**), computation of, 135
Irreducible dynamical equation, 203
Irreducible realization, 205, 217, 221
  controllable canonical form, 227
  Jordan canonical form, 229, 240
  observable canonical form, 225

## J

Jacob, J. P., 419
Johnson, C. D., 419
Jordan block, 43, 49, 57
Jordan-canonical-form matrix, 43, 46, 48, 58
Joseph, P. D., 419
Jury, E. I., 419

## K

Kaiser, J. F., 421
Kalman, R. E., 236, 420, 423
Kaplan, W., 420
Kreindler, E., 420
Kuh, E. S., 421
Kumar, S., 421
Kuo, F. F., 421
Kuo, Y. L., 253

## L

$L_2$ function space, 319
$L_p$ stability, 317
Laplace transform, 60
LaSalle, J., 421
Lefschetz, S., 421
Levinson, N., 418

Liénard–Chipart criterion, 327
Linear dependence, 13
Linear independence, of time functions, 163, 177
  of vectors, 13, 14, 40, 41
Linear mapping (*see* Linear operator)
Linear operator, 22
  matrix representation of, 23
    change of basis, 25
  nullity, 32, 33
  null space of, 32
  range of, 29
Linear space, 10, 32
  basis of, 16
  dimension of, 15, 66
Linear system, 92
Linear time-invariant system, 92
Linear transformation (*see* Linear operator)
Linear vector space (*see* Linear space)
Linearity, 22, 75
Link, 99
Luenberger, D. G., 417, 421
Lyapunov function, 343
Lyapunov theorem, 341
Lyapunov transformation, 144, 351

## M

Mantey, P. E., 421
Mason's gain formula, 151
Matrix, 8, 26
  characteristic polynomial of, 36
  eigenvalue of, 35
  eigenvector, 35
    generalized, 40
  function of, 49
  fundamental, 128
  hermitian, 339, 409
    positive definite, 340, 347
    positive semidefinite, 340, 347
  Jordan form of, 43, 46, 48, 57, 58
  nonsingular, 31, 34
  norm of, 61
  polynomial of, 49
  principal minor, 341
    leading, 341
  product of, 33
  rank of, 30
  rational, 85
    proper, 150
    strictly proper, 150
  state transition, 129
  symmetric, 339
  trace of, 157
  unitary, 412
  *See also* Linear operator
Mayne, D. Q., 421
McMillan, B., 421
Meadows, H. E., 422
Mesarovic, M., 417
Minimal polynomial, 50
Minimum-energy control, 402
Mode, 137, 213
Model, 2
Monic polynomial, 50
Morgan, B. S., Jr., 421
Morse, A. S., 423
Motor, armature-controlled dc, 95

## N

Narendra, K. S., 420, 421
Nering, E. D., 421
Neuman, C. P., 421
Newcomb, R. W., 416, 421
Niederlinski, A., 418
Nonanticipative system (*see* Causality)
Nonlinear system, 75
Norm, of matrix, 62, 63
  Euclidean, 62
  of vector, 61, 319
Normal tree, 100
Normalization, 410
Nullity, 32
Null space, 32
Nyquist plot, 370
Nyquist stability criterion, 371

## O

Observability, 183, 188, 192, 201, 203, 209, 295, 365
  differential, 186
  index, 188, 295

Observability (*continued*)
instantaneous, 187
matrix, 188, 261, 293
uniform, 187
Observable-canonical-form dynamical equation, 227, 265
Ogata, K., 421
Open-loop control system, 271
Orthogonality of vectors, 410
Orthonormal set, 19, 410
Output controllability, 206
Output equation, 90, 126
Output feedback system, 307, 365
Output function controllability, 208

**P**

Panda, S. P., 421
Parallel connection, 107, 356, 361
Parks, P. C., 422
Periodically varying system, 144
Physical system, 1
Polak, E., 422
Pole, 85
Pole-zero cancellation, 358
Pontryagin, L. S., 422
Positive definite matrix, 340
Positive semidefinite matrix, 340
Principle minor, 341
leading, 327, 341
Principle of the argument, 369
Principle of superposition, 76
Proper matrix, 150
Pulse function, 76

**Q**

Quadratic form, 412

**R**

Range, of function, 21
of linear operator, 29
Rank, of matrix, 30, 39, 44
of product of matrices, 33
Rational matrix, 85
proper, 150
strictly proper, 150
Realization, irreducible, 205, 217
minimal-dimensional, 217
of rational function, 147, 216
of rational matrix, 151, 153, 235, 240
of time-varying differential equation, 234
*See also* Irreducible realization
Reducibility, 203
Rekasius, Z. V., 422
Relaxedness, 75, 80, 154
Representation, of linear operator, 23, 25, 28
of vector, 17–19
Response, 74
impulse, 78, 121
step, 106, 121
zero-input, 132
zero-state, 132
RLC network, 82, 99, 103, 120
Rohrer, R. A., 421
Routh–Hurwitz criterion, 325
Roy, R. J., 418

**S**

Sampled transfer function, 115
Sampled transfer-function matrix, 116
Sampling instants, 403
Sampling period, 403
Sandberg, I. W., 422
Sarachik, P. E., 420
Scalar, 9
Scalar product (*see* Inner product)
Schmidt orthonormalization, 411
Schwarz, R. J., 422
Schwarz inequality, 64, 319
Separation property, 289
Sequence, input, 113
output, 113
weighting, 113
$z$-transform of, 114
Set of linear algebraic equations, 29
Shifting operator, 82, 118
Signal-flow graph, 152, 157
Silverman, L. M., 422
Similar matrices, 25
Similarity transformation, 25, 64

Simulation of dynamical equation, 92
  analog computer, 92
  digital computer, 93
Solution space, 127
Stability, asymptotic, 330, 336, 338
  bounded-input bounded-output, 313
  global, 331
  in the sense of Lyapunov (i.s.L), 330
  total (T-), 335, 338, 371, 374, 376
Stabilization, 276
State, 86–89
  definition of, 86
  equilibrium, 330
  initial, 91
  variable, 89
  vector, 89
State equation, 90, 126
State estimator, 281
  asymptotic, 283
  open-loop, 283
State feedback, 271, 288, 298, 307
State space, 89
State transition matrix, 129
Step function, 106
Step response, 106
Strictly proper matrix, 150
Subspace, 12, 30
Superposition, principle of, 76
Sylvester's inequality, 33
Symmetric matrix, 339
System, 2
  causal, 79
  continuous-time, 113
  discrete-time, 113
  linear, 75
  multivariable, 74, 277, 292, 375
  relaxed, 75
  single-variable, 74, 271, 281, 368
  time-invariant, 82
  zero-memory, 74

**T**

Tandem connection, 107, 356, 363
Thorpe, J. A., 253
Time-invariance, 82, 91
Time-varying, 82, 90
Tissi, P., 423
Total stability (T-stability), 335, 338, 371, 374, 376
Tou, J. T., 419
Toumire, E., 419
Trace of matrix, 157
Transfer function, 84
Transfer-function matrix, 85, 134
Tree, 99
  branch, 99
  normal, 100
Truncation operator, 118
Truxal, J. G., 422

**U**

Uncontrollability, 169
Uniform controllability, 175
Uniform observability, 187
Uniqueness of solution of differential equation, 90
Unitary matrix, 412
Unit-time-delay system, 80
Unobservability, 183

**V**

Vandermonde determinant, 68
Varaiya, P., 418
Vector, 10
  inner product of, 61
  linear dependence, 13
  linear independence, 13
  norm of, 61
  normalized, 410
  orthogonal, 410
  representation of, 17
Vector space, 10
  basis of, 16
  complex, 12
  dimension of, 15
  real, 12, 13
  *See also* Linear space

## W

Weighting sequence, 113
Weiss, L., 422
Wolovich, W. A., 253, 281, 423
Wonham, W. M., 419, 423
Wu, M. Y., 418
Wylie, C. R., Jr., 423

## Y

Youla, D. C., 417, 423

## Z

Zadeh, L. A., 423
Zames, G., 423
Zero, 85
Zero-input response, 132
Zero-memory system, 74
Zero-state equivalence, 140
Zero-state response, 132
$z$-transfer function, 115
$z$-transform, 114